Structural Materials and Engineering

Structural Materials and Engineering

Ference H. Hagy
Editor

Nova Science Publishers, Inc.
New York

Copyright © 2009 by Nova Science Publishers, Inc.

All rights reserved. No part of this book may be reproduced, stored in a retrieval system or transmitted in any form or by any means: electronic, electrostatic, magnetic, tape, mechanical photocopying, recording or otherwise without the written permission of the Publisher.

For permission to use material from this book please contact us:
Telephone 631-231-7269; Fax 631-231-8175
Web Site: http://www.novapublishers.com

NOTICE TO THE READER

The Publisher has taken reasonable care in the preparation of this book, but makes no expressed or implied warranty of any kind and assumes no responsibility for any errors or omissions. No liability is assumed for incidental or consequential damages in connection with or arising out of information contained in this book. The Publisher shall not be liable for any special, consequential, or exemplary damages resulting, in whole or in part, from the readers' use of, or reliance upon, this material. Any parts of this book based on government reports are so indicated and copyright is claimed for those parts to the extent applicable to compilations of such works.

Independent verification should be sought for any data, advice or recommendations contained in this book. In addition, no responsibility is assumed by the publisher for any injury and/or damage to persons or property arising from any methods, products, instructions, ideas or otherwise contained in this publication.

This publication is designed to provide accurate and authoritative information with regard to the subject matter covered herein. It is sold with the clear understanding that the Publisher is not engaged in rendering legal or any other professional services. If legal or any other expert assistance is required, the services of a competent person should be sought. FROM A DECLARATION OF PARTICIPANTS JOINTLY ADOPTED BY A COMMITTEE OF THE AMERICAN BAR ASSOCIATION AND A COMMITTEE OF PUBLISHERS.

LIBRARY OF CONGRESS CATALOGING-IN-PUBLICATION DATA

Structural materials and engineering / editors, Ference H. Hagy.
   p. cm.
Includes bibliographical references and index.
ISBN 978-1-60692-927-8 (hardcover)
1. Structural engineering. 2. Building materials. I. Hagy, Ference H.
TA637.S885 2009
624.1'8--dc22

                                                                                                 200805549

Published by Nova Science Publishers, Inc. ✝ *New York*

CONTENTS

Preface		vii
Chapter 1	Microstructure and Fracture Aspects of Short Fiber Reinforced Thermoplastics and Toughened with Elastomers *S. C. Tjong*	1
Chapter 2	Recycled Aggregate Structural Concrete: A Methodology for the Prediction of its Properties *Jorge de Brito*	43
Chapter 3	A Kinetic Model of the Oxide Growth and Restructuring on Structural Materials in Nuclear Power Plants *Iva Betova, Martin Bojinov, Petri Kinnunen, Klas Lundgren and Timo Saario*	91
Chapter 4	The Flexural Modulus of Polymer Matrix Composites *C.J.R. Verbeek, P.D. Ewart and D. Murtagh*	135
Chapter 5	Probabilistic Modeling of Cleavage Fracture n the Ductile-to-Brittle Transition Region *Xiaosheng Gao*	161
Chapter 6	Towards Performance-Based Seismic Design for Buildings in Taiwan *Qiang Xue and Cheng-Chung Chen*	179
Chapter 7	Recent Advances in the Design of Industrial Ground-Floor Slabs with Special Emphasis on Permissible Deformations *Ali A. Abbas, Milija N. Pavlović and Michael D. Kotsovos*	245
Chapter 8	Towards Efficient Analytical Models for Seismic Analysis of Multistoried Buildings *Y. Belmouden and P. Lestuzzi*	281

Chapter 9	Dynamic Stability of Beams Using a Higher Order Theory *Sebastián P. Machado and Víctor H. Cortínez*	**295**
Chapter 10	The Prediction of Survival Probabilities of Building Structures Under Transient Extreme Execution Loads *Antanas Kudzys*	**329**
Index		**359**

PREFACE

Structural materials are defined as those which are load-bearing. This new book presents the lastest research from around the globe including that on the nature of a material's physical properties based upon its microstructure and operating environment and on related structural engineering problems as well.

Chapter 1 - Thermoplastic polymer composites reinforced with short glass or carbon fibers have been widely used as structural materials in many engineering applications due to their low density, high strength and stiffness. The mechanical performances of polymer microcomposites depend on the nature of reinforcements and polymers, fiber length and volume fraction, fiber distribution, and interfacial fiber-matrix adhesion. The short fibers improve the tensile stiffness and strength of thermoplastics but sacrifice their tensile ductility and impact toughness. The introduction of elastomeric particles into these materials is considered to be an effective route to restore their toughness. The essential work of fracture approach can then be used to characterize the fracture toughness of composites toughened with elastomers. Fundamental understanding of relationships between the microstructure and the mechanical properties of short fiber reinforced thermoplastics is essential for designing microcomposites with control morphology for advanced structural applications. This chapter provides an overview on the microstructure, mechanical behavior and fracture toughness of short fiber reinforced composites with polymer matrix based on polyamide, polyester or polypropylene, and toughened with elastomers.

Chapter 2 - Structural concrete made with recycled aggregates from construction and demolition waste is an eco-efficient solution to reduce both the production of waste and the depletion of natural non-renewable materials. Even though this material shows great potential as an alternative to conventional concrete (made with primary aggregates), its large-scale use has been hampered by lack of regulation and technical documentation. Furthermore, some production procedures hinder the use of recycled aggregates under normal waste collection conditions.

In this Chapter, the specific problems of preparing concrete with recycled aggregates are addressed and the properties of recycled aggregates *versus* primary (natural stone) aggregates are compared.

A general methodology to predict the long-term performance (in terms of both mechanical properties and durability) of concrete made with recycled aggregates using the corresponding performance of equivalent conventional concrete (reference concrete) is proposed and validated. The approach employed includes an international literature review and a summation of experimental campaigns monitored / supervised by the author.

This methodology allows the early estimation of the properties of concrete made with recycled aggregates based on the properties of the aggregates mix and of young concrete, which can be used both by the structural designer and the contractor, thus effectively removing a barrier to the widespread use of this material.

Chapter 3 - A deterministic model for corrosion and activity incorporation in oxides on structural materials in nuclear power plant coolant circuits is proposed based on fundamental physico-chemical mechanisms. In order to ensure adequate modelling, uncertain or non-determined fundamental parameters are set or adjusted within the range of reasonable values by evaluating well-defined or well-controlled in-plant observations or laboratory experiments. As a result of the calculation procedure developed on the basis of the model, the kinetic and transport parameters of the growth and restructuring of the oxides on austenitic stainless steels (AISI 304 and AISI 316) and nickel based alloys (Alloys 600 and 690) in Light Water Reactors (LWRs) were determined via quantitative comparison of the equations of the Mixed-Conduction Model for oxide films with ex-situ analytical results on thickness and composition of such layers obtained from both laboratory and in-reactor exposures. The obtained parameters were used to predict film growth and restructuring, as well as corrosion release with time of exposure to the LWR coolant. Subsequently, kinetic and transport parameters for the incorporation of Co and Zn in the oxide layers on stainless steels and nickel alloys were estimated by reproducing the experimental depth profiles of these elements. The diffusion-migration equations for the non-steady state transport of such species were solved subject to the boundary conditions at the solution side set by the stability constants of the corresponding Zn and Co complexes. The calculational results are discussed in terms of the influence of incorporation of solution-originating species on the kinetics of film growth and layer restructuring.

Chapter 4 - The flexural and Young's modulus of materials are widely used material properties in design. Young's modulus is often tested either in tensile or in bending, assuming that those values are the same. Bending is often preferred due to the simplicity of the test specimens. However, when testing composite materials, the results obtained using these methods vary significantly.

Planar randomly oriented glass and wood fibre reinforced polyethylene composites as well as talc filled polyethylene were used to evaluate and compare the tensile, compression, three and four point bending modulus. It was found that the tensile modulus was consistently higher than the flexural modulus. It was concluded that because a beam is under a mixed stress state of tension and compression, the measured flexural modulus values were intermediate of the tensile and compressive modulus. The differences were small at low volume fractions reinforcement and independent of the degree of interfacial adhesion, at any volume fraction fibre. At high fibre volume fractions fibre, it was assumed that non-uniform stress distribution lead to shear deformation, causing greater differences between the various modulus values, exacerbated by debonding and inefficient fibre packing.

It was found that the glass and wood fibre reinforced composites can be successfully modelled assuming a layered micro-structure. It was concluded that the number of plies used in the model strongly affects the predicted flexural modulus and that using five plies in the model element is most representative of the material. The geometry of the reinforcement became irrelevant at a large number of plies, leading to predictions similar to the isostrain model, scaled with the adhesion coefficient.

It was found that low aspect ratio reinforcements will lead to the flexural modulus being higher than Young's modulus. It was concluded that in bending, stiffness is proportional to the arrangement or geometry of the reinforcement phase and an individual fibre's contribution to the flexural modulus is a function of the relative distance to the neutral axis. The orientation of low aspect ratio fillers is therefore not important with respect to the flexural modulus. Furthermore, the difference between the various flexural tests was found to be consistent regardless reinforcement geometry and it was concluded that shear deformation may be the most important factor contributing to the difference between these.

Chapter 5 - Transgranular cleavage fracture in the ductile-to-brittle transition (DBT) region is of a statistics nature and follows a weakest link mechanism. The Weibull stress model, derived from the failure mechanism, provides a framework to quantify the relationship between the macro and micro-scale driving forces for cleavage fracture. It has been successfully applied to predict the constraint effect on cleavage fracture and the scatter of macroscopic fracture toughness values. This chapter summarizes the research work conducted by the author and his co-workers in recent years to extend the engineering applicability of the Weibull stress model to predict cleavage fracture in structural steels. These recent efforts have introduced a threshold parameter in the Weibull stress model, developed more robust calibration methods for determination of model parameters, predicted experimentally observed constraint effect and toughness scatter, demonstrated temperature and loading rate effects on model parameters, and expanded the Weibull stress model to include the effects of plastic strain and microcrack nucleation.

Chapter 6 - Lessons learned from the devastating Chi-Chi earthquake has wakened the scientific community in Taiwan to consider new seismic design methodologies. Performance-based seismic design has been recognized as an ideal method for use in the future practice of seismic design. In this chapter, lessons learned from the Chi-Chi earthquake have been summarized. The state-of-the-art and the state-of-the-practice in performance-based seismic engineering have been reviewed. Possible misleading in engineering practice has been pointed out. The outline, design flowchart and key issues of performance-based seismic design (PBSD) methodology implemented in the performance-based seismic design draft code for buildings in Taiwan have been described. A direct displacement-based design method has been demonstrated through examples. Performance criteria regarding displacement to be applied in a direct displacement-based design method for non-ductile or less ductile structures are discussed. Suggestions on seismic performance criteria and seismic evaluation of existing buildings have been made. Feedback and difficulties encountered during the development of the draft code have been discussed. Finally, future research topics related to performance-based seismic design have been summarized. It is believed that performance-based earthquake engineering will bring a new era to engineering practices with increased confidence in, and reliability on, seismic performance and safety.

Chapter 7 - This chapter begins by briefly summarizing an extensive programme of research aimed at producing more accurate and economical guidelines for the analysis and design of ground-floor slabs (GFS), thus improving long-standing formulae that have been in use by engineers for many decades. This research addressed primarily the question of stress analysis and strength, as summarized in five recent publications. Current guidelines on GFS place emphasis on initial construction flatness and stresses (or strength) due to subsequent structural loading but, surprisingly, give very little guidance on displacements (and no guidance on slopes) under such loading, despite the fact that increasing heights of storage and forklift machinery require that serious consideration be given to such deformational matters. Most of the present chapter is, therefore, devoted to the question of serviceability under operating conditions. This is done by means of a parametric linear finite-element analysis (LFEA) of GFS resting on a Winkler subgrade. Square-patch loads with widths ranging from zero to 300mm were applied at the centre, edge and corner of the slab unit. The ensuing LFEA-based design equations and charts enable maximum deflections and slopes to be estimated for the full range of slab thicknesses and subgrade qualities found in practice. The work presently reported complements a similar study (the first of the five articles mentioned earlier) based on similar LFEA which proposes a more accurate permissible-stress analysis of GFS and offers the potential for reducing currently used thicknesses. This is illustrated by a fully worked out example in the appendix to the present chapter, where the reduced-thickness slab (obtained from the LFEA-based stress analysis) is now further checked for permissible deflection and slope criteria to ascertain the effect of thickness reduction on serviceability.

Chapter 8 - In this chapter, a novel equivalent planar-frame model with openings is presented. The proposed model was developed to assess the seismic resistance of multistoried structures without resorting to a finite element approach. The model is mainly developed for generation of capacity curves. The model deals with seismic analysis using the Pushover method for masonry and reinforced concrete buildings. It belongs to simple equilibrium analytical models. Its formulation is based on the well-known beam theory (the beam distribution method). The model allows the determination of the member forces (bending moments and shear forces) through three moment equilibrium. Each wall with openings can be decomposed into parallel structural walls made of an assemblage of piers and a portion of spandrels. As formulated, the structural model undergoes inelastic flexural as well as inelastic shear deformations. The mathematical model is based on the smeared cracks and distributed plasticity approach. Both zero moment location shifting in piers and spandrels can be evaluated. The constitutive laws are modeled as bilinear curves in flexure and in shear. A biaxial interaction rule for both axial force - bending moment and axial force – shear force are considered. The model can support any shape of failure criteria. An event-to-event strategy is used to solve the nonlinear problem. Two applications are used to show the ability of the model to study both reinforced concrete and unreinforced masonry structures. Relevant findings are compared to analytical results from experimental, simplified models and finite element models such as Drain3DX and ETABS finite element package.

Chapter 9 - The dynamic stability of thin walled beams subjected to different sets of boundary conditions is investigated in this presentation. The analysis is based on a seven-degree-of-freedom shear deformable beam theory. The theory is formulated in the context of large displacements and rotations, through the adoption of a displacement field considering moderate bending rotations and large twist. This geometrically non linear formulation is used for analyzing regions of dynamic instability of simply supported, cantilever and fixed-end beams subjected to axial and transverse periodic loads. The influence of shear deformation and inertial effects (corresponding to loading plane) on the unstable regions is analyzed, for mono- and bi-symmetric cross-section beams. This last influence is generally neglected in most of the dynamic instability studies. Such an assumption is valid to a certain extent when the exciting frequency is small in comparison with the free frequency of the loading plane.

This is the case frequently assumed for analyzing the dynamic stability of bars subjected to axial excitation. However for transverse excitation, the frequency at which a parametric resonance occurs can be the same order as the natural frequency of the loading plane vibrations.

Ritz variational method is used to reduce the governing equation, the independent displacements vector is expressed as a linear combination of given x-function vectors and unknown t-function coefficients. Regions of dynamic instability of simple and combination resonances are determined by applying Hsu's procedures to the Mathieu equation. The static load parameter and the dynamic amplitude of the excitation load are scaled with the buckling load. This critical value is obtained taking into account the effect of prebuckling deflections. The excitation frequency is also scaled with the lowest frequency value of parametric resonance.

The influence of non-conventional effects is noted in the numerical results. The interaction between the forced vibration and the parametrically excited vibrations on the regions of dynamic instability is significant in some cases. This effect depends on the closeness of the frequency values, i.e. it depends on the stiffnesses ratio between the parametrically excited mode and the vertical mode corresponding to the loading plane.

Chapter 10 - The effect of random execution (construction) loads, climate actions and their combinations on the probabilistic quality and reliability of structures is discussed. The necessity of using more attention from the probabilistic point of view to the structural reliability prediction of load-carrying members subjected to extreme action (load) effects caused by unfavourable transient site situations is discussed. The methodology of survival probability assessments and predictions of structures during the periods of building carcass erection and mixed construction work is analysed. The design features of survival probability predictions of building structures recommended by Standards SEI/ASCE 37 in the USA and ENV 1991 in Europe for their limit state design are described. The new probabilistic analysis approaches based on the concepts of conventional resistance, coincidence of extreme action effects, random sequence of member safety margins, and transformed conditional probabilities are presented. The prediction of instantaneous and time-dependent survival probabilities of members during their life cycles of carcass erection and mixed construction work is considered. The analysis of partial and total survival probabilities and reliability indices of members as quantitative parameters of their structural quality under execution loads is considered and illustrated by a numerical example. The survival probabilities and reliability indices of precast multiholowcore concrete slabs during construction of masonry walls and mosaic floors for multistorey industrial buildings are under consideration.

Chapter 1

MICROSTRUCTURE AND FRACTURE ASPECTS OF SHORT FIBER REINFORCED THERMOPLASTICS AND TOUGHENED WITH ELASTOMERS

S. C. Tjong[1]

Department of Physics & Materials Science, City University of Hong Kong,
Tat Chee Avenue, Kowloon, Hong Kong.

ABSTRACT

Thermoplastic polymer composites reinforced with short glass or carbon fibers have been widely used as structural materials in many engineering applications due to their low density, high strength and stiffness. The mechanical performances of polymer microcomposites depend on the nature of reinforcements and polymers, fiber length and volume fraction, fiber distribution, and interfacial fiber-matrix adhesion. The short fibers improve the tensile stiffness and strength of thermoplastics but sacrifice their tensile ductility and impact toughness. The introduction of elastomeric particles into these materials is considered to be an effective route to restore their toughness. The essential work of fracture approach can then be used to characterize the fracture toughness of composites toughened with elastomers. Fundamental understanding of relationships between the microstructure and the mechanical properties of short fiber reinforced thermoplastics is essential for designing microcomposites with control morphology for advanced structural applications. This chapter provides an overview on the microstructure, mechanical behavior and fracture toughness of short fiber reinforced composites with polymer matrix based on polyamide, polyester or polypropylene, and toughened with elastomers.

Keywords: Composites, glass fiber, fracture toughness, elastomer, essential work of fracture.

---

[1] E-mail: aptjong@cityu.edu.hk

INTRODUCTION

Recently the demand for polymer composites has been increasing due to the fast growing rate of their applications in automotive, building, transportation and other industrial sectors. The addition of short fibers to thermoplastics is known to improve their stiffness, strength and thermal stability. Short fiber reinforced thermoplastics (SFRT) are attractive structural materials because of their low cost, reduced weight, ease of fabrication and excellent mechanical properties. They can be mass-produced by using existing melt-compounding techniques for polymers such as extrusion and injection molding. Thus, short fiber reinforced composites have replaced many metallic materials commonly used as structural engineering components. The matrix materials commonly used for SFRT include polyamides, polyolefins and polyesters.

Polyamides, poly(butylene terephthalate) [PBT] and polyolefins are major engineering and commodity thermoplastics consumed in large quantities today. However, these thermoplastics have some deficiencies that limit their applications in industries. In the presence of a notch, these plastics are notch-sensitive and exhibit poor impact behavior, particularly at low temperatures. The crack resistance of these materials is greatly reduced, resulting in brittle failure with low fracture toughness. Therefore, large efforts have been made to improve their mechanical properties by adding either inorganic fillers, elastomers or both.

The mechanical properties of short fiber polymer composites are influenced to a greater extent by the nature of reinforcements and polymers, fiber length and volume fraction, distribution of reinforcement in polymer matrix, and reinforcement-polymer interface. The glass fiber commonly used to reinforce polymers is E-type consisting of 52-56% SiO_2, 12-16% Al_2O_3, 16 -25% CaO, and 8-13% B_2O_3. It has a density of 2.54 g/cm^3, tensile strength of 3.45 GPa, elastic modulus of 75 GPa, and a strain failure of about 3%. Carbon fibers for polymer composites are produced mainly from polyacrylonitrile (PAN). The oriented aromatic molecular chain renders the fibers exceptional strength and stiffness. The tensile strength and tensile modulus vary widely and affected by heat treatment, degree of crystallinity and the presence of defects. The processing steps for forming carbon fibers from PAN precursor involve stabilization, carbonization and graphitization. During the process, the PAN fibers are initially drawn into fine molecular fibrils followed by air oxidation at 200 -220 °C in order to stabilize them. The stabilized PAN fibers are then pyrolyzed at 1000 - 1500 °C in an inert atmosphere to form carbon ring structure. The carbonized filaments have high tensile strength (~ 3.10 to 4.07 GPa), low modulus (~ 231 to 248 GPa), and very low strain to failure ($\leq$ 1.6%). They are further heated at or above 2000 °C (graphitization stage) in vacuum or an inert atmosphere to produce highly oriented graphite layers with low strength and high modulus (~ 480 GPa) [1]. The density of PAN carbon and pitch carbon fibers is about 1.8 and 2.1 g/cm^3. Carbon fibers can also prepare from pitch through spinnerets and subsequent heat treatments somewhat similar to those of PAN-based fibers.

The load bearing capacity of short fiber composites depends greatly on the length of fibers embedded within thermoplastic matrices. However, the length of fibers may be reduced dramatically during the manufacturing process. The fiber length reduction is caused by the fiber contact with machinery surfaces, fiber-fiber interaction and fiber interaction with viscous polymer during melt-compounding. Measurement of fiber length is frequently performed on the micrographs of short fibers obtained from burning off or dissolution of polymer matrix [2]. The orientation of fibers in SFRT also plays a crucial role on their mechanical properties. Processes that can produce composites with well-aligned fibers for enhanced mechanical performance are desirable. However, it is rather difficult to fabricate a composite with all fibers oriented in one specific direction using conventional melt compounding method. For injection composites, the fibers are generally known to align with the melt flow direction near the skin section but disperse randomly in the core region.

As mentioned above, the mechanical strength of SFRT is related to the interfacial adhesion between the reinforcing phase and the matrix. Good adhesion between short fiber and polymer matrix ensures effective stress transfer from the matrix to the reinforcing phase during mechanical loading. Short glass fiber (SGF) generally have poor adhesion with polyamides and polyolefins. Addition of appropriate coupling agents and compatibilizers is needed to enhance adhesion between the fibers and thermoplastics. Organofunctional silane coupling agents (R'-Si(OH)$_3$)) are frequently used for surface treatment of glass fibers. The functional group R' of silane reacts chemically with the polymer to form an interfacial network between glass fibers and polymer matrix during the composite fabrication [3,4]. However, the lack of reactive groups on some polymers can prevent direct coupling of the fibers and polymer matrix. In this case, modification of the matrix or a modified polymer is introduced into the composite in order to enhance interfacial adhesion. Thus, compatibilizers having functional groups based on maleic anhydride or acrylic acid can be grafted to thermoplastics to improve the interfacial adhesion [5-9]. Despite the fact that enhanced interfacial bonding between the fibers and the matrix improves the tensile strength of SFRT, however, the tensile ductility and impact strength of the composites decrease dramatically. This is because strong interfacial bonding prevents excessive fiber pull-out from the polymer matrix during mechanical deformation. The energy dissipation for fiber pullout contributes mainly to the toughness of SFRT. In this regards, the interfacial bonding strength between fibers and polymer matrix cannot be very strong or too weak. There is inevitable compromise between the high strength /stiffness and the high ductility/toughness as these properties are controlled by the fiber-matrix adhesion. A balance between them must be maintained in order to achieve desired tensile strength and toughness. The optimization of the performance of SFRT through proper control of the matrix polymer, fiber reinforcement, interfacial adhesion and processing route remains major technical challenges.

FRACTURE TOUGHNESS CHARACTERIZATION

Understanding the fracture behavior of SFRT is of particularly importance because it can assist us to prevent catastrophic failure of the composites during their practical service applications. Depending on the bonding strength of interfacial adhesion, several failure processes may take place during mechanical loading of SFRT composites. These include fiber breakage, crack deflection at the fiber-matrix interface, crack bridging, fiber pullout, and matrix shear yielding. These events contribute to major energy dissipation mechanisms. The toughening mechanisms of SFRT are rather complicated as both the matrix deformation and fiber-related failure mechanisms are always involved.

Generally, the fracture behavior of polymers composites can be determined using Izod and Charpy impact testing techniques. These tests involve the measurement of the amount of energy absorbed when the pendulum strikes and fractures the sample at a notch under a high strain rate condition. The impact energy determined is often termed as impact toughness, and provides information on whether the materials fail in ductile or brittle mode. However, the impact energy is not a parameter used by engineers for designing structural components. Another drawback in using impact results for toughness characterization of ductile polymers is that the impact specimens do not fully break into two parts during the tests. To measure the materials resistance to fracture, it is more appropriate to use the fracture mechanics approach. Linear elastic fracture mechanics (LEFM) is commonly employed for the characterization of brittle materials displaying only elastic deformation or small scale yielding near the crack tip. In this concept, the applied stress distribution in the vicinity of a crack tip can be defined in terms of the stress intensity factor (K). A critical value of the stress intensity factor that causes failure of the materials under uniaxial tension is termed as the fracture toughness, K_C. The plane strain K_C is usually utilized in design analysis but the measurement is not as simple to perform as impact test. If the crack length of a material can be estimated or measured by available flaw detection techniques such as non destructive test (NDT), the fracture stress can be determined from K_C. Accordingly, the design operating stress must be smaller than fracture stress for safe operation.

For ductile materials in which a large plastic zone is developed at the crack tip, the LEFM is inappropriate to characterize their toughness. This is because the LEFM criterion requires the use of specimens with very large thickness for ductile materials, which cannot be prepared easily by injection molding. In this case, the J-integral concept [10] and the essential work of fracture (EWF) approach proposed by Broberg [11] should be used to characterize the fracture behavior of ductile materials. The determination of a critical value of J-integral (J_c) is generally carried out through the construction of the resistance curve J-Δa where Δa is the advanced crack length. The J_c value is determined at the point of intersection between the J-R curve and the blunting line (J = 2 $\sigma_y \Delta$a , where σ_y is the yield stress). The process of determination of J_c value of the ductile material is rather tedious when compared to EWF concept.

The EWF approach is based on the assumption that the total work (W_f) involved in fracture is dissipated in inner process zone (W_e) and the outer plastic zone (W_p) when a cracked ductile solid is mechanically loaded under plane-stress condition. The former is referred to as the essential work of fracture whilst the later is non-essential work. The essential work of fracture is the specific energy required to create two new surfaces and that consumed in the fracture process alone. Hence, the total fracture work (W_f) can be described as:

$$W_f = W_e + W_p \tag{1}$$

$$W_f = w_e L t + \beta w_p L^2 t \tag{2}$$

$$w_f = W_f/(Lt) = w_e + \beta w_p L \tag{3}$$

where w_f is the specific total fracture work, w_e and w_p are the specific essential fracture work and specific plastic work, respectively; L is the ligament length, t is sample thickness, and β

is a shape factor of the plastic zone. An important prerequisite of EWF approach is that crack propagates only after the ligament has fully yielded and the plastic zone is scaled with the square of the ligament length. The validity range of ligament L under plane-stress condition is given by:

(3-5) t ≤ L ≤ min (B/3, 2r_p) (4)

where B is the width of the specimen and 2r_p is the size of plastic zone. The total work of fracture can be determined from the area of the recorded load-displacement curve for each specimen. The w_f can then be obtained by dividing the integrated area value by cross-sectional area of the specimen. This approach has been recently incorporated in the European Structural Integrity Society (ESIS) test protocol for essential work of fracture under quasi-static loading conditions [12]. The ESIS protocol for EWF measurements recommends the use of double edge notched tension (DENT) specimen (Fig. 1(a)). A typical plastic zone with elliptical shape developed in PP composite reinforced with 50 wt% glass beads under EWF test is shown in Fig. 1(b) [13].

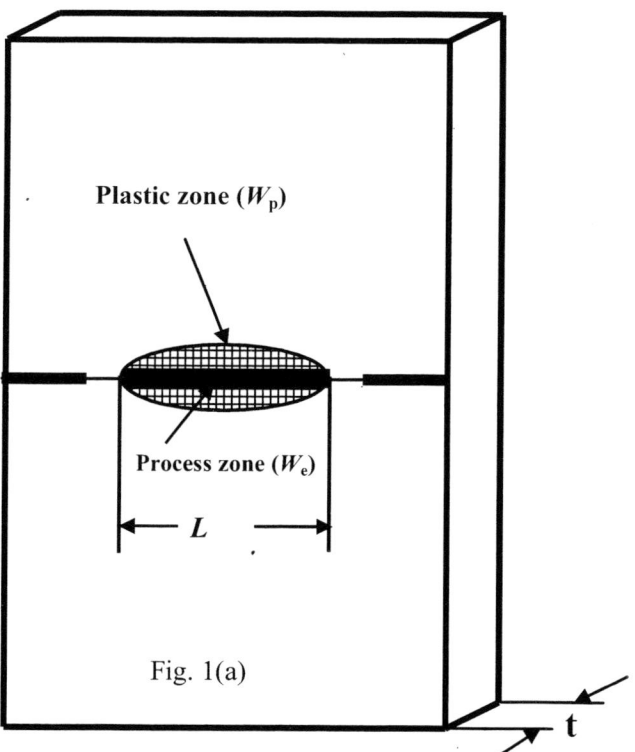

Figure 1. Continued

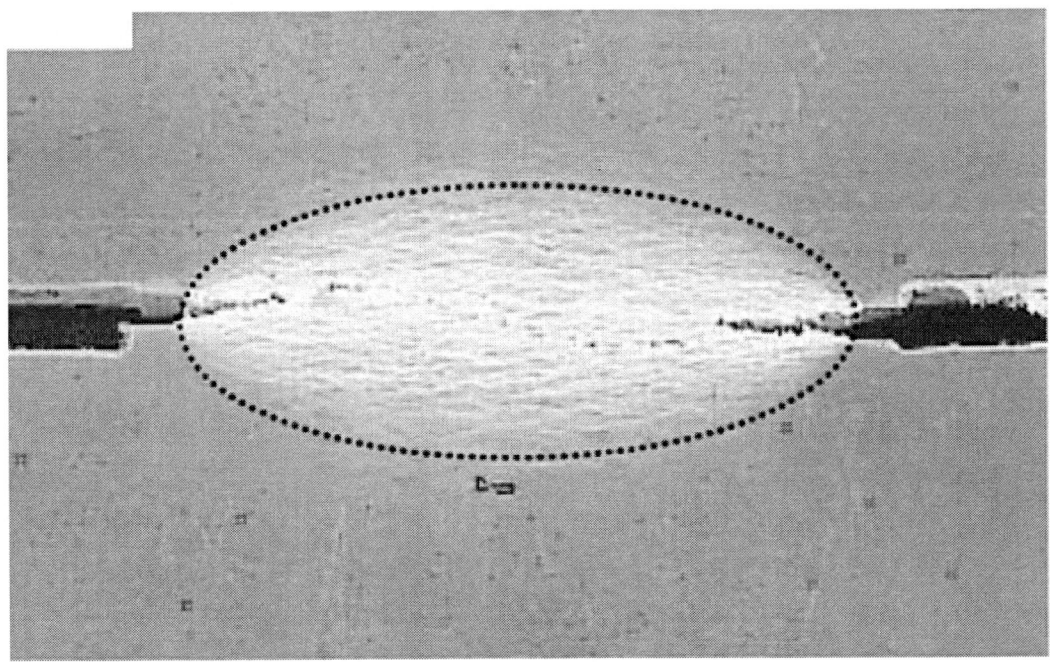

Fig. 1. (b)

Figure 1. (a) DENT specimen for EWF tests. (b) Elliptical plastic zone developed in PP composite reinforced with 50 wt% glass beads. Reprinted from [Ref. 13] with permission of Elsevier.

POLYAMIDE-BASED COMPOSITES

For many years, polyamides have been successfully reinforced with short glass fibers, carbon fibers and other inorganic fillers. Table 1 lists typical mechanical properties of injection molded nylon 6 reinforced with short glass fiber or short carbon fiber (SCF) of different contents [14]. The length of both kinds of fibers prior to compounding was 6 mm. The tensile modulus and strength of nylon 6 improve considerably by adding SGF or SCF. At fixed fiber content, the tensile modulus and strength of SCF composite are much higher than those of the SGF composite as expected. However, The SCF composites have lower elongation at break than their SGF counterparts.

Table 1. Tensile properties of nylon 6, nylon 6/SGF and nylon 6/SCF composites. Reprinted from [Ref.14] with permission of Elsevier

Sample	Young's modulus (MPa)	Tensile stress (MPa)	Elongation at break (%)
Nylon 6	1073	51.8	276.3
Nylon6/10 wt% SGF	2164	60.7	20.3
Nylon6/20 wt% SGF	3158	78.9	7.93
Nylon6/30 wt% SGF	4321	96.2	5.63
Nylon6/10 wt% SCF	3261	70.9	5.00
Nylon6/20 wt% SCF	6379	111.3	3.37
Nylon6/30 wt% SCF	8942	135.5	2.78

Fiber length generally plays an important role on the strength of the SGF reinforced composites. For SFRT, the applied load is transmitted through the matrix to the fiber by shear transfer mechanism. There exists a critical fiber length (l_c) below which stress transfer from the matrix to the fiber is inefficient. The critical fiber length can be defined as:

$$\frac{l_c}{d} = \frac{\sigma_f}{2\tau_y} \qquad (5)$$

where d is the diameter of fiber, σ_f is the fiber strength and τ_y is interfacial adhesion strength. In short fiber composites, it is necessary to maintain the fiber length above the critical value. However, melt mixing and injection molding processes often lead to a drastic reduction in fiber length. Fig. 2 shows the weight average residual fiber length (L_w) versus fiber volume fraction for nylon 6,6 prepared by melt blending followed by injection molding [15]. The initial fiber length prior to compounding is 4 mm. The fiber length reduces to less than 1 mm after compounding and injection molding processes. Fig. 2 reveals that L_w decreases almost linearly with increasing fiber concentration. This can be attributed to increase fiber-fiber and fiber-machine interaction as well as an increased apparent melt viscosity associated with increasing fiber loadings.

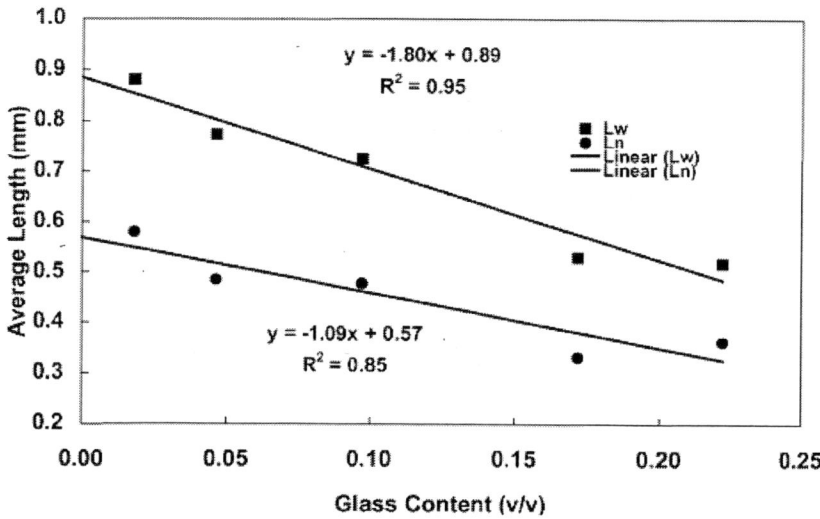

Figure 2. Weight-average fiber length (L_w) vs fiber volume fraction for nylon 6,6 prepared by melt mixing and injection molding. Number-average fiber length (L_n) vs fiber volume fraction is also shown for comparison. Reprinted from [Ref. 15] with permission of Elsevier.

The efficiency of transfer of an applied load from the matrix to the fibers depends on the state of the bonding between the matrix and the fibers. Thus, strong interfacial bond favors efficient load transfer. To ensure effective load transfer from the matrix to discontinuous fiber reinforcement, coupling agents such as silanes and titanates are used to enhance the interfacial adhesion. Fig. 3 shows schematic representation of the interfacial bonding developed between non-treated SGF, aminosilane-treated SGF, (aminosilane + maleic anhydride) treated SGF and nylon 66 matrix [4]. For non-treated SGF, only weak hydrogen bonds can be developed between SGF and nylon 66. In the case of SGF treated with aminosilane coupling agent, covalent bonds are formed due to the interaction between the amine group of coupling agent with the amine or amide groups of nylon 66. The tensile strength for the nylon 66 reinforced with 33 wt% non-treated SGF, silane treated SGF, and (silane +MA) treated SGF is 110, 143 and 193 MPa, respectively [4].

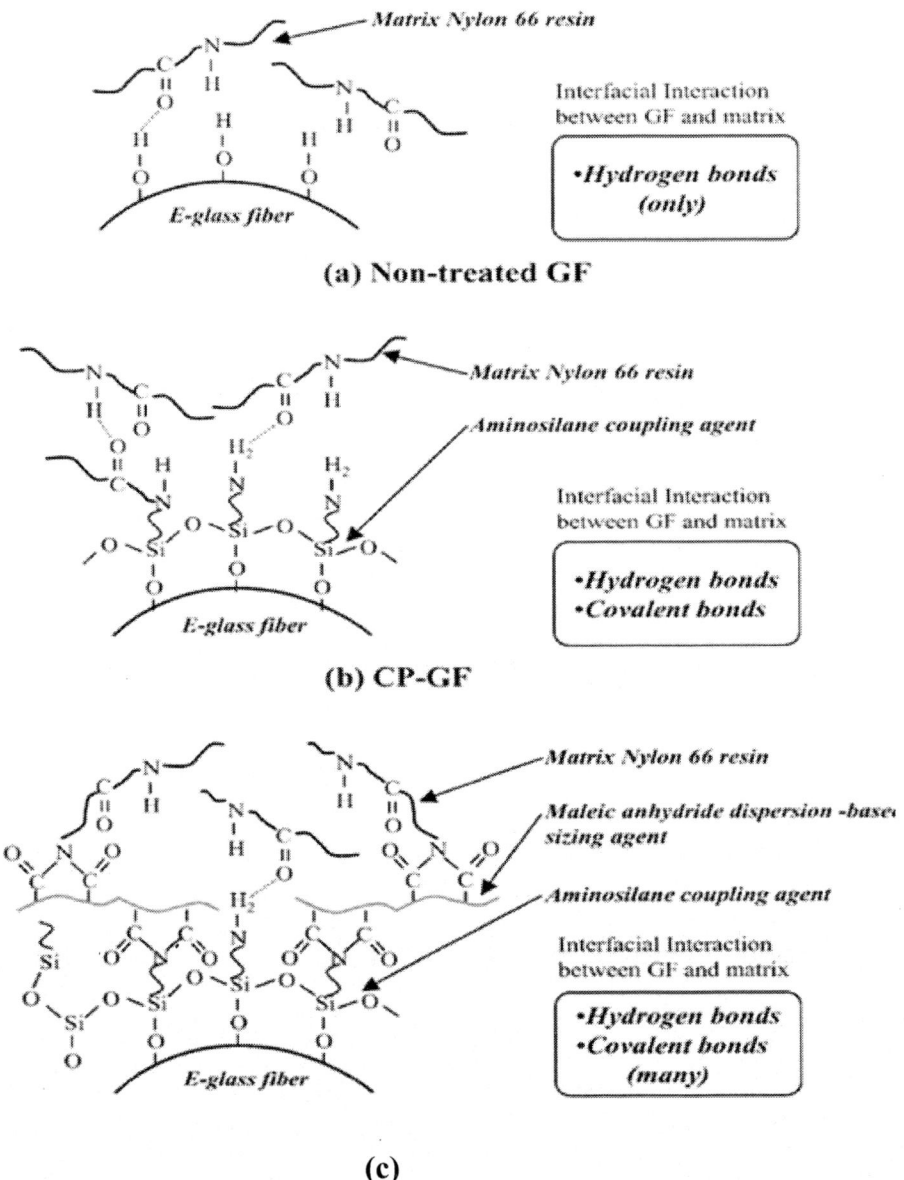

Figure 3. Schematic representation of SGF/nylon 66 interfacial models for (a) non-treated, (b) silane treated, and (c) silane and maleic anhydride treated glass fiber. Reprinted from [Ref. 4] with permission of Elsevier.

Paul and coworkers studied the effect of glass surface chemistry on the mechanical properties of SGF reinforced nylon 6 [16]. The coupling agents used include silanes having reactive functional groups based on succinic anhydride, epoxy or amine, and other functional groups such as alkoxy and octyl. Tables 2 and 3 list the mechanical properties of nylon 6 reinforced with 15 wt% and 30 wt% SGF in which the fibers have been treated with different coupling agents. Apparently, all three reactive silane coupling agents produce nearly identical property mechanical values. The octyl silane treatment yields the lowest yield strength and Izod impact strength due to the dissimilar nature of octyl group and the polar nylon 6. When 20 wt% maleated ethylene-propylene random copolymer (EPR-g-MA) elastomer is added to the polymer matrix, the yield strength and impact strength are highest for the anhydride silane, followed by epoxy silane and amine silane (Table 2).

Table 2. Mechanical properties of nylon 6 with 15 wt% glass fibers. Reprinted from [Ref. 16] with permission of Elsevier.

Silane coupling agent	Tensile modulus (GPa)	Yield strength (MPa)	Elongation at break (%)	Notched Izod impact strength (J/m)	Tensile modulus (GPa)	Yield strength (MPa)	Elongation at break (%)	Notched Izod impact strength (J/m)
	No EPR-g-MA				20 wt% EPR-g-MA			
Anhydride	5.6	117	4.2	60	4.0	80.5	4.3	292
Epoxy	5.5	110	4.5	59	4.1	79.8	3.9	265
Amine	5.7	114	4.3	58	4.2	73.7	4.3	225
Polyalkoxy	5.3	92	4.0	47	4.3	72.5	3.9	196
Octyl	5.3	81	3.4	45	4.1	53.4	2.6	148
Unreinforced materials	Neat nylon 6				Nylon 6/EPR-g-MA (80/20)			
	2.8	70	73	52	1.9	47.0	73	617

Table 3. Mechanical properties of nylon 6 reinforced with 30 wt% glass fibers. Reprinted from [Ref. 16] with permission of Elsevier.

Silane coupling agent	Tensile modulus (GPa)	Yield strength (MPa)	Notched Izod impact strength (J/m)	Tensile modulus (GPa)	Yield strength (MPa)	Notched Izod impact strength (J/m)
	No EPR-g-MA			20wt% EPR-g-MA		
Amine	9.7	155	147	6.6	73.2	173
Epoxy	9.6	153	135	7.2	102	269
Anhydride	9.6	159	163	7.7	109	306

These results are attributed to the difference in the degree of adhesion between the glass fiber and nylon 6. Paul and coworkers demonstrated that the rubber particle size and rubber type can affect the impact strength of SGF reinforced nylon 6 toughened with elastomers [17]. The types of elastomers used included EPR and maleated EPR as well as styrene-hydrogenated butadiene-styrene tribock copolymer (SEBS) and maleated SEBS. Control of rubber morphology is achieved by combining each rubber with its maleated counterpart. Figs. 4(a)-(e) are TEM micrographs showing the microstructure of nylon 6/EPR/EPR-g-MA blends reinforced with 15 wt% succinic anhydride treated SGF. The unmaleated elastomer has a coarse morphology and poor adhesion with the polyamide matrix. Moreover, the elastomer particles become smaller as the maleic anhydride is increased. This is due to the SGF addition increases polymer matrix melt viscosity. Figs. 5(a)-(b) show the effects of elastomer particle size and MA content on the Izod impact strength of nylon6/EPR/EPR-g-MA blends with or without glass fiber reinforcement. The blend of polyamides and unfunctionalized elastomer exhibits low impact toughness because the rubber particles formed during melt blending process are relatively large. For nylon 6/EPR/EPR-g-MA blends, a maximum in impact energy is observed at an average rubber particle size between 0.3 and 0.4 μm. However, there is no maximum for composites containing 15 wt% SGF; instead the impact energy increases linearly with decreasing rubber particle size. Increasing MA content in elastomer improves its adhesion with nylon matrix, thereby enhancing the impact strength of nylon 6/EPR/EPR-g-MA blends and their composites (Fig. 5(b)). A similar decreasing trend in elastomer particle size is found for the nylon 6/SEBS/SEBS-g-MA blends by increasing maleic anhydride content of the elastomer. The nylon 6/SEBS/SEBS-g-MA blends have higher impact energies than nylon 6/EPR/EPR-g-MA blends. In the presence of 15 wt% SGF, impact test results show that the nylon 6/EPR/EPR-g-MA blends are tougher than nylon 6/SEBS/SEBS-g-MA blends.

Recently, Tjong et al. studied the morphology, mechanical behavior and fracture resistance of the PA6,6/SEBS-g-MA 80/20 blend and its SGF-reinforced composites [18,19]. The mechanical properties of these materials are listed in Table 4. It is apparent that the incorporation of SEBS-g-MA elastomer into PA6,6 results in drastic reductions in both the yield stress and stiffness. However, the tensile ductility is improved considerably associated with the elastomer addition. To restore the yield stress, fibers from 5 to 30 wt% are added to the PA6,6/SEBS-gMA 80/20 blend. The tensile strength and stiffness of the PA6,6/SEBS-gMA 80/20 blend are observed to increase with increasing SGF content. PA6,6 exhibits low impact strength of 5.80 kJ/mm^2 because it is notch-sensitive under impact deformation. Incorporation of SEBS-g-MA into PA6,6 leads to a dramatic improvement in the impact strength. It is noted the PA6,6/SEBS-g-MA 80/20 is very tough, and not broken during the impact testing. The SGF additions lead to a reduction in the mpact strength of PA6,6/SEBS-g-MA 80/20 blend. Figs 6(a)-b) show the typical SEM micrographs of the fracture surfaces of (PA6,6/SEBS-g-MA)/SGF (80/20)/10 and (PA6,6/SEBS-g-MA)/SGF (80/20)/20 composites after tensile testing.

Table 4. Mechanical properties of PA6,6, PA6,6/SEBS-g-MA 80/20 blend and its SGF reinforced composites [Ref.18]

Sample	Young's modulus (MPa)	Tensile stress (MPa)	Elongation at break (%)	Impact strength (kJ/m^2)
PA6,6	1805 ± 110	51 ± 3.6	134	5.80 ± 0.28
PA6,6/SEBS-g-MA 80/20	825 ± 55	39 ± 1.6	290	151.6 ± 10.1
(PA6,6/SEBS-g-MA)/SGF (80/20)/5	1806 ± 104	39 ± 1.6	215	13.9 ± 1.41
(PA6,6/SEBS-g-MA)/SGF (80/20)/10	2442 ± 120	42 ± 3.4	71	10.25 ± 0.84
(PA6,6/SEBS-g-MA)/SGF (80/20)/15	2704 ± 123	49 ± 3.8	17	9.83 ± 0.43
(PA6,6/SEBS-g-MA)/SGF (80/20)/20	3286 ± 199	52 ± 3.5	12	10.01 ± 0.48
(PA6,6/SEBS-g-MA)/SGF (80/20)/30	4072 ± 225	65 ± 4.9	6	9.85 ± 0.33

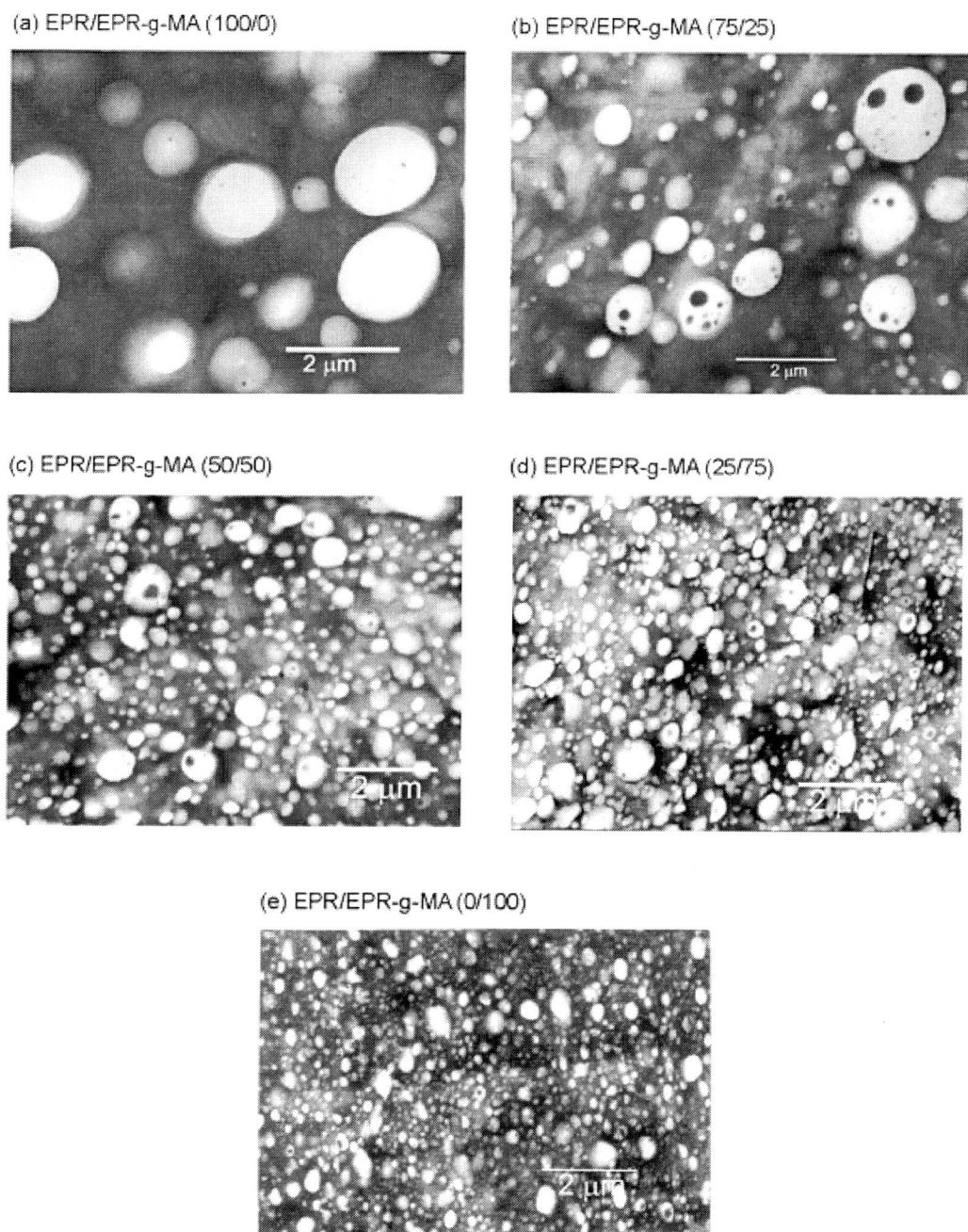

Figure 4. TEM micrographs of nylon 6/EPR/EPR-g-MA blends reinforced with 15 wt% glass fibers containing EPR/EPR-g-MA in the ratio of (a) 100/0, (b) 75/25, (c) 50/50, (d) 25/75, and (e) 0/100. The total elastomer mass (EPR + EPR-g-MA) makes up 20 wt% of the polymer mass (nylon 6 + EPR + EPR-g-MA) excluding the glass fibers. Reprinted from [Ref. 17] with permission of Elsevier.

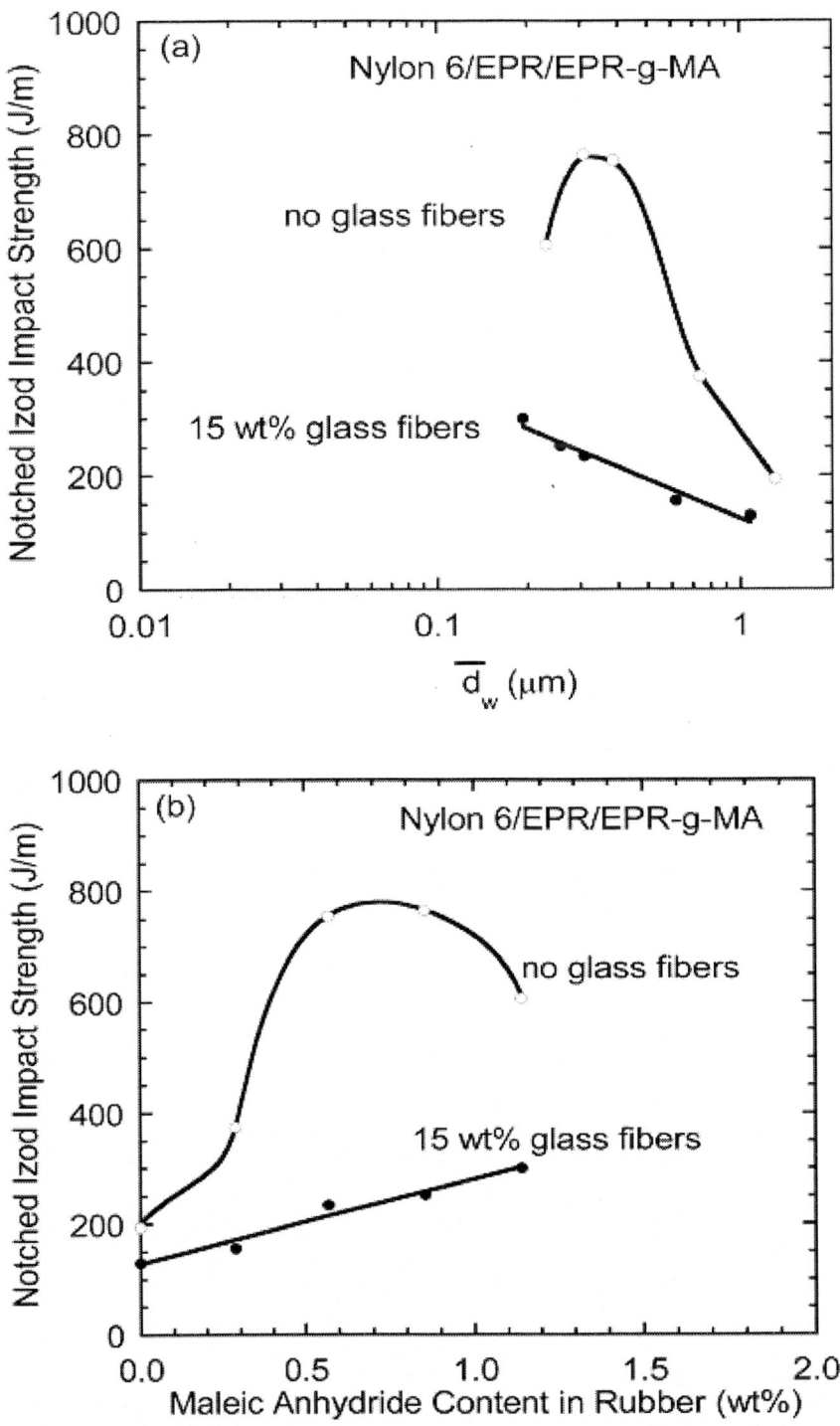

Figure 5. Notched Izod impact strength for blends of nylon 6/EPR/EPR-g-MA as a function of (a) weight average elastomer particle diameter and (b) maleic anhydride content in the elastomer phase. The total elastomer mass (EPR + EPR-g-MA) makes up 20 wt% of the polymer mass (nylon 6 + EPR + EPR-g-MA) excluding the glass fibers. Reprinted from [Ref. 17] with permission of Elsevier.

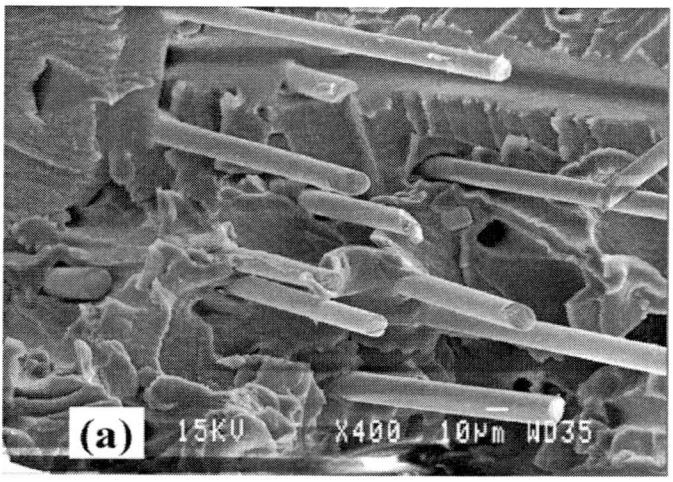

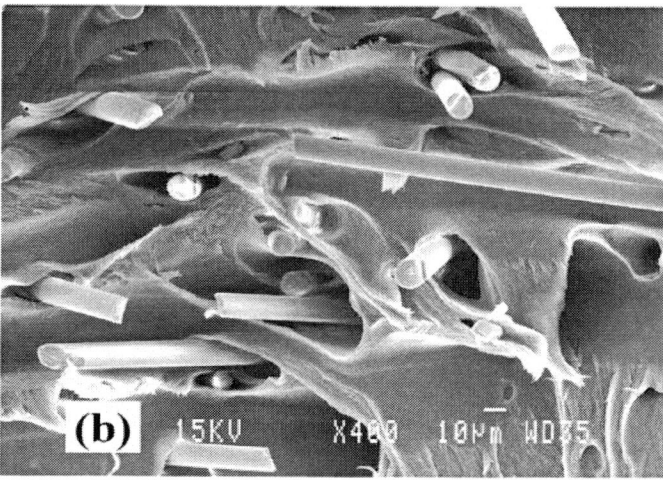

Figure 6. SEM micrographs showing the fracture surfaces of (a) (PA6,6/SEBS-g-MA)/SGF (80/20)/10 and (b) (PA6,6/SEBS-g-MA)/SGF (80/20)/20 composites after tensile test [Ref. 18].

The fiber surface is relatively clean, and cavity associated with the debonding of SGF from the matrix can be readily seen (Fig. 6(b)). The matrix of composites reinforced with 5 and 10 % SGF exhibit fibrillar morphology, indicating that plastic deformation has taken place in the matrix material during tensile test. The mechanical properties of SGF reinforced polyamide composites toughened with maleated SEBS depend on the interfacial adhesion between the matrix (PA6,6) and SEBS, between the SEBS and SGF, and between the PA6,6 and SGF. In general, anhydride functional group grafted to EB can react with hydroxyl groups on the glass fiber surfaces during compounding, thereby improving the compatibility between the SGF and SEBS. The reaction takes place between SEBS-g-MA and SGF is given as follows:

[6]

However, most of MA functional groups of SEBS react readily with the amine and amide groups of PA6,6. Consequently, only fewer remaining MA groups are available to react with the hydroxyl groups on the SGF surfaces. The interaction between the maleated SEBS and SGF is weakened accordingly, leading to the formation of clean fiber surface during tensile test (Figs. 6(a)-(b)). In the case of PA6,6 and SGF interaction, only weak hydrogen interaction can develop between them due to their polar structures (Fig. 3).

The fracture resistance of the PA6,6/SEBS-g-MA 80/20 blend and its composites is evaluated by the EWF method where the specific essential work indicates the resistance to crack initiation and the specific non-essential work implies the resistance against crack propagation. Figs. 7(a)-(d) show the typical load-displacement curves for the PA6,6/SEBS-g-MA 80/20 blend, (PA6,6/SEBS-g-MA)/SGF (80/20)/10, (PA6,6/SEBS-g-MA)/SGF (80/20)/20 and (PA6,6/SEBS-g-MA)/SGF (80/20)/30 composites. Gross yielding and necking of ligament were observed in these samples during the deformation process.

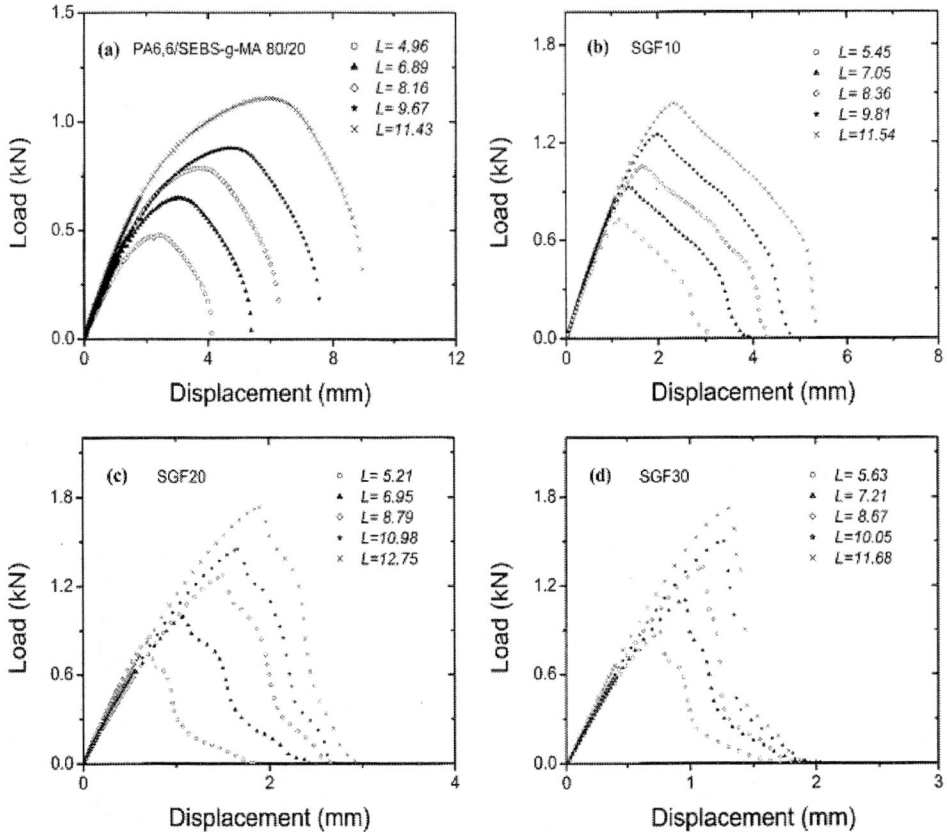

Figure 7. Load-displacement curves for (a) PA6,6/SEBS-g-MA 80/20 blend, (b) (PA6,6/SEBS-g-MA)/SGF (80/20)/10 and (b) (PA6,6/SEBS-g-MA)/SGF (80/20)/20 and (c) (PA6,6/SEBS-g-MA)/SGF (80/20)/30 composites [Ref. 18].

Moreover, similarities in the load-displacement curves for these samples having various ligament lengths are observed. Figs. 8(a)-(d) show the plots of w_f versus ligament length PA6,6/SEBS-g-MA 80/20 blend, (PA6,6/SEBS-g-MA)/SGF (80/20)/10, (PA6,6/SEBS-g-MA)/SGF (80/20)/20 and (PA6,6/SEBS-g-MA)/SGF (80/20)/30 composites.

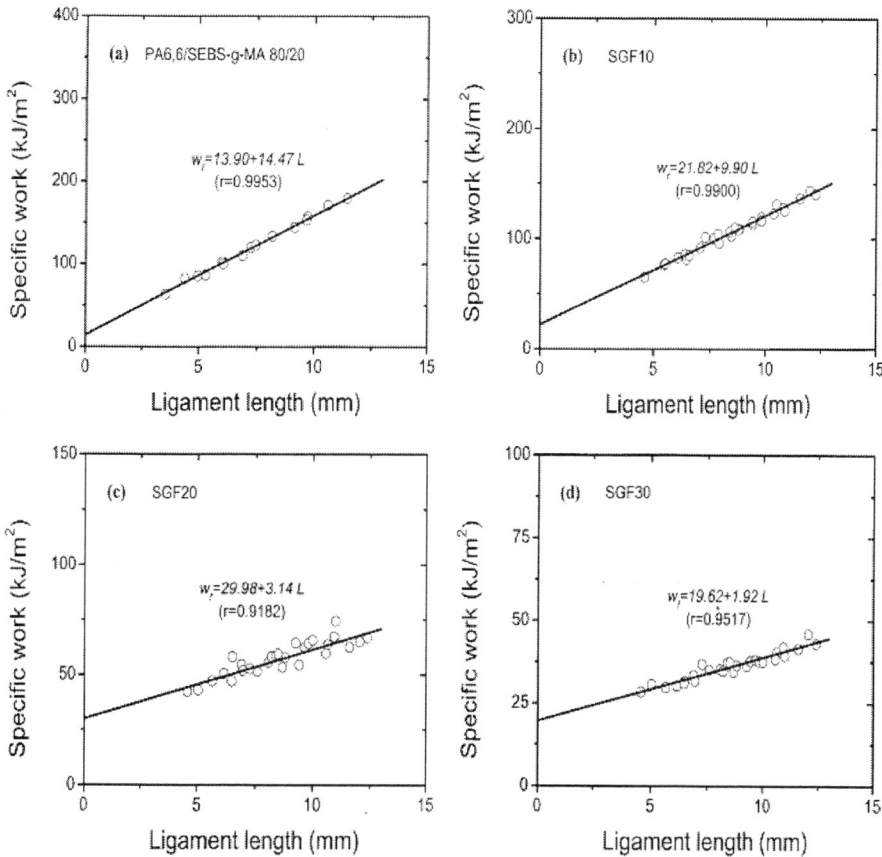

Figure 8. w_f versus L plots for (a) PA6,6/SEBS-g-MA 80/20 blend, (b) (PA6,6/SEBS-g-MA)/SGF (80/20)/10 and (b) (PA6,6/SEBS-g-MA)/SGF (80/20)/20 and (c) (PA6,6/SEBS-g-MA)/SGF (80/20)/30 composites [Ref. 18].

The specific essential work (w_e) and the specific non-essential work (βw_p) can be determined from the intercept and the slope of the regression line, respectively. For the PA6,6/SEBS-g-MA 80/20 blend, extensive energy is dissipated in the outer plastic zone, leading to a steep positive slope in the w_f and L plot. The incorporation of SGF into this blend tends to reduce the slope of regression lines. The variations of specific essential and non-essential-work of fracture for the PA6,6/SEBS-g-MA 80/20 blend and its composites are shown in Fig. 9. It can be seen that the incorporation of 5 to 10 % SGF to the PA6,6/SEBS-g-MA 80/20 blend leads to an increase in the specific essential work of fracture or fracture toughness from 13.90 kJ/m^2 to the range of 22–25 kJ/m^2. The w_e values can be increased further to 32.65 kJ/m^2 by adding 15% SGF. As the SGF content is increased to 20%, the w_e value begins to drop to 29.98 kJ/m^2. With further increasing SGF content to 30%, the w_e value drops dramatically to 19.62 kJ/m^2, but it is still much higher compared to unreinforced PA6,6/SEBS-g-MA 80/20 blend. It is noted that the specific non-essential work of fracture decreases with increasing fiber content. For the (PA6,6/SEBS-g-MA)/SGF (80/20)/20 and (PA6,6/SEBS-g-MA)/SGF (80/20)/30 composites, the presence of larger amount of SGF restricts the matrix deformation, leading to the lowest non-essential specific fracture work.

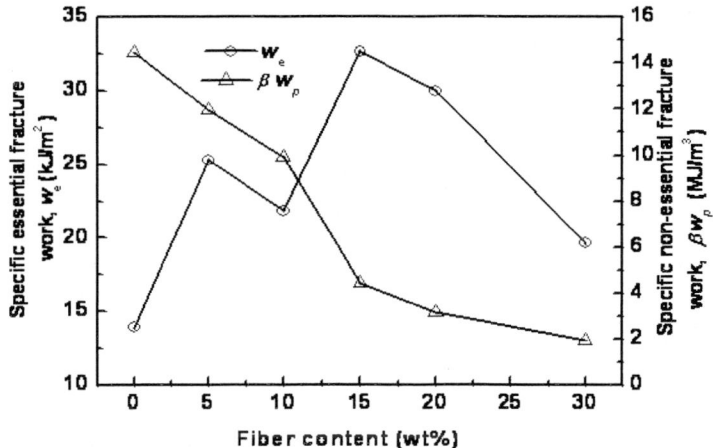

Figure 9. Specific essential and non-essential work of fracture vs SGF weight content for PA6,6/SEBS-g-MA 80/20 blend and its composites [Ref. 18].

On the basis of EWF measurements, it appears that the SGF is beneficial in increasing the fracture toughness of the PA6,6/SEBS-g-MA 80/20 blend. This implies that the additions of SGF up to 15% favor formation of supertough composites. From Table 4, the short glass fiber stiffen and strengthen the PA6,6/SEBS-g-MA 80/20 blend considerably, thus the SGF additions up to 15% provide both strengthening and toughening effects for this blend. This implies that there is a synergistic effect between glass fiber reinforcement and elastomer to enhance the fracture toughness. Figs.10(a)-(b) show SEM fractographs of (PA6,6/SEBS-g-MA)/SGF (80/20)/5 composite after tensile EWF measurements. It is evident that extensive plastic deformation occurs in the bulk matrix of these hybrids. Plastic deformation is caused by the fiber debonding and pullout processes that impose a shear stress component on the matrix, thereby promoting shear deformation of matrix in the plastic zone ahead of crack tip. Combined fiber- and rubber-toughening mechanisms contribute exclusively to the enhanced fracture toughness of these composite materials.

In a similar study, Wong et al also observed that the w_e value of 10% SGF reinforced nylon 6,6 composites and toughened with 20 vol% EPDM rubber is much higher than that of the rubber-toughened matrix without fiber reinforcements [20,21]. By increasing the fiber content above 10%, the w_e value decreases with increasing fiber content. The w_p value decreases dramatically by adding SGF (Fig. 11). As the length of glass fiber is reduced considerably during melt-compounding processes, Wong et al. used two processing routes to prepare the composites, i.e. melt mixing of all composite components followed by injection molding, and one-step injection molding without extrusion compounding. Fig. 12 shows the variations of specific essential fracture work with fiber content for the SGF reinforced rubber toughened nylon 6,6. As expected, one-step injection molded composites exhibit superior toughening in comparison with the pre-compounded composites. The decrease in toughness for the pre-compounded composites is attributed to the degradation of fiber lengths caused by the mixing process in a twin screw extruder. The in situ observation of crack opening for the rubber toughened composites containing 10-30 wt% SGF is shown in Figs. 13(a)-(c). Fiber bridging and pullout can be readily seen in these optical micrographs. These processes contribute to the major energy dissipating events.

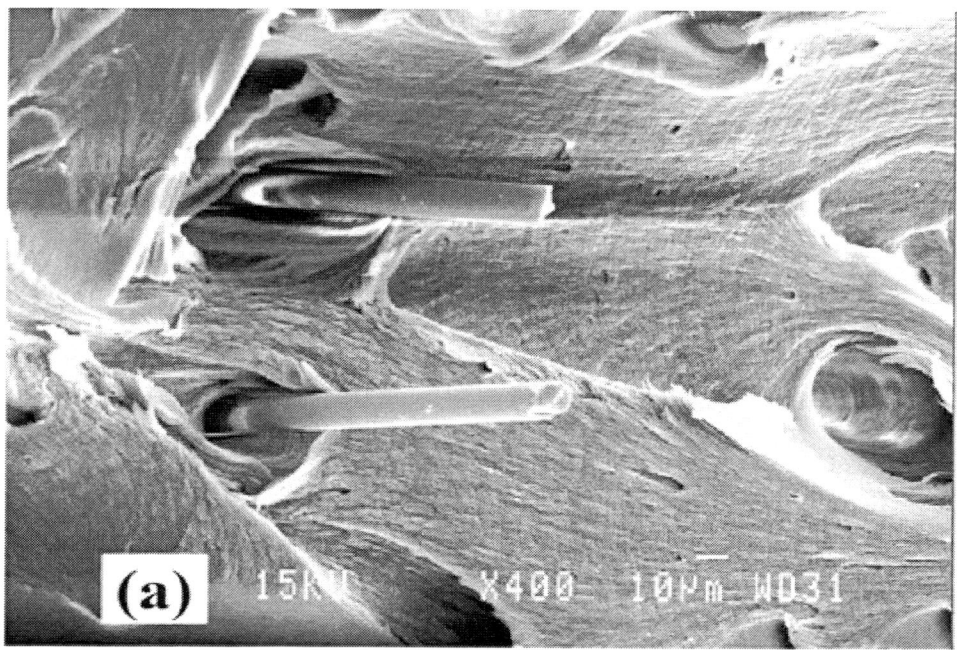

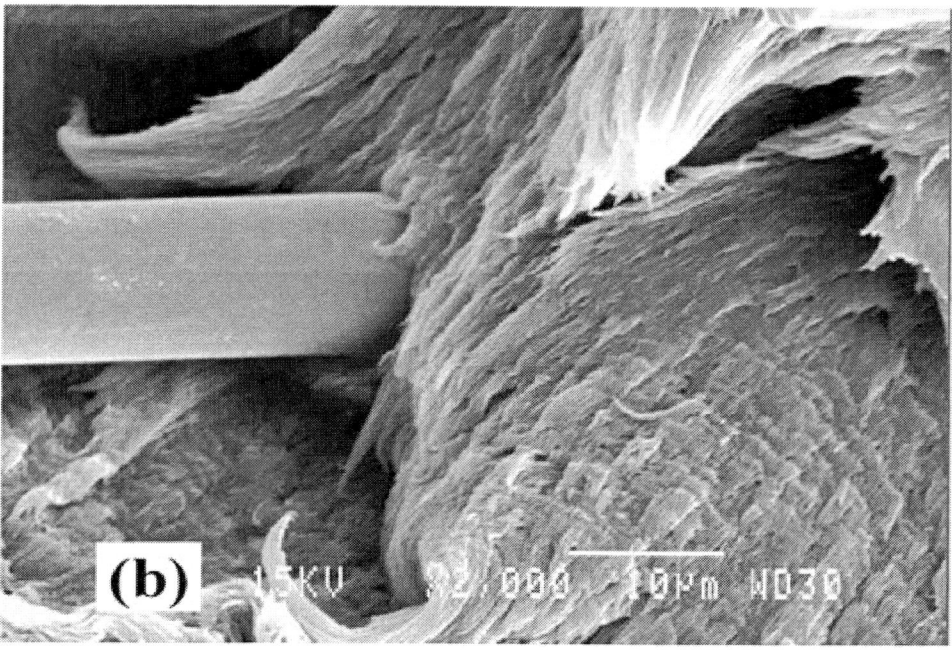

Figure 10. SEM micrographs showing the fracture surfaces of (a) (PA6,6/SEBS-g-MA)/SGF (80/20)/5 with L = 8.60 mm . (b) Higher magnification showing plastic deformation in the matrix [Ref. 18].

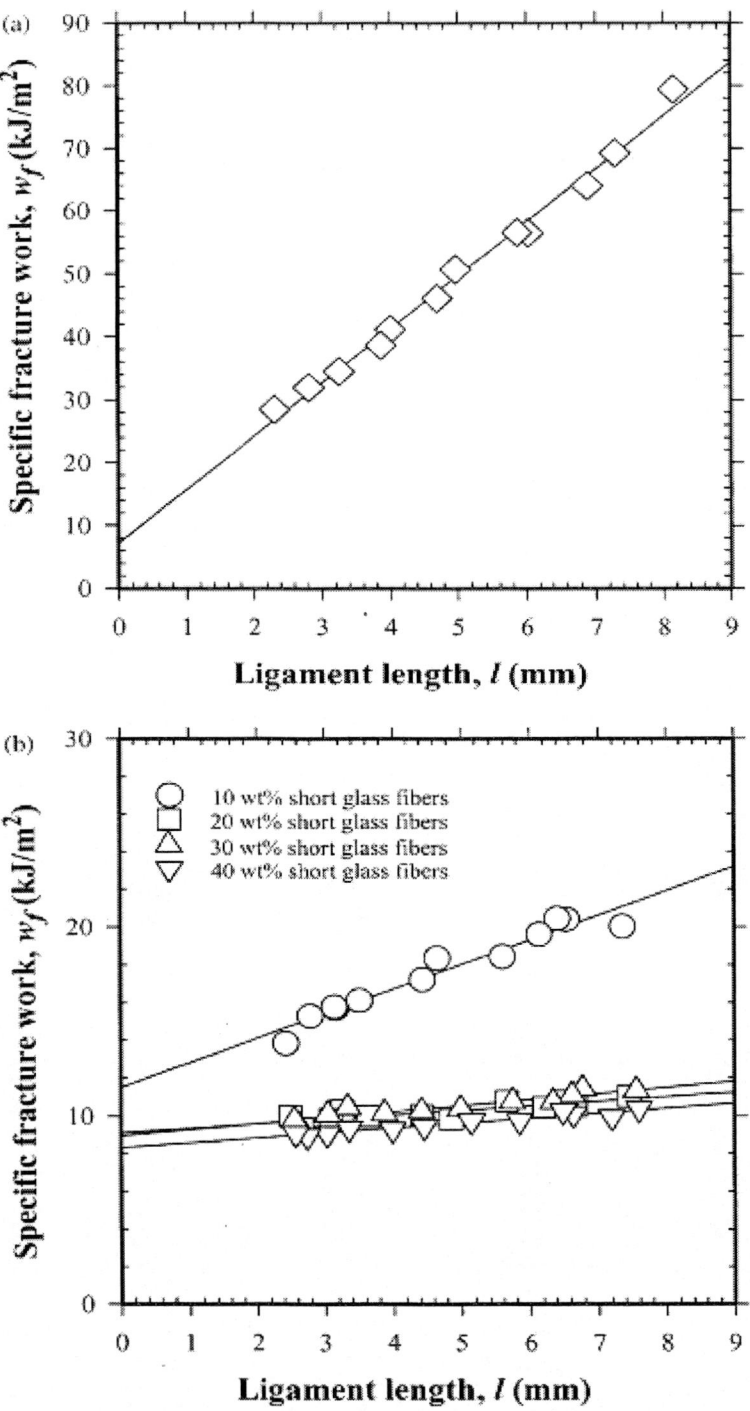

Figure 11. Specific fracture work versus ligament length for (a) unreinforced rubber toughened nylon 6,6 and (b) reinforced blends with 10, 20, 30 and 40 wt% short glass fibers. Reprinted from [Ref. 20] with permission of Elsevier.

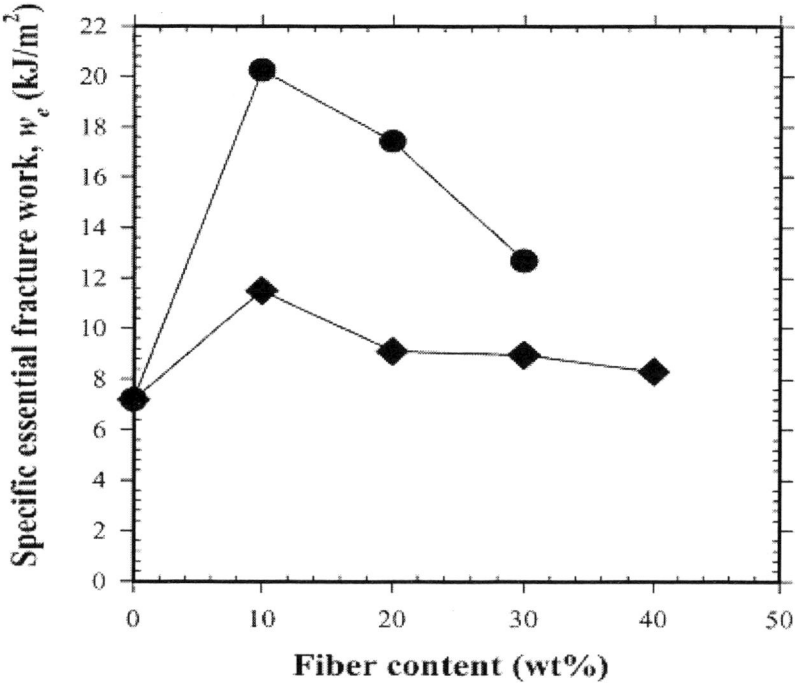

Figure 12. Specific essential fracture work for the unreinforced and fiber reinforced rubber toughened nylon 6,6 versus fiber content: (●) one-step injection molded (♦) extrusion pre-compounded. Reprinted from [Ref. 20] with permission of Elsevier.

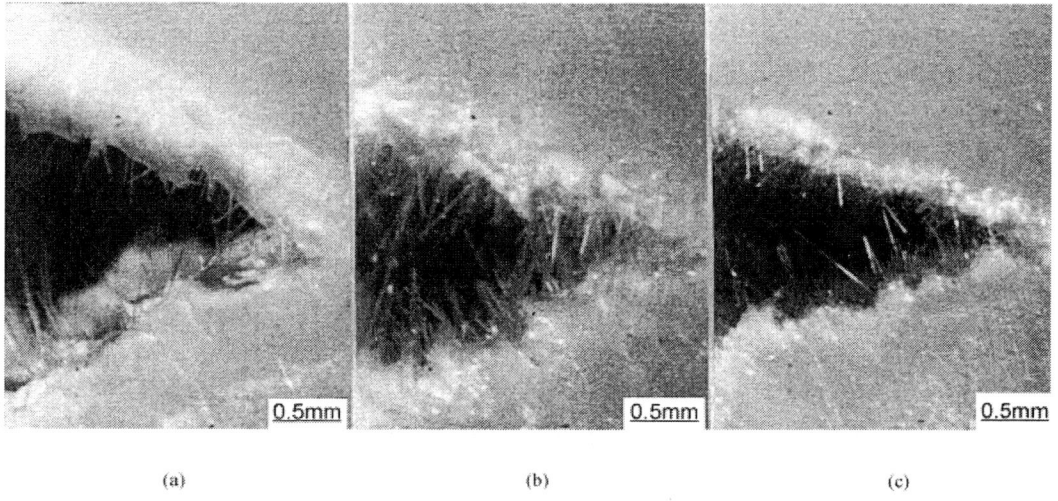

Figure 13. Photomicrographs of observations for crack opening, showing fiber bridging and pullout. (a) 10, (b) 20 and (c) 30 wt% fibers. Reprinted from [Ref. 21] with permission of Elsevier.

POLYPROPYLENE-BASED COMPOSITES

Generally, the melt-processing conditions affect the mechanical performances of SFRT considerably. This is because the processing conditions can influence the fiber orientation, fiber length, matrix degradation, etc of the resulting composites. Thus, processing parameters such as barrel temperature and injection speed must be optimized in order to obtain the composites with desired mechanical properties [22 -24]. Figs. 14(a)-(b) shows the effects of barrel temperature profile on the tensile strength and elastic modulus of injection molded PP, PP/SGF 80/20 and PP/SGF 70/30 composites, respectively.

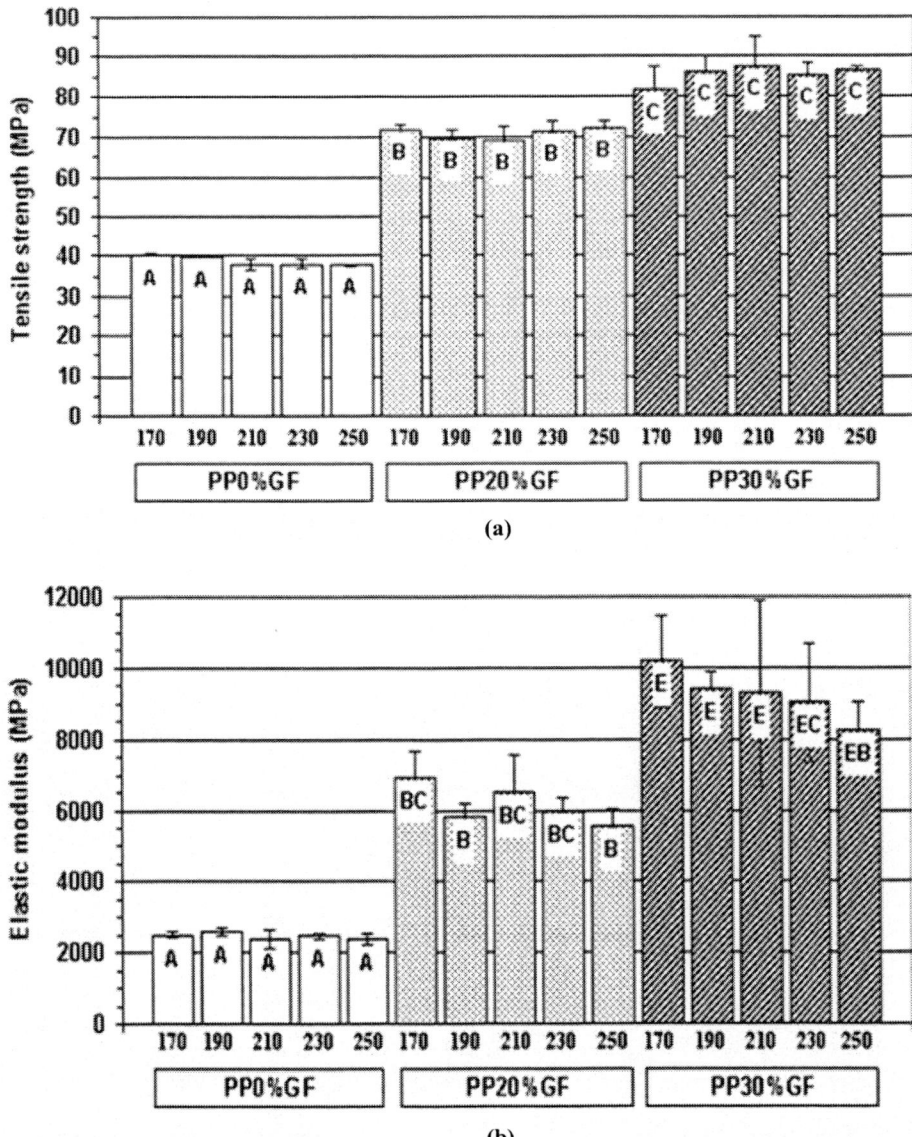

Figure 14. Variations of (a) tensile strength and (b) elastic modulus with injection temperature for PP and PP-glass fiber reinforced composites. Reprinted from [Ref. 23] with permission of Elsevier.

The temperature profiles of injection molder used for preparing these specimens are: (a) 165, 168, 170, 170 °C, (b) 185, 188, 190, 190 °C, (c) 205, 208, 210, 210 °C, (d) 225, 228, 230, 230 °C, and (e) 245, 248, 250, 250 °C. These heating profiles are simplified and denoted as 170, 190, 210, 230 and 250 °C in Fig. 16. For pure PP, the tensile strength decreases with increasing temperature due to a reduction of its molecular weight. The tensile strength of PP remains nearly unchanged by adding 20 wt% SGF possibly resulting from delayed degradation of the matrix due to reinforcement. However, the composites show a tendency of decreasing modulus with injection temperature. The increase of injection temperature drastically alters the flow behavior of the polymer melt, thereby affecting the orientation of fibers in the composites.

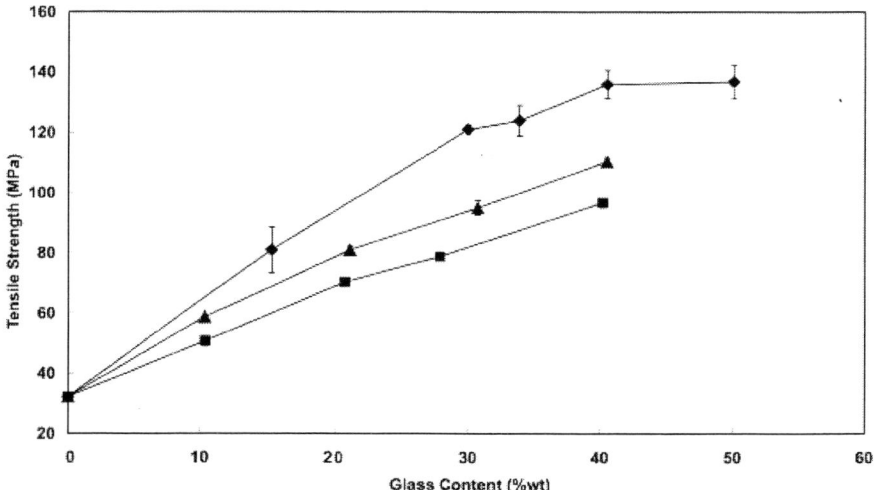

Figure 15. Tensile strength vs fiber content for long and short fiber reinforced PP composites (♦: LF 19, ■: SF19, ▲: SF14). Reprinted from [Ref. 25] with permission of Elsevier.

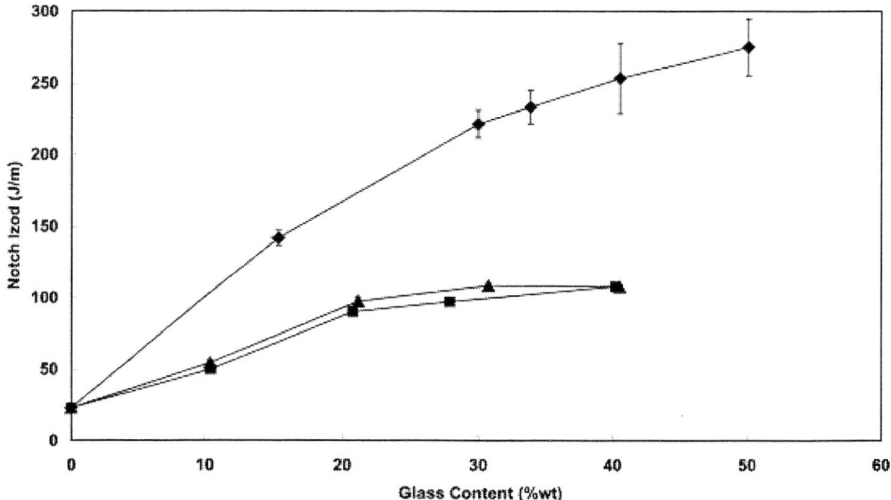

Figure 16. Notched Izod impact energy vs fiber content for long and short fiber reinforced PP composites (♦: LF 19, ■: SF19, ▲: SF14). Reprinted from [Ref. 25] with permission of Elsevier.

Thomason studied the effect of fiber length and content on the mechanical properties of glass fiber reinforced PP composites prepared by melt-mixing and injection molding [25]. Surface modified glass fibers with respective diameter of 19 and 14 μm were used to produce short fiber composites having an average residual fiber length of less than 1 mm. They were denoted as SF19 and SF14, respectively Moreover, long fiber (19 μm diameter; denoted as LF19) reinforced composites having an average residual fiber length of 4-5 mm were also fabricated for comparison purposes. Fig. 15 shows the tensile strength vs fiber content for the PP composites investigated. The tensile strength of long fiber and short fiber composites increases with increasing fiber content. The long fiber composites give higher tensile performance as expected. A lower diameter yield higher performance in short fiber composites. The tensile elongation of long and short fiber composites decreases markedly with increasing fiber content (not shown). However, the notched Izod impact strength of both composites improves with increasing fiber content (Fig. 16). The presence of fibers increases the amount of energy required to propagate an existing crack via several mechanisms such as fiber pullout and matrix deformation.

Polypropylene is a commodity polymer in which its properties can be enhanced and extended into the applications of high performance engineering materials by adding both short glass fibers and elastomers. PP is nonpolar and thus lacks interaction with SGF. Accordingly, grafted form of PP with polar functionality (e.g. PP-g-MA) is commonly added to PP in order to improve the interfacial adhesion between SGF and PP matrix. Tjong et al. studied the compatibilizing effect of MA functional group on the mechanical performances of SGF reinforced composites and toughened with SEBS elastomer [26-28]. MA can be either grafted to PP to form PP-g-MA or SEBS to form SEBS-g-MA. Table 5 summarizes the chemical compositions and tensile properties of unmaleated and maleated SGF/SEBS/PP composites.

Table 5. Compositions (wt. %) and tensile properties of PP composites [Ref. 26 and 27]

Sample	PP	PP-g-MA	SEBS	SEBS-g-MA	SGF	σ_y (MPa)	E (GPa)	ε_b(%)
SGF/SEBS/PP	61.5	---	15.4	---	23.1	33	4.25	3.9
SGF/SEBS-g-MA/PP	61.5	---	---	15.4	23.1	54	4.75	5.0
SGF/SEBS/mPP	58.4	3.1	15.4	---	23.1	51.5	4.22	7.8
SGF/SEBS-g-MA/mPP	58.4	3.1	---	15.4	23.1	49.7	4.15	7.3

* σ_y Yield stress; E Young's modulus; ε_b Elongation at break

The composites were prepared by an initial melt-mixing of the blend/composite components in a twin screw extruder followed by injection molding. Fig. 17 shows the cross-sectional optical micrograph of the SGF/SEBS/mPP composites view along the mold filling direction (MFD). A typical skin/core structure is observed. The fibers align preferentially along MFD in the skin layer and orient randomly in the core region. For the SGF/SEBS/PP composite containing no MA group, the interfacial adhesion between SGF and PP matrix is poor, thus this material has the lowest yield strength of 33 MPa. The yield stress of composites can be improved dramatically by adding PP-g-MA or SEBS-g-MA due to effective stress transfer across the fiber-matrix interface

Figure 17. Cross-sectional optical micrograph of the SGF/SEBS/mPP composite view along the mold filling direction [Ref. 26].

Figs. 18(a)-(b) show SEM fractographs of the SGF/SEBS/PP and SGF/SEBS-g-MA/PP composites subjected to Charpy impact test at a speed of 3 m/s. The fiber surfaces of SGF/SEBS/PP composite are very clean, indicating that the bonding between the SGF and the matrix is relatively weak. The fibers can be pullout easily from the matrix during mechanical deformation. The interaction between SGF and PP is limited because SGF has a polar surface but PP is non-polar polyolefin. In contrast, the fiber surfaces of SGF/SEBS-g-MA/PP composite are coated with a thin layer of deformed matrix material, demonstrating good interfacial adhesion between the fibers and the polymer matrix. A similar polymer coated SGF morphology is observed for the SGF/SEBS/mPP and SGF/SEBS-g-MA/mPP composites.

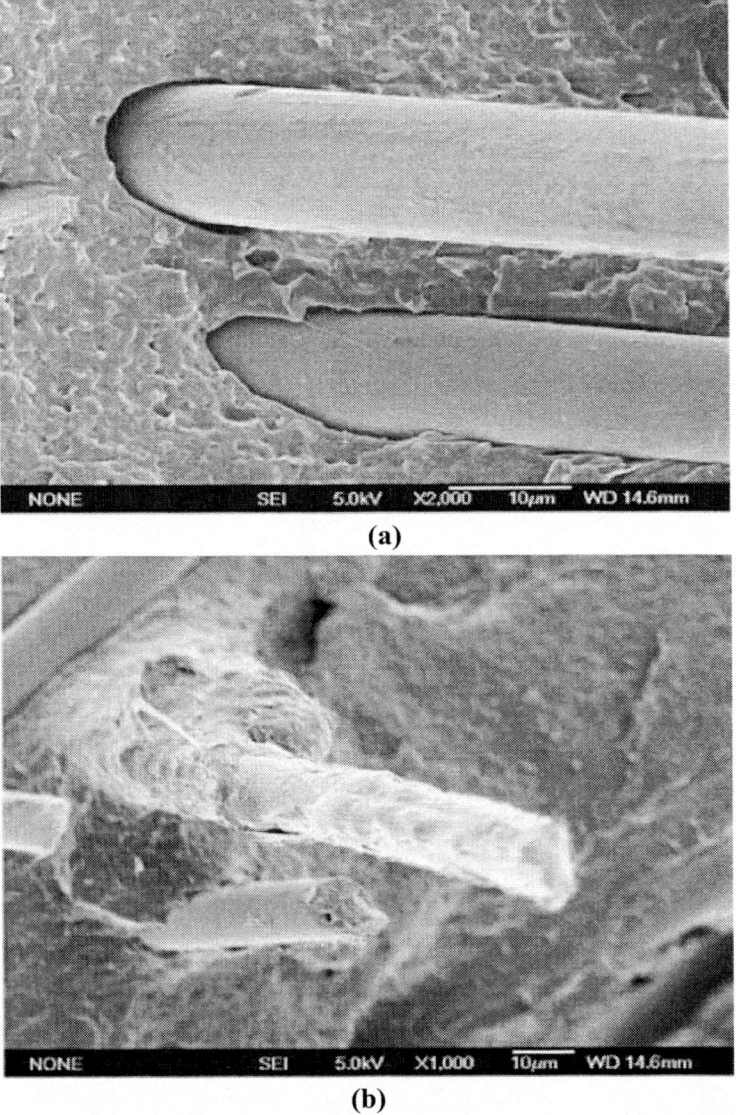

Figure 18. SEM micrographs showing the fracture surface morphology of (a) SGF/SEBS/PP and (b) SGF/SEBS-g-MA/PP composites tested at an impact speed of 3 m/s [27].

Figures 19(a)-(b) show the load-displacement curves for the SGF/SEBS/PP and SEBS/SEBS-g-MA/PP composites. The plots of w_f versus ligament length for SGF/SEBS/PP and SEBS/SEBS-g-MA/PP composites are shown in Fig. 20(a)-(b). The SGF/SEBS/PP composite exhibits the highest specific work of fracture value (29.2 kJ/m^2) as a result of extensive fiber pull-out. The fiber pullout contributes to the main energy dissipation process for SGF/SEBS/PP composite. Replacing SEBS elastomer with SEBS-g-MA in the composite leads to a sharp decrease in fracture toughness to 11.55 kJ/m^2. The incorporation of PP-g-MA into the polymer matrix of composite also results in a reduction in fracture toughness (Table 6). Clearly, strong interfacial bonding between the fibers and the polymer matrix restricts the fiber pull out, thereby yielding low fracture toughness.

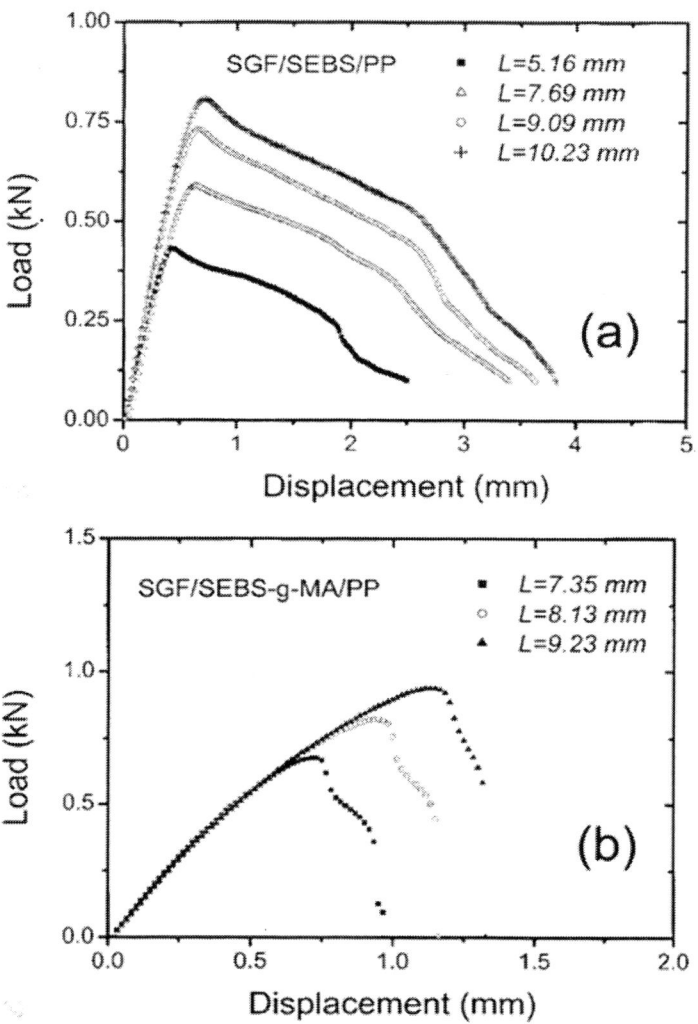

Figure 19. Load-displacement curves for (a) SGF/SEBS/PP and (b) SGF/SEBS-g-MA/PP composites [Ref. 27].

Table 6. Specific essential work of fracture and specific plastic work for PP composites toughened with elastomer [Ref 26 and 27]

Sample	w_e (kJ/m^2)	βw_p (MJ/m^3)
SGF/SEBS/PP	29.20	2.43
SGF/SEBS-g-MA/PP	11.55	1.28
SGF/SEBS/mPP	7.81	2.92
SGF/SEBS-g-MA/mPP	7.85	2.11

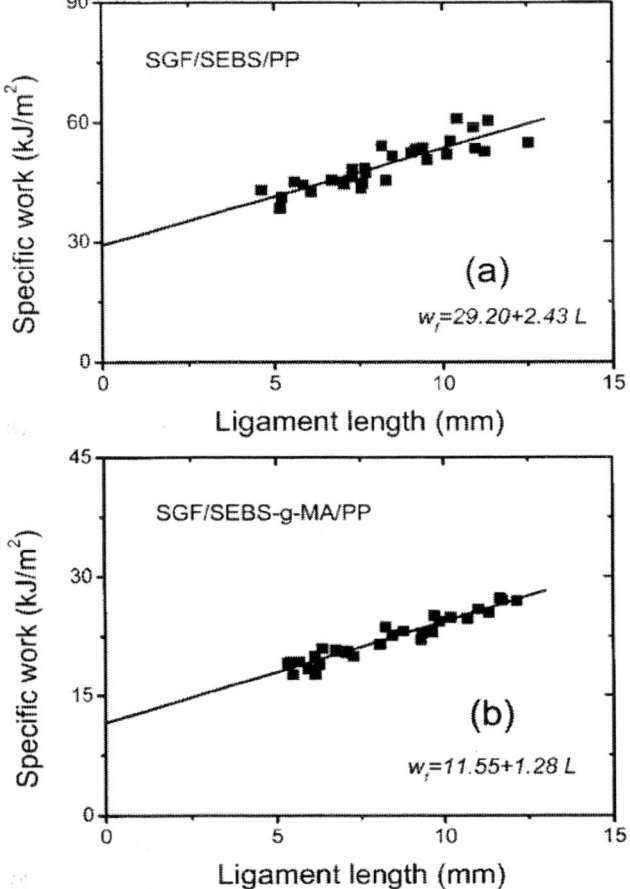

Figure 20. w_f versus L plots for (a) (a) SGF/SEBS/PP and (b) SGF/SEBS-g-MA/PP composites [Ref. 27].

The effect of short carbon fiber (SCF) additions on the mechanical performance of PP composites is now considered. Fu et al. prepared the PP/SCF and PP/SGF composites via melt-mixing followed by injection molding [29, 30]. Figs. 21(a)-(b) show number-average residual fiber length vs fiber content for PP/ SGF and PP/SCF composites. The residual length of both fibers decreases with increasing fiber content due to greater fiber-fiber interaction. Moreover, the mean carbon length is less than the mean glass fiber length owing to the carbon fibers are more brittle and fracture more easily during processing.

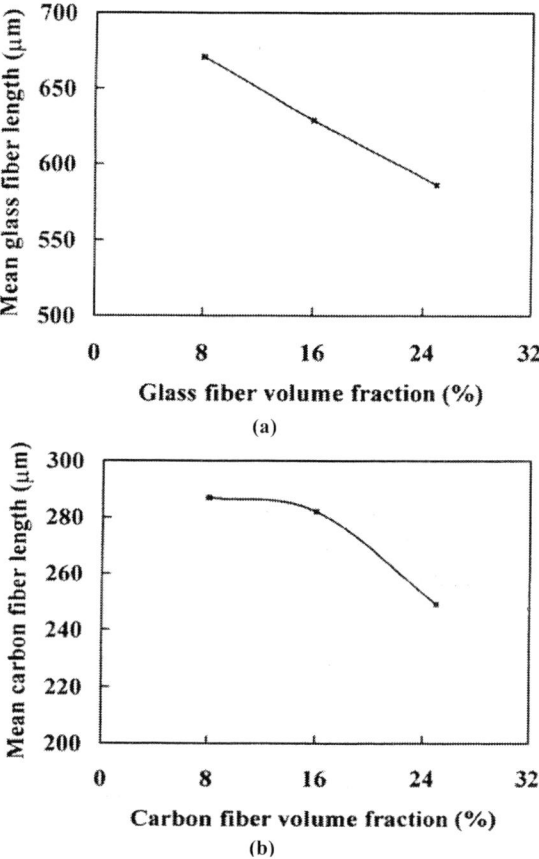

Figure 21. Number-average fiber length vs fiber content for (a) PP/SGF and (b) PP/SCF composites. Reprinted from [Ref. 29] with permission of Elsevier.

Fig. 22 shows the variation of tensile strength with fiber content for PP/SGF and PP/SCF composites. The additions of glass and carbon fibers are beneficial to enhance the tensile strength of PP. The tensile strength of PP/SCF composites is higher than that of PP/SGF composites because carbon fibers exhibit a much higher strength than glass fibers. Furthermore, the tensile strength of both composites increases slightly with increasing fiber content. This can be explained in terms of the competition between fiber content and mean fiber length. The increase in tensile strength caused by the increase in fiber content is offset by the decrease in composite strength due to the reduction in mean fiber length. On the other hand, the elastic modulus of both composites increases markedly with increasing fiber content (Fig. 23). In this case, the elastic modulus is more dependent on fiber content than on fiber length. The incorporation of high modulus carbon fiber to PP leads to the PP/SCF composites exhibit remarkable high stiffness values.

As mentioned previously, carbon fibers are very strong, stiff and expensive reinforcements, while glass fibers are relatively much cheaper and have lower stiffness. Combining the advantages of both kinds of fibers, one can design composites containing hybrid reinforcements with desired mechanical properties. In this context, Fu et al also prepared PP hybrids reinforced with both short glass and carbon fibers by injection molding [31]. The total glass and carbon content is fixed at 25%. Fig. 24 shows the variations of

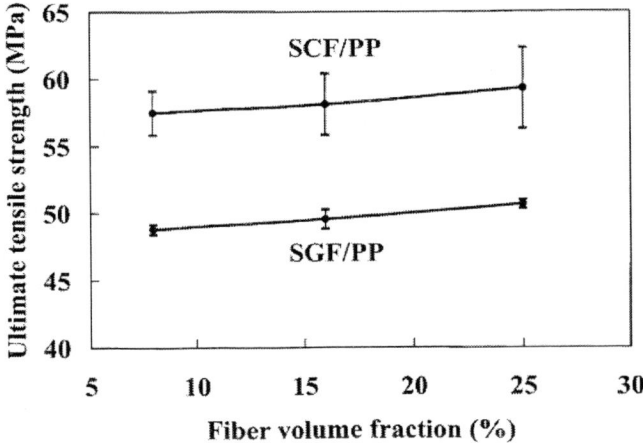

Figure 22. Tensile strength vs fiber volume content for PP/SGF and PP/SCF composites Reprinted from [Ref. 30] with permission of Elsevier.

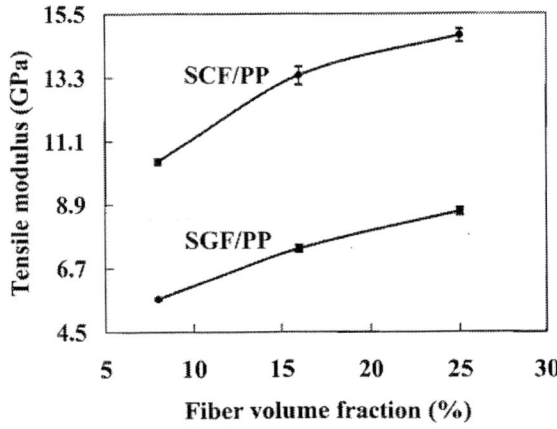

Figure 23. Tensile modulus vs fiber volume content for PP/SGF and PP/SCF composites Reprinted from [Ref. 30] with permission of Elsevier.

fracture toughness with relative carbon fiber content for the PP hybrids. The fracture toughness was determined using compact tension (CT) specimens. In the measurements, L-crack specimen having crack propagated parallel to mold filling direction whilst T-cracked specimen having crack perpendicular to MFD were used. The schematic illustrations showing orientation of fibers and pre-crack notch direction for the composite and hybrid is shown in Fig. 25. It can be seen from Fig. 24 that the T-cracked specimens exhibit higher fracture toughness than the L-cracked specimens for all the composites. This is because the fiber alignment in the T-cracked and L-cracked specimens with respect to MFD is different. Fibers aligned perpendicular to the notch direction obstructs the crack propagation, leading to zigzag path mode. In contrast, the crack propagates more easily in a straightforward mode when fibers aligned parallel to the notch direction (Fig. 26). From Fig. 24, The fracture toughness of the hybrid PP/(SCF+SCF) hybrids is higher than that for both the single PP/SGF and PP/SCF composites. Furthermore, the values of fracture toughness of the hybrids are higher than those predicted by the rule of hybrid mixtures (the dotted lines), showing synergistic toughening effect of both kinds of fibers.

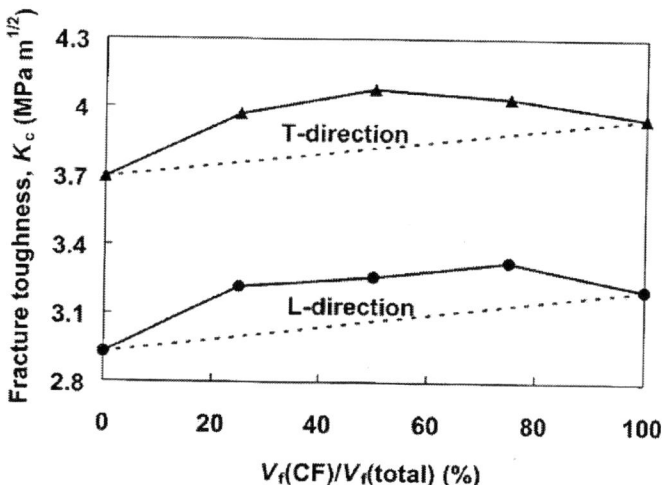

Figure 24. Fracture toughness vs relative carbon fiber content for PP hybrids. Reprinted from [Ref. 31] with permission of Elsevier.

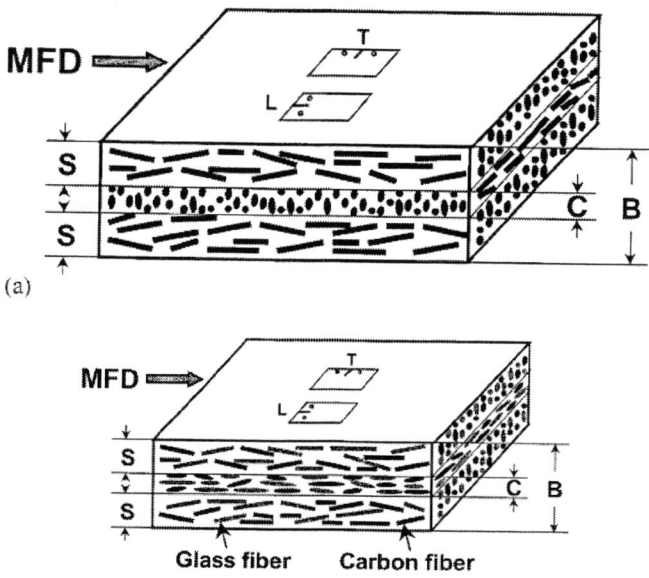

Figure 25. Schematic illustrations showing skin-core structure and notch orientation with respect to MFD in compact tension specimens for (a) single and (b) hybrid composites. Reprinted from [Ref. 31] with permission of Elsevier.

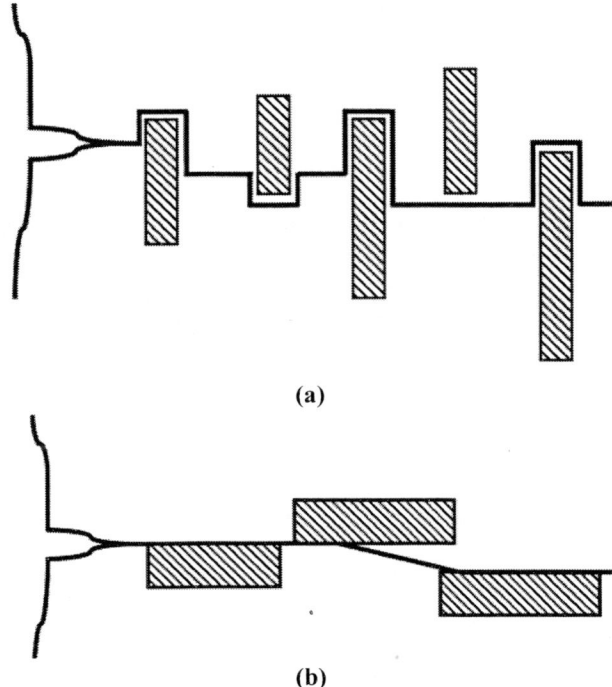

Figure 26. Schematic illustrations showing fibers (a) perpendicular and (b) parallel to crack propagation.

REINFORCED PA/PP BLENDS

Polymer blending is an effective route to make new polymer materials at low cost and to combine the performances of the corresponding pure polymers. Recently, considerable efforts have been devoted to tailor the mechanical properties of SFRT by varying the composition of matrix blends. The PA6,6/PP blend represents a wide range of obtainable matrix materials prepared by changing the PA6,6/PP ratio. In general, maleated SEBS has been used by several researchers to toughen and to compatibilize the PA6,6/PP and PA/PP blends [32,33]. It was shown that the maximum fracture toughness at a given PA6,6/PP ratio can be obtained when the ratio of maleated SEBS to un-maleated SEBS is 50/50 [32]. Fu et al. studied the effects of PA6,6/PP ratio on the mechanical properties of SGF reinforced and rubber-toughened PA6,6/PP blends. The polymer blends containing various PA6,6/PP ratios, plus a mixture of 10 wt% SEBS and 10 wt% SEBS-g-MA. Glass fiber was added to the blends at a weight ratio of 60/40 polymer/SGF. The composites were prepared via melt mixing followed by injection molding process [34,35]. Fig. 27 shows the stress-strain curves for SGF (12 μm) reinforced PA6,6/PP composites. The tensile strength of the composites at the 100/0 and 75/25 PA6,6/PP ratios exhibit relatively high values. The results of tensile and Charpy impact measurements are listed in Table 7. This Table reveals that the composite strength increases while the elastic modulus decreases with increasing PA6,6/PP ratio. Impact results indicate that the composites at intermediate PA6,6/PP ratios (75/25 or 50/50) exhibit higher impact strength than that of composites with the 100/0 or 0/100 PA6,6/PP ratio.

Table 7. Mean fiber aspect ratio (l/d), critical fiber ratio (l_c/d), interfacial adhesion strength and mechanical properties of SGF reinforced and rubber-toughened PA6,6/PP blends. Reprinted from Ref. [34] with permission of Elsevier

PA6,6/PP	l/d	l_c/d	τ (MPa)	τ/τ_m	Tensile strength (MPa)	Elastic modulus (GPa)	Failure strain (%)	Notched Charpy impact energy (kJ/m^2)
100/0	18.23	26.02	38.43	1.92	78.25 ± 1.50	6.41 ± 0.17	1.95 ± 0.30	7.29 ± 0.16
75/25	20.90	27.58	36.26	2.27	74.57 ± 0.73	6.70 ± 0.15	3.22 ± 0.10	11.35 ± 4.32
50/50	26.86	43.01	23.25	2.22	47.28 ± 0.57	6.94 ± 0.20	3.02 ± 0.31	12.09 ± 0.41
25/75	32.44	65.00	15.38	2.06	47.36 ± 0.55	7.57 ± 0.55	2.62 ± 0.12	11.51 ± 0.68
0/100	33.22	95.06	10.52	1.06	34.55 ± 0.46	8.01 ± 0.28	1.27 ± 0.05	9.01 ± 1.36

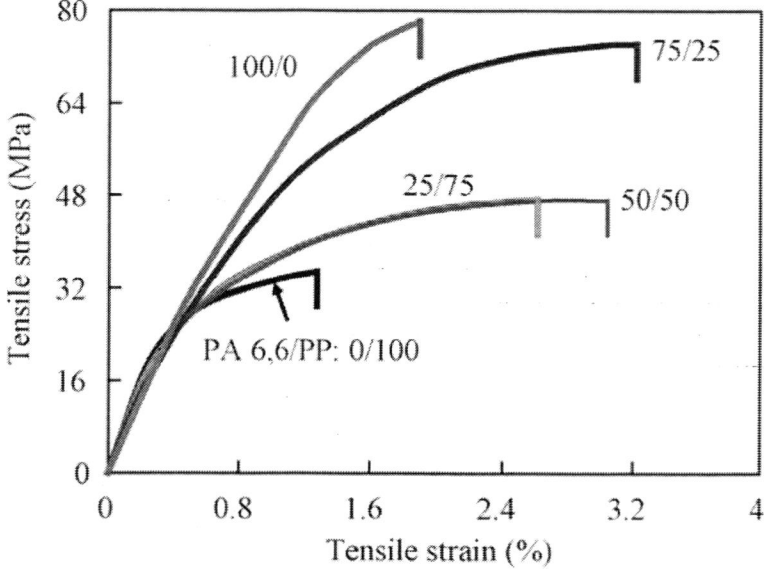

Figure 27. Stress-strain curves for SGF reinforced PA6,6/PP composites. Reprinted from [Ref. 34] with permission of Elsevier.

Fig. 28 shows the SEM fractographs of SGF reinforced and rubber toughened PA6,6/PP blends after impact testing. It can be seen that the fibers are sheathed with deformed matrix materials for the composites with medium PA6,6/PP ratios (Figs. 28(c) and (d)). Fu et el. determined the interfacial adhesion strength using Equation (5), matrix shear yield strength (τ_m; equals half of the matrix tensile yield strength), mean fiber length of all composites investigated. The results are also listed in Table 7. It can be seen that the ratios of interfacial adhesion sthength to matrix shear yield strength are higher at medium PA6,6/PP ratios than at the 100/0 and 0/100 PA6,6/PP ratios. Consequently more matrix shear deformation takes place in the vicinity of fiber surfaces for the medium PA6,6/PP ratios. Therefore, control of the PA6,6/PP ratio is considered to be an effective way to produce composites with optimal mechanical properties.

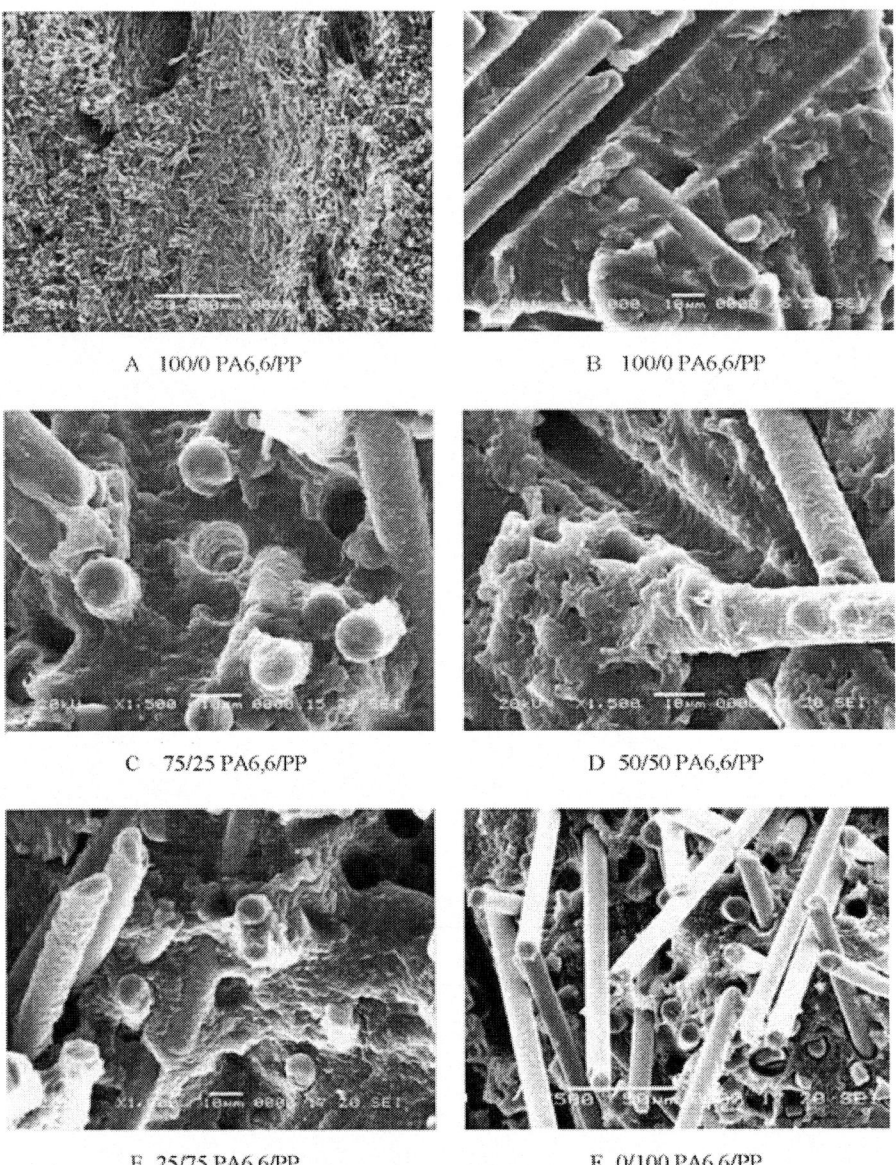

Figure 28. SEM fractographs of SGF reinforced PA6,6/PP composites after impact testing. Reprinted from [Ref. 35] with permission of Elsevier.

In another study, Wong and Mai used EWF and J-integral methods to determine the fracture toughness of the SGF reinforced 75/25 PA6,6/PP blend and toughened with 20 wt% maleated SEBS elastome [36]. The MA contents in the blend were controlled by varying SEBS/SEBS-g-MA from 0 to 10 wt.% SEBS-g-MA at five different levels. The SGF was added at a ratio of 80/20 polymer/SGF. The composite was prepared via melt mixing followed by injection molding into 6 mm thickness plaques. Single edge-notched bend (SENB) specimens were used for the EWF evaluation. The SENB specimens with a larger thickness (6 mm) generally meet the thickness criterion for plane-strain condition. It has been demonstrated by Wu and Mai [37] that the plane-strain specific essential work of fracture

(w_{Ie}) is equivalent to the plane-strain J_{IC} value obtained via ASTM E813-81. In other words, the w_{Ie} is a plane strain fracture toughness value provided that the required specimen thickness for plane strain is satisfied. The specific essential work of fracture determined under plane-strain is considerably lower than that obtained under plane-stress condition. The plane-strain specific essential work of fracture is a material constant parameter, independent of sample geometry used (DENT or SENB). Figs. 29(a)-(b) show the plain strain fracture toughness of the SGF reinforced 75/25 PA6,6/PP blend with SEBS-g-MA (MA content: 0.37 %) determined from EWF and J-integral method. The toughness value of the composite obtained from EWF agrees reasonably with that determined from the multiple specimen J-integral method and straight line extrapolation. From Fig. 29(a), two data points lying below the regression line are excluded from the toughness evaluation. The deviation from linear line is believed to be associated with a skin-core effect of the injection molded composite. The fracture toughness values for composite specimens with different MA contents are summarized in Table 8. Apparently, the fracture toughness of composite specimens depends on the MA-grafted SEBS content to a greater extent. In other words, the MA-grafted SEBS content affects the degree of interfacial adhesion between various blend components (i.e. PA66, PP and SEBS), and the size of dispersed phases. In the PA66/PP blends containing SEBS-g-MA, the non-polar olefinic rubber can mix well with the PP domains and the grafted anhydride groups exhibit a strong affinity for the amine-end groups on nylons, which then establish an imide linkage with SEBS at the interphase [38]. The MA-grafted SEBS content also controls the interfacial bonding between these blend components and glass fibers [34]. Thus the compatibilizing effect of MA group in the SGF reinforced composites having multiple phases is a complicated issue. Poor interfacial adhesion (MA content = 0) and strong interfacial bonding (MA content = 1.47) do not favor formation of tough polymer composites. At an appropriate MA content of 0.37, composite with high fracture toughness can be produced. This clearly demonstrates that the interfacial bonding strength between the fibers and the polymer matrix can be neither very strong nor too weak. From Table 8, it is noted that the ASTM E813-89 procedure defined by the 0.2 mm offset line yields higher fracture toughness values for all composite specimens. The discrepancy is attributed to the extra energy dissipation in this procedure by a 0.2 mm crack growth. The $J_{0.2}$ value is determined from the intersection of J-R curve with 0.2 mm offset line parallel line to the blunting line. Thus $J_{0.2}$ is not a true fracture toughness value for crack initiation.

Table 8. Fracture toughness value of SGF reinforced 75/25 PA6,6/PP blends with SEBS-g-MA. Reprinted from Ref [36] with permission of John Wiley & Sons

MA content (%)	ASTM E813-89, $J_{0.2}$ (kJ/m^2)	Linear Extrapolation for J_{IC} (kJ/m^2)	Specific essential fracture work (kJ/m^2)	βw_p (MJ/m^3)
0	Not available	Not available	4.4	0.47
0.37	12.1	10.8	9.6	1.01
0.92	8.0	7.6	6.2	0.90
1.47	4.8	4.4	3.6	0.86

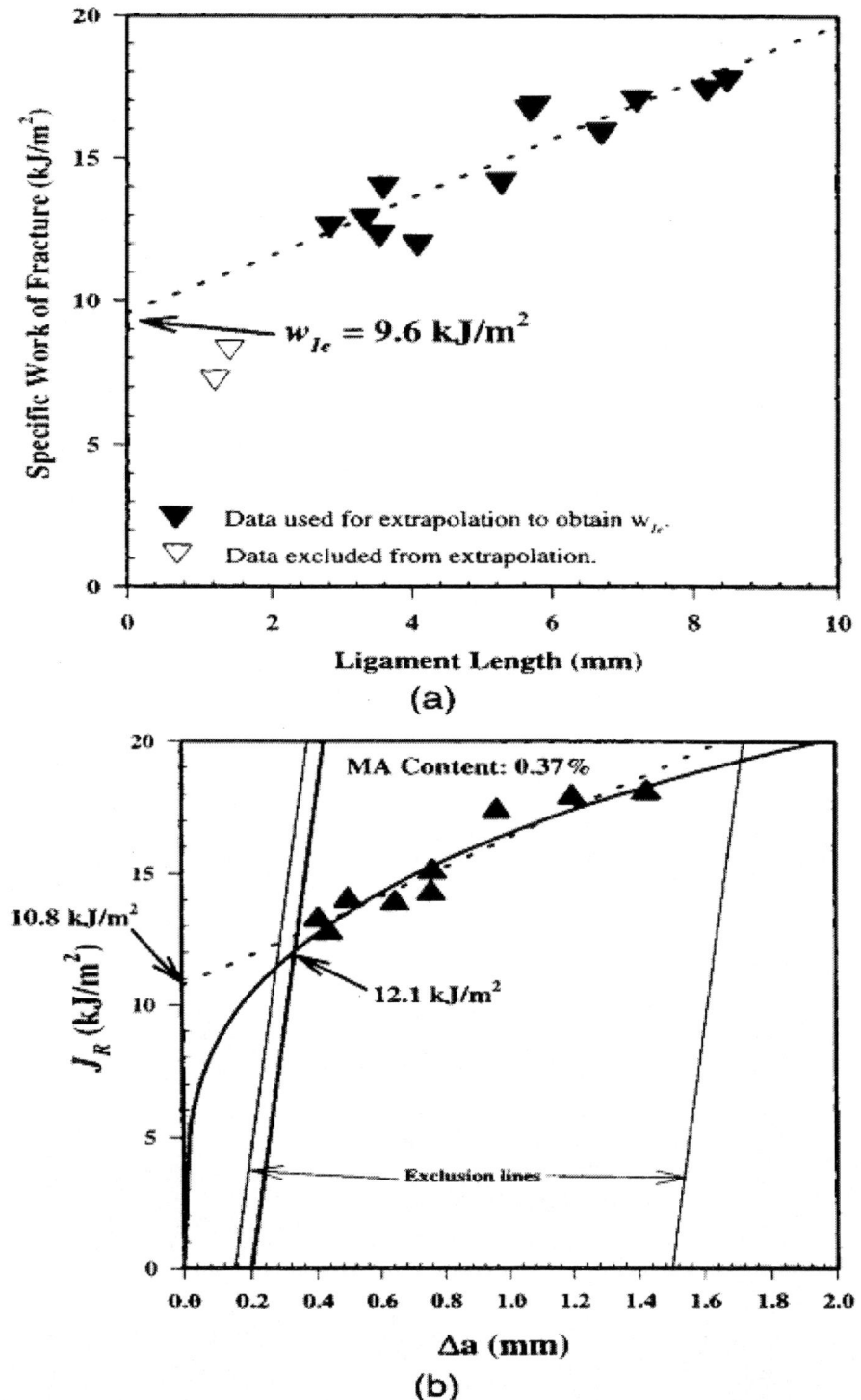

Figure 29. Plain strain fracture toughness of the SGF reinforced 75/25 PA6,6/PP blend with SEBS-g-MA (MA content: 0.37 %) determined from (a) EWF method and (b) J-R curve. Reprinted from [Ref. 36] with permission of John Wiley & Sons.

POLYESTER BASED COMPOSITES

PBT is an aromatic polyester widely used to produce injection molded articles due to its relatively high crystallization rate, good chemical and physical properties as well as good processability. However, PBT exhibits low ductility at high deformation rates and at low temperatures. It can be improved by adding impact modifiers such as acrylonitrile-butadiene styrene (ABS), core-shell rubber, acrylate-styrene-acrylonitrile, etc. However, PBT with super-toughness can be achieved by blending with functionalized rubbers. The most common functionalized groups used for grafting rubbers are maleic anhydride and epoxy. During melt blending, the carboxylic or hydroxyl end groups of PBT can react with functionalized rubber to produce grafted copolymers that compatiblize the blend. Typical functionalized rubbers include MA-grafted EPR, MA-grafted EPDM and MA-grafted poly(ethylene-octene) [39-41].

PBT is susceptible to hydrolysis when subjected to hydrothermal aging at high temperatures. Polymers used in engineering applications are often subjected to moistures or high temperatures for prolonged periods of time. Thus, the investigation of the environment effects on the mechanical and fracture properties of PBT-based composites is becoming increasingly important as the SGF reinforced PBT composites are used in electronic, communication and automobile applications. Mohd Ishak et al. investigated the effects of thermal aging on the tensile properties of SGF reinforced PBT composites [42]. Some fibers were treated with a silane coupling agent, i.e. 3-aminopropyltriethoxysilane (3-APE).The composite specimens were hygrothermally aged (HA) by immersing them in water at 80 °C and 100 °C. Once the water uptake in the specimens reached a saturation limit, tensile tests were performed subsequently. Some HA specimens were also subjected to further re-drying (RD) treatment in a vacuum at 80 °C for 72 h. Fig. 30 shows the effects of thermal aging at 100 °C on the tensile properties of PBT and its composites reinforced with untreated SGF of different volume fractions. The tensile modulus and strength of the as-received (AR) composites increase with increasing fiber content at the expense of their ductility. HA treatment reduces the tensile strength, stiffness and ductility of the composites significantly. PBT and its composites become very brittle with elongation at break of less than 1.3 %. This implies that the moisture degrades the PBT matrix of the composites considerably, leading to poor mechanical performances. Fig. 31 shows the effect of silane coupling addition on the tensile performances of PBT/15% SGF composite in the as received, HA and RD conditions. For untreated PBT/15% SGF composite, HA treatment at 100 °C degrades the tensile properties more seriously compared to HA at 80 °C. However, the silane agent improves the stiffness and strength of the composite due to enhanced interfacial bonding. This leads to an increase in the efficiency of stress transfer from PBT matrix to the SGF during tensile loading. The coupling agent is also beneficial in improving the retention in the tensile properties of PBT/15% SGF composite subjected to HA at 80 °C and 100 °C.

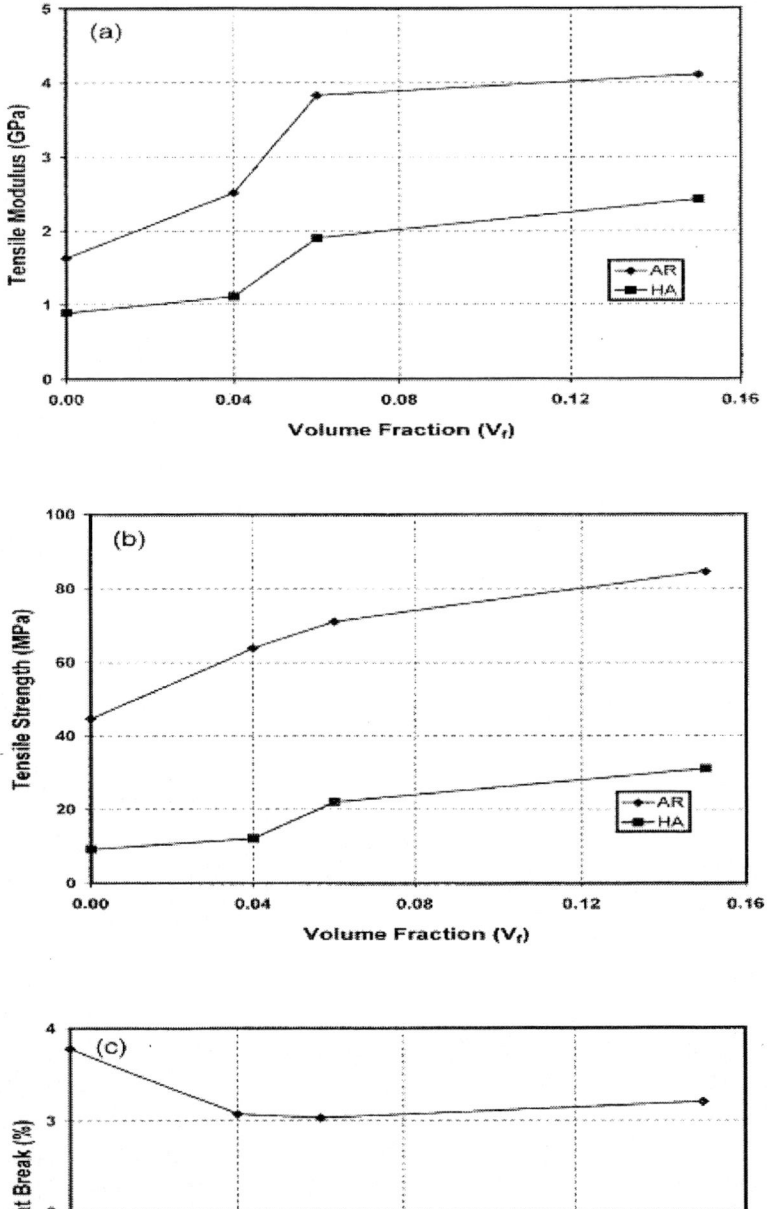

Figure 30.

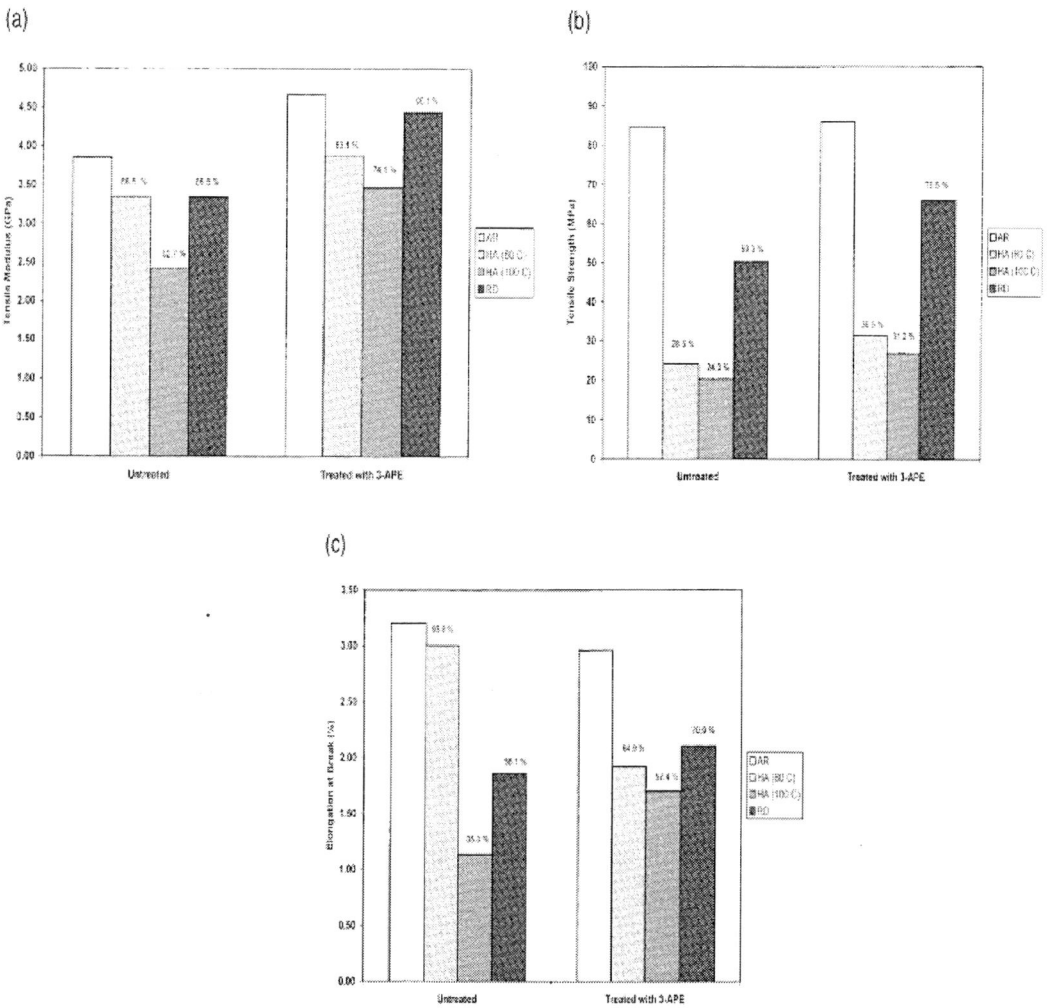

Figure 31. Effects of hydrothermal aging (HA) at 100 °C on (a) tensile modulus, (b) tensile strength and (c) elongation at break of PBT and its SGF composites. Reprinted from [Ref. 42] with permission of Elsevier.

In another study, Mohd Ishak et al. studied the effect of HA treatment on the mechanical and fracture properties of PBT/SGF composites and toughened with 20 wt% core-shell rubber (CSR) [43]. Table 9 summarizes the tensile properties of PBT, PBT/30% SGF and PBT/30% SGF composite toughened with 20 wt% CSR under AR, HA and RD conditions. HA treatment at 90 °C leads to drastic reduction in tensile strength and ductility of PBT and its composites. All specimens become very brittle with tensile elongation ≤ 1.3%. Fig. 33 shows the variations of critical fracture toughness (K_c) and strain energy release rate (G_c) of PBT/CSR/SGF composite. Both fracture toughness values increase with increasing fiber content as a result of a combination of fiber bridging and pullout. The variation of K_c with temperature for pure PBT, PBT/30% SGF and PBT/CSR/SGF composites is shown in Fig. 34. Apparently, addition of 30 wt% SGF to PBT leads to higher fracture toughness at all testing temperatures as compared to that of PBT-CSR matrix. The ineffectiveness of CSR rubber to toughen the composite is related to the nature of impact modifier consisting of rigid

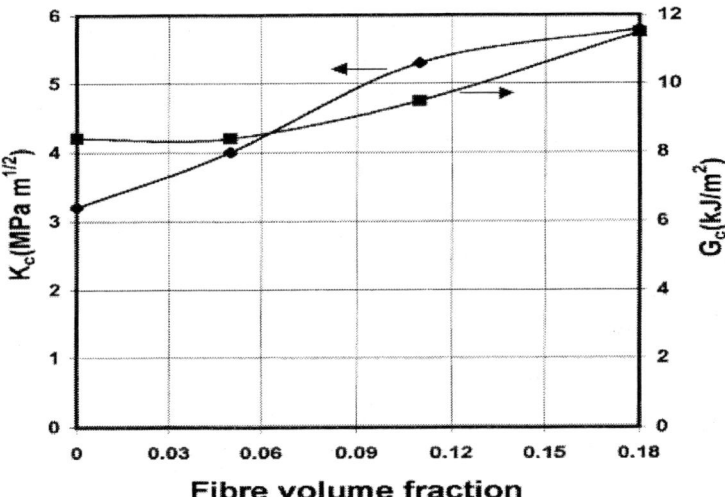

Figure 32. Effect of coupling agent on (a) tensile modulus, (b) tensile strength and (c) elongation at break of PBT/15%SGF composite in AR, HA and RD conditions (% values indicate percentage retention). Reprinted from [Ref. 42] with permission of Elsevier.

SAN shell and soft acrylate core. The rigid elastomer restrains the matrix deformation, leading to embrittlement of PBT. However, the K_c value of PBT/CSR/SGF composite approaches to that of PBT/30% SGF at 80 °C. This is due to the pullout of matrix resin sheathed fibers above the glass transition temperature of the matrix. Fig. 35 shows the effect of immersion temperature on the K_c of pure PBT, PBT/30% SGF and PBT/CSR/SGF composites. HA treatment at 90 °C leads to a marked reduction in the fracture toughness of PBT. This is because HA treatment at 90 °C reduces the tensile strength and ductility of PBT (Table 9). Accordingly, pure PBT becomes very brittle and incapable of supporting the applied stress. In contrast, PBT/30% SGF and PBT/CSR/SGF composites exhibit much higher fracture toughness than pure PBT when subjected to HA treatment at 90 °C. The PBT/CSR/SGF composite is even tougher than PBT/30%SGF. This demonstrates that the SGF and CSR act synergistically to produce a positive effect on fracture toughness.

CONCLUSIONS

This chapter gives an overview of the microstructure and the toughening mechanisms of the SFRT with polymer matrix based on PA, PBT or PP. Short fiber reinforced polymer microcomposites offer a wide range of attractive structural applications because of their enhanced stiffness and tensile strength as well as low specific gravity properties. The mechanical properties of SFRT can be enhanced and extended into the applications of high performance engineering materials either by adding elastomers and controlling the blend compositions of polymer matrix, or by hybridization of glass fibers with high modulus carbon fibers. Further development of such microcomposites depends on comprehensive understanding of their processing, structure and correlation relationship between these

Table 9. Tensile properties of PBT, PBT-glass fiber (30 wt% glass fiber) and PBT-CSR-glass fiber (20 wt% CSR + 30 wt% glass fiber) before AR and after hygrothermal aging (HA; RD)[a] Reprinted from Ref. [43] with permission of Elsevier.

Material	Unaged (AR)[b]		Aged at 30°C				Aged at 90°C			
			HA[b]		RD[b]		HA		RD	
	Tensile strength (MPa)	Tensile strain (%)	Tensile strength (MPa)	Tensile strain (%)	Tensile strength (MPa)	Tensile strain (%)	Tensile strength (MPa)	Tensile strain (%)	Tensile strength (MPa)	Tensile strain (%)
PBT	55.9	8.0	50.6	16.1	54.4 (97.3)[c]	11.4 (142.8)	11.6	0.8	19.52 (34.93)	1.1 (13.4)
PBT-glass fiber	109.2	3.5	100.0	3.0	108.7 (99.5)	3.3 (94.2)	38.9	1.1	59.6 (55.5)	1.5 (44.1)
PBT-CSR-glass fiber	97.6	3.3	87.5	2.9	95.6 (98.0)	2.9 (88.9)	46.1	1.3	77.8 (79.7)	2.1 (61.5)

[a] Testing conditions: 20°C, 1 mm/min.
[b] AR, as-received; HA, hygrothermally aged; RD, redried.
[c] Percentage recovery with respect to AR samples shown in parentheses.

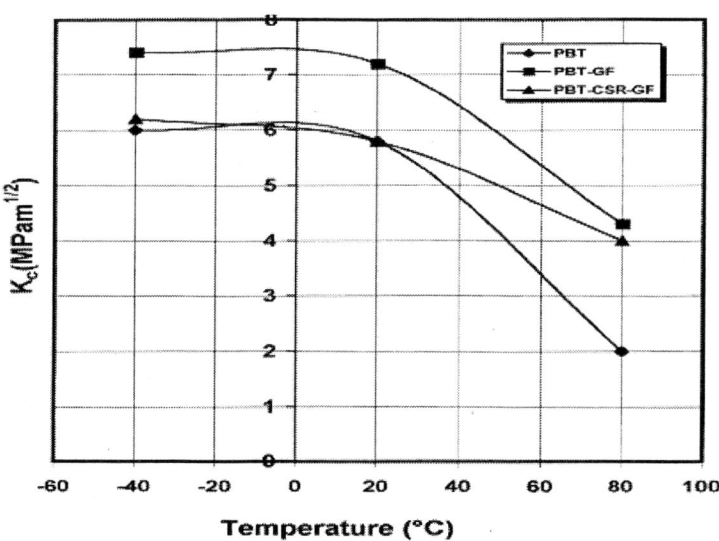

Figure 33. The variations of K_c and G_c with glass fiber loading for (PBT/CSR)/SGF (80/20)/30 composite. Reprinted from [Ref. 43] with permission of Elsevier.

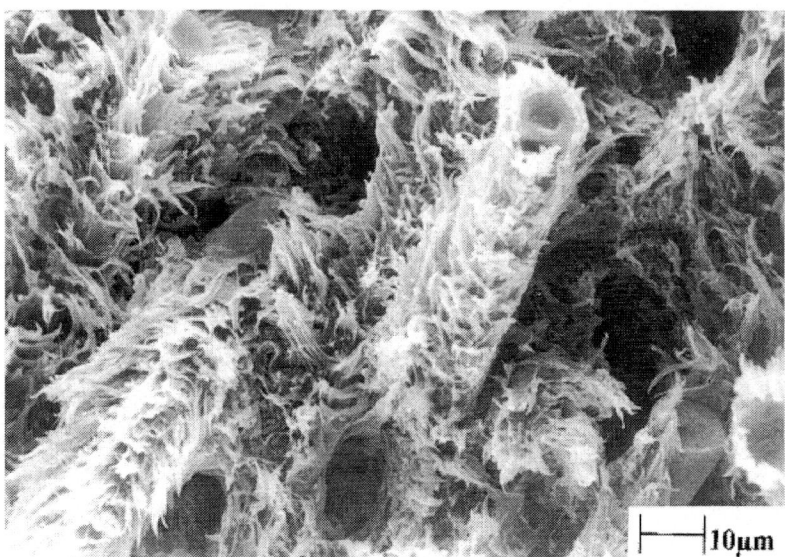

Figure 34. The effect of testing temperature on the K_c of PBT, PBT/30% SGF and (PBT/CSR)/SGF (80/20)/30 composite. Reprinted from [Ref. 43] with permission of Elsevier.

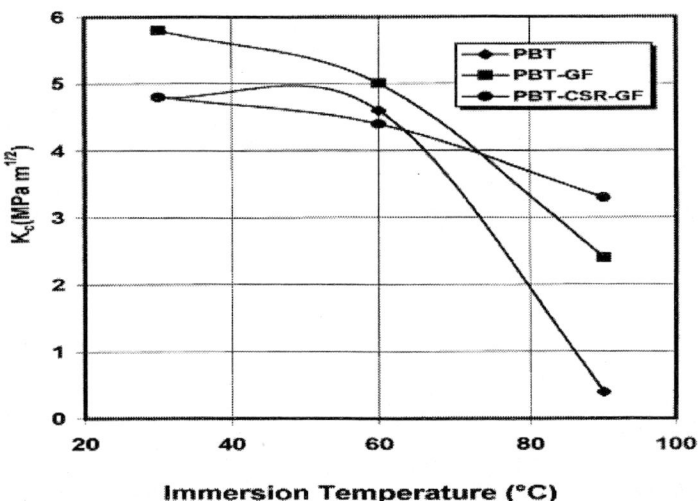

Figure 35. The effect of immersion temperature of HA treatment on the K_c of PBT, PBT/30% SGF and (PBT/CSR)/SGF (80/20)/30 composite. Reprinted from [Ref. 43] with permission of Elsevier.

properties. SFRT are commonly fabricated by means of the melt-compounding processes due to its versatility, low cost and mass production capability. However, the length of fibers decreases markedly during processing, thereby reducing the reinforcing effect of short fibers. For SFRT, the applied load is transmitted through the matrix to the fiber by shear transfer mechanism. The interfacial bond between the fibers and the matrix is a crucial factor controlling the reinforcing efficiency of SFRT. However, interfacial bonding strength between the fibers and the polymer matrix can be neither very strong nor too weak. Strong interfacial adhesion prevents fiber pullout. And the fracture toughness of SFRT derives from the energy dissipation during fiber pullout. A balance between them must be maintained in order to achieve desired tensile strength and toughness. Super-tough composites based on polyamides and toughened with 20% elastomer can be prepared by adding short glass fibers up to 15%. Combined fiber- and rubber-toughening mechanisms are responsible for enhanced fracture toughness of these composite materials. The essential work of fracture approach has been used successfully to characterize the fracture toughness of SFRT toughened with elastomers. Finally, moisture and temperature play crucial roles on the mechanical behavior and fracture toughness of PBT-based composites.

REFERENCES

[1] Inagaki, M. New Carbons: Control of Structure and Functions, Elsevier, Oxford, 2002.
[2] Fu, S.Y.; Mai, Y.W.; Ching, E.C.; Li, K.Y. Composites Part A 2002, vol 33, 1549 - 1555.
[3] Dibenedetto, A.T. Mater Sci Eng A 2001, vol 302, 74-82.
[4] Noda, K.; Tsuji, M.; Takahara, A.; Kajiyama, T. Polymer 2002, vol 43, 4055 -4062.
[5] Pak, S.H.; Caze, C. J Appl Polym Sci 1997, vol 65, 143 -153.
[6] Zhou, X.; Dai, G.; Guo, W.; Lin, Q. J Appl Polym Sci 2000, vol 76, 1359 -1365
[7] Roux, C.; Denault, J.; Champagne, M.F. J Appl Polym Sci 2000, vol 78, 2047 -2060.
[8] Ozkoc G.; Bayram, G.; Bayramli, E. Polymer 2004, vol 45, 8957 -8966.
[9] Bikiaris, D.; Matzinos, P.; Prinos, J.; Flaris, V.; Larena, A.; Panayiotou, C. J Appl Polym Sci 2001, vol 80, 2877 - 2888.
[10] (a) ASTM Standard E813-81, Standard Test Method for J_{IC}, A Measure of Fracture Toughness in 1981 Annual Book of ASTM Standards, American Society for Testing and Materials, Philadelphia, PA, 1981, p.810.
(b) ASTM Standard E813-89, Standard Test Method for J_{IC}, A Measure of Fracture Toughness in 1989 Annual Book of ASTM Standards, American Society for Testing and Materials, Philadelphia, PA, 1989, p. 700.
[11] Broberg, K. B. Int J Frac 1968, vol 4, 11- 16.
[12] Clutton, E. In Fracture Mechanics Testing Methods for Polymers, Adhesives and Composites; Moore, D. R.; Pavan. A.; Williams, J. G.; Eds.; ESIS Publication 28; Elsevier, New York, 2001, 177 -196.
[13] Arencon, D.; Velasco, J.I.; Realinho, V.; Antunes, M.; Maspoch, M.L. Polym Test 2007, vol 26, 761-769.
[14] Wu, S.H.; Wang, F.Y.; Ma, C.C.; Chang, W.C.; Kuo, C.T.; Kuan, H.C.; Chen, W. J. Mater Lett 2001, vol 49, 327 -333.
[15] Thomason, JL Compos Sci Technol 2001, vol 61, 2007 -2016.
[16] Laura, D.M.; Keskkula, H.; Barlow, J.W.; Paul, D. R. Polymer 2002, vol 43, 4673 - 4687.
[17] Laura, D.M.; Keskkula, H.; Barlow, J.W.; Paul, D.R. Polymer 2003, vol 44, 3347 - 3361.
[18] Tjong, S.C.; Shi, S.A.; Li, R.K.; Mai, Y.W. Compos Sci Technol 2002, vol 62, 2017- 2027.
[19] Tjong, S.C.; Shi, S.A.; Mai, Y.W. Mater Sci Eng A 2003, vol 347, 338-345.
[20] Sui, G.X.; Wong, S.C.; Yue, C.Y. Compos Sci Technol 2001, vol 61, 2481-2490.

[21] Sui, G.X.; Wong, S.C.; Yue, C.Y J Mater Process Technol 2001, vol 113, 167-171.
[22] Jancar, J. J Mater Sci. 1996, vol 31, 3983 -3987.
[23] Ota, W.N.; Amico, S.C.; Satyanarayana, K.G. Compos Sci Technol 2005, vol 65, 873 - 881.
[24] Gullu, A.; Ozdemir, A.; Ozdemir, E. Mater Design 2006, vol 27, 316 -323.
[25] Thomason, JL. Composites Part A 2002, vol 33, 1641-1652.
[26] Tjong, S.C.; Shi, S.A.; Li, R.K.; Mai, Y.W. Compos Sci Technol 2002, vol 62, 831-840.
[27] Tjong, S.C.; Shi, S.A.; Li, R.K.; Mai, Y.W. Polym Int 2002, vol 51, 1248-1255.
[28] Tjong, S.C.; Shi, S.A.; Mai, Y.W. J Appl Polym Sci 2003, vol 88, 1384-1392.
[29] Fu, S.Y.; Lauke, B.; Mader, E.; Hu, X.; Yue, C. Y. J Mater Proc Technol 1999, vol 89-90, 501-507.
[30] Fu, S.Y.; Lauke, B.; Mader, E.; Yue, C. Y; Hu, X. Composites Part A 2000, vol 31, 1117 -1125.
[31] Fu, S.Y.; Mai, Y.W.; Lauke, B.; Yue, C. Y. Mater Sci Eng A 2002, vol 323, 326-335.
[32] Wong, S.C.; Mai, Y.W. Polymer 1999, vol 40, 1553-1566.
[33] Kim, G. M.; Michler, G.H.; Rosch, J.; Mulhaupt, R. Acta Polymer 1998, vol 49, 88-95.
[34] Fu, S.Y.; Lauke, B.; Zhang, Y.H.; Mai, Y.W. Composites Part A, 2005, vol 36, 987 -994.
[35] Fu, S.Y.; Lauke, B.; Zhang, Y.H.; Mai, Y.W. Composites Part B, 2006, vol 37, 182-190.
[36] Wong, S.C.; Mai, Y.W. Polym Eng Sci 1999, vol 39, 356-364.
[37] Wu, J.: Mai, Y. W. Polym Eng Sci 1996, vol 36, 2275 -2288.
[38] Wong, S.C.; Mai, Y.W. Polymer 1999, vol 40, 1533 -1566.
[39] Cecere, A.; Greco, R.; Ragosta, G.; Scarinzi, G.; Taglialatela, A. Polymer 1990, vol 31,. 1239-1244..
[40] Arostegui, A.; Gaztelumendi, M.; Nazabal, J. Polymer 2001, vol 42, 9565-9574.
[41] Matin, P., Devaux, J., Legras, R., van Gurp, M., van Duin, M. Polymer 2001, vol 42, 2463 -2478.
[42] Mohd Ishak, Z. A.; Ariffin, A.; Senawi, R. Eur Polym J. 2001, vol 37, 1635 -1647.
[43] Mohd Ishak, Z. A.; Ishiaku, U. S.; Karger-Kocsis, J. Compos Sc. Technol 2000, vol 60, 803-815.

In: Structural Materials and Engineering
Editor: Ference H. Hagy

ISBN 978-1-60692-927-8
© 2009 Nova Science Publishers, Inc.

Chapter 2

RECYCLED AGGREGATE STRUCTURAL CONCRETE: A METHODOLOGY FOR THE PREDICTION OF ITS PROPERTIES

Jorge de Brito

Department of Civil Engineering and Architecture, Instituto Superior Técnico, Technical University of Lisbon, Portugal

ABSTRACT

Structural concrete made with recycled aggregates from construction and demolition waste is an eco-efficient solution to reduce both the production of waste and the depletion of natural non-renewable materials. Even though this material shows great potential as an alternative to conventional concrete (made with primary aggregates), its large-scale use has been hampered by lack of regulation and technical documentation. Furthermore, some production procedures hinder the use of recycled aggregates under normal waste collection conditions.

In this Chapter, the specific problems of preparing concrete with recycled aggregates are addressed and the properties of recycled aggregates *versus* primary (natural stone) aggregates are compared.

A general methodology to predict the long-term performance (in terms of both mechanical properties and durability) of concrete made with recycled aggregates using the corresponding performance of equivalent conventional concrete (reference concrete) is proposed and validated. The approach employed includes an international literature review and a summation of experimental campaigns monitored / supervised by the author.

This methodology allows the early estimation of the properties of concrete made with recycled aggregates based on the properties of the aggregates mix and of young concrete, which can be used both by the structural designer and the contractor, thus effectively removing a barrier to the widespread use of this material.

ABBREVIATIONS

CDW	construction and demolition waste;
CPA	coarse primary (natural) aggregates;
CRA	coarse recycled aggregates;
FPA	fine primary (natural) aggregates;
FRA	fine recycled aggregates;
MRA	mixed (concrete and masonry) recycled aggregates;
PA	primary (natural) aggregates;
RA	recycled aggregates;
RAC	recycled aggregate concrete;
RC	reference concrete (i.e. without recycled aggregates);
RCA	recycled concrete aggregates;
RMA	recycled (brick and/or stone) masonry aggregates.

INTRODUCTION

The Construction and Demolition Waste Problem

The construction industry, even though essential to the progress of Society, is also a serious contributor to major environmental impacts. One of the most visible aspects of these impacts is construction and demolition waste (CDW), which represents from 20% to 30% (including soils) of all the solid waste produced each year.

Even though innocuous in comparison with other types of waste, the absence of a viable reuse for CDW contributes to the depletion of natural non-renewable - even if abundant - resources (stone turned into gravel and sand), to the dramatic decrease of the landfills where they are dumped instead of other more dangerous industrial waste and, when dumped illegally as frequently happens, to an unacceptable onslaught on the environment and each person's individual freedom to enjoy Nature.

Therefore, this problem needs to be addressed and, as with other similar waste materials, the best solution is to reduce their production (e.g. by building constructions that last much longer). When that is not possible, the second-best strategy is to find and implement viable uses (from a technical and economic point of view) for this material. The construction elements may be reused as they are in other locations and with similar functions and, when that is not possible either (as happens with most of bulk CDW), the materials may be recycled.

This last option can be divided into three levels: *upcycling*, when the recycled use of the materials is of greater value than its original one (this is not usually possible with CDW); *cycling* or *recycling*, when the quality demand of the new use is similar to the original one; and *downcycling*, when the materials are used for a less demanding function than the original one, i.e. there is very little profit from their intrinsic value (in terms of cost or potential properties). This is the commonest way in which CDW is being recycled at the moment. Since most of bulk CDW consists of inert materials (concrete, brick masonry and mortars), CDW is largely recycled as aggregate in roads, streets and the floors of buildings, used in the landscaping of spent mining infrastructures, or as drainage layers of landfills and in concrete production.

Unfortunately, in many countries none of these options is chosen (though this is happening increasingly less often) and the CDW ends up in landfills or dumped beside remote roads.

Production of CDW in Europe

Table 1 shows the average values of CDW production in various European countries, taken from a paper presented by the European Union in 2003 (Muthmann 2003) to which the average annual growth rates determined from Eurostat Environmental Statistics were added, as well as an estimation of *per capita* production. These statistics have the drawback of including data from the 1990s, probably outdated now, and also of resulting from different criteria for estimating the amount in each country (e.g. whether soil is taken into account or not, the use of enquiries or intelligent guesses).

Table 1. CDW production in the EU by country (adapted from (Muthmann 2003) in (Gonçalves 2007))

Country	Average CDW production (1000 tonnes)	Average annual growth	Time scale	Population in 2005 (millions of inhabitants)	Per capita production (kg/person)
Belgium	6 559	n/a	1994	10.4	631
Denmark	2 787	6.05%	1992-2000	5.4	516
Germany	238 580	2.07%	1996-2000	82.5	2 892
Greece	1 898	3.90%	1996-2000	11.1	171
Spain	22 000	n/a	1991	43.0	512
France	24 300	-0.05%	1991-1997	59.9	406
Ireland	2 012	27.19%	1995-1998	4.1	491
Italy	26 226	-4.20%	1991-1999	58.5	448
Luxembourg	4 359	42.16%	1997-1999	0.5	8 717
Netherlands	15 604	4.33%	1990-2001	16.3	957
Austria	27 500	n/a	1999	8.2	3 354
Finland	33 545	4.12%	1997-1999	5.2	6 451
United Kingdom	70 625	0.39%	1990-1999	60.0	1 177
Norway	1 840	-3.74%	1990-2000	4.6	400
Switzerland	6 393	n/a	1998	7.5	852
Cyprus	555	-2.34%	1990-1999	0.7	793
Czech Republic	8 486	16.55%	1998-2001	10.2	832
Estonia	294	16.08%	1995-2000	1.3	226
Latvia	39	n/a	2001	2.3	17
Lithuania	231	10.00%	2000-2001	3.4	68
Malta	970	-2.29%	1990-2001	0.4	2 424
Poland	668	2.94%	1998-2001	38.2	17
Romania	623	27.68%	1995-2000	21.7	29
Slovakia	477	-6.80%	1998-2000	5.4	88
Slovenia	427	35.64%	1995-2001	2.0	213
Croatia	290	n/a	2000	4.4	66
Total	497 285			467.2	1 064

The result is that there are some dubious values, too low or too high, suggesting a figure of 1064 kg per person annually, which is most probably higher than the real one. Therefore, these results should be viewed with caution especially where they differ most from a previous estimate, put together in the so-called Symonds report (1999).

In Table 2 some more updated statistics for some European countries are presented, also containing their CDW recycling ratio. These values clearly represent a positive evolution compared with existing literature. For example, in the Symonds report (1999) it is stated that 180 million tonnes are produced in Europe every year, leading to a *per capita* yearly production of 480 kg, less than half the figure in Table 1, which may be explained by soil not being considered in the Symonds report and possibly also by an increasing trend.

The origin of this CDW, according to the Waste Centre of Denmark (WCD) (2005), is as follows: 5%-10% from new construction works, 20%-25% from rehabilitation works and 70%-75% from demolition sites. These percentages may vary widely from country to country, reflecting their level of development / stagnation (in less developed countries the relative influence of new construction is dramatically higher).

Table 2. Reuse of CDW in selected countries in Europe (UEPG 2006)

Country	CDW produced (M tonnes)	Recycling M tonnes	%	Dumping M tonnes	%	Source
Netherlands	25.8	24.6	95.3	0.9	3.5	Survey 2002-2003, 2005
United Kingdom	91.0	81.8	89.9	9.2	10.1	QPA, 2003
Germany	213.9	186.3	87.1	27.6	12.9	Monitoring-Bericht, 2002
Belgium	14.0	12.0	85.7	2.0	14.3	WS and FDERECO, 2005
France	309.0	195.0	63.1	114.0	36.9	Survey FNTP et Ademe, 2001

In terms of composition, CDW varies a lot and the first thing that matters is whether it results from new construction, rehabilitation work or demolition (Figure 1). In the last case, the age of the construction demolished has a great influence on the materials but there are major differences in buildings of similar age where different structural materials have been used (e.g. a reinforced concrete frame with solid slabs *versus* structural brick masonry walls supporting timber pavements). Statistics from different countries exhibit this heterogeneity (compare Table 3 for Denmark with Figure 1 for Brazil).

Table 3. CDW composition in Denmark in 2003 (DEPA 2005)

Materials	%
Concrete	25
Coil and rocks	22
Asphalt	19
Roof tiles	6
Other construction waste	11
Other waste	8
Other recyclable waste	6
Non-incineratable waste	3

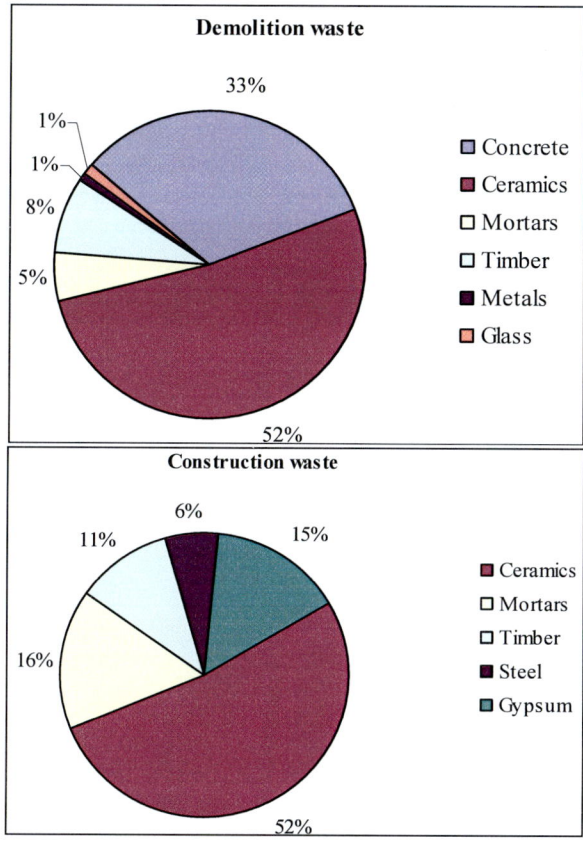

Figure 1. Composition of CDW from a demolition and a new construction site (Angulo 1998)

Production of Aggregates in Europe

In terms of aggregates, various countries have been increasing the reuse of CDW for that effect. Table 4 shows the production of aggregates by type in each country while Figures 2 and 3 indicate the relative importance of recycled aggregates (RA) production in the overall aggregates market, which covers 3 million tonnes per year (Arab 2006).

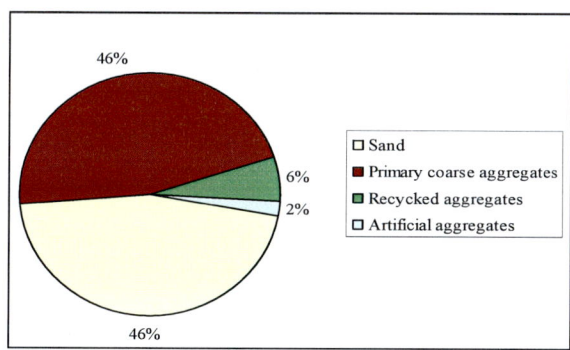

Figure 2. Aggregates produced by type in the EU in 2005 (UEPG 2006)

Table 4. Production of aggregates in the EU by country in 2005 (UEPG 2006)

Country	Aggregate production (Mt)				
	Sand	Primary coarse aggregates	Recycled aggregates	Artificial aggregates	Total
Austria	66.0	32.0	3.5	3.0	104.5
Belgium	13.9	38.0	12.0	1.2	65.1
Czech Republic	25.5	38.0	3.4	0.3	67.2
Finland	53.0	45.0	0.5	n/a	98.5
France	170.0	223.0	10.0	7.0	410.0
Germany	263.0	174.0	46.0	30.0	513.0
Ireland	54.0	79.0	1.0	0.0	134.0
Italy	225.0	145.0	4.5	3.0	377.5
Netherlands	24.0	4.0	20.2	n/a	48.2
Norway	15.0	38.0	0.2	n/a	53.2
Poland	104.3	37.7	7.2	1.6	150.8
Portugal	6.3	82.0	n/a	n/a	88.3
Slovakia	8.9	16.9	0.2	0.3	26.3
Spain	159.0	300.0	1.3	0.0	460.3
Sweden	23.0	49.0	7.9	0.2	80.1
Switzerland	46.5	5.3	5.3	n/a	57.1
United Kingdom	124.0	85.0	56.0	12.0	277.0
Total	1 381.4	1 391.9	179.2	58.6	3 011.1

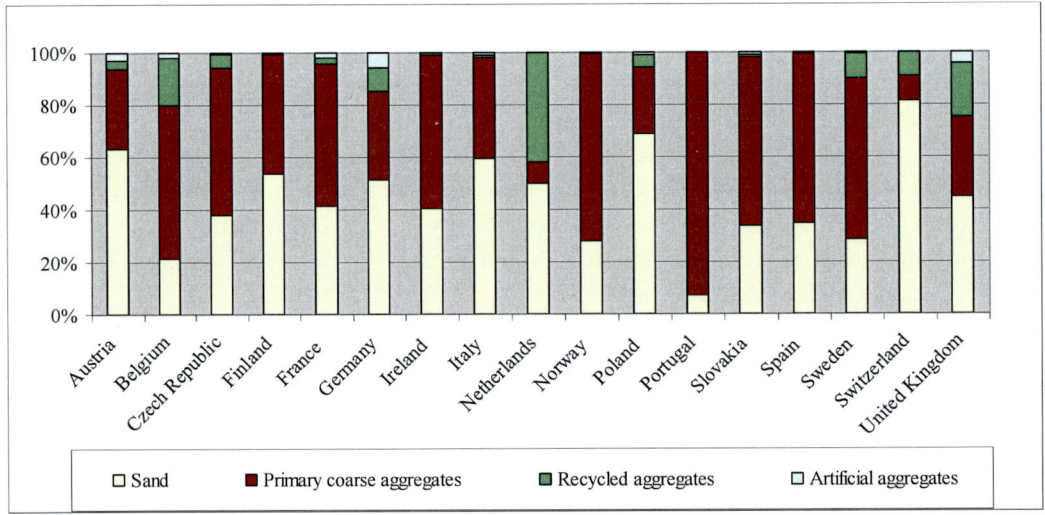

Figure 3. Aggregates produced by type and country in the EU in 2005 (UEPG 2006)

RECYCLED AGGREGATES FROM CONSTRUCTION AND DEMOLITION WASTE

Classification of Recycled Aggregates

To refer to "recycled aggregates" (RA) without specifying their origin is as vague as mentioning "steel" or "timber" without describing their production process and class in the first case and the species and quantity of defects in the second case. As a matter of fact and taking into account the various possible uses of AR, there are fundamentally the following types rated in decreasing order of their quality (measured in terms of water absorption and density and the consequent potential of the construction materials / elements in which they are used; e.g. in concrete the quality of the aggregates is reflected in its mechanical and durability-related performance), as it is generally reported in the literature (de Brito 2005):

- (Coarse) stone aggregates recovered after washing fresh ready-mixed concrete which for some reason has not been used (this material is virtually identical to the original primary aggregates (PA) and is currently reused for new ready-mixed concrete);
- The coarse and/or fine fractions resulting from the demolition of architectural concrete elements (from buildings, special structures and the waste from the precast industry), when there is a reasonable assurance that no other materials will enter the mixture;
- The coarse and/or fine fractions resulting from the demolition of architectural concrete elements already made with RA (multiple recycling), where the RA may have several sources but it is likely that they will only be coarse recycled concrete; this type of RA is not likely to be found in large quantities until some decades from now;
- The coarse and/or fine fractions resulting from the demolition of rendered / coated concrete elements similar to the previous ones but where the resulting RA are made of a mixture of stone, mortar and possibly ceramics; this is today the most common CDW in construction sites;
- The coarse and/or fine fractions resulting directly from more or less undifferentiated CDW, where the stone, mortars and ceramic elements may be joined by small quantities of other materials such as timber, plastics, metals, glass and so on, leading to a mixture with worse characteristics and less quality control; one of the purposes of selective demolition (as an alternative to conventional demolition) is exactly to separate the materials at the building site in order to minimize this contamination;
- The coarse and/or fine fractions resulting from the ceramic industry waste or from the demolition of ceramic elements (bricks and tiles) with or without mortars, leading to mostly ceramic RA with worse characteristics for producing concrete.

This classification is generally reflected in international legislation. In his comparative study of 17 norms for the use of RA in concrete production, from 14 different countries, Gonçalves (2007) concluded that the following three main types of RA were considered with variable quantities of other materials allowed:

- RCA: recycled concrete aggregates;
- RMA: recycled (brick and/or stone) masonry aggregates;
- MRA: mixed (concrete and masonry) recycled aggregates.

In the following sections of this Chapter the main differences in terms of the properties of the concrete mixes made with the different types of AR described above will be described. To understand these characteristics, the main differences between PA and RA are presented next:

- Greater water absorption (both for the RA obtained from concrete elements and those with incorporation of ceramic elements, for the coarse fraction as well as the fine fraction), with various consequences in terms of the workability and, water / cement ratio of fresh concrete and of the mechanical and durability-related properties of hardened concrete;
- Lower compactness (especially for the ceramic RA) and consequent lower crushing resistance and modulus of elasticity, which has an impact on both the immediate and long-term deformability of the concrete elements, and to a lesser degree on its mechanical strength.

Recycled Concrete Aggregate

In the case of the RA from uncoated (or almost uncoated) concrete elements, there are two types of RA: the coarse fraction (CRA) and the fine fraction (FRA), whose differences are more qualitative than quantitative.

What distinguishes these CRA from the equivalent coarse primary aggregates (CPA) is the hardened paste (mortar) that adheres to the former original CPA. This paste is made of primary fines and cement (there is no consensus in the scientific community on whether the non-hydrated part of this cement provides the paste with pozzolanicity).

Their high water absorption and reduced density are responsible for the lower performance of these RA. In other words, the greater the proportion of hardened paste adhering to the original PA, the higher the differences in performance between the CRA and the CPA. Hansen & Narud (1983) measured the percent volume of paste adhering to the PA and concluded that it increases as the size of the CRA decreases, leading to poor expectations for concrete made with FRA.

Since CRA recovered from fresh ready-mixed concrete and washed almost straightaway have hardly any adhering paste they are practically identical to CPA.

This paste is also the reason why the crushing process of the concrete elements (in terms of crusher unit, free space between the blades and number of crushing operations) can strongly affect the characteristics of the RA thus obtained. In an experimental campaign, Matias and de Brito (2005) obtained similar preliminary results in mechanical terms from concrete mixes made with coarse RCA and CPA when the crushing process was the same for the primary and the recycled aggregates, i.e. it was the aggregate crushing process and not the type of aggregates that affected the concrete performance. In another experimental campaign, Gonçalves (2002) obtained different results when the CRA were obtained in a small laboratory unit from those when the crusher unit was on an industrial scale.

Finally, it is also to be expected that the higher the replacement ratio between CPA and CRA in the production of concrete the greater the differences in terms of the respective mixes' performance. This point was unanimous in all the literature reviewed and is proven beyond doubt in the sections which deal with the analysis of this review.

The FRA obtained from crushing concrete elements display exactly the same trends as the CRA in terms of water absorption and density, only more markedly (which is emphasized by an unavoidably greater contamination rate). In practical terms this means that theoretically a lower replacement ratio of FPA with FRA is enough to lead to the same variation in behaviour of the respective RAC as a given replacement ratio of CPA with CRA. The literature review presented later identified several instances where this is not true, in particular the research by Evangelista (2007) with concrete FRA where the results in terms of mechanical strength with replacement ratios of up to 100% were very promising, and better than for equivalent ratios CPA/CRA. One of the reasons for this trend (better than for durability-related characteristics) may have been the non-consensual pozzolanicity of the concrete fines.

A question that frequently comes up is the hypothetical influence of the original concrete grade on the performance of the resulting RA and RAC (naturally if the difference in grade of two original concrete mixes results from the characteristics of the aggregates themselves, this will certainly have an influence on the resulting RA; but that is not the issue here). In their ongoing research Santos et al (2004) concluded that this influence does exist if the gap between the grades is sufficiently wide, but its importance is greater in some properties (modulus of elasticity) of the resulting RAC than in others (compressive strength). Hansen & Narud (1983) reached similar conclusions.

Multiple (repetitive) recycling, i.e. the possibility of crushing concrete elements in whose production RA (either from concrete or ceramics) have already been used is nowadays practically an academic question. Nevertheless, Gonçalves (2002) consecutively tested three concrete mixes with concrete CRA in which the original PA were always the same with promising results in terms of compressive strength, i.e. it did not decrease after the first cycle. She also concluded that cyclic recycling progressively increased the mass of adherent paste and therefore the water absorption kept growing and the density decreasing, though not linearly. In other words, the repetition of the process helped to "worsen" the CRA characteristics, and those of the resulting RAC (recycling RCA where ceramic RA have been used would further lower the quality of the resulting second-cycle RA). This was the reason why, in the classification above, the RA resulting from the first recycling of concrete elements were considered in principle to be of a higher quality than those resulting from multiple recycling.

Recycled Ceramic Aggregate

By comparison with stone aggregates, ceramic aggregates exhibit much higher water absorption (so high that these aggregates always need to be pre-saturated before being used in concrete production) and lower density. It seems obvious that ceramic aggregates, either recycled or not, are "worse" than stone (primary or concrete) aggregates, because their unfavourable characteristics affect all their volume (unlike concrete aggregates where only the adhering paste has these characteristics). Therefore, the higher the percentage of ceramics in the RA mix the greater the differences in terms of performance for the resulting RAC, thus explaining the rating proposed above in terms of the relative quality of the RA. The use of ceramic RA in concrete production has been studied only in terms of the coarse fraction (Rosa 2002) and the fine fraction has been limited to its use in mortars (Silva 2006).

Expected Trends in Recycled Aggregate Concrete

Any estimation of the future performance of an RAC is impossible without knowing the origin and size distribution of the RA and their relative weight in the aggregate mixture. In other words, each RA composition has unique characteristics that will affect the performance of the concrete mix made with it. Based on the considerations set forth earlier, the following general trends are to be expected (the results of the literature review presented further on corroborate most of them) (de Brito 2005):

- Similar or worse performance of every RAC relative to the corresponding RC (reference concrete, with exactly the same characteristics as the RAC except for the use of RA, i.e. 100% PA);
- A greater replacement ratio of PA with RA (either fine or coarse, recycled concrete or ceramic) implies a similar or worse performance of the RAC;
- Worse performance of RAC made with FRA and CRA when compared with that of RAC whose RA are only CRA (this trend may depend on the class strength intended for the RAC and its dependence on the aggregates themselves, i.e. compressive strength of low grade concretes is less dependent on the coarse aggregates than on the cement and fines paste);
- Similar or worse performance of RAC with multiply recycled RA compared with RAC whose RA are being recycled for the first time (with the same replacement ratio);
- Similar or worse performance of RAC with RMA than of RAC with RCA only (with the same replacement ratio).

TECHNICAL BARRIERS TO THE USE OF RA IN GENERAL AND IN RAC PRODUCTION

There are various technical barriers to the use of RA from CDW. They are going to be presented in two parts: those that limit the use of RA in general (regardless of their application) and those that specifically concern RAC production. As a matter of fact, even in the countries that maximize the reuse of CDW and minimize CDW dumping (e.g. the Netherlands and Denmark), the production of RAC is residual and that question needs to be addressed.

Most applications of CDW as RA (in roads, streets and buildings' floors, in landscaping of worked-out mining infrastructures, or as drainage layers of landfills), in many cases using crushed concrete, can be classified as *downcycling* (as defined above), since the materials' characteristics are not fully profited from in these generally less-demanding applications.

Even though this concept is not consensual among researchers and authorities, it is the Author's opinion that only the use of RCA (especially recycled concrete) as RA in concrete production configures a *cycling* operation without a loss of potential of the material, i.e. sustainable.

General Barriers to the Use of CDW RA

The main barriers are related to the general management of CDW production and disposal. Depending on each country's legislative system, this problem is efficiently taken care of (e.g. in the Netherlands, Denmark, the United Kingdom, Germany), incipiently considered because the market has not yet risen to the opportunities presented by new legislation (e.g. in Portugal, Spain, Italy) or almost totally ignored (e.g. most countries from Eastern Europe).

These issues are addressed in the following ways:

- Classification of CDW, in particular addressing the problem of contamination with impurities and toxic waste;
- Reliable estimations of the CDW volume and average composition within each district to gain a global perspective of the needs for recycling plants and dependant post-processing industries;
- Appropriate legislation addressing the problems of compulsory selective demolition (without which CDW recycling is neither technically nor economically viable), environmental taxes and landscaping demands at CDW dumping sites and at stone quarries, state / municipal supervision of illegal dumping and CDW production ("environmental police"), fines for failing to comply with the legislation, among others; the price issue is fundamental, especially in regions where PA are particularly cheap and CDW RA thus become less competitive;
- Development of appropriate technical procedures (contract specifications) for selective demolition to minimize the economic gap relative to conventional demolition (always less expensive and time-consuming) and maximize the environmental impact advantages (in terms of safety, disturbance, air, water and soil impact, and - especially - reuse ratio); some of these techniques must also be used for new construction and rehabilitation sites; the waste from certain industries (e.g. precast concrete, ceramic artefacts) must also be governed by adequate processing specifications to maximize their potential for reuse;
- Tax (or other) incentives for the creation and adequate location of dedicated CDW dumping / sorting sites (recycling plants) and of firms specializing in reusing materials and components from CDW; failing to take care of this part of the problem leads inevitably to laws being broken because the CDW producers are too far from the recycling plants or pay too much for CDW disposal and the plants have no market to sell the processed materials;
- Technical specifications (quality control) for the reuse of CDW materials as raw materials for all the possible future applications, as well as for the specialized equipment (e.g. crushers, magnetic units, air blowers) and procedures adequate for

recycling plants and related industries; in various regions this and the other problems have been addressed in various pilot plants / projects.

Barriers to the Use of CDW RA in RAC Production

As noted above, the use of CDW RA in concrete production faces additional barriers that have so far prevented their widespread use, especially in structural applications. All other things being equal, PA are always preferred to RA by all the agents involved: quarry owners because of their profit; concrete manufacturers because they lack confidence in RA and the characteristics of conventional concrete tend to be better than those of RAC; structural designers because they have a deep-seated but mistaken notion that conventional concrete is less prone to execution errors than RAC and because their calculations would have to be adapted to the RA's actual characteristics and incorporation rate; contractors because there is an additional step in the process (screening and disposing of CDW), and building owners because they do not see any advantage in having to worry about and pay for selective demolition. The lack of technical specifications and of incentives to recycle CDW in concrete production is another major problem.

These issues are addressed thus:

- Each RA composition and PA/RA replacement ratio leads to the resulting concrete mix having different long-term properties; even though the first part of this problem also applies to conventional concrete, its impact is much greater in RAC, making it hard to devise specifications for structural RAC; other critical factors in terms of the performance of RAC are: target strength class of the RAC, water / cement used, cement content and type used in the RAC, size distribution and shape of the RA, percentage of cement and sand adhering to the PA for RC, compacting and curing methods used, type and quantity of plasticizers and fillers;
- In various norms (e.g. German DIN 4226-100, Hong Kong WBTC No. 12/2002, British BS 8500-2:2002, Portuguese LNEC E 471, Dutch CUR-VB-1994, and Swiss Ot 70085), the problem is solved by imposing limits on the PA/RA replacement ratio and the type of the RA used which, when complied with, lead to RAC with a performance sufficiently similar to that of conventional concrete to be considered for practical purposes as equivalent; the problem with this approach is that this strongly limits the incorporation of RA in RAC production due to the low values of the limit replacement ratios imposed and the high demands in terms of the quality control and source of the CDW; quite often only the coarse fraction of RA is accepted in structural concrete;
- Other norms (e.g. Brazilian NBR 15.116, RILEM TC 121-DRG, Belgian PTV 406) allow corrective factors to be applied to the properties of conventional concrete and thus obtain the properties of RAC; these coefficients are made to depend upon the use of pre-determined maximum replacement ratios and on the classification of the RA used; this is a generalization of the first approach where, by allowing the RAC's performance to be different (worse) than that of the equivalent conventional concrete, the use of CDW as RA is clearly enhanced; however, there are still stringent

limitations on this use, notably the need to use pre-determined types of RA which, since most CDW obtained from real situations and with undetermined origin, do not comply with; both these approaches may also include limitations in terms of the maximum strength class of RAC or other restrictions in terms of the field of application;
- The many properties that define fresh and hardened concrete performance (workability, density, compressive and tensile strength, modulus of elasticity, abrasion resistance, shrinkage, creep, water absorption, carbonation and chloride penetration) have evolve differently in RAC and conventional concrete (a very promising step forward is the study by Xiao et al (2006), where constitutive laws are provided for RAC); since most of the norms mentioned above only deal with compressive strength, they may deviate significantly from RAC's expected performance, especially in terms of durability; other properties such as the thermal, hydro-thermal and acoustic behaviour further complicate the possibility of viable predictions of the performance of RAC;
- The best approach, if possible, would be to make these predictions depend on easily obtained early results (such as characterization tests of the aggregate mixture used, including of the PA and/or RA or of concrete at early age) and to use corrective factors of the most important properties of a reference concrete to find the corresponding properties of RAC, without having to know the origin of the CDW or having to classify the resulting RA in standard classes; the next section of this Chapter presents a methodology that renders such an approach viable.

PREDICTION OF RAC'S LONG-TERM PROPERTIES

The methodology presented here, the subject of the Portuguese patent PT 103756 (2007), is based on the concept that trends that are found in conventional concrete also occur in RAC, specifically a strong dependency (and consequent correlation) of the former long-term performance on two characteristics of the aggregates (density and water absorption) and one characteristic of young concrete (7-day compressive strength).

"Aggregates" here is the mixture of fine and coarse aggregates in the composition of the RAC produced, i.e. taking into account the relative weight / replacement ratio that the RA represent in the mixture. In other words, if the replacement ratio PA/RA is small, the difference between the density and water absorption of the mixture of aggregates in the corresponding RAC is also small in comparison with those of the equivalent RC. Therefore, both the origin of the RA (through their intrinsic characteristics - density and water absorption) and the replacement ratio are simultaneously taken into account.

Methodology Proposed

This weighed value of the density of the mixture of all aggregates in the RAC composition is calculated using the density of each individual aggregate (RA and PA, fine and coarse) and the composition of each concrete mix. The following equation is used to obtain the weighed density value (density of the mixture of aggregates):

$$D_{mix} = \frac{FA}{100} \times \left[\frac{subst_{FRA} \times D_{FRA} + (100 - subst_{FRA}) \times D_{FNA}}{100} \right] + \frac{(100 - FA)}{100} \times \left[\frac{subst_{CRA} \times D_{CRA} + (100 - subst_{CRA}) \times D_{CNA}}{100} \right] \quad (1)$$

where: D_{mix} - weighed density of the mixture of aggregates in the concrete mix;
FA - percentage of fine aggregates used in the mix;
$subst_{FRA}$ - replacement ratio of FRA by FPA;
$subst_{CRA}$ - replacement ratio of CRA by CPA;
D_{FRA} - density of the FRA;
D_{FNA} - density of the FPA;
D_{CRA} - density of the CRA;
D_{CNA} - density of the CPA.

The weighed value of the water absorption of the mixture of aggregates (wa_{mix}) used in the mix is calculated using a similar equation where the density values are replaced with the water absorption values for each of the different aggregates used.

In each experimental campaign, the values of D_{mix} and wa_{mix} can be determined for each RAC composition and for the RC using the replacement ratios FPA/FRA and CPA/CRA and the individual values of the density and water absorption of each type of aggregate (RA and PA). When available, the value of the 7-day compressive strength of hardened concrete (f_{c7}) for each RAC composition and for the RC can also be taken into account. By dividing the values of these parameters for each RAC by the corresponding values of the RC, non-dimensional factors are obtained that allow the comparison between different experimental results regardless of the source of the RA and the replacement ratios used.

For each property of hardened concrete determined within an experimental campaign, graphs can be prepared where the abscissas contain the non-dimensional values described above (one at a time) and the ordinates contain the ratio between the value of the property for each RAC mix and the value of the same property for the corresponding RC. By gathering the results of various campaigns it is possible to have a statistically significant number of results and to determine a linear regression trend-line that is valid for any RAC, no matter what the origin of the RA used and their replacement ratio.

In the data collection (international and Portuguese literature review) described next, the establishment of these linear correlations between the concrete mixes' properties and the aggregate mixture's density and water absorption, as well as the compressive strength of concrete at the age of 7 days, a graphic analysis methodology was created. It involved the following steps:

- Analysis and organization of available data from each experimental campaign, including information on the test results for the properties of the aggregates used to make concrete;

- Calculation of the exact value of the density and water absorption of the mixture of aggregates used in the mix, through the mix proportions of the concretes (with primary aggregates PA only and with recycled aggregates RA) and the individual density and water absorption of the aggregates (natural and recycled);
- Graphical analysis of the relationship between the replacement ratio of PA by RA and each property of concrete;
- Graphical analysis of the variation of the ratio between the properties of the concrete with RA (RAC) and the one with PA only (conventional reference concrete RC) and the replacement ratio of NA by RA;
- Graphical analysis of the variation of the ratio between the properties of the RAC and the RC and the ratio between the weighed value of the density of the mixture of aggregates (D_{mix}) in the RAC and the RC;
- Graphical analysis of the variation of the ratio between the properties of the RAC and the RC and the ratio between the weighed value of the water absorption of the mixture of aggregates (wa_{mix}) in the RAC and the RC;
- Graphical analysis of the variation of the ratio between the properties of the RAC and the RC and the ratio between the compressive strength at 7 days (f_{c7}) of the RAC and the RC;
- Superposition of the graphical results of each concrete property for the various campaigns analyzed and determination of the linear regression lines with the respective correlation coefficient;
- Correction of the linear regression lines obtained, so that they represent the physical behaviour under analysis, forcing them to pass through the point that corresponds to the RC, with the nuisance of lowering the respective correlation coefficient;
- Compilation of the data in a table, including the slope of the linear regression line and the respective correlation coefficient.

In order to analyse this graphical summary for each property, qualitative criteria were established so that it would be possible to classify the corresponding correlation value. This classification is presented in Table 5.

Table 5. Qualitative classification of the correlation values obtained in the graphical analysis of the experimental campaign literature review (Robles 2007) (Alves 2007)

Classification	Values range
Very good	$R^2 \geq 0.95$
Good	$0.80 [R^2 < 0.95$
Acceptable	$0.65 [R^2 < 0.80$
Not acceptable	$R^2 < 0.65$

Literature Reviews

To implement the methodology described above, two literature reviews were conducted, each resulting in a Masters Thesis:

- An international review by Robles (2007);
- A Portuguese review by Alves (2007).

These reviews were guided by the criteria below:

- Availability of the tests results performed on RA and PA, especially in terms of water absorption and density;
- Availability of the experimental values relative to the greatest possible number of concrete properties in the fresh and hardened states (mechanical and durability-related), especially the 7-day compressive strength;
- The greatest possible PA/RA replacement ratio options;
- The greatest possible number of fixed parameters (e.g. water / cement ratio, grading curve, workability, curing method) during the experimental production of RAC;
- Existence of a reference concrete, without which the corresponding campaign was eliminated from the survey.

The criterion related to the fixed parameters is of paramount importance. As a matter of fact, since the whole purpose of this research is to isolate the impact of the direct replacement of PA with RA, it is essential that all other factors that may disturb the understanding of this influence are eliminated. The following factors are the most important (but not the only) ones since, when they change, almost every concrete property can be affected thus masking the influence of the replacement PA/RA and jeopardizing the objectives of the study:

- Effective water / cement ratio (differentiating the total amount of water introduced into the mix from that which effectively contributes to the hydration of the cement and the workability of fresh concrete); this property has a strong influence on the concrete's mechanical performance, but even more on its durability;
- Workability (this property must be maintained by using plasticizers or increasing the total amount of water without increasing the effective water / cement ratio, for example, by pre-saturating the recycled aggregates); for practical purposes, concrete mixes with different workability levels may not have the same range of applications and therefore should not be directly compared;
- Grading curve (size distribution) of the aggregates (when replacing PA with RA this curve should be kept exactly constant because any change leads to uncontrolled shifts in almost every relevant property of concrete);
- In the case of RA from recycled concrete, the origin / nature of the original PA should not differ drastically from that of the PA used in to make the RAC and RC.

Surprisingly, since the number and quality of the references accessed was much greater than for the Portuguese literature review, it was more difficult to find international literature references that complied with these criteria. The main reasons for this were (Robles 2007):

- Significant differences at the level of procedures and of organization / presentation of the results;
- Unavailability of data due to existing information being excluded from the test

description or because relevant data were simply not determined (absence of data on the aggregates' properties, both RA and PA, or on the composition of the concrete mixes tested, lack of specification on whether the water / cement ratio specified was the global or the effective w/c ratio);
- Variability of individual factors of each campaign.

Since most of the campaigns on RAC in Portugal took place at the same location (IST - Instituto Superior Técnico in Lisbon) and under the same supervision, the criteria used in the tests and in the presentation of the results were basically the same, which was crucial to the outcome of the literature review.

In both cases, a database (much too large to be presented here) was prepared where all the information collected was organized so as to facilitate perception of the main points analyzed in each experimental campaign. The main points recorded in the database were (Robles 2007):

- The origin of the RA: concrete, ceramic, the two types, and undifferentiated;
- The grading of the aggregates replaced: coarse, fine or both;
- The fixed parameters, i.e. those that remained constant in the production of the various families of concrete mixes (e.g. water / cement ratio, grading curve, compressive strength, mass or volume proportions, workability, cement content);
- The parameters that varied, i.e. the criteria that defined the various RAC families produced that were correlated with the variation of the concrete properties in the fresh and hardened states (e.g. replacement ratio of PA with RA, water / cement ratio, added flying ash or synthetic fibre content, type of curing used, RA density, RA age at crushing, type of cement, surface treatment of the RA, age at testing);
- The tests performed on aggregates (e.g. density, water absorption, abrasion resistance, retained water content, grading curve, fineness module, crushing resistance, bulk density, frost resistance, shape coefficient);
- The tests performed on fresh concrete (e.g. workability, density, bleeding, and air content); the test standard is also provided when known;
- The tests performed on hardened concrete (e.g. compressive strength, flexural and splitting tensile strength, creep, shrinkage, abrasion resistance, chloride and carbonation penetration resistance, water absorption, density, porosity, permeability); the test standard is also provided when known.

Short Description of the Experimental Campaigns Selected

Given the difficulties mentioned above, only a relatively small number of experimental campaigns were selected for analysis in each literature review. Their results are analyzed in the next section and the resulting conclusions presented for each property of hardened concrete.

The following experimental campaigns that complied with the criteria defined above were analysed in the international literature review:

- Leite (2001), at the Federal University of Rio Grande do Sul (UFRGS), tested replacement ratios of 0, 11.5, 50, 88.5 and 100% of coarse and fine natural aggregates with coarse and fine concrete and ceramic aggregates to determine the compressive and tensile strength and modulus of elasticity of hardened concrete;
- Soberón (2002), at the Technical University of Catalonia (UTC), tested replacement ratios of 0, 15, 30, 60 and 100% of coarse natural aggregates with coarse concrete aggregates to determine the porosity, density, permeability, compressive and tensile strength, modulus of elasticity, shrinkage, creep and water absorption of hardened concrete;
- Katz (2003), at the Israel Institute of Technology (IIT), tested replacement ratios of 0 and 100% of coarse and fine natural aggregates with coarse and fine concrete aggregates to determine the compressive and tensile strength, modulus of elasticity, shrinkage, water absorption and carbonation penetration of hardened concrete;
- Kou et al (2004), at the Hong Kong Polytechnic University (HKPU), tested replacement ratios of 0, 20, 50 and 100% of coarse natural aggregates with coarse undifferentiated aggregates to determine the compressive strength, modulus of elasticity, shrinkage and chloride penetration of hardened concrete;
- Carrijo (2005), at the São Paulo University (USP), tested replacement ratios of 0 and 100% of coarse basaltic aggregates with coarse concrete and ceramic aggregates to determine the compressive strength, water absorption by immersion and modulus of elasticity of hardened concrete;
- Cervantes et al (2007), at the University of Illinois (UI), tested replacement ratios of 0, 50 and 100% of coarse natural aggregates with coarse concrete aggregates to determine the compressive and tensile strength, shrinkage and modulus of elasticity of hardened concrete.

The following experimental campaigns that complied with the criteria defined above were analysed in the Portuguese literature review:

- Rosa (2002), at the Technical University of Lisbon (IST), tested replacement rates of 1/3, 2/3 and 3/3 of coarse limestone aggregates with coarse recycled ceramic aggregates to determine the compressive and tensile strength, abrasion resistance, water absorption by capillarity and immersion of hardened concrete;
- Rocha and Resende (2004), at the University of Aveiro (UA), tested replacement rates of 50 and 100% of coarse granite aggregates with coarse recycled concrete aggregates to determine the density, compressive and tensile strength and modulus of elasticity of hardened concrete;
- Figueiredo (2005), at the Faculty of Engineering of Porto University (FEUP), tested replacement rates of 50 and 100% of coarse granite aggregates with coarse recycled concrete and white ceramic aggregates to determine the compressive and tensile strength, modulus of elasticity, abrasion resistance, shrinkage, water absorption by capillarity and immersion, carbonation and chloride penetration of hardened concrete;
- Matias and de Brito (2005), at IST, tested replacement rates of 25, 50 and 100% of fine limestone aggregates with fine recycled concrete aggregates (with a focus on the

use of super-plasticizers) to determine the compressive and tensile strength, abrasion resistance, shrinkage, water absorption by capillarity and immersion, carbonation and chloride penetration of hardened concrete;
- Evangelista (2007), at IST, tested replacement rates of 10, 20, 30, 50 and 100% of fine limestone aggregates with fine recycled concrete aggregates to determine the compressive and tensile strength, modulus of elasticity, abrasion resistance, shrinkage, water absorption by capillarity and immersion, carbonation and chloride penetration of hardened concrete;
- Gomes (2007), at IST, tested various combinations of replacement rates of coarse limestone aggregates with coarse recycled concrete, ceramic and mortar aggregates to determine the compressive and tensile strength, modulus of elasticity, shrinkage, water absorption by capillarity and immersion, carbonation and chloride penetration of hardened concrete;
- Ferreira (2007), at IST and Polytechnic University of Barcelona (UPC), tested replacement rates of 20, 50 and 100% of coarse limestone aggregates with coarse recycled concrete aggregates (with a focus on the pre-saturation of recycled aggregates) to determine the density, compressive strength, modulus of elasticity, shrinkage and water absorption by capillarity and immersion of hardened concrete.

RESULTS AND DISCUSSION

General Remarks

The methodology proposed above made it possible to compare directly results from different experimental campaigns with different test procedures and norms, different origins and sizes of RA, and various other factors that led to the dispersion of results (e.g. curing method, addition of flying ash or synthetic fibres, different strength classes, different w/c ratios, different crushing ages of recycled concrete aggregates).

Therefore, the linearity (measured by the correlation factor R^2) exhibited by each individual experimental campaign for each property naturally decreased when other experimental campaigns' results were juxtaposed in the same graph. There was a further reduction of the correlation factors obtained when the linear regression lines in each graph were corrected to pass through the point that corresponds to the RC, to make them representative of the physical behaviour under analysis.

As a consequence of these two situations the correlation factors obtained for each of the properties of hardened concrete with the three parameters selected (density and water absorption of the mixture of aggregates used in the concrete mix and 7-day compressive strength of the concrete mix) were not always high enough to prove the accuracy of the methodology beyond doubt. However, it must be stated that similar rates of dispersion of the results would be obtained for conventional concrete (i.e. without recycled aggregates) and that RAC therefore does not exhibit higher scatter than conventional concrete.

In the following points all the more relevant properties of hardened concrete (from a mechanical and a durability point of view) are analyzed using the methodology proposed, and the results are described and explained based on the intrinsic properties of RA and their effects on hardened concrete. These properties include density, compressive strength, modulus of elasticity, splitting and flexural tensile strength, abrasion resistance, shrinkage, creep, water absorption by capillarity and immersion, carbonation and chloride penetration resistance.

Most of these properties showed a poorer performance for RAC than for the corresponding RC, as expected and in agreement with the international and Portuguese studies cited above, and others mentioned in the references.

Density

Only the Portuguese literature review collected results on RAC's density.

Figures 4, 5 and 6 show the relationship of the ratio between the hardened concrete densities and the ratio between the densities and the water absorptions of the mixture of aggregates and the 7-day compressive strengths of the concrete, respectively. The correlation coefficients determined (see Table 5) are respectively good, acceptable and not acceptable.

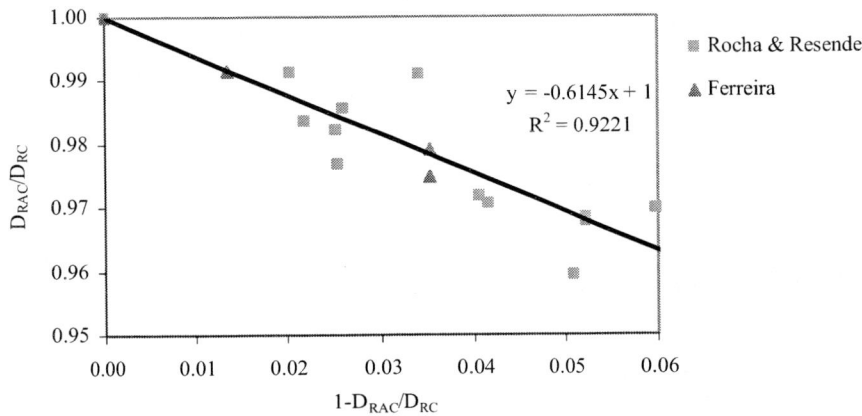

Figure 4. Ratio between hardened concrete densities *versus* the ratio between the densities of the mixture of aggregates (Alves 2007)

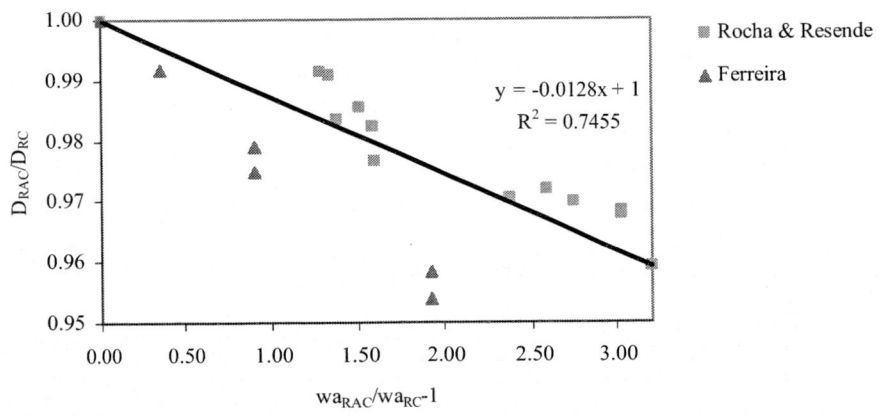

Figure 5. Ratio between hardened concrete densities *versus* the ratio between the water absorptions of the mixture of aggregates (Alves 2007)

For every parameter used as an indicator, the concrete density decreases with the replacement ratio. This is because of the lower density of the recycled aggregates, due to their lower intrinsic porosity (in ceramic and mortar aggregates) or the mortar adhering to the original aggregates (in recycled concrete aggregates).

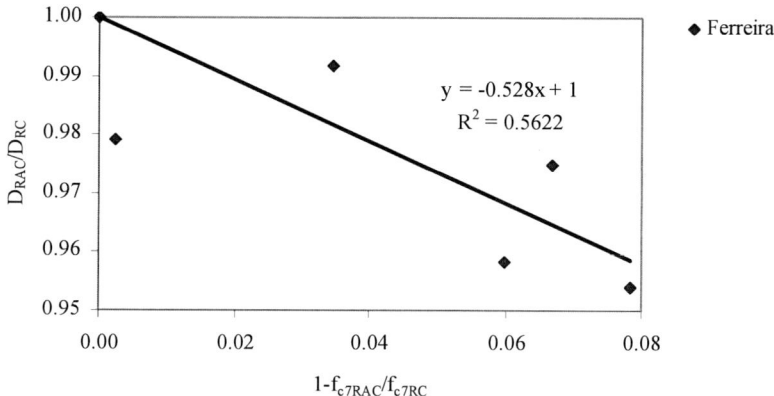

Figure 6. Ratio between hardened concrete densities *versus* the ratio between the concrete's 7-day compressive strengths (Alves 2007)

Compressive Strength

The international and the Portuguese literature reviews both contain many results on RAC's compressive strength, since it is the most defining property of hardened concrete. Other references analyzed but not included in the graphs are: Angulo (1998), Angulo (2005), Azzouz et al (1998), Barra and Vázquez (1998), Buttler (2003), Buyle-Bodin and Hadjieva-Zaharieva (2002), Chen et al (2003), Corinaldesi (2003), Etxeberria et al (2004), Etxeberria et al (2006), Fraaij et al (2002), Juan and Gutiérrez (2004), Larrañaga (2004), Latterza and Machado (2003), Levy (2001), Levy and Helene (2004), Limbachiya et al (2000), Lin et al (2004), Khatib (2004), Kimura et al (2004), Knights (1998), Mendes et al (2004), Merlet and Pimienta (1993), Muller (2004), Oliveira et al (2004), Roos (1998), Sagoe-Crentsil (2001), Sugiyama (2004), Tsujino et al (2007), Vieira et al (2004), and Wainwright et al (1994).

Figures 7, 8 and 9 respectively show the relationship of the ratio between the compressive strengths of hardened concrete and the ratio between the densities and the water absorptions of the mixture of aggregates and the 7-day compressive strengths of concrete, respectively. The correlation coefficients are considered respectively good, not acceptable and not acceptable.

Even though the correlation coefficients were disappointingly low in two of the cases (due to the large number and variety of studies juxtaposed in the same graph), it is still possible to identify a clear trend towards a decrease in compressive strength as the replacement ratio increases, explained by the lower mechanical characteristics of the ceramics and of the mortar adhering to the natural aggregates (in recycled concrete).

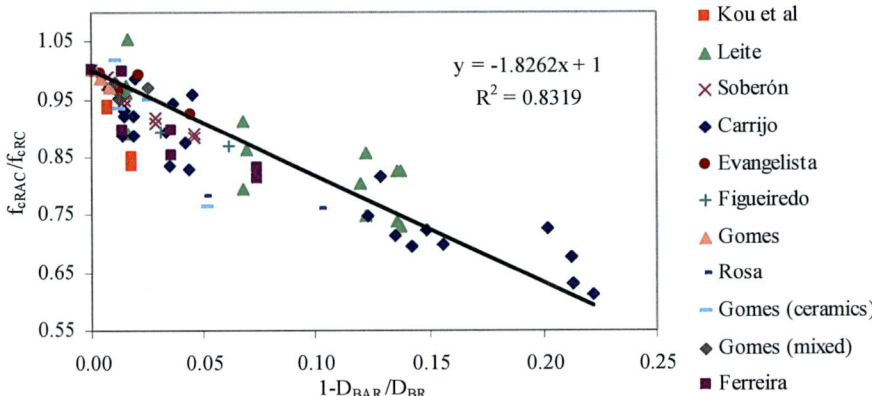

Figure 7. Ratio between concrete compressive strengths *versus* the ratio between the densities of the mixture of aggregates (Robles 2007) (Alves 2007)

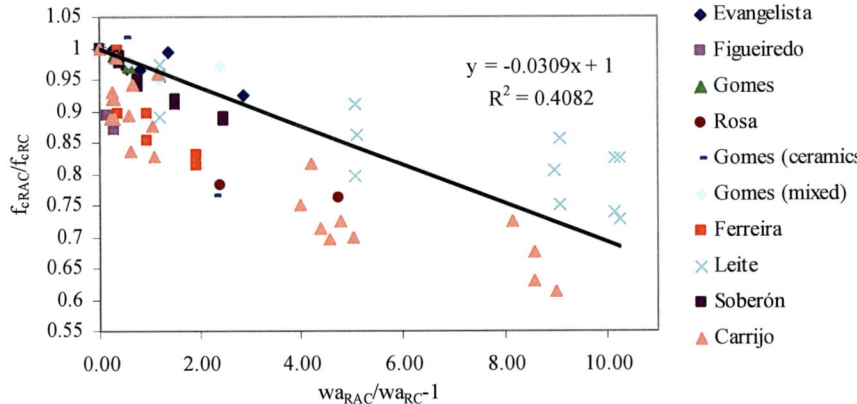

Figure 8. Ratio between concrete compressive strengths *versus* the ratio between the water absorptions of the mixture of aggregates (Robles 2007) (Alves 2007)

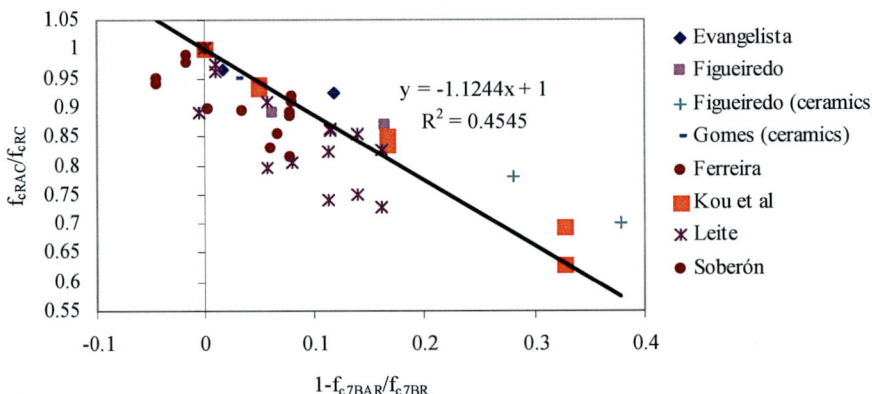

Figure 9. Ratio between concrete compressive strengths *versus* the ratio between the concrete 7-day compressive strengths (Robles 2007) (Alves 2007)

Modulus of Elasticity

Both the international and the Portuguese literature reviews contain many results on RAC's modulus of elasticity, another of the most defining properties of hardened concrete. Other references analyzed but not included in the graphs are: Angulo (2005), Azzouz et al (1998), Barra and Vázquez (1998), Buttler (2003), Chen et al (2003), Etxeberria et al (2004), Etxeberria et al (2006), Juan and Gutiérrez (2004), Larrañaga (2004), Latterza and Machado (2003), Levy (2001), Khatib (2004), Mendes et al (2004), Merlet and Pimienta (1993), Oliveira et al (2004), Roos (1998), Sugiyama (2004), and Tsujino et al (2007).

Figures 10, 11 and 12 display the relationship of the ratio between the modules of elasticity of hardened concrete and the ratio between the densities and the water absorptions of the mixture of aggregates and the 7-day compressive strengths of concrete, respectively. The correlation coefficients are considered respectively acceptable, acceptable and not acceptable.

Again there is a clear trend towards a fall in the mechanical property of hardened concrete (in this case the modulus of elasticity) as the replacement ratio increases. The reason is the same as for the compressive strength, even though this time it is related to the lower stiffness of the recycled aggregates. A very interesting conclusion is that the slope of the correlation lines is higher for the modulus of elasticity than for the compressive strength in every parameter, confirming a trend of higher loss of stiffness of RAC reported in the literature review undertaken within the study.

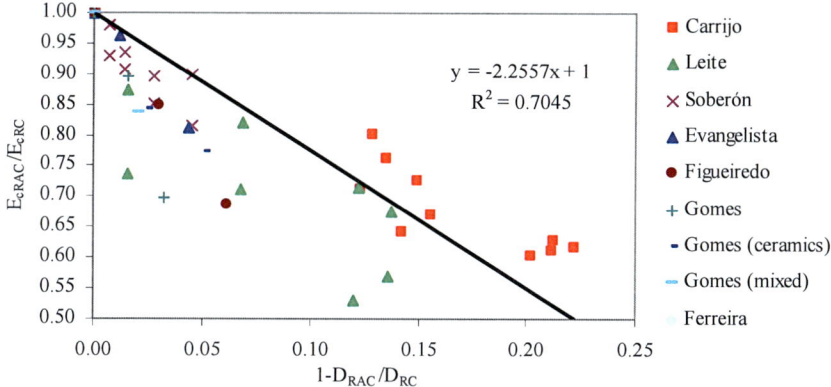

Figure 10. Ratio between concrete modules of elasticity *versus* the ratio between the densities of the mixture of aggregates (Robles 2007) (Alves 2007)

Splitting Tensile Strength

Both the international and the Portuguese literature reviews contain results on RAC's splitting tensile strength. Other references analyzed but not included in the graphs are: Angulo (1998), Buttler (2003), Corinaldesi (2003), Etxeberria et al (2004), Juan and Gutiérrez (2004), Larrañaga (2004), Latterza and Machado (2003), Merlet and Pimienta (1993), Sagoe-Crentsil (2001), and Tsujino et al (2007).

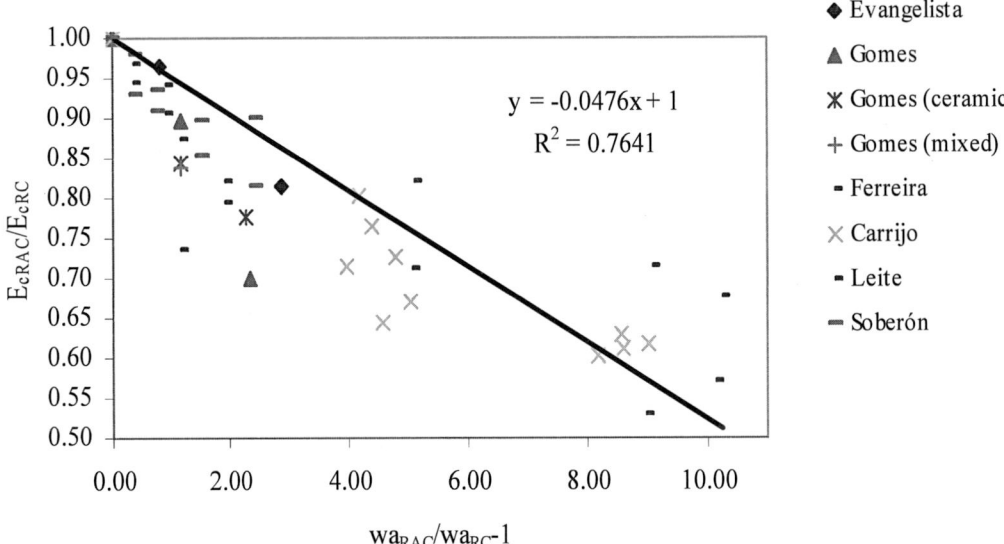

Figure 11. Ratio between concrete modules of elasticity *versus* the ratio between the water absorptions of the mixture of aggregates (Robles 2007) (Alves 2007)

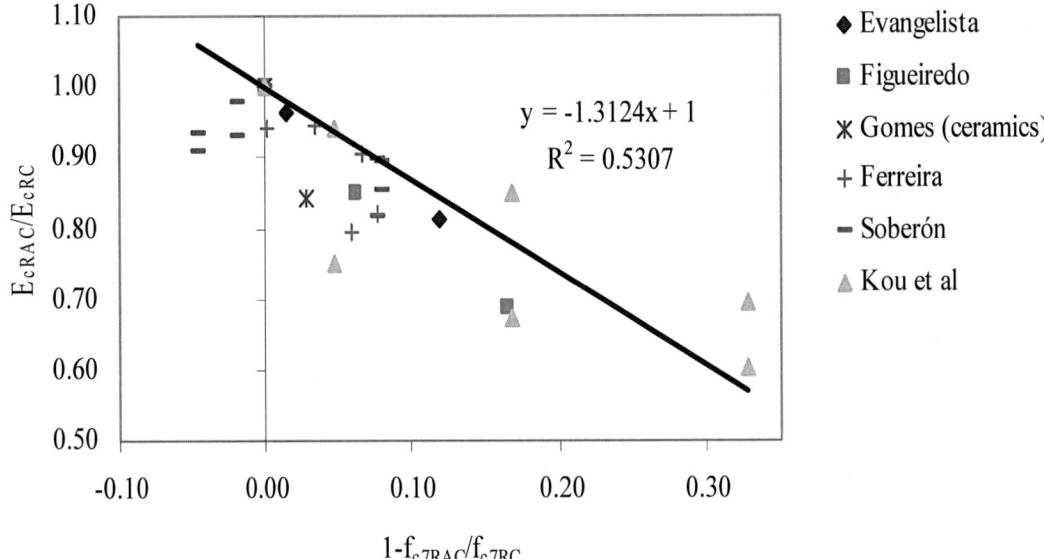

Figure 12. Ratio between concrete modules of elasticity *versus* the ratio between the concrete 7-day compressive strengths (Robles 2007) (Alves 2007)

Figures 13, 14 and 15 display the relationship of the ratio between the splitting tensile strengths of hardened concrete and the ratio between the densities and the water absorptions of the mixture of aggregates and the 7-day compressive strengths of concrete, respectively. The correlation coefficients are considered respectively acceptable, not acceptable and not acceptable.

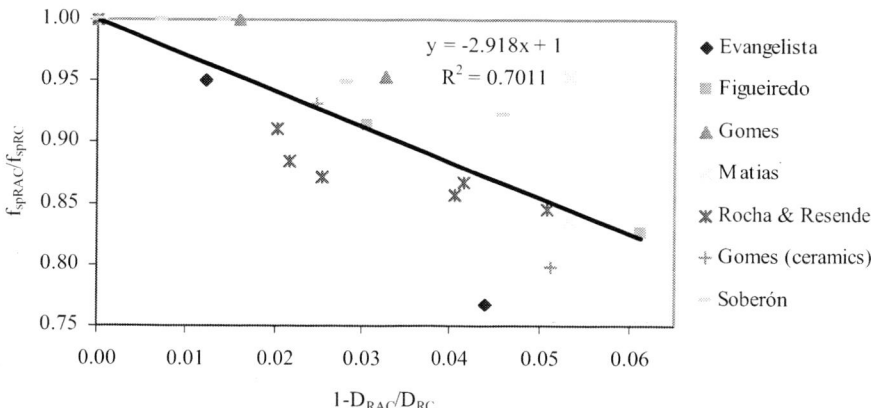

Figure 13. Ratio between concrete splitting tensile strengths *versus* the ratio between the densities of the mixture of aggregates (Robles 2007) (Alves 2007)

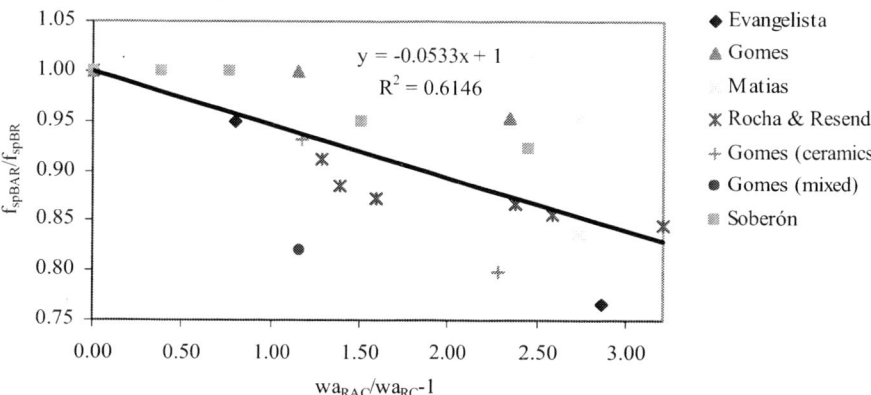

Figure 14. Ratio between concrete splitting tensile strengths *versus* the ratio between the water absorptions of the mixture of aggregates (Robles 2007) (Alves 2007)

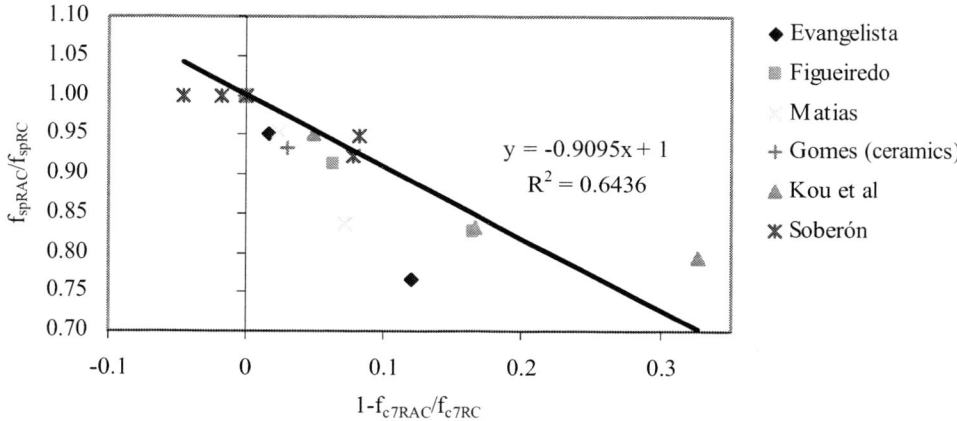

Figure 15. Ratio between concrete splitting tensile strengths *versus* the ratio between the concrete 7-day compressive strengths (Robles 2007) (Alves 2007)

Once again, the lower mechanical properties of RA when compared with PA lead to a decrease in the splitting tensile strength of the former as the replacement ratio increases. Notwithstanding the relatively low values of the correlation factors, the slope of the correlation lines is generally higher than for the compressive strength and even for the modulus of elasticity (except for the correlation with the 7-day compressive strength of concrete). This contradicts existing literature that states that the RAC is slightly less sensitive to the inclusion of these aggregates for tensile strength than for compressive strength and even more than for the modulus of elasticity. A possible explanation for this is the well known scatter that is usually registered in the results for the splitting tensile strength of concrete, both RAC and conventional.

Flexural Tensile Strength

Both the international and the Portuguese literature reviews contain results on RAC's splitting tensile strength, though very few (no correlations between this property and the 7-day compressive strength of concrete were found in the Portuguese literature review). Other references analyzed but not included in the graphs are: Chen et al (2003), Etxeberria et al (2006), and Latterza and Machado (2003).

Figures 16, 17 and 18 show the relationship of the ratio between the flexural tensile strengths of hardened concrete and the ratio between the densities and the water absorptions of the mixture of aggregates and the 7-day compressive strengths of concrete, respectively. The correlation coefficients are considered respectively acceptable, good and good.

Conclusions are the same as for splitting tensile strength, except that this property seems to be less sensitive to the inclusion of recycled aggregates (the exception is the correlation with the 7-day compressive strength of concrete). However, these conclusions are only provisional due to the small number of valid results obtained and experimental campaigns analysed.

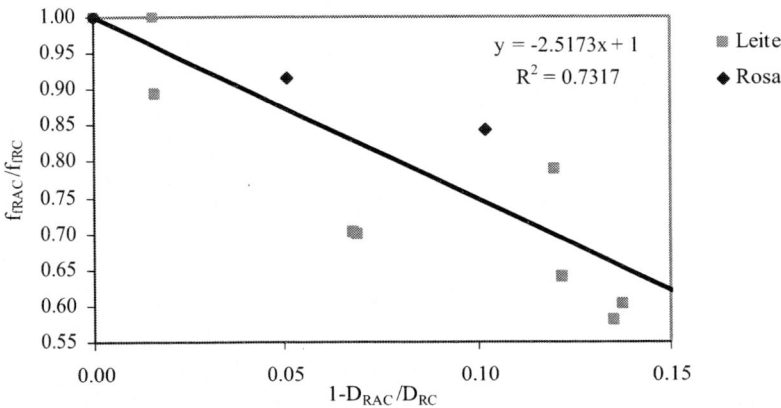

Figure 16. Ratio between concrete flexural tensile strengths *versus* the ratio between the densities of the mixture of aggregates (Robles 2007) (Alves 2007)

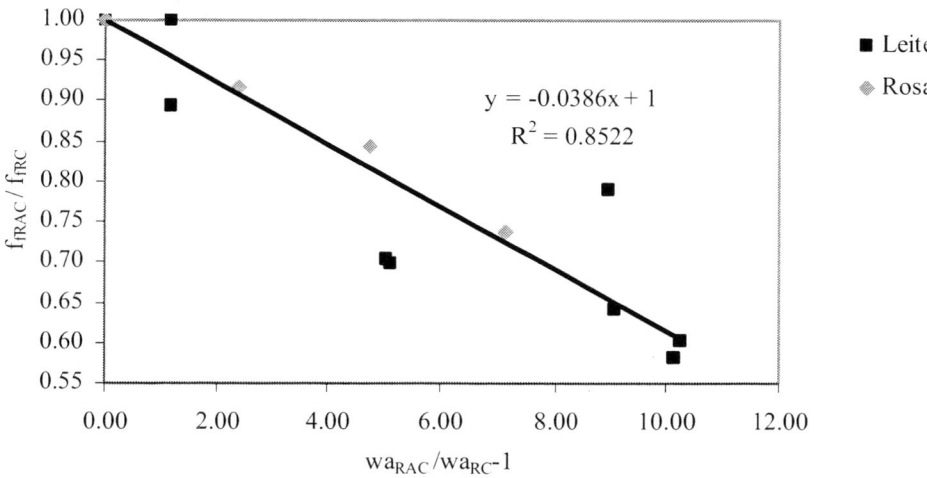

Figure 17. Ratio between concrete flexural tensile strengths *versus* the ratio between the water absorptions of the mixture of aggregates (Robles 2007) (Alves 2007)

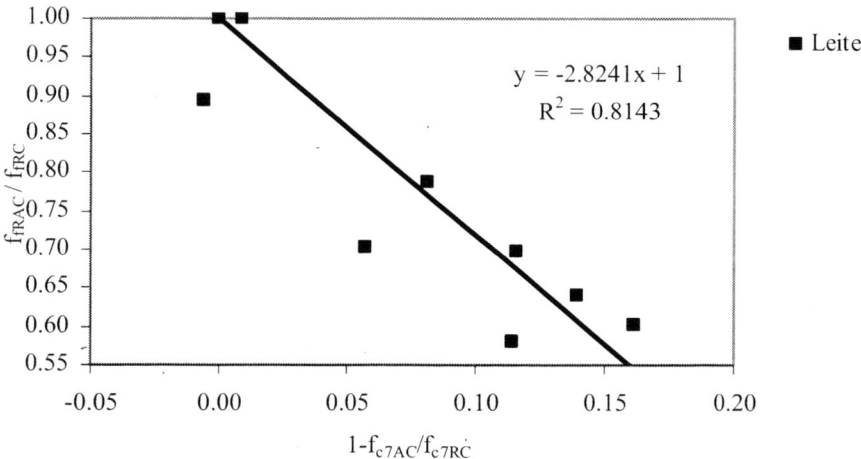

Figure 18. Ratio between concrete flexural tensile strengths *versus* the ratio between the concrete 7-day compressive strengths (Robles 2007)

Abrasion Resistance

Only the Portuguese literature review contains results on RAC's abrasion resistance, a relatively unimportant property in terms of concrete's most common applications. Other references analyzed but not included in the graphs are: Latterza and Machado (2003), Limbachiya et al (2000), and Sagoe-Crentsil (2001).

Figures 19, 20 and 21 display the relationship of the ratio between the abrasion resistances (inversely proportional to the loss of mass) of hardened concrete and the ratio between the densities and the water absorptions of the mixture of aggregates and the 7-day compressive strengths of concrete, respectively. The correlation coefficients are considered respectively good, good and not acceptable.

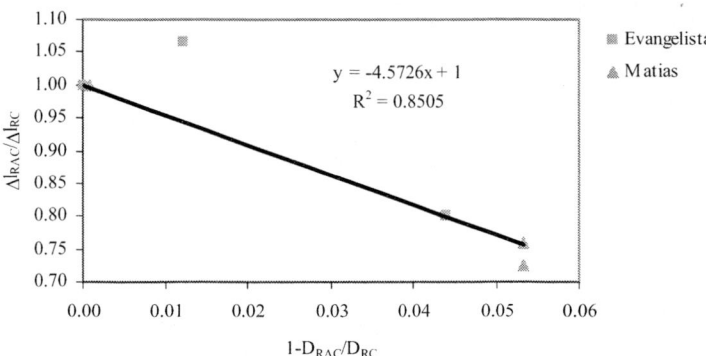

Figure 19. Ratio between concrete abrasion mass losses *versus* the ratio between the densities of the mixture of aggregates (Alves 2007)

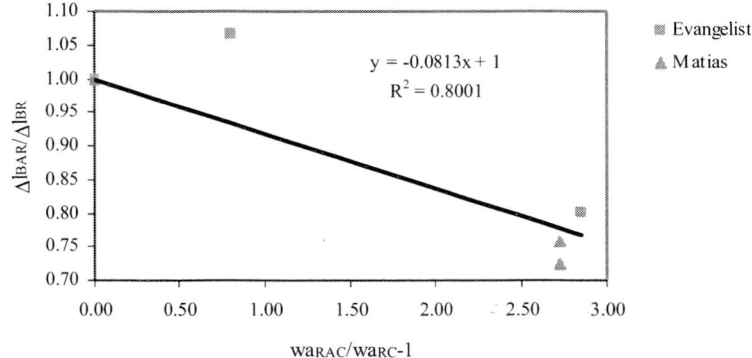

Figure 20. Ratio between concrete abrasion mass losses *versus* the ratio between the water absorptions of the mixture of aggregates (Alves 2007)

The abrasion resistance of RAC tends to increase (less loss of mass) as the replacement ratio increases, which is explained in the meagre existing literature by greater adherence between particles provided by the more porous surface of recycled aggregates. This is the only property analysed where RAC performs markedly better than conventional concrete.

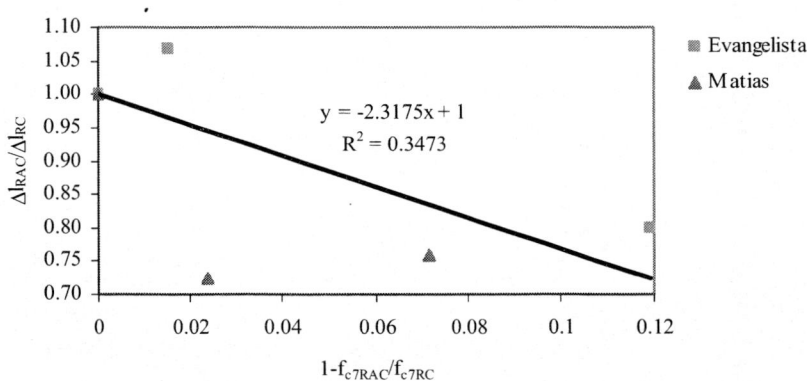

Figure 21. Ratio between concrete abrasion mass losses *versus* the ratio between the concrete 7-day compressive strengths (Alves 2007)

Shrinkage

Both the international and the Portuguese literature reviews contain results on RAC's shrinkage, a particularly defining characteristic of hardened concrete. Other references analyzed but not included in the graphs are: Azzouz et al (1998), Buyle-Bodin and Hadjieva-Zaharieva (2002), Corinaldesi (2003), Fraaij et al (2002), Juan and Gutiérrez (2004), Limbachiya et al (2000), Khatib (2004), Kimura et al (2004), Merlet and Pimienta (1993), Roos (1998), and Sagoe-Crentsil (2001).

Figures 22, 23 and 24 show the relationship of the ratio between the shrinkages of hardened concrete and the ratio between the densities and the water absorptions of the mixture of aggregates and the 7-day compressive strengths of concrete, respectively. All the correlation coefficients are considered not acceptable, which may be explained by the great influence that curing conditions have on shrinkage results and by the fact that it could not be known in detail which curing conditions were used in each experimental campaign.

Shrinkage of hardened concrete is greatly affected by the porosity of the aggregates and their stiffness and, regardless of the poor correlation factors determined, shows a clear trend towards increasing with the PA/RA replacement ratio. This was expected, based on the relative properties of the RA and on the literature review. Another common conclusion in the literature is that shrinkage is the most sensitive property in terms of the inclusion of recycled aggregates, and this is confirmed in the present study by the considerably higher slopes of the correlation lines, even higher than those for the modulus of elasticity.

Creep

Creep is one of the least studied properties of concrete because testing is both costly and time-consuming. Therefore, only the international literature review contains results on RAC's creep. Other references analyzed but not included in the graphs are: Fraaij et al (2002), Limbachiya et al (2000), Kimura et al (2004), Mendes et al (2004), and Roos (1998).

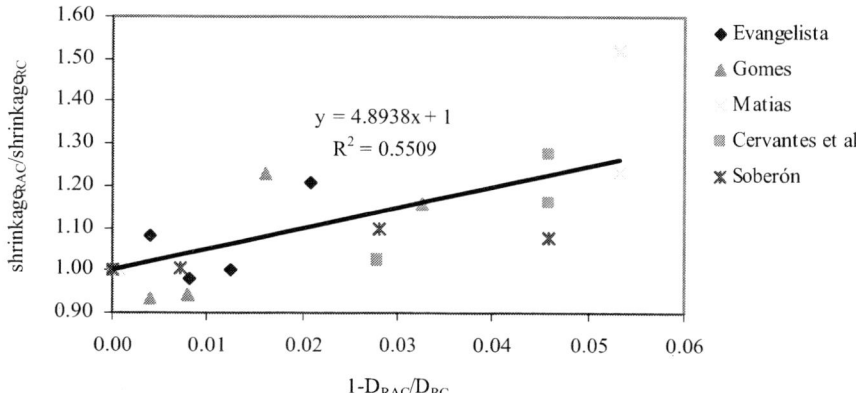

Figure 22. Ratio between concrete shrinkages *versus* the ratio between the densities of the mixture of aggregates (Robles 2007) (Alves 2007)

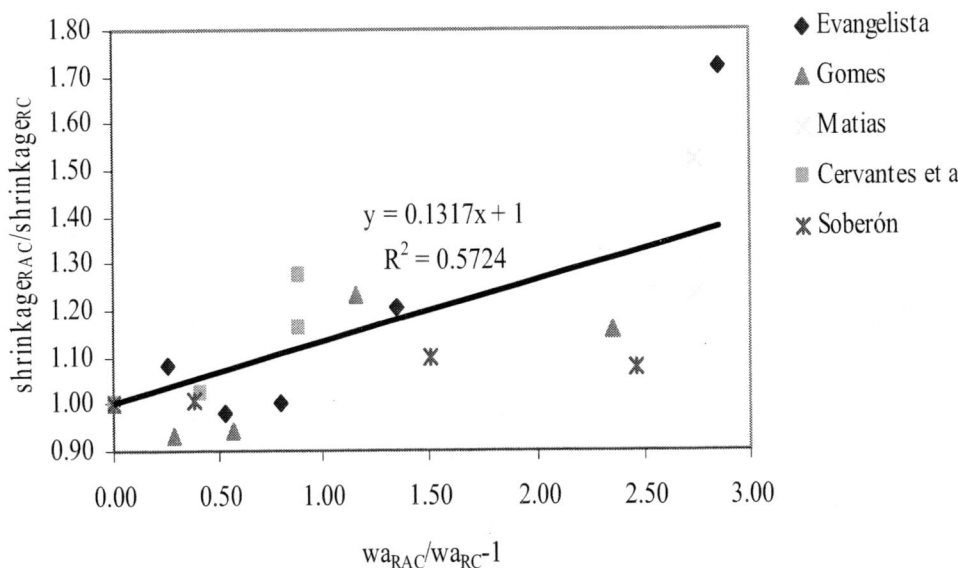

Figure 23. Ratio between concrete shrinkages *versus* the ratio between the water absorptions of the mixture of aggregates (Robles 2007) (Alves 2007)

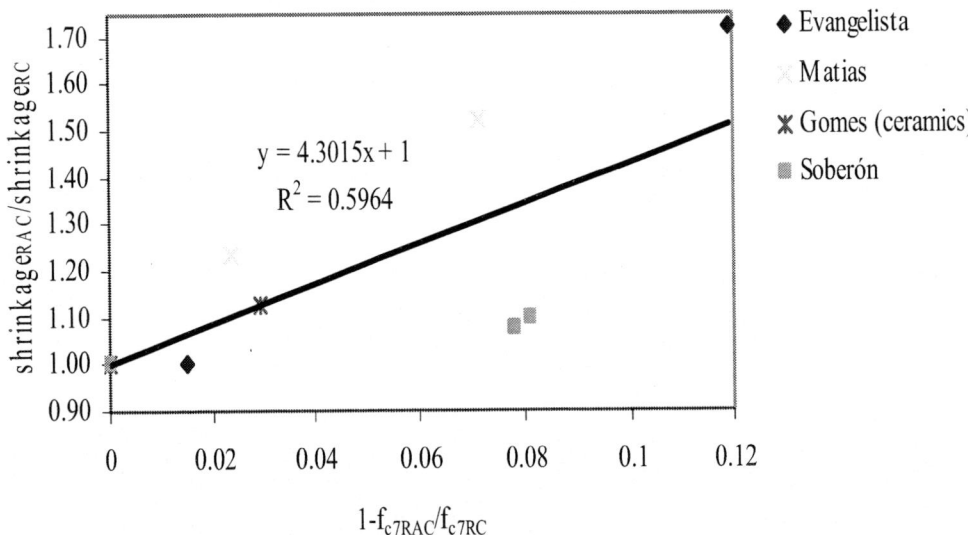

Figure 24. Ratio between concrete shrinkages *versus* the ratio between the concrete 7-day compressive strengths (Robles 2007) (Alves 2007)

Figures 25, 26 and 27 display the relationship of the ratio between the creep of hardened concrete and the ratio between the densities and the water absorptions of the mixture of aggregates and the 7-day compressive strengths of concrete, respectively. All the correlation coefficients are considered very good, which is mostly due to the fact that only one experimental campaign (Soberón 2002) was used for the graphs.

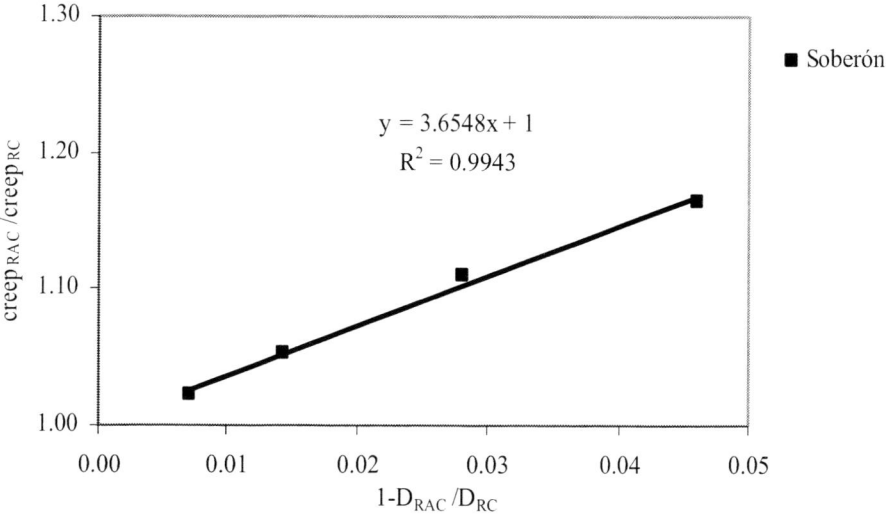

Figure 25. Ratio between concrete creep *versus* the ratio between the densities of the mixture of aggregates (Robles 2007)

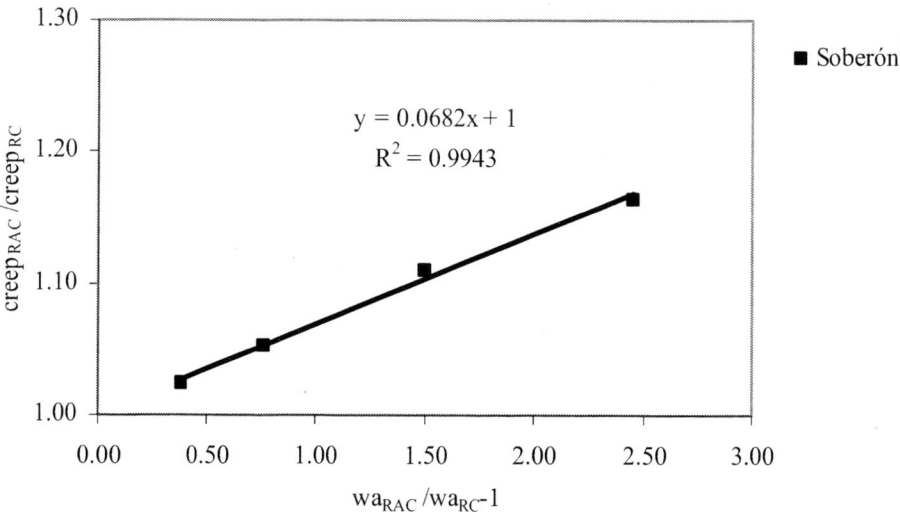

Figure 26. Ratio between concrete creep *versus* the ratio between the water absorptions of the mixture of aggregates (Robles 2007)

The lower stiffness of RA when compared with PA contributes to higher creep with increasing replacement ratio of PA with RA, as seen in Figures 25 to 27. Also, a hypothetical increment of the w/c ratio, to balance the higher water absorption of RA compared with PA, can contribute to higher values of creep in the concrete made with RA. The few studies listed above on creep in RCA suggest a slightly better behaviour than for shrinkage. However, the slopes of the correlation lines obtained from the Soberón study (2002) are slightly higher than those in Figures 25 and 26 (the exception is Figure 27, in relation to the 7-day compressive strength of concrete). The small number of valid results in terms of both shrinkage and creep does not allow definitive quantitative conclusions.

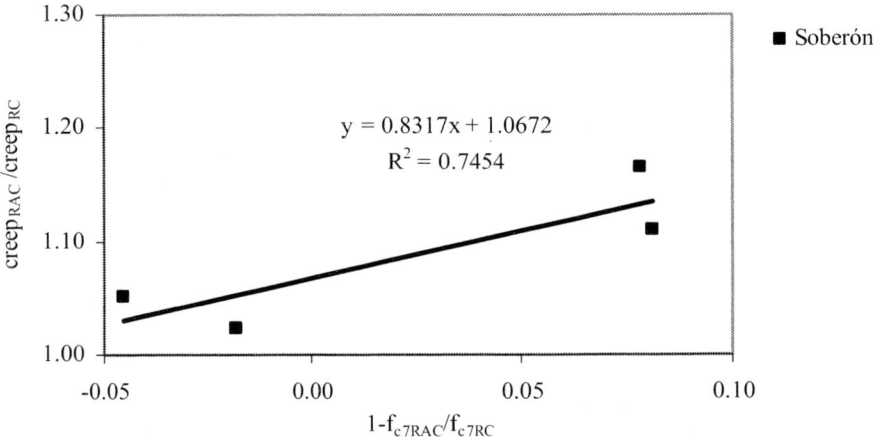

Figure 27. Ratio between concrete creep *versus* the ratio between the concrete 7-day compressive strengths (Robles 2007)

Water Absorption by Capillarity

Only the Portuguese literature review contains results on RAC's water absorption by capillarity. Only one international reference was found for this property in RCA (Limbachiya et al 2000), but was not used to build the graphs.

Figures 28, 29 and 30 display the relationship of the ratio between the capillary water absorptions of hardened concrete and the ratio between the densities and the water absorptions of the mixture of aggregates and the 7-day compressive strengths of concrete, respectively. All the correlation coefficients are considered not acceptable and are very low, which may be an indication that for this particular property the methodology proposed is not appropriate for quantitative estimations.

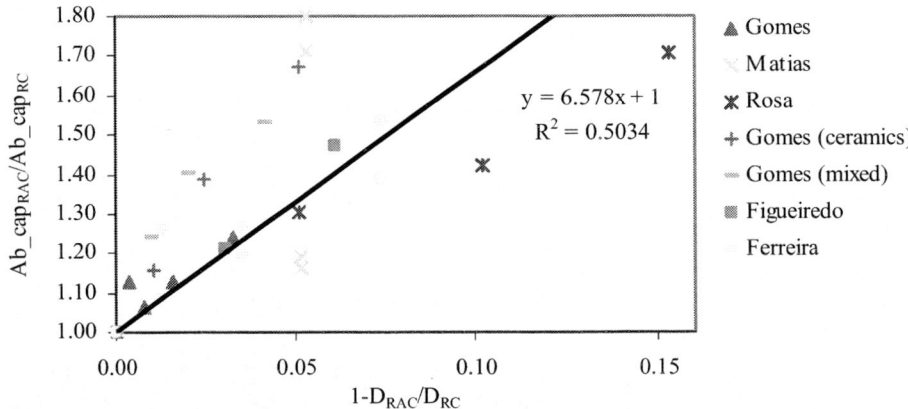

Figure 28. Ratio between concrete capillary water absorptions *versus* the ratio between the densities of the mixture of aggregates (Alves 2007)

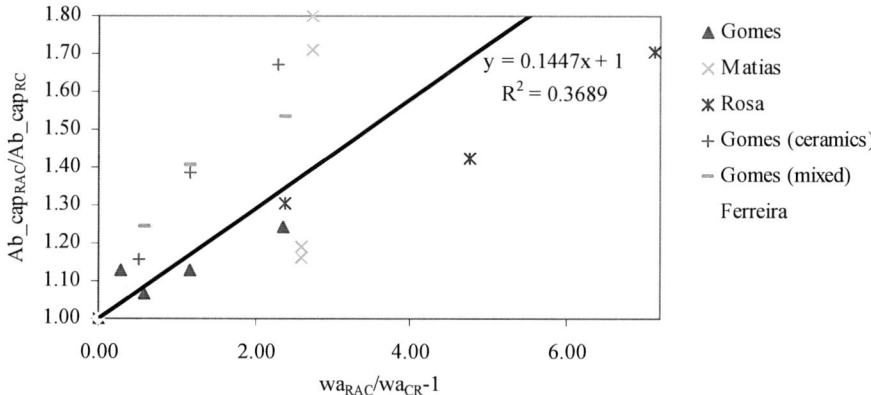

Figure 29. Ratio between concrete capillary water absorptions *versus* the ratio between the water absorptions of the mixture of aggregates (Alves 2007)

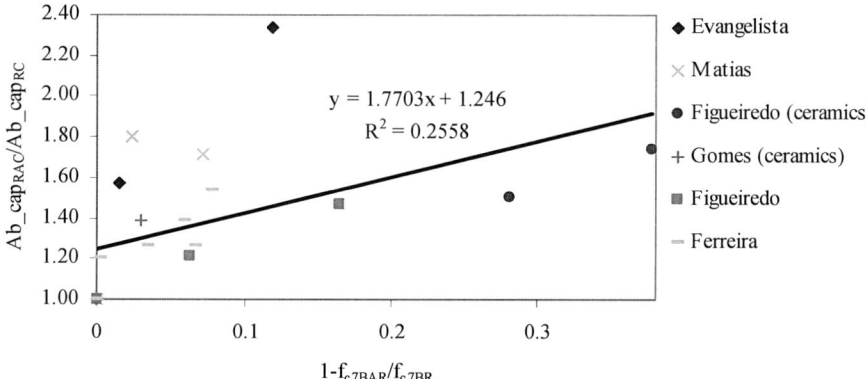

Figure 30. Ratio between concrete capillary water absorptions *versus* the ratio between the concrete 7-day compressive strengths (Alves 2007)

Even though the correlation obtained is poor, the trend is very clear for all the parameters and it shows the strong impact of the higher porosity of the RA on the water absorption by capillarity, an impact that is even greater than that found for shrinkage and creep (except for the correlation found between shrinkage and the 7-day compressive strength of concrete). Again, this confirms the literature, which singles out durability, along with volume stability, as the performance aspects of RCA exhibiting the greatest losses.

Water Absorption by Immersion

Both the international and the Portuguese literature reviews contain results on RAC's water absorption by immersion. Other references analyzed but not included in the graphs are: Angulo (2005), Azzouz et al (1998), Barra and Vázquez (1998), Buttler (2003), Buyle-Bodin and Hadjieva-Zaharieva (2002), Levy (2001), Levy and Helene (2004), Knights (1998), Merlet and Pimienta (1993), Oliveira et al (2004), Olorunsogo and Padayachee (2001), Sagoe-Crentsil (2001), and Wirquin et al (2000).

Figures 31, 32 and 33 show the relationship of the ratio between the water absorptions by immersion of hardened concrete and the ratio between the densities and the water absorptions of the mixture of aggregates and the 7-day compressive strengths of concrete, respectively. The correlation coefficients are considered respectively not acceptable, acceptable and good.

The conclusions are similar to those for water absorption by capillarity, with similar slopes (higher in two cases) in terms of correlation lines, proving once again the great susceptibility of concrete made with RA to absorb water.

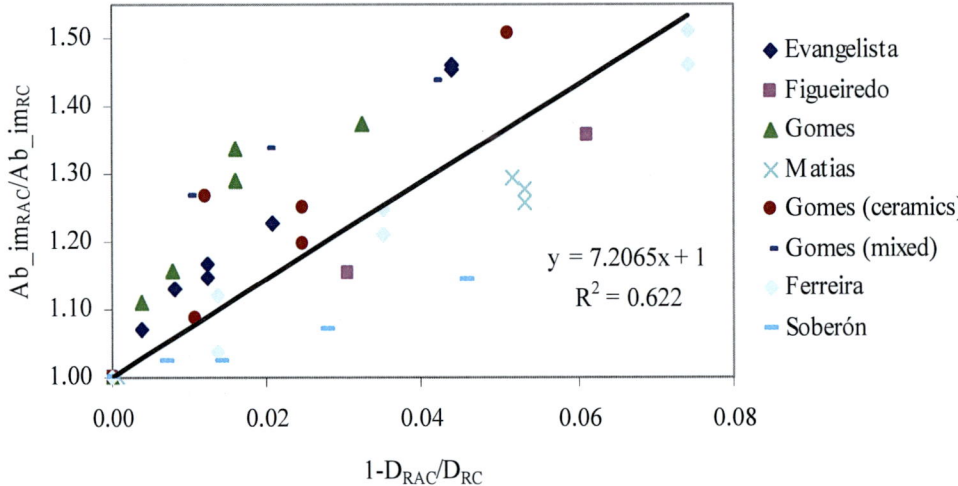

Figure 31. Ratio between concrete water absorptions by immersion *versus* the ratio between the densities of the mixture of aggregates (Robles 2007) (Alves 2007)

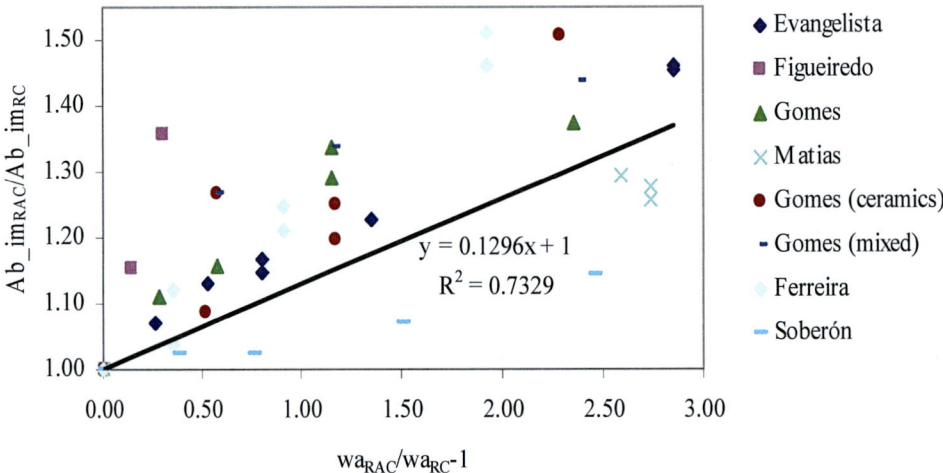

Figure 32. Ratio between concrete water absorptions by immersion *versus* the ratio between the water absorptions of the mixture of aggregates (Robles 2007) (Alves 2007)

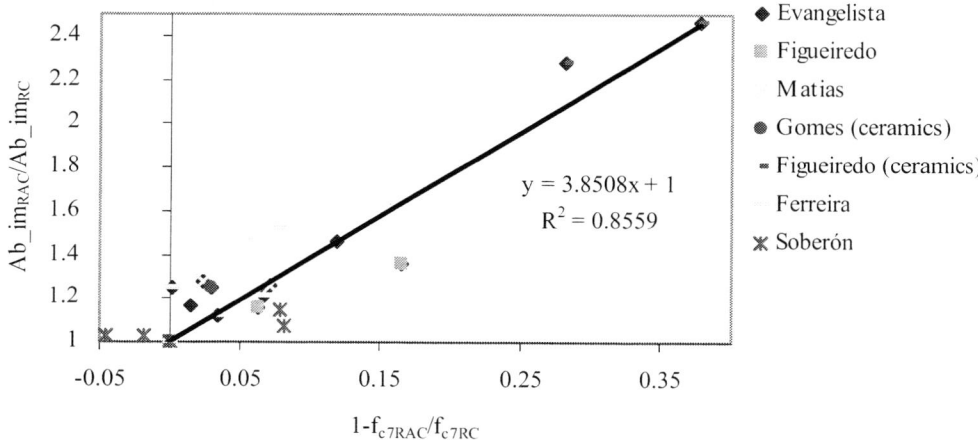

Figure 33. Ratio between concrete water absorptions by immersion *versus* the ratio between the concrete 7-day compressive strengths (Robles 2007) (Alves 2007)

Carbonation Resistance

Both the international and the Portuguese literature reviews contain results on RAC's carbonation resistance, a vital characteristic in terms of reinforcement corrosion. Other references analyzed but not included in the graphs are: Buyle-Bodin and Hadjieva-Zaharieva (2002), Corinaldesi (2003), Levy (2001), Levy and Helene (2004), Kimura et al (2004), Merlet and Pimienta (1993), and Sagoe-Crentsil (2001).

Figures 34, 35 and 36 show the relationship of the ratio between the carbonation depths of hardened concrete and the ratio between the densities and the water absorptions of the mixture of aggregates and the 7-day compressive strengths of concrete, respectively. The correlation coefficients are considered respectively acceptable, acceptable and not acceptable.

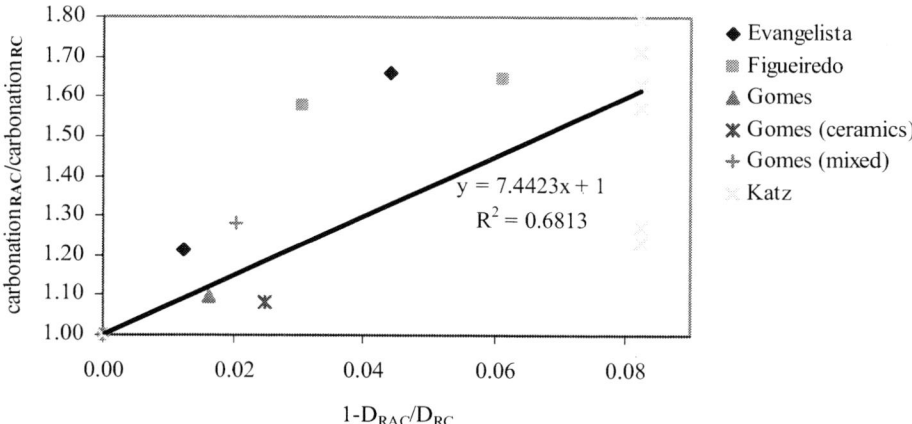

Figure 34. Ratio between concrete carbonation depths *versus* the ratio between the densities of the mixture of aggregates (Robles 2007) (Alves 2007)

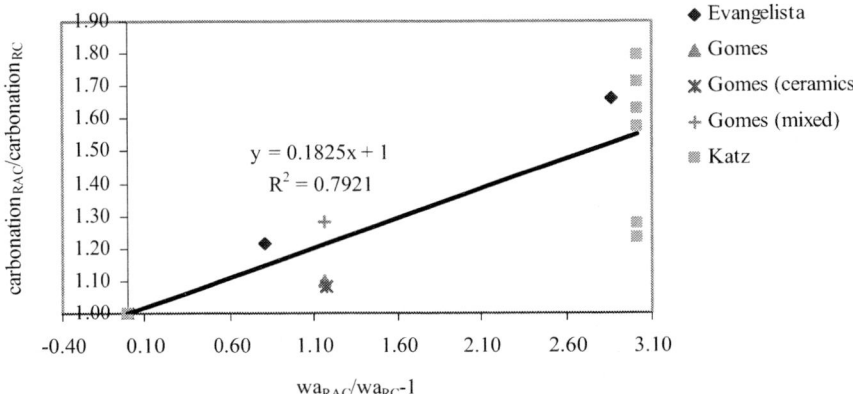

Figure 35. Ratio between concrete carbonation depths *versus* the ratio between the water absorptions of the mixture of aggregates (Robles 2007) (Alves 2007)

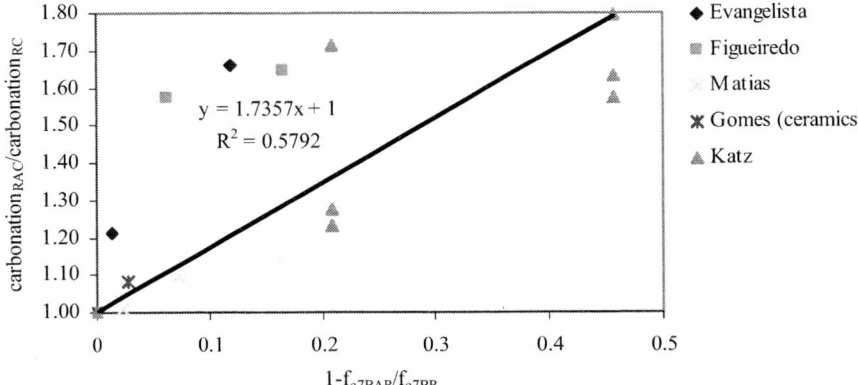

Figure 36. Ratio between concrete carbonation depths *versus* the ratio between the concrete 7-day compressive strengths (Robles 2007) (Alves 2007)

The graphs exhibit a very strong influence of the replacement ratio on the depth the carbonation front reaches within concrete, stronger than for any other characteristic so far (except for the correlation with the 7-day compressive strength of concrete, which exhibits stronger trends for some other properties). This has been identified in the literature review as one of the most vulnerable aspects of the inclusion of RA in concrete production and is explained, as for the other durability-related characteristics (e.g. water absorption) by the greater porosity of these aggregates compared with that of PA.

Chloride Penetration Resistance

Both the international and the Portuguese literature reviews contain results on RAC's chloride penetration resistance, a defining characteristic for concrete's durability in coastal areas. Other references analyzed but not included in the graphs are: Corinaldesi (2003), Fraaij et al (2002), Levy (2001), Limbachiya et al (2000), and Olorunsogo and Padayachee (2001).

Figures 37, 38 and 39 display the relationship of the ratio between the chloride penetration of hardened concrete resistance (in terms of the diffusion coefficients or electric charge measured) and the ratio between the densities and the water absorptions of the mixture of aggregates and the 7-day compressive strengths of concrete, respectively. The correlation coefficients are considered respectively good, not acceptable and acceptable.

The slope of the correlation lines is similar to that for carbonation (in one case slightly higher and in the other two lower) and therefore the trend of increasing vulnerability to chloride penetration arising from the inclusion of RA is also very clear. The reasons for this phenomenon remain the same (i.e. greater porosity of the RA).

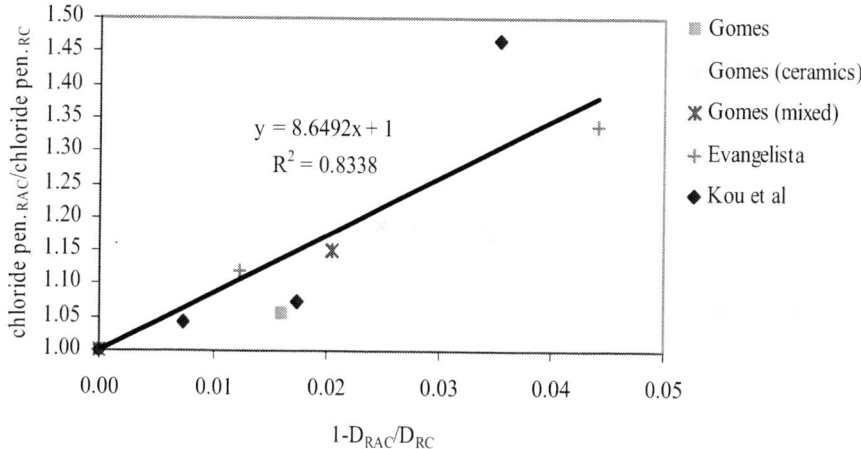

Figure 37. Ratio between concrete chloride penetration depths *versus* the ratio between the densities of the mixture of aggregates (Robles 2007) (Alves 2007)

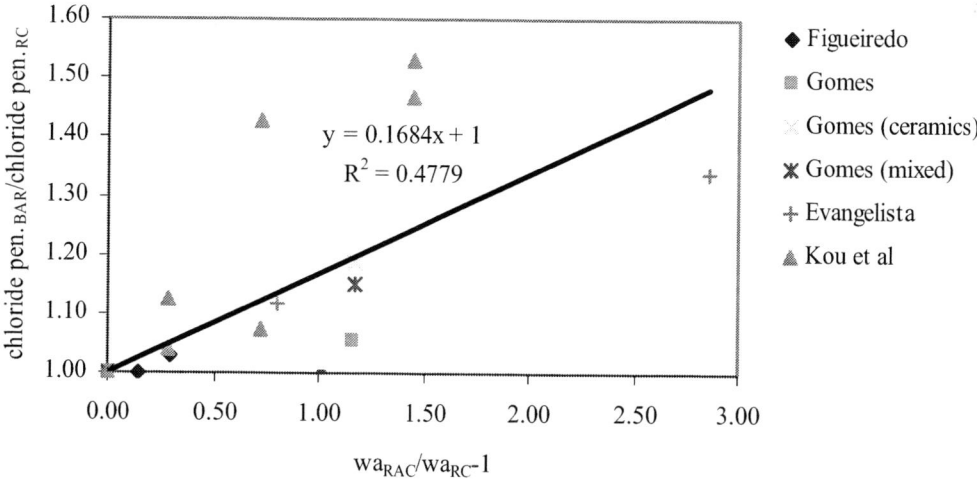

Figure 38. Ratio between concrete chloride penetration depths *versus* the ratio between the water absorptions of the mixture of aggregates (Robles 2007) (Alves 2007)

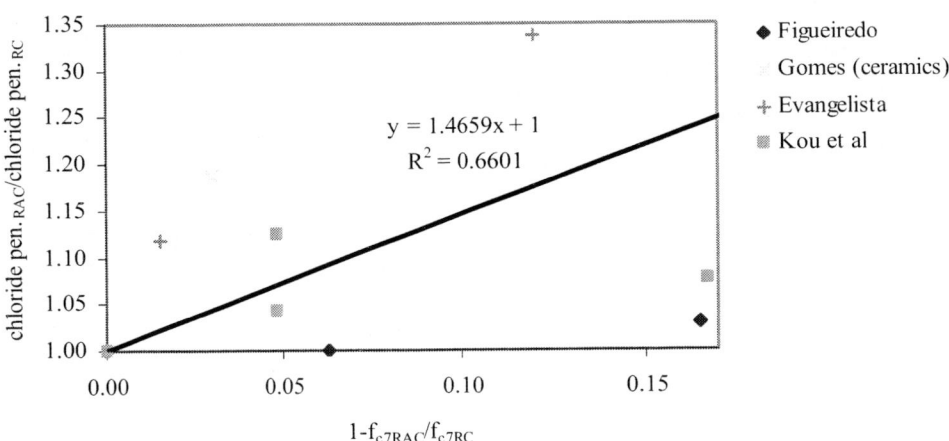

Figure 39. Ratio between concrete chloride penetration depths *versus* the ratio between the concrete 7-day compressive strengths (Robles 2007) (Alves 2007)

Summary of the Results

Table 6 summarizes the correlation coefficients for the relationship of each hardened concrete property with the weighed densities and water absorptions of the mixture of aggregates and the 7-day compressive strengths of concrete. The regression lines' slopes are also presented.

The correlation coefficient classification is indicated by different colours.

Based on the thirteen experimental campaigns on RAC chosen (six international and seven Portuguese), it was possible to analyze twelve different properties of hardened concrete. The relationship between these properties and the density and water absorption of the mixture of aggregates and the 7-day compressive strength of concrete allowed the following conclusions to be drawn (according to Table 6):

- With very few exceptions, it is possible to establish a linear relationship for the variation of the ratio between the concrete properties and the ratio of the three parameters mentioned;
- Generally, the density of the mixture of aggregates showed higher correlation coefficients with the hardened concrete properties, in the graphical analysis;
- The 7-day compressive strength of concrete seems to be the least adequate parameter to estimate the long-term concrete properties, since the lowest correlation coefficients were, in general, obtained for this property; this behaviour may be due to the influence of the variation of mixture procedures from one campaign to another and to the higher scatter of results for young concrete; an additional reason is the relative rate of growth of compressive strength for RCA *versus* RC (in some of the Portuguese campaigns, it was faster and in other others it was slower, and therefore it needs further research);

Table 6. Summary of the correlation trend lines between the different concrete properties and the density and water absorption of the mixture of aggregates and the 7-day compressive strength of concrete (Robles 2007) (Alves 2007)

Property	Experimental studies	Aggregates' density R^2	Aggregates' density Slope	Aggregates' water absorption R^2	Aggregates' water absorption Slope	Concrete's 7-day compressive strength R^2	Concrete's 7-day compressive strength Slope
Density	Rocha and Resende/ Ferreira	0.9221 (0.9225)	-0.6145	0.7455 (0.7555)	-0.0128	-	-
	Ferreira	-	-	-	-	0.5622 (0.6208)	-0.528
Compressive strength	Evangelista/ Figueiredo/ Gomes/ Rosa/ Ferreira/ Carrijo/ Kou/ Leite/ Soberón	0.8319 (0.8490)	-1.8262	-	-	-	-
	Evangelista/ Figueiredo/ Gomes/ Rosa/ Ferreira/ Carrijo/ Leite/ Soberón	-	-	0.4082 (0.5484)	-0.0309	-	-
	Evangelista/ Figueiredo/ Gomes/ Ferreira/ Kou/ Leite/ Soberón	-	-	-	-	0.4545 (0.6254)	-0.8568
Modulus of elasticity	Evangelista/ Figueiredo/ Gomes/ Ferreira/ Carrijo / Leite / Soberón	0.7045 (0.7641)	-2.2557	-	-	-	-
	Evangelista/ Gomes/ Ferreira/ Carrijo / Leite / Soberón	-	-	0.7641 (0.8166)	-0.0476	-	-
	Evangelista/ Figueiredo/ Ferreira/ Kou/ Leite/ Soberón	-	-	-	-	0.5307 (0.7022)	-1.3124
Splitting tensile strength	Evangelista/ Figueiredo/ Gomes/ Matias/ Rocha and Resende	0.7011 (0.7012)	-2.918	-	-	-	-
	Evangelista/ Gomes/ Matias/ Rocha and Resende/ Soberón	-	-	0.6146 (0.6149)	-0.0533	-	-
	Evangelista/ Figueiredo/ Gomes/ Matias/ Kou/ Soberón	-	-	-	-	0.6436 (0.7211)	-0.9095
Flexural tensile strength	Rosa/ Leite	0.7317 (0.7388)	-2.5173	0.8522 (0.8556)	-0.0386	-	-
	Leite	-	-	-	-	0.8143 (0.8477)	-2.8241
Abrasion resistance	Evangelista/ Matias	0.8505 (0.8856)	-4.5726	0.8001 (0.8313)	-0.0813	0.3473 (0.3855)	-2.3175
Shrinkage	Evangelista/ Gomes/ Matias/ Cervantes/ Soberón	0.5509 (0.5606)	4.8938	0.5724 (0.5819)	0.1317	-	-
	Evangelista/ Gomes/ Matias/ Soberón	-	-	-	-	0.5964 (0.5972)	4.3015
Creep	Soberón	0.9943 (0.9943)	3.6548	0.9943 (0.9943)	0.0682	(0.7454)	0.8317
Water absorption by capillarity	Figueiredo/ Gomes/ Matias/ Rosa/ Ferreira	0.5034 (0.5769)	6.578	-	-	-	-
	Gomes/ Matias/ Rosa/ Ferreira	-	-	0.3689 (0.5052)	0.1447	-	-
	Evangelista/ Figueiredo/ Gomes/ Matias/ Ferreira	-	-	-	-	(0.2558)	1.7703

Table 6. Summary of the correlation trend lines between the different concrete properties and the density and water absorption of the mixture of aggregates and the 7-day compressive strength of concrete (Robles 2007) (Alves 2007)

Property	Experimental studies	Aggregates' density		Aggregates' water absorption		Concrete's 7-day compressive strength	
		R^2	Slope	R^2	Slope	R^2	Slope
Water absorption by immersion	Evangelista/ Figueiredo/ Gomes/ Matias/ Ferreira /Soberón	0.622 (0.6657)	7.2065	0.7329 (0.7332)	0.1296	0.8559 (0.8656)	3.8508
	Evangelista/ Figueiredo/ Gomes/ Matias/ Katz	-	-	-	-	0.5792 (0.6276)	1.7357
Carbonation resistance	Evangelista /Figueiredo/ Gomes/ Katz	0.6813 (0.6919)	7.4423	-	-	-	-
	Evangelista / Gomes/ Katz	-	-	0.7921 (0.7923)	0.1825	-	-
Chloride penetration resistance	Evangelista/ Figueiredo/ Matias/ Gomes/ Kou	-	-	-	-	0.5792 (0.6276)	1.7357
	Evangelista/ Gomes/ Kou	0.8338 (0.8398)	8.6492	-	-		
	Evangelista/ Figueiredo/ Gomes/ Kou	-	-	0.4779 (0.4953)	0.1684	-	-

correlation coefficient acceptable ($0.65 \leq R^2 < 0.80$)
correlation coefficient good ($0.80 \leq R^2 < 0.95$)
correlation coefficient very good ($R^2 \geq 0.95$)

Note: the values of R2 between parentheses correspond to the correlation lines that do not pass through the point corresponding to the RC.

- The lowest results were obtained for water absorption (especially by capillarity), which can be justified by the greater variability of this property compared with compressive strength, for example (a trend shared with conventional concrete);
- For every property of hardened concrete analyzed, except abrasion resistance, the performance of RCA is worse than that of the corresponding RC and the difference increases with the value of the replacement ratio;
- Based on the slope of the correlation lines for either of three reference parameters - density and water absorption of the mixture of aggregates and 7-day compressive strength of concrete - the vulnerability of RCA with the inclusion of RA instead of PA grows from mechanical properties (compressive strength, splitting and flexural tensile strength and modulus of elasticity) to those related to its rheologic behaviour (shrinkage and creep) and from these to the durability-related properties (water absorption, carbonation and chloride penetration).

An Application of the Methodology

In order to illustrate the potential of the methodology, a theoretical example of its application is now described. Even though this is not a real situation, all the values suggested are feasible and in agreement with past experience in experimental campaigns on RAC.

Let us suppose that the following types of aggregates are available: coarse primary (limestone) aggregates - CPA, fine primary (river sand) aggregates - FPA, coarse recycled concrete aggregates - CRA(RCA), coarse brick masonry aggregates - CRA(RMA), and fine recycled (concrete) aggregates - FRA. Their main characteristics (oven-dry density and water absorption) are provided in Table 7.

Table 7. Density and water absorption of the aggregates

	CPA	FPA	CRA(RCA)	CRA(RMA)	FRA
Oven-dry density (kg/m^3)	2650	2550	2350	2050	1950
Water absorption (%)	1.0	1.0	6.0	12.0	13.0

The composition of a C30/37 reference concrete (RC) was determined using the Faury method (leading to 35% of fine and 65% of coarse aggregates). Four types of RAC are being examined (Table 8). RAC1 is a composition which, according to various norms for RAC, is equivalent to a conventional concrete. The other three RAC compositions are more or less within the pragmatic limits for the maximization of the application of CRA(RCA), FPA and CRA(RMA) without unacceptable loss of performance. For every RAC composition, the following factors are expected to remain the same as for the RC: aggregates (fine and coarse fractions) grading curve, effective water / cement ratio, and slump value. Several specimens of each RAC and of the RC were tested for 7-day compressive strength and the corresponding results are given in Table 8, together with the weighed values for density (in accordance with equation (1) presented above) and water absorption of the mixture of aggregates used in each concrete mix.

Table 9 presents the corrective factors for the hardened concrete properties of all the RAC mixes, according to the correlation lines determined in the research described, as summarized in Table 6. These factors were obtained as an average of the factors provided by each of the parameters: D_{RAC}/D_{RC}; wa_{RAC}/wa_{RC}; f_{c7RAC}/f_{c7RC}.

This study shows that RAC1 in fact performs very much like the RC: practically the same density, losses of mechanical performance limited to 3% (with a 5% gain in abrasion resistance), increased deformability limited to 6%, and losses in durability-related aspects limited to 8% (not negligible, but small). Therefore, it shows that small percentages of coarse recycled concrete aggregates can in fact be used to make structural concrete without affecting its properties.

Table 8. PA/RA replacement ratios and reference parameters for all concrete mixes

	Replacement ratios (%)			D_{mix} (kg/m³)	wa_{mix} (%)	f_{c7} (MPa)	D_{RAC}/D_{RC}	wa_{RAC}/wa_{RC}	f_{c7RAC}/f_{c7RC}
	CPA/ CRA(RCA)	CPA/ CRA(RMA)	FPA/ FRA						
RC	20	0	0	2576.0	1.65	26.8	0.9851	1.65	0.9926
RAC1	100	0	0	2420.0	4.25	26.0	0.9254	4.25	0.9630
RAC2	0	0	25	2562.5	2.05	26.5	0.9799	2.05	0.9815
RAC3	0	25	0	2356.3	2.79	26.0	0.9011	2.7875	0.9630
RAC4	0	0	0	2615.0	1.00	27.0	-	-	-

Contradicting most norms that bar the use of FRA, the RAC3 composition (25% replacement) is the one with the next best performance: practically the same density as the RC, losses of mechanical performance limited to 5% (with a 7% gain in abrasion resistance), increases in deformability limited to 11% (not negligible, but small), and losses in durability-related aspects limited to 13% (not negligible, but small). This is good news for the environment since in some countries (e.g. Portugal) the greatest impact from aggregate production is related to the fine fraction (obtained from river beds, estuaries and sandy beaches).

Table 9. Corrective factors for the hardened concrete properties of all RAC mixes

	RAC1	RAC2	RAC3	RAC4
D_{RAC}/D_{RC}	0.99	0.96	0.99	0.97
f_{cRAC}/f_{cRC}	0.98	0.91	0.97	0.91
E_{cRAC}/E_{cRC}	0.98	0.88	0.96	0.88
f_{spRAC}/f_{spRC}	0.99	0.93	0.98	0.93
f_{fRAC}/f_{fRC}	0.97	0.86	0.95	0.86
$\Delta i_{RAC}/\Delta i_{RC}$	0.95	0.77	0.93	0.77
$\Delta\varepsilon_{RAC}/\Delta\varepsilon_{RC}$	1.06	1.32	1.11	1.29
$\varphi_{RAC}/\varphi_{RC}$	1.03	1.17	1.05	1.17
Ab_cap_{RAC}/Ab_cap_{RC}	1.07	1.34	1.11	1.33
Ab_im_{RAC}/Ab_im_{RC}	1.07	1.37	1.12	1.36
$carb._{RAC}/carb._{RC}$	1.08	1.40	1.12	1.13
$chlo._{RAC}/chlo._{RC}$	1.08	1.42	1.13	1.41

Legend: D_{RAC}/D_{RC} - concrete density; f_{cRAC}/f_{cRC} - compressive strength; E_{cRAC}/E_{cRC} - modulus of elasticity; f_{spRAC}/f_{spRC} - splitting tensile strength; f_{fRAC}/f_{fRC} - flexural tensile strength; $\Delta i_{RAC}/\Delta i_{RC}$ - loss of mass due to abrasion; $\Delta\varepsilon_{RAC}/\Delta\varepsilon_{RC}$ - extension due to shrinkage; $\varphi_{RAC}/\varphi_{RC}$ - creep coefficient; Ab_cap_{RAC}/Ab_cap_{RC} - water absorption by capillarity; Ab_im_{RAC}/Ab_im_{RC} - water absorption by immersion; $carb._{RAC}/carb._{RC}$ - carbonation depth; $chlo._{RAC}/chlo._{RC}$ - chloride diffusion coefficient.

The RAC2 and RAC4 compositions have similar levels of performance: practically the same density as the RC, losses of mechanical performance limited to 12% (with a 23% gain in abrasion resistance), increases in deformability up to 32% (liable to create problems in service), but, especially, losses in durability-related aspects up to 42% (which may have to be countered by increasing the reinforcement cover, in this case around 20%). These are not excellent results but they are acceptable in contexts where the importance of recycling large amounts of CDW outweighs the needs for exceptional concrete performances.

GENERAL CONCLUSIONS

The search for international experimental results for this study revealed great differences in the test procedures and organization of the published information. Most of the campaigns accepted the variation of more than one parameter, including the w/c ratio, making the analysis of the effect of the replacement ratio impracticable. These obstacles excluded a great number of campaigns from the graphical analysis undertaken.

The Portuguese literature review, even though it had a much shorter campaign domain, did not suffer the same difficulties, largely thanks to the scientific supervision, which was common to most of the works.

Notwithstanding the variability of factors introduced by each researcher in the experimental procedures, it was still possible to validate this methodology for the estimation of the properties of the concrete with recycled aggregates. Again, it must be pointed out that the scatter found in the relationships between the various properties of hardened RAC and the reference parameters is no greater that that found in similar relationships concerning conventional concrete, i.e. RAC are no less trustworthy in this aspect than conventional concrete.

The major advantage of this procedure is related to its low cost and the short time needed to obtain the results required to estimate the long-term properties of hardened concrete. The generalization of this methodology can, in the future, allow construction developers to decide, in an economic and fast way, cheaply and quickly about the use of RA in the building of new concrete structures.

Structural designers may also use acceptable estimations of the expected RAC properties in their designs, based on data provided by the contractor, the aggregates supplier or the CDW recycling plant. They can also provide the contractors with limits of application of RA in terms of the three parameters proposed here: D_{RAC}/D_{RC} (ratio between weighed values of the density of the mixture of aggregates used in the concrete mix); wa_{RAC}/wa_{RC} D_{RC} (ratio between weighed values of the water absorption of the mixture of aggregates used in the concrete mix); f_{c7RAC}/fD_{c7RC} (ratio between 7-day compressive strengths). As long as the contractors comply with the limits established by the structural designer the latter's calculations are valid.

REFERENCES

Alves, F. Concrete with coarse recycled concrete aggregates - The national experimental 'state-of-the-art' summary *(in Portuguese)*, *Masters Thesis in Civil Engineering, Instituto Superior Técnico*, Technical University of Lisbon, Portugal, 2007.

Angulo, S. Production of concrete with recycled aggregates *(in Portuguese)*, *Diploma Thesis in Civil Engineering*, State University of Londrina, Brazil, 1998.

Angulo, S. Characterization of aggregates from construction and demolition waste and the influence of their characteristics on the performance of concrete *(in Portuguese)*, *PhD Thesis in Civil Engineering*, São Paulo University, São Paulo, Brazil, 2005.

Arab, H. B. Study - Aggregates from construction and demolition waste in Europe, *European aggregates association*, UEPG, Brussels, 2006.

Azzouz, L.; Kenai, S.; Debied, F. Mechanical properties and durability of concrete made with coarse and fine recycled aggregates, *International Symposium on sustainable concrete construction*, University of Dundee, Scotland, 1998, pp. 383-392.

Barra, M., Vázquez, E. Properties of concretes with recycled aggregates: influence of properties of the aggregates and their interpretation, *International Symposium on the use of recycled concrete aggregate*, University of Dundee, Scotland, p. 19-30, 1998.

Buttler, A. Concrete with coarse recycled concrete aggregates - Influence of the age of recycling on the properties of the recycled aggregates and concrete *(in Portuguese)*, *Masters Dissertation in Structural Engineering*, University of São Paulo, Brazil, 2003.

Buyle-Bodin, F.; Hadjieva-Zaharieva, R. Influence of industrially produced recycled aggregates on flow properties of concrete, *Materials and Structures 35(8)*, 2002, pp. 504-509.

Cervantes, V.; Roesler, J.; Bordelon, A. Fracture and drying shrinkage properties of concrete containing recycled concrete aggregate, *Technical note CEAT*, University of Illinois, 2007.

de Brito, J. Recycled aggregates and their influence on concrete properties *(in Portuguese)*, *Public Lecture within the Full Professorship in Civil Engineering Pre-Admission Examination, Instituto Superior Técnico*, Technical University of Lisbon, Portugal, 2005.

de Brito, J. Methodology to estimate the properties of recycled aggregate concrete *(in Portuguese)*, *Patent PT 103756*, Lisbon, Portugal, 2007.

Carrijo, P. Analysis of the influence of the density of coarse aggregates from construction and demolition waste on the mechanical performance of concrete *(in Portuguese)*, *Master Dissertation in Civil Engineering*, University of São Paulo, São Paulo, 2005.

Chen, H., Yen, T., Chen, K. Use of building rubbles as recycled aggregates, *Cement and Concrete Research 33(1)*, 2003, pp.125-132.

Corinaldesi, V. Influence of recycled aggregate on the concrete performance, *Training course on technology road mapping for building materials in developing countries and promotion of related projects*, Italy, 2003.

DEPA. Waste statistics 2003 [online], Environmental Protection Agency, cited June 2006. Available from: *http://www.mst.dk/homepage/default.asp?Sub=http://www.mst.dk/udgiv/Publications/2005/87-7614-585-9/html/default_eng.htm*.

Etxeberria, M.; Vázquez, E.; Marí, A.; Barra, M.; Hendriks, F. The role and influence of recycled aggregate in recycled aggregate concrete, *International RILEM Conference on the Use of Recycled Materials in Buildings and Structures*, Barcelona, 2004, pp. 665-674.

Etxeberria, M.; Vázquez; E., Marí, A. Recycled concrete aggregate as a structural material, *Materials and Structures* DOI 10.1617/s11527-006-9161-5, 2006.

Evangelista, L. Concrete made with fine recycled concrete aggregates *(in Portuguese)*, *Masters Thesis in Construction, Instituto Superior Técnico*, Technical University of Lisbon, Portugal, 2007.

Ferreira, L. Structural concrete with coarse recycled concrete aggregates - Influence of the pre-saturation *(in Portuguese)*, *Masters Dissertation in Civil Engineering, Instituto Superior Técnico*, Technical University of Lisbon, Lisbon, Portugal, 2007.

Figueiredo, F. Integrated management of construction and demolition waste *(in Portuguese)*, *Report of research project POCTI/ECM/43057/2001*, Faculty of Engineering of Porto University, Porto, Portugal, 2005.

Fraaij, A. L., Pietersen, H. S., de Vries, J. Performance of concrete with recycled aggregates, *International Conference on sustainable concrete construction*, Dundee, 2002, Scotland, pp.187-198.

Gomes, M. Structural concrete made with coarse ceramic and recycled concrete aggregates *(in Portuguese)*, *Masters Dissertation in Construction, Instituto Superior Técnico*, Technical University of Lisbon, Lisbon, Portugal, 2007.

Gonçalves, A. P. Analysis of the performance of concrete made with aggregates resulting from construction waste *(in Portuguese)*, *Masters Thesis in Construction, Instituto Superior Técnico*, Technical University of Lisbon, Portugal, 2002.

Gonçalves, P. Recycled aggregate concrete - Commented Analysis of existing legislation *(in Portuguese)*, *Masters Thesis in Civil Engineering, Instituto Superior Técnico, Technical University of Lisbon*, Portugal, 2007.

Hansen, T.; Narud, H. Strength of recycled concrete made from crushed concrete coarse aggregate, *Concrete International 5(1), American Concrete Institute,* Detroit, 2003, pp. 79-83.

Juan, M. S.; Gutiérrez, P. A. Influence of recycled aggregate quality on concrete properties, *International RILEM Conference on the Use of Recycled Materials in Buildings and Structures, Barcelona,* 2004, pp. 545-553.

Katz, A. Properties of concrete made with recycled aggregate from partially hydrated old concrete, *Cement and Concrete Research, V. 33 (5),* 2003, pp. 703-711.

Khatib, J. M. Properties of concrete incorporating fine recycled aggregate, *Cement and Concrete Research 35(4),* 2004, p. 763-769.

Kimura, Y., Imamoto, K., Nagayama, M., Tamura, H., High quality recycled aggregate processed by decompression and rapid release, *RILEM International Symposium on Environment-Conscious Materials and Systems for Sustainable Development,* Koriyama, Japan, 2004, pp. 163-170.

Knights, J. Relative performance of high quality concretes containing recycled aggregates and their use in construction, *International Symposium on sustainable concrete construction,* University of Dundee, Scotland, 1998, pp. 275-286.

Kou, S. C.; Poon, C. S.; Chan, D. Properties of steam cured recycled aggregate fly ash concrete, *International RILEM Conference on the Use of Recycled Materials in Buildings and Structures*, Barcelona, 2004, pp. 590-599.

Larrañaga, M. E. Experimental study on microstructure and structural behaviour of recycled aggregate concrete, *PhD Thesis*, Polytechnic University of Catalonia, Barcelona, 2004.

Latterza, L.; Machado, E. Concrete with coarse recycled aggregate: properties in the fresh and hardened states and application in light precast elements *(in Portuguese)*, *Engineering and Structures Journal, No. 21*, 2003, pp. 27-58.

Leite, M. Evaluation of the mechanical properties of concrete made with recycled aggregates from construction and demolition waste *(in Portuguese)*, *PhD Thesis in Civil Engineering, Federal University of Rio Grande do Sul*, Porto Alegre, 2001.

Levy, S. Contribution to the study of the durability of concrete made with concrete and masonry waste *(in Portuguese)*, *PhD Thesis in Civil Engineering, Polytechnic School of São Paulo, São Paulo, Brazil*, 2001.

Levy, S. M.; Helene, P. Durability of recycled aggregates concrete: a safe way to sustainable development, *Cement and Concrete Research 34(11), 2004*, pp. 1975-1980.

Limbachiya, M.C.; Dhir, R. K.; Leelawat, T. Use of recycled concrete aggregate in high-strength concrete, Materials and Structures, V. 33, 2000, pp 574-580.

Lin, Y., Tyan, Y., Chang, T., Chang, C. An assessment of optimal mixture for concrete made with recycled concrete aggregates, *Cement and Concrete Research 34(8), 2004*, pp. 1373-1380.

Matias, D.; de Brito, J. Concrete with coarse recycled concrete aggregates resorting to plasticizers - Experimental study performed at IST *(in Portuguese)*, *Report ICIST DTC 3/05, Instituto Superior Técnico*, Technical University of Lisbon, Portugal, 2005.

Mendes, T.; Morales, G.; Carbonari, G. Study on ARC's aggregate utilization recycled of concrete, *International RILEM Conference on the Use of Recycled Materials in Buildings and Structures*, Barcelona, 2004, pp. 629-635.

Merlet, J. D.; Pimienta, P. Mechanical and physico-chemical properties of concrete produced with coarse and fine recycled concrete aggregates, *Demolition and reuse of concrete and masonry, RILEM proceedings* 23, Odense, 1993, pp. 343-353.

Muller, A. Lightweight aggregates from masonry rubble, *International RILEM Conference on the Use of Recycled Materials in Buildings and Structures*, Barcelona, 2004, pp. 97-106.

Muthmann, R. Waste generated and treated in Europe - Data 1991-2001 [online], *European Communities*, Luxembourg, 2003, cited April 2006. Available from: http://epp.eurostat.ec.europa.eu/cache/ITY_OFFPUB/KS-55-03-471/EN/KS-55-03-471-EN.PDF.

Oliveira, M. de, de Assis, C., Terni, A. Study on compressed stress, water absorption and modulus of elasticity of produced concrete made by recycled aggregate, *International RILEM Conference on the Use of Recycled Materials in Buildings and Structures*, Barcelona, 2004, pp. 636-642.

Olorunsogo, F. T.; Padayachee, N. Performance of recycled aggregate concrete monitored by durability indexes, *Cement and Concrete Research*, V. 32, 2001, pp. 179-185.

Robles, R. Prediction of the behaviour of concrete with coarse recycled concrete aggregates - The international experimental 'state-of-the-art' summary *(in Portuguese)*, *Masters Thesis in Civil Engineering, Instituto Superior Técnico*, Technical University of Lisbon, Portugal, 2007.

Rocha, B.; Resende, C. Properties of concrete made with recycled aggregates *(in Portuguese)*, *Diploma Thesis in Civil Engineering, Aveiro University*, Aveiro, Portugal, 2004.

Roos, F. Verification of the dimensioning values for concrete with recycled aggregates, *International Symposium on the use of recycled concrete aggregate*, University of Dundee, Scotland, 1998, pp. 309-319.

Rosa, A. S. Use of coarse ceramic aggregates in concrete production *(in Portuguese), Masters Thesis in Construction, Instituto Superior Técnico,* Technical University of Lisbon, Portugal, 2002.

Sagoe-Crentsil, K. K., Brown, T., Taylor A. H. Performance of concrete made with commercially produced coarse recycled concrete aggregate, *Cement and Concrete Research 31(5), 2001*, pp.707-712.

Santos, J. R.; Branco, F. A.; de Brito, J. Almeida, N. Concrete with coarse concrete aggregates: compressive strength and modulus of elasticity *(in Portuguese)*, Conference Construction 2004, Porto, Portugal, 2004, pp. 357-362.

Silva, J. Incorporation of red-brick waste in cementitious mortars *(in Portuguese)*, Masters Thesis in Construction, Instituto Superior Técnico, Technical University of Lisbon, Portugal, 2006.

Soberón, J. G. "Porosity of recycled concrete with substitution of recycled concrete aggregate, an experimental study", *Cement and Concrete Research, V. 32 (8)*, 2002, pp. 1301-1311.

Sugiyama, M. The compressive strength of concrete containing tile chips, crushed scallop shells, or crushed roofing tiles, *International RILEM Conference on the Use of Recycled Materials in Buildings and Structures*, Barcelona, 2004, pp. 658-664.

Symonds. Construction and demolition waste management practices and their economic impact" *Report to DGXI, European Commission*, Brussels, 1999.

Tsujino, M.; Noguchi, T.; Tamura, M.; Kanematsu, M.; Maruyama, I. Application of conventionally re-cycled coarse aggregate to concrete structure by surface modification treatment, *Journal of Advanced Concrete Technology 5(1), 2007*, pp 13-25.

UEPG. The European aggregates industry - Statistics 2005 [online], *European Aggregates Association,* Brussels, 2006, cited April 2006. Available from: *http://www.uepg.eu/index.php?pid=122.*

Vieira, G., Molin, D., Lima, F. de Strength and durability of concrete produced with recycled aggregates from construction and demolition waste *(in Portuguese)*, *Civil Engineering Journal 19*, Minho University, 2004, pp. 5-18.

Wainwright P. J.; Trevorrow, A.; Yu, Y.; Wang, Y. Modifying the performance of concrete made with coarse and fine recycled concrete aggregates, *Demolition and reuse of concrete - RILEM Proceedings 23,* E & FN Spon, 1994, pp. 319-330.

Wirquin, E.; Buyle-Bodin, F.; Hadjieva-Zaharieva, R. Utilisation de l'absorption d'eau des bétons comme critères de leur durabilité - application aux bétons de granulats recyclés, *Materials and Structures 33(6), 2000*, pp. 403-408.

Xiao, J. Z.; Li, J. B.; Zhang, C. On relationships between the mechanical properties of recycled aggregate concrete - an overview, *Materials and Structures 39(6), 2006*, pp. 655-664.

Chapter 3

A KINETIC MODEL OF THE OXIDE GROWTH AND RESTRUCTURING ON STRUCTURAL MATERIALS IN NUCLEAR POWER PLANTS

Iva Betova[1], Martin Bojinov[2], Petri Kinnunen[3], Klas Lundgren[4] and Timo Saario[3]

[1]Department of Chemistry, Technical University of Sofia, Sofia, Bulgaria
[2]Department of Physical Chemistry, University of Chemical Technology and Metallurgy, Sofia, Bulgaria
[3]VTT Materials and Building, VTT Technical Research Centre of Finland, Espoo, Finland
[4]ALARA Engineering AB, Skultuna, Sweden

ABSTRACT

A deterministic model for corrosion and activity incorporation in oxides on structural materials in nuclear power plant coolant circuits is proposed based on fundamental physico-chemical mechanisms. In order to ensure adequate modelling, uncertain or non-determined fundamental parameters are set or adjusted within the range of reasonable values by evaluating well-defined or well-controlled in-plant observations or laboratory experiments. As a result of the calculation procedure developed on the basis of the model, the kinetic and transport parameters of the growth and restructuring of the oxides on austenitic stainless steels (AISI 304 and AISI 316) and nickel based alloys (Alloys 600 and 690) in Light Water Reactors (LWRs) were determined via quantitative comparison of the equations of the Mixed-Conduction Model for oxide films with ex-situ analytical results on thickness and composition of such layers obtained from both laboratory and in-reactor exposures. The obtained parameters were used to predict film growth and restructuring, as well as corrosion release with time of exposure to the LWR coolant. Subsequently, kinetic and transport parameters for the incorporation of Co and Zn in the oxide layers on stainless steels and nickel alloys were estimated by reproducing the experimental depth profiles of these elements. The diffusion-migration equations for the non-steady state transport of such species were solved subject to the boundary conditions at the solution side set by the stability constants of the corresponding Zn and Co complexes. The calculational results are discussed in terms of the influence of incorporation of solution-originating species on the kinetics of film growth and layer restructuring.

ABBREVIATIONS

D_{Co}	diffusion coefficient of Co in the inner layer, cm^2 s^{-1}
$D_{Co,OL}$	formal diffusion coefficient of Co in the outer layer, cm^2 s^{-1}
D_{Cr}	diffusion coefficient of Cr in the inner layer, cm^2 s^{-1}
$D_{Cr,OL}$	formal diffusion coefficient of Ni in the inner layer, cm^2 s^{-1}
$D_{Ni,OL}$	formal diffusion coefficient of Ni in the outer layer, cm^2 s^{-1}
D_{Zn}	diffusion coefficient of Zn in the inner layer, cm^2 s^{-1}
$D_{Zn,OL}$	formal diffusion coefficient of Zn in the outer layer, cm^2 s^{-1}
$\vec{E}$	field strength in the inner layer, V cm^{-1}
F	Faraday number, 96487 C mol^{-1}
K	equilibrium constant of surface complexation reactions
$K_{enr,Co}$	enrichment factor of Co at the inner layer/electrolyte interface
$K_{enr,Zn}$	enrichment factor of Zn at the inner layer/electrolyte interface
$k_{1,Cr}$	rate constant of oxidation of Cr at the alloy/inner layer interface, cm^4 mol^{-1} s^{-1}
$k_{1,Fe}$	rate constant of oxidation of Fe at the alloy/inner layer interface, cm^4 mol^{-1} s^{-1}
$k_{1,Ni}$	rate constant of oxidation of Ni at the alloy/inner layer interface, cm^4 mol^{-1} s^{-1}
$k_{3,Cr}$	rate constant of dissolution of Cr at the inner layer/electrolyte interface, cm s^{-1}
$k_{3,Fe}$	rate constant of oxidation of Fe at the inner layer/electrolyte interface, cm s^{-1}
$k_{3,Ni}$	rate constant of oxidation of Ni at the inner layer/electrolyte interface, cm s^{-1}
L	inner layer thickness, cm
L(t=0)	initial film thickness, cm
R	universal gas constant, 8.314 J mol^{-1} K^{-1}
T	temperature, K
$V_{m,MO}$	molar volume of the phase in the inner layer, cm^3 mol^{-1}
X	formal valency state of Fe in the inner layer
y_{Co}	mass fraction of Co in the oxide
y_{Cr}	mass fraction of Cr in the oxide
y_{Fe}	mass fraction of Fe in the oxide
y_{Ni}	mass fraction of Ni in the oxide
y_{Zn}	mass fraction of Zn in the oxide
$y_{Co,a}$	mass fraction of Co in the alloy substrate
$y_{Cr,a}$	mass fraction of Cr in the alloy substrate
$y_{Fe,a}$	mass fraction of Fe in the alloy substrate
$y_{Ni,a}$	mass fraction of Ni in the alloy substrate
$y_{Zn,a}$	mass fraction of Zn in the alloy substrate
α_1	transfer coefficient of the oxidation reaction at the alloy/inner layer interface

INTRODUCTION

Importance of the Problem

The two main issues in having a safe and economic operation of a nuclear power plant are to minimise the risk and occurrence of stress corrosion cracking, and to minimise the exposure of the staff and other people to ionising radiation. Although these two issues are very disparate phenomena, the common and most important factor is the integrity of the oxide films on the structural materials in the nuclear power plants.

All materials used in the nuclear power plants rely on a passivating oxide film. This means that the oxide film formed from the corrosion attack of water on the material will protect the underlying material against further attack. The material in itself would, according to the thermodynamics, decompose into a mixture of metal oxides without this protective layer. This deterioration of the material is also observed if the passivating film is destroyed for a certain reason. The protective effect of the oxide layer originates from a low solubility of the oxide itself and a slow reaction rate for any chemical interaction between the oxide film and the surrounding environment, thecoolant water in light water reactor (LWR). The protective effect is, however, never perfect. This means that corrosion of the base material is always occurring to some extent. Furthermore, changes in the chemical composition or pH of the surrounding water, or an increase in temperature, could increase the reaction rate and hence diminish the protective ability of the oxide film.

During the almost 50 years of commercial use of LWRs, a huge experience has been gained about the materials behaviour in addition to the basic mechanical properties. Yet, the increasingly faster change of the environmental conditions has made it difficult to test all possible combinations of materials and environments in order to assess the long-term behaviour of the materials, or rather the behaviour of the protective oxides on the materials, in these new combinations of environments.

The corrosion behaviour of a certain material is always depending on the environment, which influences the release of corrosion products from the already existing oxide surfaces. The behaviour of oxide films in the primary system of a nuclear power plant is hence a complex function of their behaviour in coupled localised areas of the system depending on each other and all local environmental factors.

The inference is that a full assessment of the materials behaviour in LWRs cannot be gained only through materials testing in simulated or even real LWR environments, since the oxide film stability and protectiveness will depend on a large number of factors, and so will the materials integrity. We need a more fundamental understanding for each kind of oxide film and the impact of environmental factors on the film behaviour. We also need a theory or a modelling tool that can take into account the interaction between the various local oxides with their local chemistry in order to have an integral description and understanding of the stress corrosion cracking and the activity build-up in any LWR.

Oxide Films, Localised Corrosion and Stress Corrosion Cracking

The protective ability of the oxide film is hindering significant corrosion of the base alloy and hence the passive film assures that no significant component thinning is occurring during the exposure to the surrounding coolant in the LWR. Even very small localised attacks on the oxide film could, however, be detrimental for the component. Although the localised attack would only affect a small area or part of the material, it could still be threatening the safe operation of the LWR through stress corrosion cracking (SCC). The passive behaviour of the oxide film could then be important in two ways.

The first is for the localised corrosion attack itself. To have SCC, in addition to a tensile stress applied to the material, the material must be susceptible to localised attack, e.g. sensitised, and there must be an aggressive environment causing the attack. Although the actual chemistry is rather different in Boiling Water Reactors (BWRs) compared to Pressurized Water Reactors (PWRs) or Water-Water Energy Reactors (WWERs), still the same factors are needed. If the stressed oxide film is prone to localised attacks via chemical interaction with the coolant water, so called pitting could develop on the material surface. The pits, formed where the oxide film is locally the least protective, will establish a local aggressive chemistry in the bottom of the pit. The weakest spots of the oxide film will hence be the locations where further attacks to the underlying, no longer well-protected, base alloy occur. When a certain depth, normally claimed to be approximately 20 μm or slightly deeper, has been penetrated in the base alloy, the pit or incipient attack can develop into a propagating crack if sufficient stress is applied. The weakest spot in the oxide film is thus transformed into the weakest spot for cracking of the materials. Chlorides and other halides are known to cause localised attacks. Many more chemical impurities and combinations of chemical impurities could, however, be the possible cause for such a localised attack.

The second aspect is the re-passivation kinetics, i.e. the ability to quickly reform a protective passive layer, which is also important for the development of SCC. If the material does not re-passivate, the localised attack would be more extended and actually open up the attack area to allow wash-out of the local chemistry. The re-passivation hence allows for the slightly less stressed areas of exposed metal to rebuild an oxide which will protect them against further attack, confining the attack to a small area. This forces the attack to grow into the material instead of extending laterally and stabilising the local chemistry in a more and more confined location. In this way the localised attack continues to penetrate the material and keeps the intergranular or transgranular crack growing by continuously re-establishing the aggressive chemical conditions until the crack has propagated through the material.

The re-passivation kinetics is always important what comes to SCC in LWR environments. The weak spots in the film are only important when the tensile stresses do not immediately crack the brittle oxide film but a localised chemical attack is needed to start the cracking behaviour. Just as for the general corrosion and the impact of the environment on the passivation behaviour in general, the environment will be of decisive importance what comes to the localised attack and the re-passivation kinetics.

Oxide Films and Activity Incorporation

The activity build-up in nuclear power plants is influenced by the oxide films in several ways. To have activity build-up a certain corrosion release from the materials is needed to allow corrosion products to deposit in the core region and become activated. Without corrosion products from the non-core structural materials, basically only the cladding and its oxide would be the source for any activity distribution. The oxide formed on the zirconium material, does, however, almost exclusively consist of zirconium oxide, which has a very low solubility in BWR conditions, but also markedly low solubility in PWR and WWER environments. This means that without deposited corrosion products in the core, the activity release would be almost exclusively due to flaking of oxide on fuel cladding. The activation products of zirconium, i.e. Zr-95 and Nb-95 and their derivatives, have short half-life (e.g. 64 days for Zr-95) and the activation build-up problem would be virtually non-existent in LWRs.

In reality, the materials in the LWR primary systems do release a significant amount of products to the coolant through corrosion. Flow assisted corrosion enhances the corrosion release from some parts of the system and hence provides even higher contributions to the primary coolant flow. The corrosion release in itself will affect the oxide film build-up of any materials downstream the releasing material.

The neutron activation of the core crud will depend on the amount of crud and the residence time for the deposit on the core. These aspects are indirectly related to the integral corrosion release from all the materials in contact with the coolant before the core passage. The activity released from the core is then partly in form of transmuted corrosion products, but still to a large extent the transmuted nuclides are of the same or similar elements, i.e. cobalt (Co-58 and Co-60), iron (Fe-59), manganese (Mn-54), zinc (Zn-65), and chromium (Cr-51). In addition some elements formed from only trace amounts in the coolant flow are of importance, such as antimony (Sb-122, Sb-124) and silver (Ag-110m). Still, most of the released core crud will be non-transmuted, and hence the core will be a sink and source for the corrosion products being distributed by the coolant. The activated corrosion products will, however, be responsible for any activity build-up.

The activity build-up in the nuclear power plant is either resulting from particulate matter mainly depositing in low flow locations, or, which is one of the main focus points in this review and the planned model development, resulting from incorporation of dissolved corrosion products with the oxide films in contact with the coolant carrying the ionic activity. The ionic activity can interact with the oxide film and be incorporated, all as a function of the oxide growth, release, and restructuring behaviour.

Objective and Structure of the Chapter

The main objective of the present paper is to give the fundament, both in knowledge and in structure, for the development of a theoretical approach and eventually a predictive model of the oxide film behaviour on structural materials in nuclear power plants in relation to activity build-up and corrosion phenomena.

The basic limitation for the oxide film growth rate on such materials and hence the oxide thickness in each instance, is the solid state transport rate within the inner, also called barrier, oxide film layer. Models for such transport have been developed for many years, but only recently has the modelling reached a sufficient sophistication to allow a reasonably good quantitative correlation between the modelled values and the growth rates of oxide films on structural materials in real LWR environments. Such a quantitative model, called the Mixed-Conduction Model (MCM) is described to some detail in the Theory section of the paper. The associated calculation procedure that enables the estimation of the kinetic and transport parameters using a quantitative comparison of the model equations with experimental data stemming mainly from surface and in-depth characterisation of the oxide films on structural materials by ex-situ analytical techniques is also outlined. The physical significance and the relevance of the parameter estimates is described at length in the Results and Discussion section, in which a systematic review of the calculational results for oxide films formed both in laboratory and during plant operation is presented. Finally, the limitations of the proposed approach are summarized and some trends for future research in the field are proposed.

Composition of the Structural Materials Discussed in the Chapter

The typical compositions of the structural materials, the oxidation mechanism of which is subject to discussion in the present work, are summarized in Table 1 (stainless steels) and Table 2 (nickel-based alloys).

Table 1. Composition of stainless steels (weight-%).

Material	C	Cr	Cu	Fe	Mn	Ni	S	Si	Mo	Other
AISI 304	0.08	19		bal.	2	9.3	0.03	1		N : 0.1
AISI 316L(NG)	0.015	16.5	0.26	bal.	1.73	10.5	0.002	0.54	2.55	N : 0.056
AISI 316L	0.03	17		bal.	2	12	0.03	1	2.5	N:0.1

Table 2. Compositions of nickel-based alloys (weight-%).

Material	C	Cr	Cu	Fe	Mn	Ni	S	Si	Other
Alloy 600	<0.15	16	≤0.5	8	≤1.0	bal.	0.015	≤0.5	
Alloy 690	<0.05	30	≤0.5	9	0.31	bal.	0.006	≤0.5	

THEORY

The Mixed-Conduction Model

The MCM is at the conceptual level quite similar to the Point Defect Model (PDM) proposed and developed by Macdonald and co-workers during the last three decades [[1]-[4]]. However, some basic assumptions made in these models differ to a certain extent. Thus it is worth summarising the assumptions made within the frames of the MCM:

1. The film consists mainly of oxide and the amount of hydroxide in the film is negligible. Thus the possible doping of the layer with incorporated hydrogen e.g.

playing the role of electron donor is not taken into account. The stoichiometric composition of the oxide varies gradually with potential without abrupt structural changes.
2. Possible ionic charge carriers in the film are anion, i.e. oxygen, vacancies, interstitial cations and cation vacancies.
3. A certain concentration of oxygen vacancies is maintained in the film, in order to explain the growth of the film via the inward-motion of oxygen ions.
4. The electric field strength in the film is assumed to be homogeneous and independent of potential, while the film thickness is assumed to be proportional to the applied potential. However, very recent thickness measurements indicate that the latter assumption may not be fulfilled at high temperatures.
5. For the transport of ionic species, the high-field approximation is used at room temperature and the low-field approximation at high temperatures, i.e. the electric field strength is assumed to decrease as the film becomes thicker and/or more defective at high temperatures.
6. An exponential dependence on the local potential drops is assumed at both interfaces for the interfacial rate constants. However, the potential drop at the alloy/oxide interface is supposed to be independent on the applied potential, which is in turn distributed between the oxide bulk and the oxide/electrolyte interface. In other words, the rates of the reactions at the alloy/oxide interface are not adjusted by the applied potential but rather by the transport of defects through the oxide.
7. Each ionic point defect in the film plays the role of an electron donor or acceptor. This means that it is accompanied by an electronic defect that contributes to the electronic conductivity in the film.
8. The main contribution to the measurable film resistance comes from the part of the film in which the defect concentration and thus also the conductivity of the film is at its minimum.

In the MCM the impedance of the metal/film/solution system is a combination of the impedances of the interfaces and that of the bulk film. For the passive film in the steady state the interfacial impedances can be usually neglected, because the reactions at the interfaces are either very fast compared with the transport in the oxide film, or the capacitances corresponding to these reactions are very large so that the impedances are not distinguishable. The film impedance consists of the electronic and ionic contributions. Using the general transport equations given by Fromhold and Cook [[5]] and defining the appropriate boundary conditions, the concentration profiles for different ionic charge carriers in the film can be defined. Further, the electronic conductivity is proportional to the concentration of defects, and equations describing the electronic conductivity can be derived for both low and high potentials according to the type of charge carriers having the strongest effect. The transfer function for the electronic contribution to the total conductivity and to the film impedance depending on the film thickness and potential can be derived from an explicit physico-chemical model, being mathematically equivalent to the formal model of Young [[6]] for insulating layers with defect induced conductivity. This makes the bridge to the chemiconductor model at the conceptual level [[7]]. The ionic part of the film impedance, on the other hand, often reduces to a resistance, if the apparent diffusion coefficient is high, or to a Warburg impedance in the case of a low diffusion coefficient.

The MCM has been successfully used previously to explain the features of several metals and alloys in the passive region at room temperature [[8]]-[[11]]. A comprehensive description of the model for structural materials in high temperature aqueous electrolytes such as LWR coolants is given below.

Taking into account the above assumptions, the basic kinetic scheme of the MCM is shown in Figure 1 for the passive film on Fe as an example.

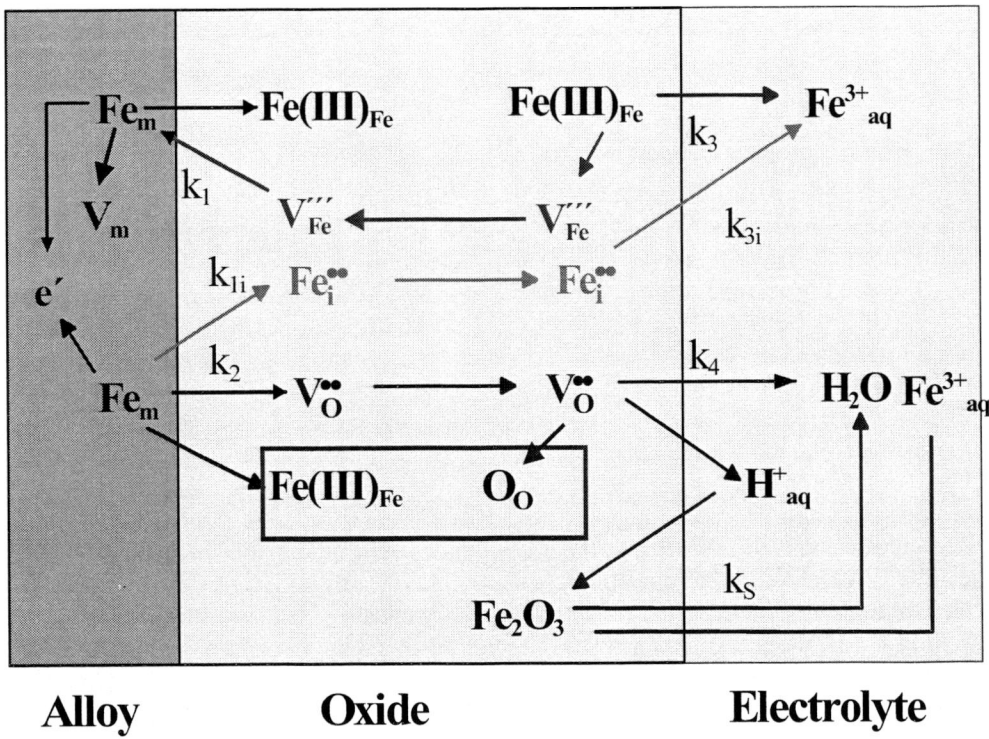

Figure 1. Basic kinetic scheme of the MCM illustrated for the passive film on iron.

As stated above (assumption 5) because of the increase of their defectiveness, the films at high temperature are not likely to support the high-field conditions. Thus the steady-state solutions of the generalised transport equation given by Fromhold and Cook [[5]] with a low-field approximation results in equations of the type

$$J_O(x) = -D_O \frac{\partial c_O(x)}{\partial x} - \frac{D_O}{a} c_O(x) \left[\frac{zFaE(x)}{RT} \right] \tag{1}$$

These equations together have to be solved with the appropriate boundary conditions in order to calculate the profiles of defects within the film. In this case, the steady state concentration profiles of anion and cation vacancies, as well as interstitial cations, can be written as follows:

$$c_O(x) = \chi k_2 \left[\left(\frac{1}{k_4} - \frac{1}{2KD_O} \right) e^{-2Kx} + \frac{1}{2KD_O} \right]$$

$$c_M(x) = k_3 \left[\left(\frac{1}{k_1} - \frac{1}{2KD_M} \right) e^{-3K(L-x)} + \frac{1}{3KD_M} \right] \quad (2)$$

$$c_i(x) = k_{1i} \left[\left(\frac{1}{k_{3i}} - \frac{1}{2KD_i} \right) e^{-2Kx} + \frac{1}{2KD_i} \right]$$

where $K = F\vec{E}/RT$. For a detailed explanation of the symbols used, the reader is referred to the Nomenclature section.

The obtained expressions for the concentration profiles of positive (e.g. oxygen vacancies) and negative (e.g. cation vacancies) defects are illustrated in Figure 2. They contain exponential terms that depend on the distance within the film and the applied potential. The latter dependence is included in the dependence of the interfacial rate constants on potential. This means that, according to the model, the difference between the distribution of defects in films formed at room and at a high temperature is only quantitative.

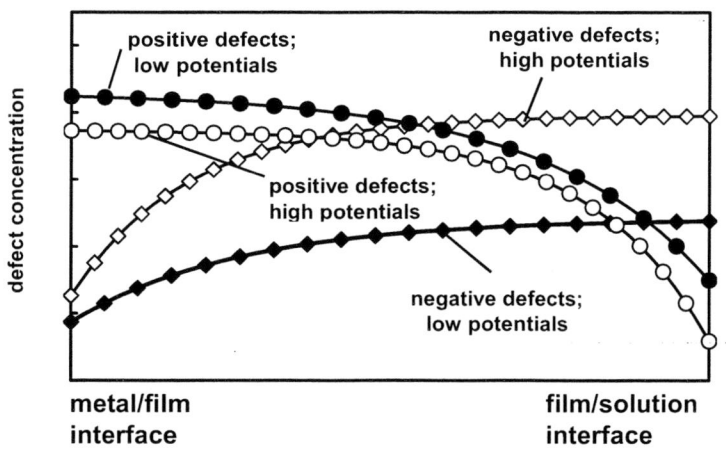

Figure 2. A scheme of simulated profiles of defect concentrations through the oxide film at high and at low potentials.

Inner Layer Growth According to the Proposed Approximation of the MCM

According to the MCM [[8]-[17]], the inner layer growth proceeds via the sequence of reactions involving the generation of normal cation positions and injection of oxygen vacancies at the alloy/film interface, their transport via a diffusion-migration mechanism and subsequent consumption at the film/electrolyte interface via a reaction with adsorbed water. In parallel to that process, metal cation dissolution through the film, involving either the generation of cation vacancies at the outer interface, their transport and consumption at the inner interface, or generation, transport and release of interstitial cations, respectively, is also expected to occur. Due to the low solubility of the constituent metals in the high-temperature electrolyte, the dissolved cations are mainly re-deposited to form the outer layer. Such a situation is illustrated schematically in Figure 3 for film growth on stainless steel. In this figure, the oxidation and dissolution of nickel has been omitted for the sake of simplicity. As will be pointed out later in this chapter, in addition to the inner and outer oxide layers resulting from the release of cations from the oxide, also a deposited layer formed via adsorption of solution originating particles onto the oxide surface can exists.

The goal of the present work is to elaborate a calculational procedure for the determination of kinetic and transport parameters of individual alloy constituents. Such a procedure is intended to be validated by comparing the predictions of the model with experimentally measured depth profiles of such constituents in the oxide, normalised to its total cation content to exclude the influence of the oxygen profile. In order to reduce the number of variables, we intend to use a simplified approach scheme presented in Fig. 4. Although it has been presumed that Cr and Ni are transported through the oxide via vacancies and Fe via interstitial cation sites, at this point we do not distinguish between the types of the defects via which the respective metallic constituent is transported through the inner layer of oxide. Thus e.g. the reaction rate constant at the alloy/inner layer interface for Fe (k_{1Fe}) can be regarded as a sum of the reaction rate constants of oxidation of Fe producing Fe interstitials (k_{1iFe}), producing normal Fe positions and oxygen vacancies (k_{2Fe}) and consuming cation vacancies with the simultaneous production of normal Fe positions (k_{1FeM}), the same being the case for Cr and Ni and also for the inner layer/electrolyte interface in the pores of the outer layer. On the other hand, the diffusion coefficients D_{Cr}, D_{Fe} and D_{Ni} can be regarded as characterising the overall transport of cations through the inner layer resulting either from its growth via generation, transport and consumption of oxygen vacancies or the dissolution of these cations through the inner layer via a vacancy or interstitial mechanism.

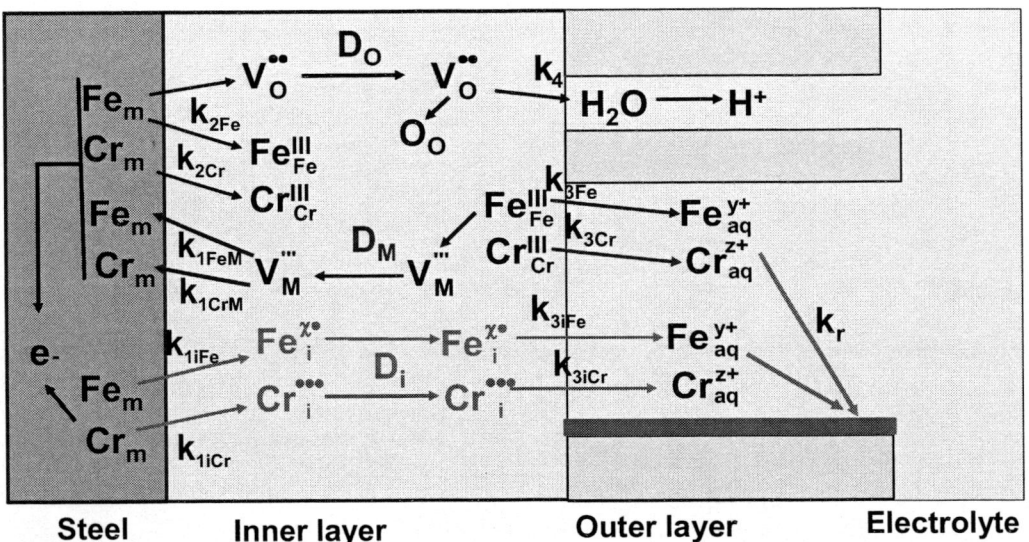

Figure 3. A scheme of the growth of the inner layer, dissolution of cations through it and subsequent precipitation of the outer layer of the corrosion film on stainless steel according to the MCM.

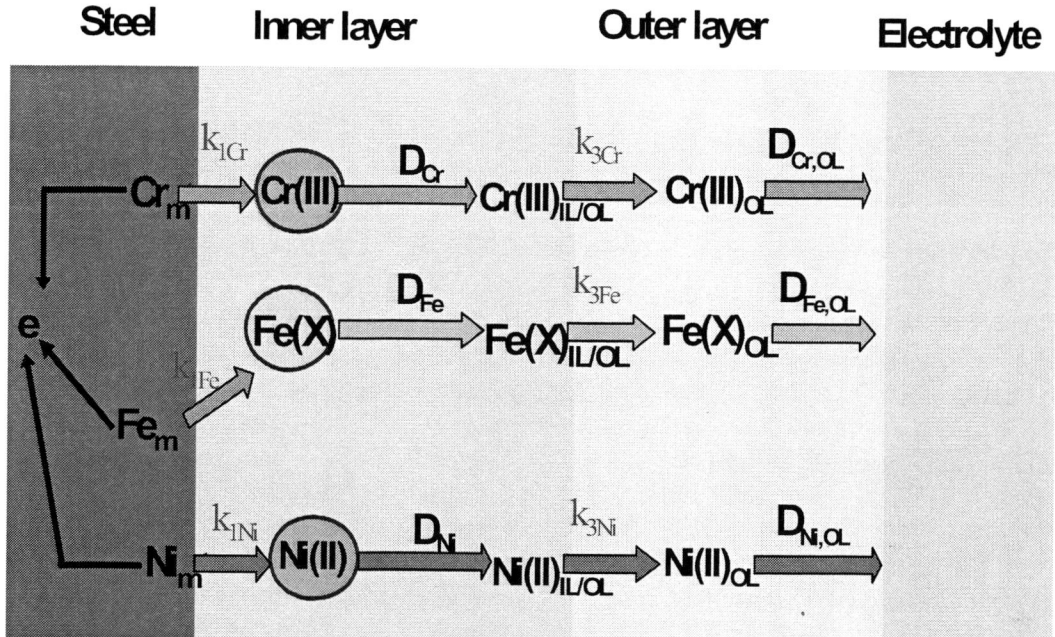

Figure 4. A simplified scheme of the growth of the inner and outer layers of the film formed on a structural material according to the proposed simplified approach. For details see text.

Mathematically, we treat the problem in conceptual analogy to the approach presented in Ref [18], where the depth profile of W in a binary Ni-W alloy has been calculated on the basis of the PDM. The depth profile of a metallic oxide constituent j = Fe, Cr or Ni, can be expressed as the dependence of its molar fraction, $y_j = c_j V_{m,MO}$, on the distance within the inner layer, where c_j is its molar concentration and $V_{m,MO}$ the molar volume of the phase in the layer. Respectively the transient diffusion-migration equations for each component

$$\frac{\partial y_{Fe}}{\partial t} = D_{Fe} \frac{\partial^2 y_{Fe}}{\partial x^2} + \frac{XF\vec{E}D_{Fe}}{RT} \frac{\partial y_{Fe}}{\partial x}$$

$$\frac{\partial y_{Cr}}{\partial t} = D_{Cr} \frac{\partial^2 y_{Cr}}{\partial x^2} + \frac{3F\vec{E}D_{Cr}}{RT} \frac{\partial y_{Cr}}{\partial x} \quad (3)$$

$$\frac{\partial y_{Ni}}{\partial t} = D_{Ni} \frac{\partial^2 y_{Ni}}{\partial x^2} + \frac{2F\vec{E}D_{Ni}}{RT} \frac{\partial y_{Ni}}{\partial x}$$

where X stands for the nominal valency of Fe in the oxide, which is close to 2.67 in the passive region and close to 3 in the transpassive region, have to be solved. The boundary conditions at the alloy/film and film/electrolyte interfaces, as well as the initial conditions can be written as

$$y_{Fe}(x,0) = y_{Fe,a}, y_{Cr}(x,0) = y_{Cr,a}, y_{Ni}(x,0) = y_{Ni,a}$$

$$y_{Fe}(0,t) = y_{Fe,a}, y_{Cr}(0,t) = y_{Cr,a}, y_{Ni}(0,t) = y_{Ni,a}$$

$$y_{Fe}(L,t) = \frac{k_{1Fe} y_{Fe,a}}{V_{m,MO}} \left[\frac{1}{k_{3Fe}} + \frac{RT}{XF\vec{E}D_{Fe}} \right] \quad (4)$$

$$y_{Ni}(L,t) = \frac{k_{1Ni} y_{Ni,a}}{V_{m,MO}} \left[\frac{1}{k_{3Ni}} + \frac{RT}{2F\vec{E}D_{Ni}} \right]$$

$$y_{Cr}(L,t) = \frac{k_{1Cr} y_{Cr,a}}{V_{m,MO}} \left[\frac{1}{k_{3Cr}} + \frac{RT}{3F\vec{E}D_{Cr}} \right]$$

where x = 0 at the alloy/film interface and x = L is the film/electrolyte or inner layer/outer layer interface.

At this stage of the modelling, we do not consider the existence of a transition layer between the bulk alloy and the alloy/oxide interface in which the composition of the underlying alloy is modified. In other words, the respective mass fractions at the alloy / film interface are taken as equal to the fractions in the bulk alloy. Albeit admittedly a crude simplification, this assumption facilitates the computations markedly. The introduction of the transition layer at a later stage of this work is, however, conceptually straightforward. Another simplification introduced at this point is that at t=0, there is no oxide on the surface, which means that the thin barrier layer formed via the transformation of the airborne oxide during heat-up is neglected. The boundary conditions at the film/electrolyte interface are given by the steady-state concentrations of the respective metallic constituents at that interface obtained by the steady-state solution of the transport equations.

An Approach to the Outer Layer Growth

The outer layer is presumed to grow via the precipitation of material that is dissolved from the substrate through the inner layer of oxide. This is admittedly a simplification with regard to the real plant conditions, in which the composition of the coolant water adjacent to the outer layer plays an important role. The growth of the outer layer is a multistep process that involves the dissolution reaction at the inner layer/solution interface (already described above by the reaction steps $k_{3,i}$ (i = Fe, Cr and Ni), liquid phase transport through the pores of the outer layer, supersaturation and re-precipitation of cations both at the pore walls and at the outer layer/electrolyte interface and hence outer layer crystal growth. In the absence of representative information on the pore diameter and tortuousity, and also of reliable data on the crystallite size of the outer layer, the growth of this layer can be formally treated as a diffusion process in a matrix constituted of the outer layer crystals and the electrolyte in between. Similar treatments for a porous oxide layer have already been presented [[19]] and discussed also by some of us in relation to incorporation of foreign species into the growing film on construction materials in nuclear power plants [[20]]. It has to be stressed, however, that the approach to porous layer growth outlined in the present paragraph is purely formal and it can not be considered analogous to the solid-state growth mechanism of the inner layer discussed in previous paragraphs. In order to devise a mechanistic model of outer layer growth, quantitative information on the kinetic constants at the outer layer/electrolyte boundary is obviously needed. Estimates of the values of such constants could be obtained e.g. by considering the kinetics of adsorption and surface complexation of electrolyte originating cations on the already grown oxide. Experimental data on these processes are currently being gathered and will be used for modelling purposes in the near future.

Since the outer layer is not continuous, but rather represents a system of discrete crystallites with electrolyte in the pores, the role of the potential gradient in this layer can be considered negligible with respect to the concentration gradient. Thus, to calculate the depth profile of a certain cation in the outer layer, the following system of equations has to be solved:

$$\frac{\partial y_{Fe,OL}}{\partial t} = D_{Fe,OL} \frac{\partial^2 y_{Fe,OL}}{\partial x^2}$$

$$\frac{\partial y_{Cr,OL}}{\partial t} = D_{Cr,OL} \frac{\partial^2 y_{Cr,OL}}{\partial x^2} \qquad (5)$$

$$\frac{\partial y_{Ni,OL}}{\partial t} = D_{Ni,OL} \frac{\partial^2 y_{Ni,OL}}{\partial x^2}$$

The boundary conditions at the inner interface of the outer oxide layer are identical to those used as outer boundary conditions at the inner layer / electrolyte interface, which ensures the continuity of the composition of the whole film as usually found by ex-situ analytical techniques. It also reflects the implicit assumption that the material dissolving at the inner layer/solution boundary is effectively re-deposited to form the outer layer. In the actual in-plant case, on the contrary, we probably will not have a confined system which will possibly affect the model by introducing additional terms for solution-originating Fe, Cr and Ni cations that are able to incorporate in the outer layer.

Incorporation of Solution-Originating Cations in the Oxide

In addition, in order to predict quantitatively the depth profile of a solution-originating cation (e.g. Zn) that is incorporated into the inner layer, we need to add the associated diffusion-migration equation for the non-steady state transport of Zn cation

$$\frac{\partial c_{Zn}}{\partial t} = D_{Zn} \frac{\partial^2 c_{Zn}}{\partial x^2} - \frac{2F\vec{E}D_{Zn}}{RT} \frac{\partial c_{Zn}}{\partial x} \qquad (6)$$

to the system of equations (3) and solve the extended system subject to the boundary conditions (4). The boundary condition for Zn at the inner layer/outer layer interface is given by the enrichment factor of Zn defined as the ratio between the concentration of Zn at that interface and the Zn concentration in the water: $y_{Zn}(L,t) = K_{enr,Zn} c_{Zn}(sol)$. A reflective boundary condition is used for Zn at the alloy / inner layer interface since there is no Zn present in the alloy substrate.

If, in addition, the solution-originating cation is present in the alloy substrate, such being the case with Co, the boundary condition at the inner layer/electrolyte interface is written as

$$y_{Co}(L,t) = K_{enr,Co} c_{Co}(sol) + \frac{k_{1Co} y_{Co,a}}{V_{mo}} \left[\frac{1}{k_{3Co}} + \frac{RT}{2F\vec{E}D_{Co}} \right]$$

Within the frames of the formal model for the outer layer growth, the depth profile of e.g. Zn in the outer layer can be predicted by solving the system of equations (5) extended with a corresponding equation for Zn:

$$\frac{\partial y_{Zn,OL}}{\partial t} = D_{Zn,OL} \frac{\partial^2 y_{Zn,OL}}{\partial x^2} \qquad (7)$$

The boundary condition at the outer layer/water interface is set by the corresponding enrichment factor at that interface.

Description of the Calculation Procedure

First, the position of the metal/film interface is determined by a sigmoidal fit of the experimental profile of the atomic concentration of oxygen obtained by the respective analytical technique - Auger Electron Spectroscopy (AES), X-ray Photoelectron Spectroscopy (XPS), Glow Discharge Optical Emission Spectroscopy (GDOES) or Secondary Ion Mass Spectroscopy (SIMS). Then, the profiles of the mass fractions of main metallic elements are normalised to the total metallic content of the film and the respective profiles are recalculated with the estimated position of the metal/film interface taken as zero. Whenever evidence of the presence of an outer layer has been found in the experimental profiles, the approximate position of the inner layer/outer layer interface is calculated by sigmoidal fits of the experimental profiles of the main metallic constituents of the alloy – Fe and Cr for stainless steels, or Ni and Cr for nickel-based alloys.

As a first step of the calculation, the system of equations (3) is solved subject to the initial and boundary conditions (4) using a Crank-Nicholson method in order to obtain the compositional profiles in the inner layer. In the cases when an outer layer is present, the inner layer thickness has been subtracted and the inner/outer layer boundary is taken as a new zero. Then the system of equations (5) is solved with the respective boundary conditions at the inner layer/outer layer interface similar as those used in (4). The boundary conditions at the outer layer / electrolyte interface are at this point assumed to be equal to the experimentally found stoichiometry of the outermost layer of oxide as determined by the respective ex-situ technique. In principle, the surface composition has been found to be in reasonable agreement with thermodynamic calculations of the solubility of the respective oxide phases discussed by other authors [[21]]. As a final step, the calculated profiles for the normalised mass fractions of the inner and outer layers are combined in order to compare the model predictions with the experimental results in a convenient way. It has to be stressed that the outlined procedure does not represent an optimisation method per se, but rather a manual search for the best possible match between the experimental and calculated profiles by a trial-and-error method. In that respect, the possibility of another set of parameters furnishing an equally good fit to the experimental data can not be excluded. Proofs of the reliability of the estimated kinetic and transport parameters are sought a posteriori by computing further characteristics of the systems under study. The adopted procedure of validation of the method is described below, using results from ex-situ characterisation of oxides formed on stainless steels and nickel-based alloys in both real (in-reactor) and simulated PWR and BWR coolant conditions.

RESULTS AND DISCUSSION

Simulated and in-Reactor PWR Conditions

Effect of Exposure Time and Zn Addition on the Kinetic Parameters of Film Growth on AISI 304 Stainless Steel

The experimental and calculated depth profiles of the mass fractions of Fe, Cr and Ni in the oxides formed on AISI 304 stainless steel in simulated PWR water without Zn addition at 260 °C for 5000 and 10 000 h are shown in Figure 5. The corresponding profiles obtained in simulated PWR water with the addition of 30 ppb of Zn after 1000 and 10 000 h of oxidation are presented in Figure 6. Double-layer oxides are formed in both environments, the inner layers being slightly enriched in Cr whereas the outer layers contain mostly Fe and Ni. The oxides formed in the presence of Zn are significantly thinner than those in the absence of Zn, and significant amounts of Zn are incorporated in the outer part of the inner layer. The profile of Zn in the inner layer follows to a certain extent that of Cr, and the enrichment of Cr in the Zn-doped inner layers is more significant than that in the inner layers formed in the absence of Zn. The outer layer is significantly thinner than the inner layer in the presence of Zn, especially for a 10 000 h of exposure.

The calculated profiles (shown in the figures with solid lines) match sufficiently well the experimental ones, except for the profile of Zn in the vicinity of the alloy/inner layer interface. This could be due to sputtering effects since no Zn is expected to be present in the base alloy. The values of the kinetic parameters calculated from the simulation are shown in Figure 7 (rate constants at the alloy/inner layer and inner layer/electrolyte interfaces) and Figure 8 (diffusion coefficients in the inner layer and formal diffusion coefficients of the growth of the outer layer, respectively).

The profiles obtained after 5000 and 10000 h of exposure in an electrolyte without Zn are reproduced with a rather homogeneous set of parameters, this being especially true for the rate constants at both interfaces. The rate constant for oxidation of Cr at the inner interface is the highest whereas the corresponding rate constant for dissolution of Cr at the outer interface is the lowest, which results in the observed Cr enrichment in the outer part of the inner layer. The opposite is true for Fe which is preferentially leached from the inner layer and subsequently redeposits to form the outer layer, as observed in the profiles.

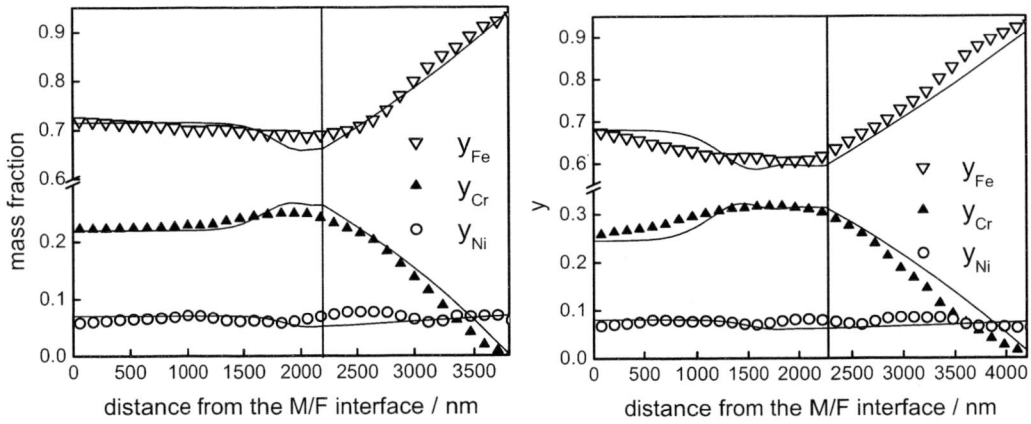

Figure 5. Experimental (points) and calculated (solid lines) XPS depth profiles of the mass fractions of Fe, Cr, and Ni in the films formed on AISI 304 during a 5000 h (left) and 10 000 h (right) exposure to a simulated PWR water at 260 °C. The inner layer / outer layer boundary indicated with a vertical line. Experimental data from Ref. [22].

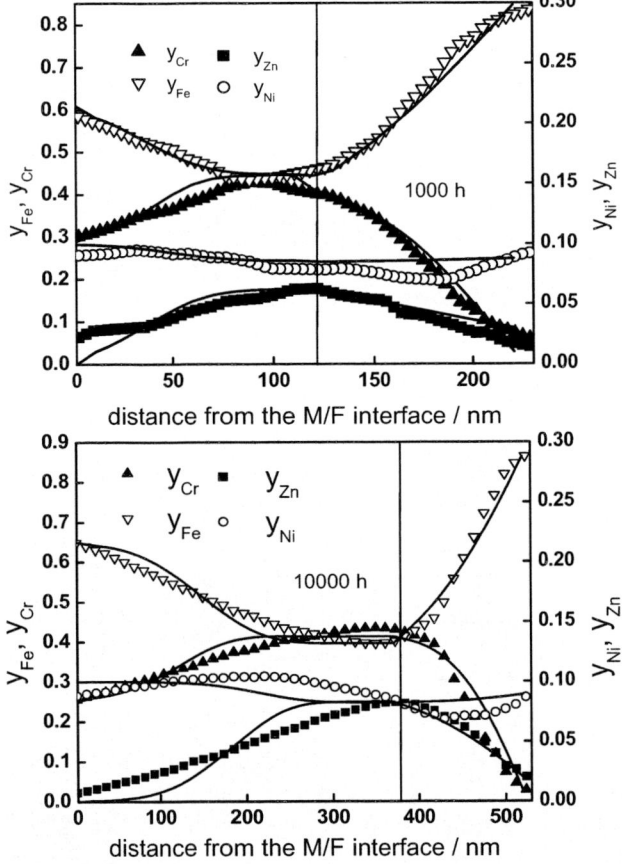

Figure 6. Experimental (points) and calculated (solid lines) XPS depth profiles of the mass fractions of Fe, Cr, Ni and Zn in the films formed on AISI 304 during a 1000 and 10,000 h exposure to a simulated PWR water with the addition of 30 ppb Zn at 260 °C. The inner layer / outer layer boundary indicated with a vertical line. Experimental data from Ref. [23].

The effect of Zn on the kinetic and transport parameters is significant, especially on the diffusion coefficients in the inner layer and the formal diffusion coefficients depicting the growth of the outer layer of oxide. In other words, when the film grows in an electrolyte containing Zn, the transport rate in the inner layer of oxide has been reduced by a factor of 3-4, and the growth of the outer oxide is almost totally suppressed, especially for longer oxidation times (Figure 8).

The decreasing effect of Zn on the kinetics of the interfacial reactions is comparatively weaker, with one notable exception – in solutions containing Zn, the rate constant of Fe dissolution from the inner layer is somewhat larger than that in electrolytes without Zn addition (Figure 7). It is tempting to assume that Zn is substituted for divalent Fe, thus expulsing the latter from the oxide, as already discussed by several authors [[23]-[25]].

The following line of reasoning is proposed to explain these observations. If the incorporation of Zn at the inner layer/electrolyte interface occurs via filling of available empty cation interstices V_i and/or cation vacancies $V_M^{'''}$:

$$V_i + Zn^{2+}_{aq} \rightarrow Zn_i^{\bullet\bullet} \tag{8}$$

$$V_M^{'''} + Zn^{2+}_{aq} \rightarrow Zn_M^{'} \tag{9}$$

then the concentration of these defects at the interface will decrease. In order their steady-state concentration to be maintained, an additional amount of Fe (the main element of the alloy) is dissolved according to the reactions

$$Fe_i^{\bullet\bullet} \rightarrow Fe^{x+}_{aq} + V_i + (x-2)e' \tag{10}$$

$$Fe_{Fe}^{X} \rightarrow Fe^{x+}_{aq} + V_M^{'''} + (x-X)e' \tag{11}$$

According to the model described above, the sum of the rate constants of the reactions (10) and (11) is equal to the rate constant k_{3Fe}, which is observed to increase with the addition of Zn to the electrolyte. Further incorporation of Zn in the bulk of the inner layer will lead to a decrease of the concentration of defects, and also their mobilities, within the oxide, which in turn explains the much smaller values of the diffusion coefficients of inner layer constituents when the film is grown in solutions that contain Zn.

The sum of the reactions (8)-(11) can be written as an equilibrium exchange between a Fe cation in the inner layer of oxide and a Zn aquoion in the electrolyte, in full analogy with what has been proposed in Ref. [23]:

$$Fe_{inner\ layer} + Zn^{2+}_{aq} \rightleftarrows Fe^{x+}_{aq} + Zn_{inner\ layer} \tag{12}$$

Alternatively, the first step of Zn incorporation in the oxide can be written as a surface complexation reaction [[26]]

$$\equiv Fe-OH^0_{1/2} + Zn^{2+}_{aq} \rightleftarrows \equiv Fe-OZn^{1/2+} + H^+ \qquad (13)$$

The equilibrium constant of this reaction on magnetite has been estimated to be 41.7 at 280 °C using the constant capacitance model to quantitatively interpret high-temperature titration data on magnetite surfaces in simulated PWR water [[27]]. The equilibrium constant can be also estimated from the present results using data on the mass fractions of Zn and Fe at the inner layer/electrolyte interface (Figure 6), the surface site concentration on magnetite estimated earlier from titration data and the speciation of soluble Zn in the electrolyte calculated using literature data [[27]]. A value of 60±10 is obtained, in reasonable agreement with the surface complexation calculations. It can be concluded that the first step of Zn incorporation into the oxide is described adequately by the surface complexation model and equation (13).

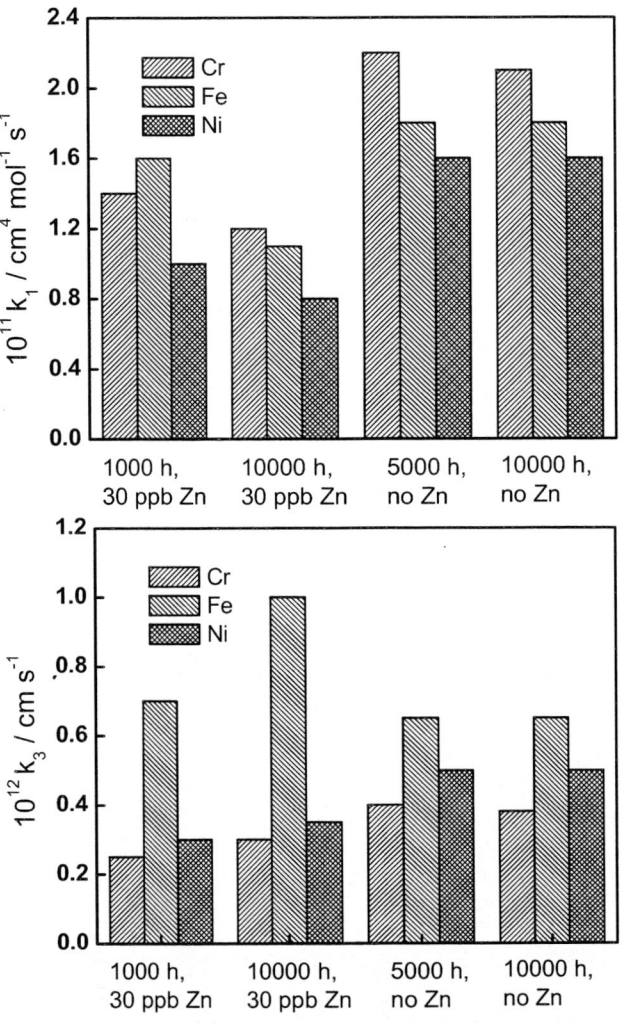

Figure 7. Dependence of the rate constants for inner layer constituents at the alloy/inner layer (left) and inner layer/electrolyte (right) interfaces on time of exposure and Zn addition.

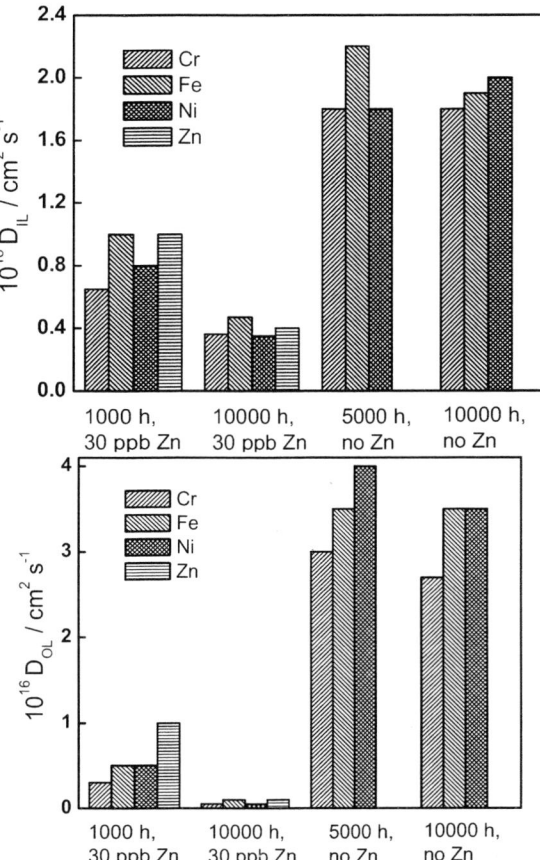

Figure 8. Dependence of the diffusion coefficients of inner layer constituents (left) and formal diffusion coefficients depicting the growth of the outer layer (right) on time of exposure and Zn addition.

Effect of Zn on in-Reactor Co Incorporation in the Oxide on Stainless Steel

The effect of Zn on the incorporation of radioactive matter (e.g. Co) from the coolant into the oxides formed on stainless steels and nickel-based alloys has been thoroughly investigated by the Halden Reactor project [[28]-[30]]. Both fresh and preoxidised samples have been exposed to typical PWR in-plant conditions (1000 ppm B, 3 ppm Li, 320 °C, $pH_{300°C}$ = 7.1, soluble Zn either 0 or ca. 50 ppb, soluble Co ca. 0.05 ppb), the duration of each exposure phase being ca. 100 Effective Full Power Days (EFPD). The SIMS depth profiles for Fe, Cr, Ni, Co and Zn in the films obtained after the first and third exposure phase on fresh AISI 304 sample in the presence of 50 ppb Zn in the water are shown in Figure 9 together with the profiles calculated using the present model. The kinetic and transport parameters estimated from the calculation are collected in Figure 10. The following conclusions can be drawn on the basis of the parameter values:

- The set of parameters used to simulate the in-reactor film growth and restructuring on AISI 304 is fully compatible with that used for the simulation of the growth and restructuring on the same steel in laboratory conditions (see previous paragraph). This demonstrates the feasibility of the present model also for the interpretation of in-reactor data.

- The equilibrium constant of Co incorporation via a reaction analogous to (13) was estimated to be close to 1, which agrees by order of magnitude with the value calculated on the basis of high-temperature titration data and the surface complexation model [[27]], whereas the corresponding value for Zn incorporation was again of the order of 50, confirming the earlier calculations and demonstrating once again the higher affinity of Zn towards the surface oxide.
- The main effects of Zn are on the values of the diffusion coefficients in both layers and to a certain extent on the values of the rate constants at the outer interface. This can be interpreted as a modification of both the oxide surface and the bulk oxide by Zn addition.
- Oxide growth/restructuring and incorporation of Co in the oxide are retarded by Zn incorporation. This could be related to the formation of new Zn-containing phases in both the inner and outer layers, as discussed also by other authors [[23]-[25],[30]].

The SIMS depth profiles for Fe, Cr, Ni, Co and Zn in the films obtained after in-reactor exposure of a preoxidised AISI 304 sample to PWR water in the presence or absence of Zn are shown in Figure 11 together with the corresponding profiles calculated by the model. The kinetic and transport parameters estimated from the calculation are collected in Figure 12. On comparing the parameter values for the fresh and preoxidised samples, it can be concluded that:

- Incorporation of Zn in already existing oxides is slower, and the layer restructuring is less pronounced. The diffusion coefficients in the inner layer formed in the presence of Zn on the preoxidised samples are ca. 50% lower than those formed on the fresh sample, and when the Zn-doped oxide is once again exposed to Zn-free PWR water, their initial values are restored. This means that the incorporation of Zn in pre-existing oxides is to a great extent reversible, thus less effective in suppressing further oxide growth and incorporation of solution-originating Co.
- There is no appreciable difference between the values of the apparent diffusion coefficients in the outer layer in both cases, with the notable exception of the diffusion coefficient for Co - when Zn is added, $D_{Co,OL}$ decreases to a half of its value before Zn addition. Thus Zn addition suppresses incorporation of Co in the outer layer as well.
- The value of the rate constant k_{3Fe} is once again much higher in the presence of Zn which could be interpreted as if the exchange reaction of Zn for Fe being efficient enough for preoxidised samples as well and its mechanism being essentially unaltered.

A summary plot of the estimated values of the field strength in the inner layer of the oxide on AISI 304 is given in Figure 13. The values of the field strength in the inner layers formed or restructured in the presence of Zn in the water are somewhat higher than those in the absence of Zn, and a slight decrease of this parameter with time of exposure is observed. On the overall, the effect of both time of exposure and Zn addition to the electrolyte is small and the field strength can be considered essentially constant, its values being in agreement with what has been calculated earlier by us in the same temperature range [[17]]. The values of the field strength are in general rather low and confirm the validity of the low-field approximation used to derive the model equations in the Theory section.

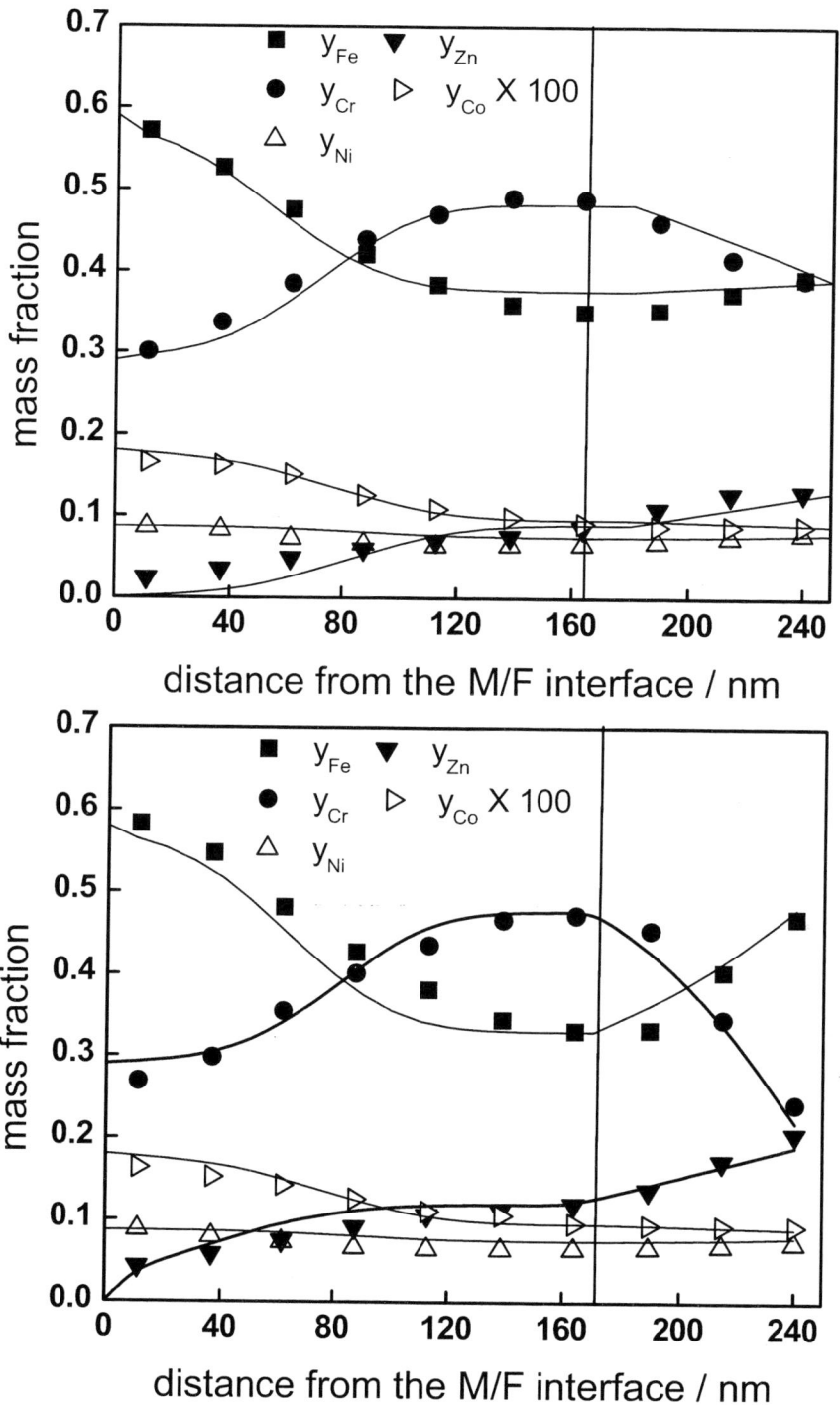

Figure 9. Experimental (points) and calculated (solid lines) depth profiles of the mass fractions of Fe, Cr, Ni, Co and Zn in the films formed on AISI 304 after the exposure to ca. 100 and 300 EFPD (125 and 336 days) in the Halden Reactor, PWR coolant conditions. The inner layer / outer layer boundary indicated with a vertical line. Experimental data taken from Refs [28]-[30].

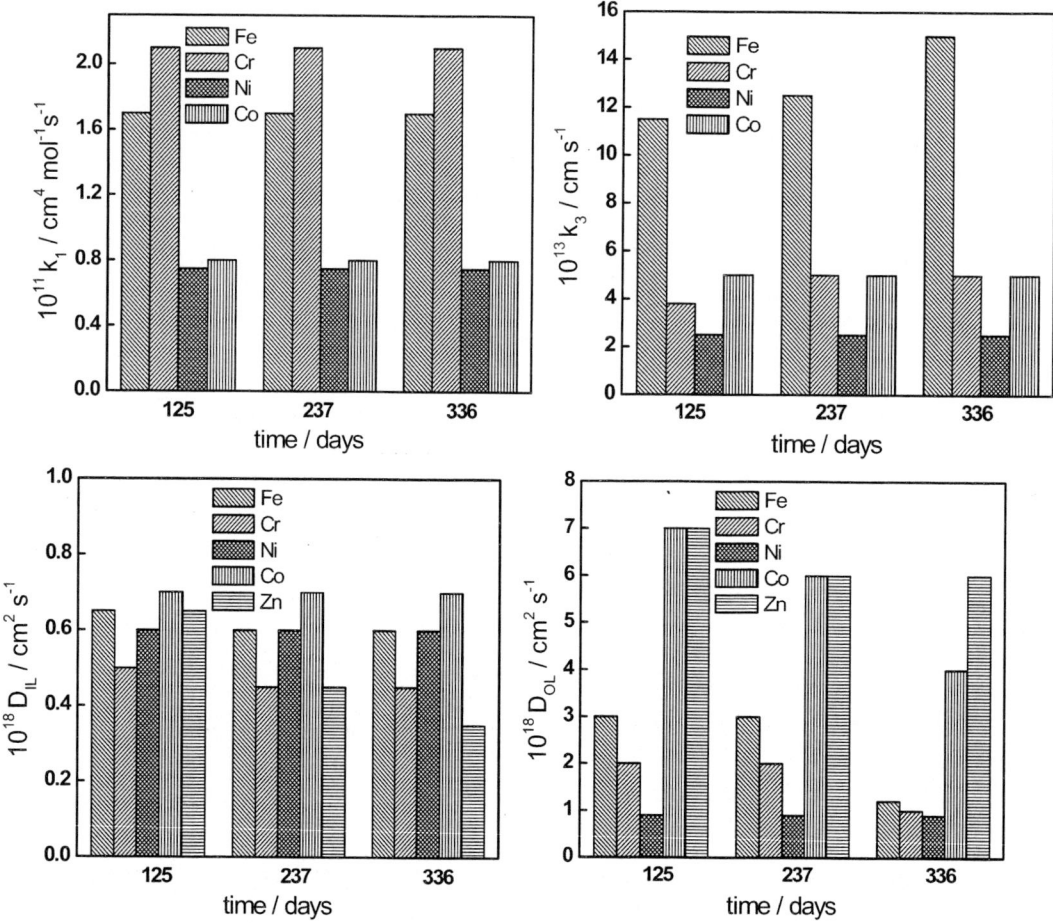

Figure 10. Dependence of the rate constants at the steel/inner layer interface (above left), the inner layer/electrolyte interface (above right), diffusion coefficients of inner layer constituents (below left) and formal diffusion coefficients depicting the growth of the outer layer (below right) on time of exposure. 50 ppb Zn added from the beginning of exposure.

Further proof for the validity of the model comes from Figure 14, in which the inner layer thicknesses for the oxides formed on AISI 304 in both simulated and in-reactor PWR water in the absence or presence of Zn are compiled as depending on exposure time (data from Refs. [22]-[33]). In the figure, experimentally determined thickness values are compared with the thickness calculated according to the relationship proposed within the frames of the PDM and MCM [[3],[17]]

$$L(t) = L(t=0) + \frac{1}{b}\ln\left[1 + V_{m,MO}(k_{1,Cr}y_{Cr,a} + k_{1,Fe}y_{Fe,a} + k_{1,Ni}y_{Ni,a})be^{-bL(t=0)}t\right]$$
$$b = \frac{3\alpha_1 F\vec{E}}{RT}$$
(14)

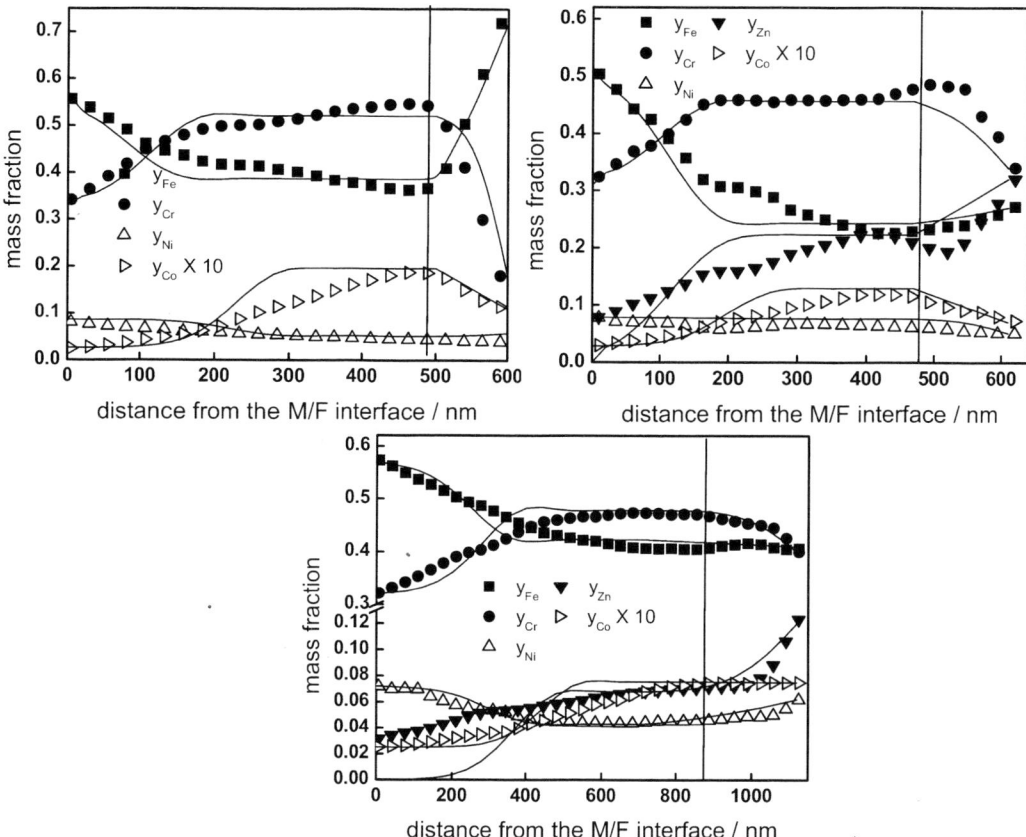

Figure 11. Experimental (points) and calculated (solid lines) depth profiles of the mass fractions of Fe, Cr, Ni, Co and Zn in the films formed on AISI 304 during a 100 day exposure with no Zn addition (above left), followed by a 237 day exposure to 50 ppb Zn (above right) and another 230 day exposure with no Zn addition (below). The inner layer / outer layer boundary indicated with a vertical line. Experimental data taken from Refs [28]-[30].

subject to the assumption that the transfer coefficients for all the three reactions at the metal film interface are similar and equal to α_1 and the values of the rate constants are taken from Figure 7, Figure 10 and Figure 12. Notwithstanding the fact that the compilation of data is stemming from exposure to slightly different experimental conditions, the solid lines shown in Figure 14 demonstrate the fair agreement between the experimentally estimated thicknesses and model predictions, which is encouraging taking into account the fact that no further adjustment has been made.

Effect of Exposure Time and Zn Addition on the Kinetic Parameters of Film Growth on Nickel-Based Alloys

Knowledge of the oxidation processes on Ni alloys in a PWR coolant is of major importance for at least two practical reasons: (i) the radioactivity of the primary circuit is primarily due to cations released by corrosion of the steam generator tubes made mainly of nickel-based alloys and (ii) the oxidation process is important in the mechanisms of the initiation of intergranular stress corrosion cracking in Alloys 600, 82 and 182. Understanding

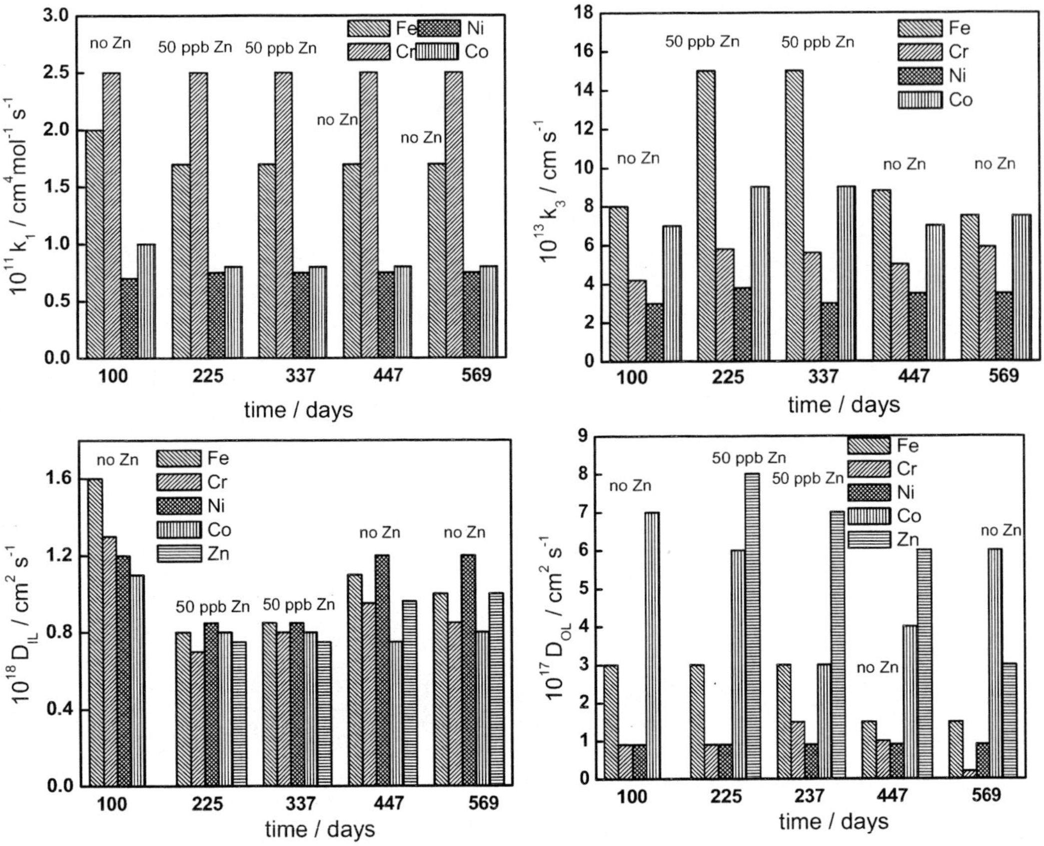

Figure 12. Dependence of the rate constants at the steel/inner layer interface (above left), the inner layer/electrolyte interface (above right),, diffusion coefficients of inner layer constituents (below left) and formal diffusion coefficients depicting the growth of the outer layer (below right) on time of exposure of AISI 304 to PWR water. Successive periods without and with Zn as indicated.

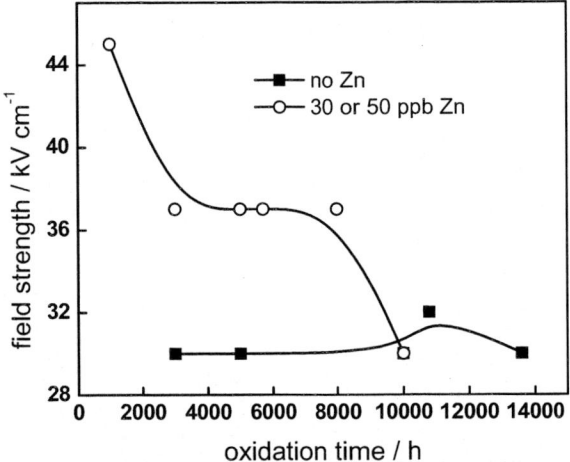

Figure 13. Summary of the field strength values for the inner layer on AISI 304 calculated from the comparison of the model equations to the data in Figure 5, Figure 6, Figure 9 and Figure 11.

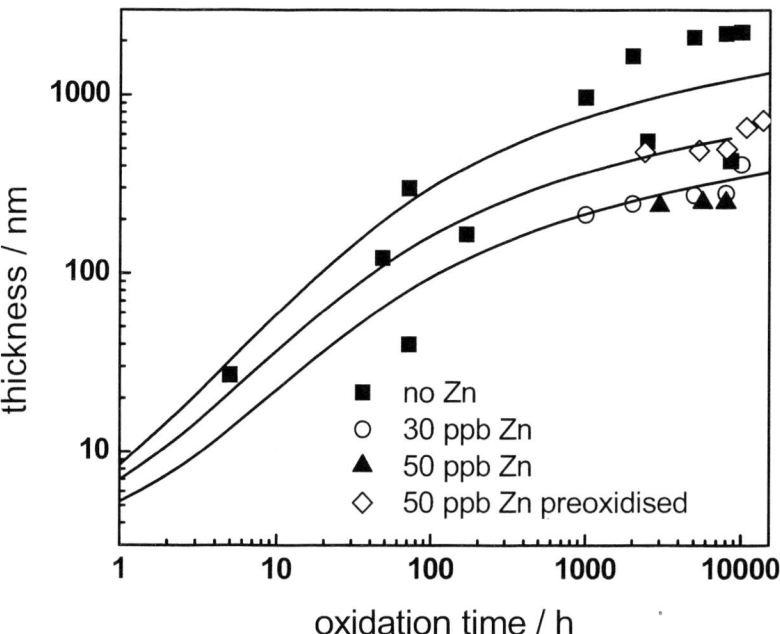

Figure 14. Inner layer thickness vs. time data for AISI 304 in simulated and in-reactor PWR water with or without Zn addition at 260-300 °C (symbols) and calculated curves according to the model (solid lines).

the root causes of this cracking should also allow the safety margins offered by the Alloys 690, 52 and 152, which have replaced the former ones, to be evaluated more accurately. Very recently, unique experiments and new data on the initial stages of passive oxide film growth on nickel-based Alloys 600, 690 and 800 in a simulated PWR coolant using a micro-autoclave technique combined with *ex situ* XPS and STM characterisations have been reported [[34]-[36]] and a preliminary treatment of these data in terms of the MCM using a set of parameters calculated from fitting of electrochemical impedance spectroscopic data on binary Ni-15% Cr and Ni-20%Cr alloys has been presented and discussed [[16]]. Further treatment of these and associated data to extract the full set of model parameters as depending on exposure time and also Zn addition to both simulated and in-reactor PWR water is given in the present chapter.

Figure 15 shows XPS depth profiles of the mass fractions of Ni, Cr and Fe in the oxides formed on Alloy 600 in simulated PWR water (2 ppm Li, 1200 ppm B, partial pressure of H_2, 0.3×10^5 Pa) for 20 and 100 h at 325 °C [[34]] together with the corresponding profiles calculated by the model. XPS depth profiles for the oxides on Alloy 600 stemming from experiments involving much longer oxidation times in simulated PWR water with and without the addition of 30 ppb of soluble Zn at 260 °C [[37],[38]] are presented in Figure 16 and Figure 17, respectively. The parameters used to reproduce the experimental profiles are collected in Figure 18. On the other hand, XPS depth profiles of the mass fractions of Ni, Cr and Fe in the oxides formed on Alloy 690 in simulated PWR water for 20, 50 and 100 h at 325 °C are summarized in Figure 19, and the corresponding parameters used to calculate the profiles are collected in Figure 20-Figure 21.

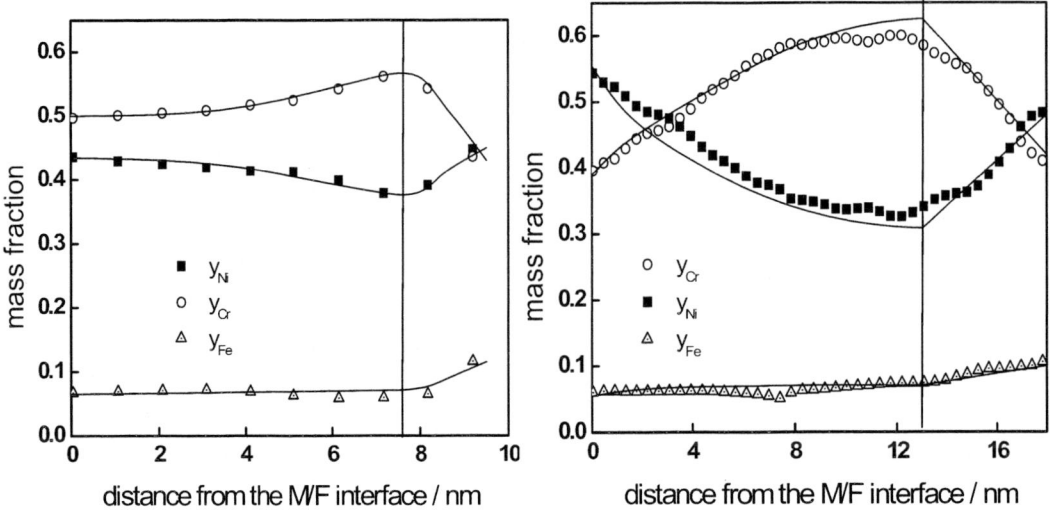

Figure 15. Experimental (points) and calculated (solid lines) depth profiles of the mass fractions of Fe,Cr, and Ni in the films formed on Alloy 600 in simulated PWR water for 20 h (left) and 100 h (right). The inner layer / outer layer boundary indicated with a vertical line. Experimental data taken from Ref. [34].

Concerning in-reactor exposure, SIMS depth profiles for Fe,Cr, Ni, Co and Zn in the films obtained on a fresh Alloy 690 sample after the first and third exposure phase of the Halden reactor project experiment discussed above in the presence of 50 ppb Zn in the water are shown in Figure 22 together with the profiles calculated using the present model. The kinetic and transport parameters estimated from the calculation are collected in Figure 23 and Figure 24.

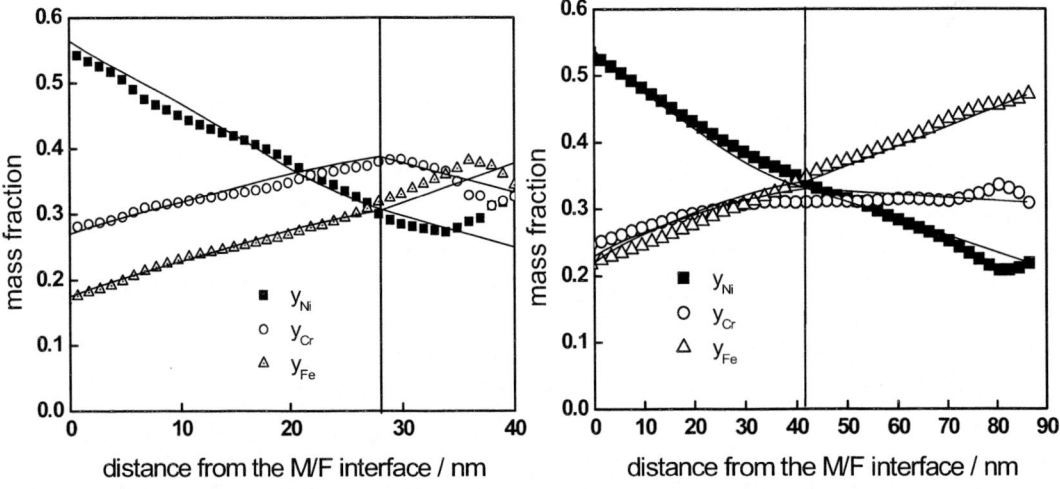

Figure 16. Experimental (points) and calculated (solid lines) depth profiles of the mass fractions of Fe,Cr, and Ni in the films formed on Alloy 600 in simulated PWR water for 5000 h (left) and 10000 h (right). The inner layer / outer layer boundary indicated with a vertical line. Experimental data taken from Ref.[37].

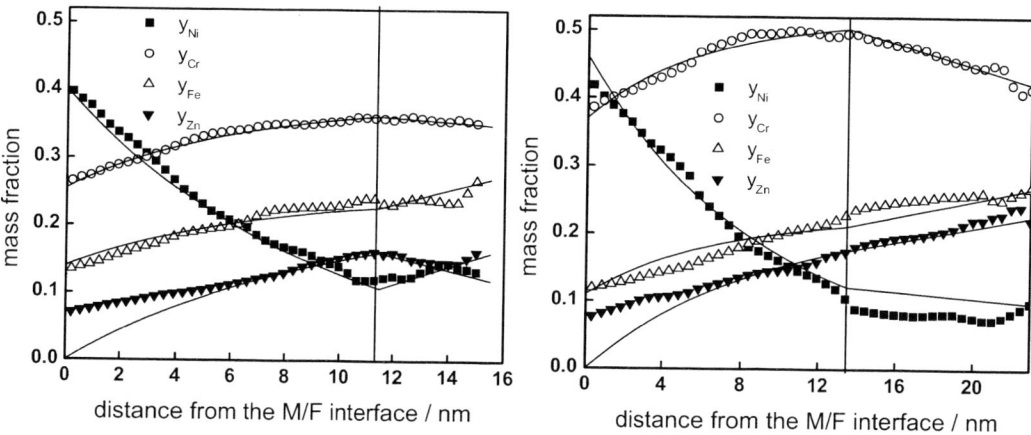

Figure 17. Experimental (points) and calculated (solid lines) depth profiles of the mass fractions of Fe, Cr, Ni, and Zn in the films formed on Alloy 600 in simulated PWR water with 30 ppb Zn for 5000 h (left) and 10000 h (right). The inner layer / outer layer boundary indicated with a vertical line. Experimental data taken from Ref. [38].

Several differences can be noticed between the estimated parameter values for the oxides on stainless steels and nickel-based aloys when comparing Figure 7, Figure 8, Figure 10, Figure 12, Figure 18, Figure 20, Figure 21, Figure 23 and Figure 24:

- The differences between parameters for oxides on Alloys 600 and 690 is not very significant, the only important difference being observed for the diffusion coefficients in the inner layer (Figure 21, left). In general it can be stated that the values of the parameters on Alloy 690 are somewhat larger and their time evolution more pronounced. The lower diffusion coefficients in the inner layer on Alloy 600 can be tentatively explained by the unavailability of Cr from the alloy layer underneath the oxide, which retards film growth, as suggested in Refs. [34]-[36].
- The diffusion coefficients of individual constituents of the inner layer are significantly (more than an order of magnitude) smaller for the oxides formed on nickel-based alloys, which explains the much slower growth and restructuring of the corrosion layers on these materials when compared to stainless steels. This fact can be in general explained by the composition of the inner layer, which is close to chromium oxide (Cr_2O_3) for nickel-based alloys, whereas it can be considered as close to the normal spinel chromite ($FeCr_2O_4$) on stainless steels. The diffusion coefficients in spinel structures are in general much larger than those in corundum structures as that of Cr_2O_3.
- There is also a very significant difference between the formal diffusion coefficients of the outer layers formed on stainless steels and nickel-based alloys which amounts to more than two orders of magnitude in certain cases. A tentative explanation for this observation could be that the growth mechanism of the outer layer on nickel-based alloys (which consists of NiO or $Ni(OH)_2$, or a mixture of the two and $NiFe_2O_4$) is closer to a solid-state process, whereas the outer layer on stainless steels (consisting mainly of trevorite $NiFe_2O_4$ doped with Cr) grows via a dissolution-precipitation reaction resulting in comparatively large crystals with electrolyte in between.

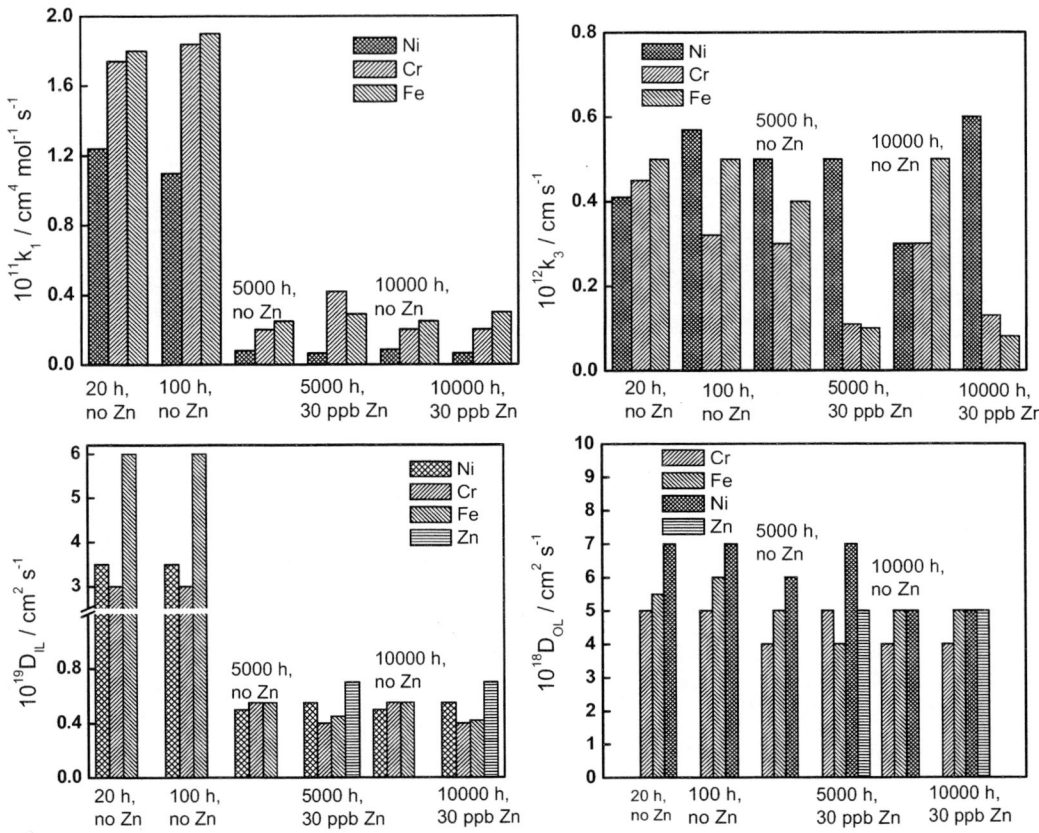

Figure 18. Dependence of the rate constants at the alloy 600 / inner layer interface (above, left), inner layer/electrolyte interface (above, right), diffusion coefficients of inner layer constituents (below left) and formal diffusion coefficients for the growth of the outer layer (below right) on time of exposure to simulated PWR water with or without Zn addition.

- On the other hand, the differences between the interfacial rate constants at the inner and outer interfaces is much smaller, although there is some tendency for a decrease of the rate constants at the inner layer /electrolyte interface from the oxide on stainless steels to that on nickel-based alloys. Thus the mechanism of the interfacial reactions does not seem to be altered remarkably by the transition of the main element in the alloy from Fe to Ni.

Concerning the effect of Zn on the oxide growth and restructuring, it can be stated that the effect is much smaller on the oxides formed on Ni-based alloys than on those formed on stainless steels, even if the amounts of incorporated Zn especially in the outer part of the inner layers on both types of materials are not very different. It can be argued that the incorporation of Zn in a Cr_2O_3 type structure does not lead to such a large alteration of the properties of such phase when compared to the incorporation of Zn in $FeCr_2O_4$ type oxides.

Indeed, by examining the values of the diffusion coefficients of inner layer constituents of the oxide on nickel-based alloys, it can be concluded that the time of exposure has a much larger influence on the values than the addition of Zn to the electrolyte. In order to explain the effect of film aging on its properties, the effect of microstructure has to be considered.

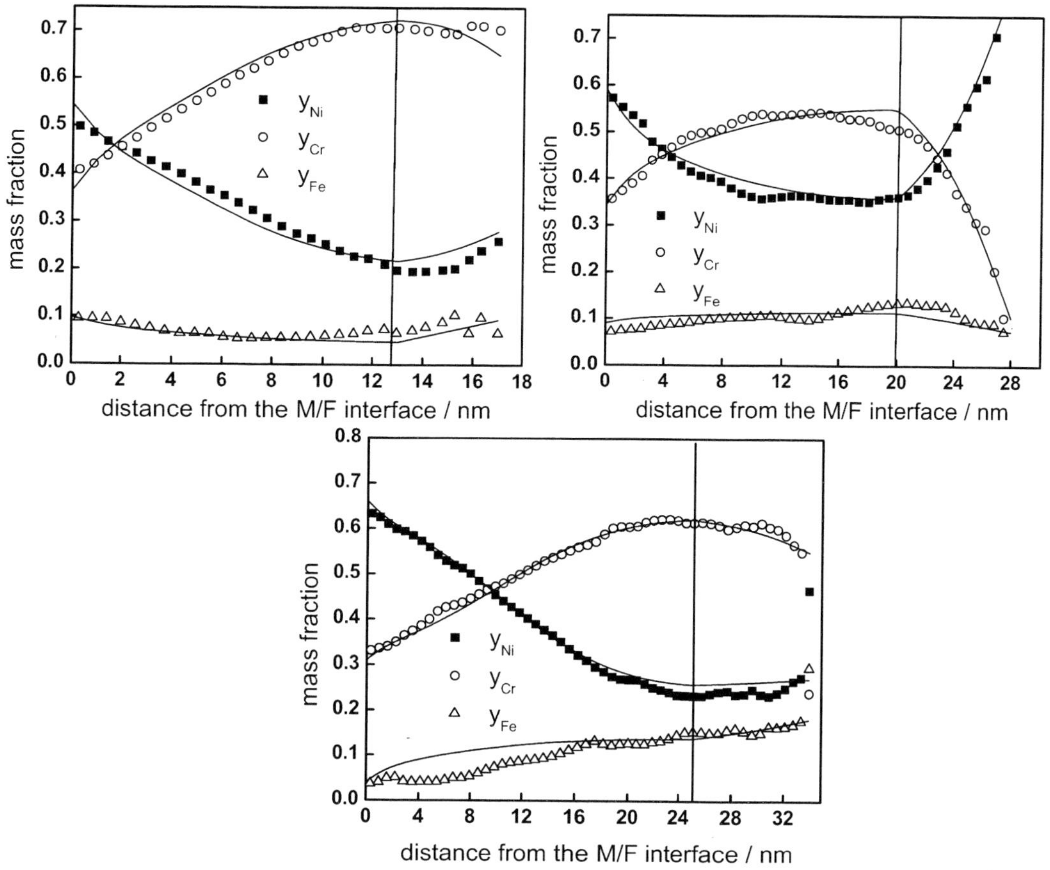

A Kinetic Model of the Oxide Growth and Restructuring on Structural Materials...119

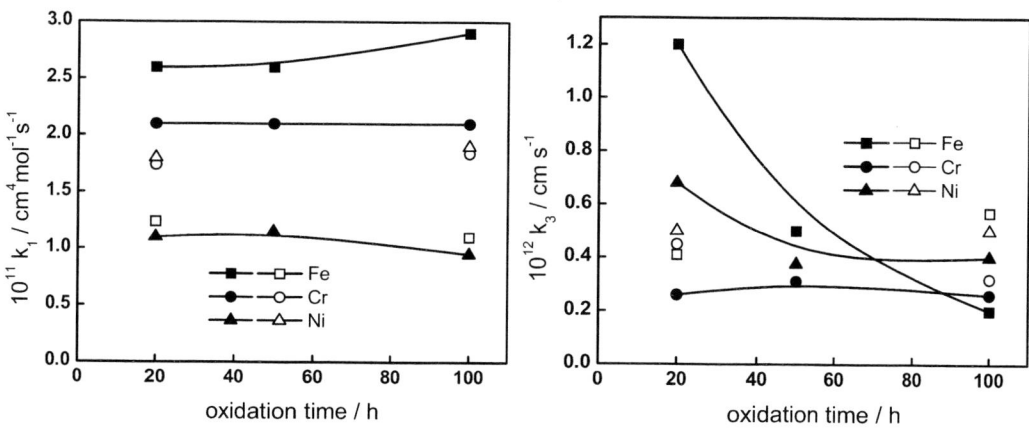

Figure 19. Experimental (points) and calculated (solid lines) depth profiles of the mass fractions of Fe, Cr, and Ni in the films formed on Alloy 690 in simulated PWR water at 325 for 20 h (above, left), 50 h (above, right) and 100 h (below). The inner layer / outer layer boundary indicated with a vertical line. Experimental data taken from Ref. [34].

Figure 20. Dependence of the rate constants at the alloy 690 / inner layer interface (left) and the inner layer/electrolyte interface (right) on the time of exposure of Alloy 690 to simulated PWR water (full symbols). The values for the oxide on Alloy 600 are shown with open symbols for comparison.

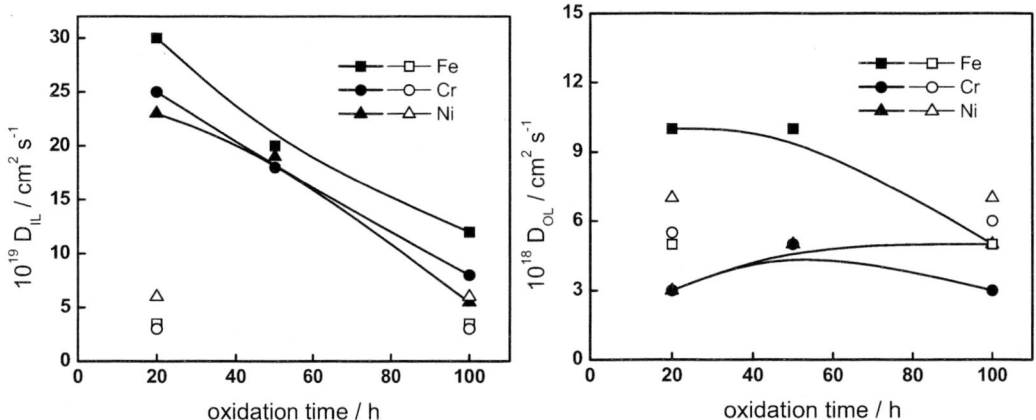

Figure 21. Dependence of the diffusion coefficients of inner layer constituents (left) and formal diffusion coefficients for the growth of the outer layer (right) on the time of exposure of Alloy 690 to simulated PWR water(full symbols). The values for the oxide on Alloy 600 are shown with open symbols for comparison.

In the simplest treatment, the effective diffusion coefficient can be expressed as a combination of the diffusion coefficients in the grain interior and at the grain boundary

$$D_{eff} = (1 - f_{gb})D_{gi} + f_{gb}D_{gb} \tag{15}$$

where for the fraction of grain boundaries the following equation holds as a first approximation

$$f_{gb} \approx \frac{3\delta_{gb}}{\Phi} \tag{16}$$

where δ_{gb} is the grain boundary width (of the order of 1 nm) and Φ is the grain size of the oxide in question. An additional explanation for enhanced conduction at boundaries is related to the formation of space charge regions in the grain areas adjacent to the boundaries. As charged species and defects tend to segregate to the grain boundaries to lower the strain and electrostatic energy of the system, the boundary charges are compensated by the formation of space charge in the adjoining grain areas. If a bulk defect with a high mobility is accumulated in the space charge region, the overall conductivity of the solid should increase. The width of the space charge region can be linked to the space charge screening length

$$L_D = \sqrt{\frac{\varepsilon\varepsilon_0 kT}{e^2 N}} \tag{17}$$

Using typical values for $\varepsilon=20$, T=573 K and a bulk concentration of the high mobility carrier N = 5 ×10^{19} cm^{-3}, we arrive at L_D = 5 nm, which can be substituted for δ_{gb} in equation (16). Using data from Refs. [34]-[38], an increase of grain size of the oxide on Alloy 600 from 10 to 30 nm is adopted for oxidation between 20 and 5000 h, and a corresponding decrease of the fraction of grain boundaries from 0.3 (at 20 h) to 0.1 (at 5000 h) is obtained from equations (16)-(17). Using the calculated value of the diffusion coefficient e.g. for Ni after 20 and 5000 h of oxidation (Figure 18), the grain interior diffusion coefficient is estimated as 10^{-20} cm^2 s^{-1} in good agreement with an extrapolation of the diffusion coefficient of Ni for dry oxidation in the intermediate temperature range [[39]]. Although this can be considered as a proof for the validity of the present approach to the heterogeneity of the inner layer of oxide as a transport medium, the availability of grain size data for the inner corrosion layers in high-temperature water is not sufficient to generalize this treatment for other structural materials.

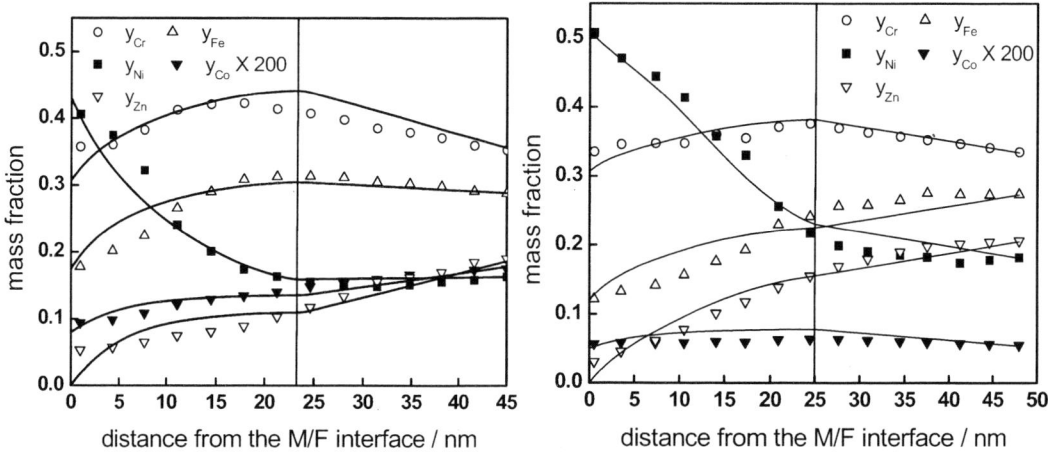

Figure 22. Experimental (points) and calculated (solid lines) depth profiles of the mass fractions of Fe, Cr, Ni, Co and Zn in the films formed on Alloy 690 during the first and second exposure to ca. 100 Effective Power Days in the Halden Reactor, PWR coolant conditions. The inner layer / outer layer boundary indicated with a vertical line. Experimental data taken from Refs [28]-[30].

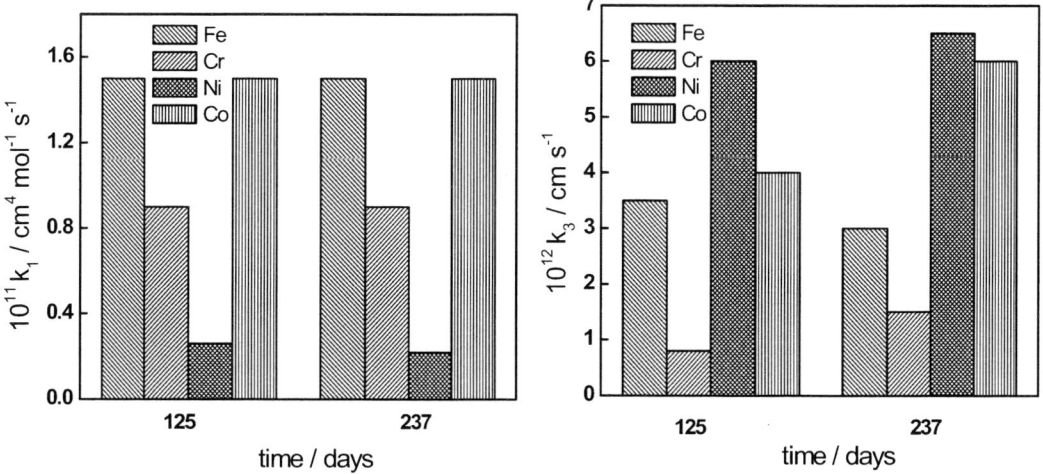

Figure 23. Dependence of the rate constants at the Alloy 690 / inner layer interface (left) and the inner layer/electrolyte interface (right) on the time of exposure of Alloy 690 to PWR water containing 50 ppb of Zn.

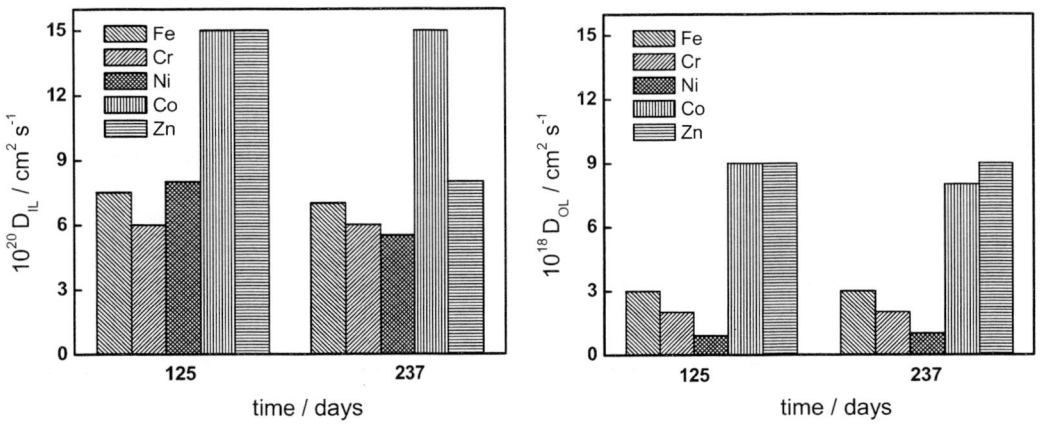

Figure 24. Dependence of the diffusion coefficients of inner layer constituents (left) and formal diffusion coefficients for the growth of the outer layer (right) on the time of exposure of Alloy 690 to PWR water with 50 ppb of Zn.

A summary plot of the estimated values of the field strength in the inner layer of the oxide on Alloys 600 and 690 is given in Figure 25. The values of the field strength in the inner layers formed or restuctured in the presence of Zn in the water are somewhat higher than those in the absence of Zn, and a significant decrease of this parameter with time of exposure is observed. It is also worth noticing that the field strength in the inner layers on nickel-based alloys is considerably higher than that on stainless steels at short exposure times, which is in agreement with our previous calculations based on quantitative treatment of electrochemical impedance spectroscopic data [[15],[16]]. If it is assumed that the space charge regions near grain boundaries determine the conduction through the oxide via a diffusion-migration mechanism, the decrease of the field strength in the inner layer with time of exposure can be tentatively attributed to the increase of the space charge screening length with time of exposure, which is tantamount of a decrease of the density of current carriers in the space charge regions.

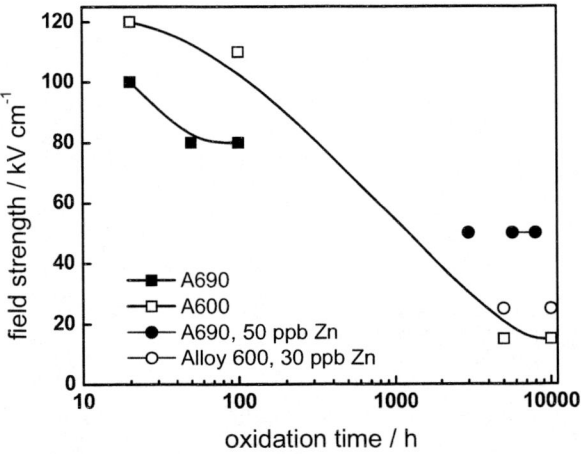

Figure 25. Summary of the field strength values for the inner layer on A600 and A690 calculated from the comparison of the model equations to the data in Figure 15, Figure 16, Figure 17 and Figure 22.

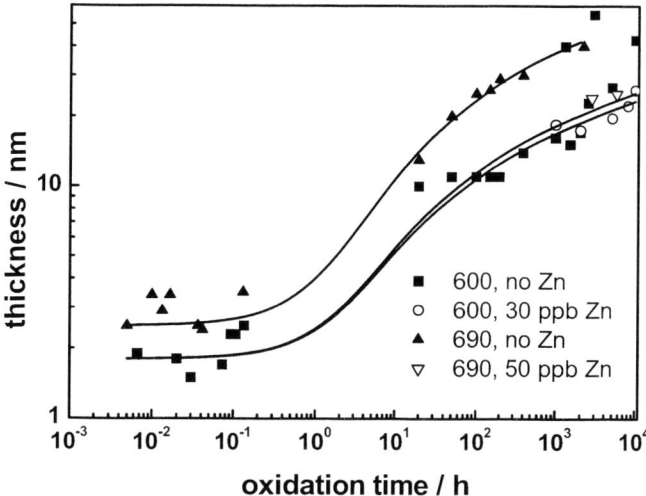

Figure 26. Inner layer thickness vs. time data for nickel-based alloys in simulated and in-reactor PWR water with or without Zn addition at 260-325 °C (symbols) and calculated curves according to the model (solid lines).

The inner layer thicknesses for the oxides formed on Alloys 600 and 690 in both simulated and in-reactor PWR water in the absence or presence of Zn in the water are compiled in Figure 26 as depending on exposure time (data from Refs. [28]-[30],[34]-[38],[40],[41]). In the figure, the experimentally determined thickness values are compared with the thickness calculated according to equation (14). Once again a reasonable agreement is obtained without any further adjustment of parameters, which demonstrates the ability of the model to predict the kinetics of growth of the oxide on nickel-based alloys as well.

Simulated and in-Reactor BWR Conditions

In-Reactor Exposure of Stainless Steels to BWR Conditions

Figure 27 shows Glow Discharge Optical Emission Spectroscopic (GDOES) depth profiles of oxygen and iron in the oxide films on AISI 304 and 316 stainless steels exposed for 7200 h to a BWR coolant (conductivity less than 0.12 μScm^{-1}, dissolved oxygen 200 ppb) at the Olkiluoto 1 plant [[42]]. The film on AISI 304 stainless steel appears to be significantly thicker than that on AISI 316. As customary for both in-reactor and simulated BWR conditions [[43]-[47]], the films comprise an outer iron-rich layer and an inner layer containing more Cr. It is worth noting that Cr is impoverished in the films with respect to its content in the alloy substrate due to its transpassive dissolution as Cr(VI) in BWR conditions as demonstrated by wall-jet ring-disk electrode measurements [[48]]. A plateau in Ni concentration is detected close to the metal / film interface in the oxide formed on AISI 304, whereas a well-pronounced maximum of Ni concentration exists in the inner layer formed on AISI 316 (Figure 27). In general, it can be stated that the difference in thickness between the films formed on 304 and 316 stainless steels is largely due to the inner layer being thicker on 304 steel.

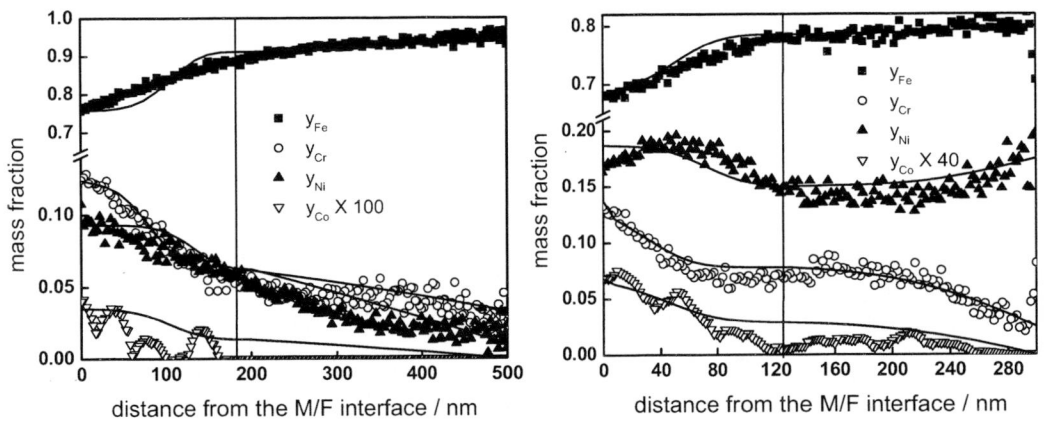

Figure 27. Experimental (points) and calculated (solid lines) depth profiles of the mass fractions of Fe, Cr, Ni, and Co in the films formed on AISI 304(left) and AISI 316 (right) after a 7200 h exposure to BWR reactor coolant at Olkiluoto 1 power plant. The inner layer / outer layer boundary indicated with a vertical line.

The calculated profiles of Fe, Cr, Ni and Co in the films are also shown in Figure 27 with solid lines and demonstrate the ability of the model to account for BWR data as well. The estimated values of the rate constants and diffusion coefficients are collected in Figure 28 and Figure 29, respectively. The values for the rate constants and diffusion coefficients in the inner layer are rather similar for both alloys being, however, somewhat smaller for the oxide formed on AISI 316 than for that formed on AISI 304, thus explaining the lower thickness and hence the slower growth of the inner layer on the former steel. A somewhat conjectural explanation of this fact is that the more pronounced Ni-enrichment in the inner layer formed on AISI 316 acts as a diffusion barrier preventing further growth of the inner oxide on that steel. The fact that also the formal diffusion coefficients of the growth of the outer layer on AISI 316 are significantly smaller can also be explained by the larger concentration of Ni in this phase which may hinder its growth. Indeed, the outer layer on AISI 304 is composed almost exclusively of iron (more than 90%), whereas that on AISI 316 contains ca. 15% of nickel.

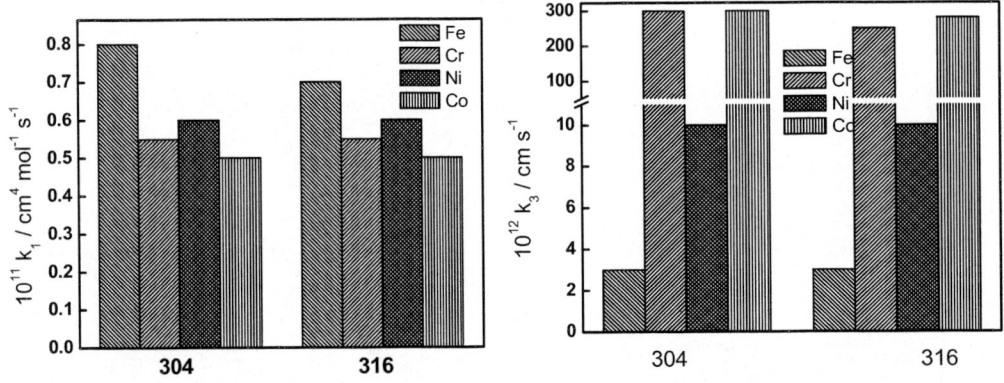

Figure 28. Dependence of the rate constants at the steel / inner layer interface (left) and the inner layer/electrolyte interface (right) on the steel type.

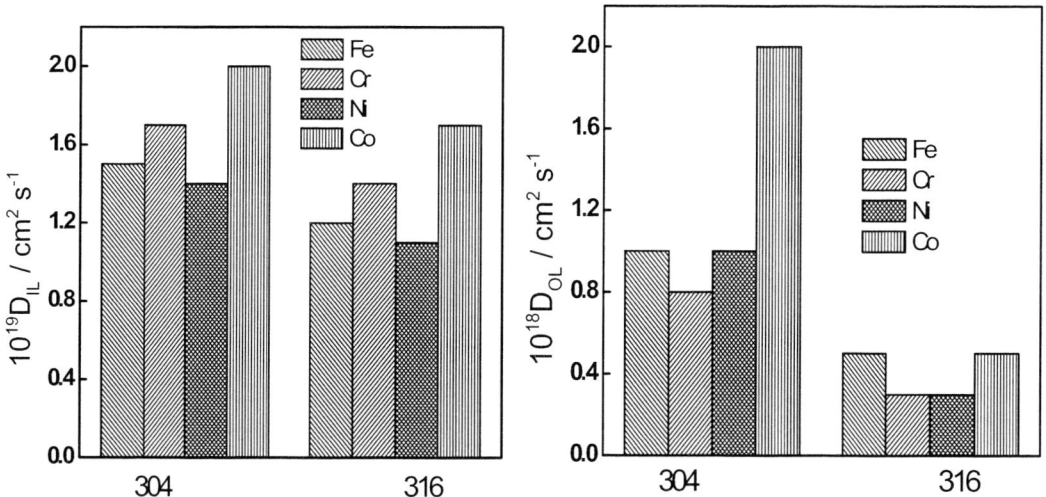

Figure 29. Dependence of the diffusion coefficients of inner layer constituents (left) and formal diffusion coefficients for the growth of the outer layer (right) on the steel type.

On comparing the sets of parameters used to reproduce the depth profiles of AISI 304 in PWR and BWR water (e.g. Figure 7-Figure 8 and Figure 28-Figure 29), it can be concluded that the most important differences are found between the rate constants at the inner layer/water interface, the diffusion coefficients in the inner layer and the formal diffusion coefficients of the growth of the outer layer. Notably, the rate constants of Cr and Co dissolution at the outer interface are orders of magnitude higher for films formed in BWR water than those for films formed in PWR water, which reflects the transpassive dissolution condition of the steels. On the basis of this result it can be tentatively assumed that Co substitutes Cr in the phase in the inner layer and is dissolved transpassively together with Cr. On the other hand, the diffusion coefficients in the inner layer are smaller in BWR water than in PWR water, which may reflect a change in the phase composition of the inner layer of oxide towards a phase containing mainly Fe(III) in which the concentration and mobility of Fe(II) interstitials is lower in comparison to those in the phase formed in PWR conditions. The fact that also the formal diffusion coefficients of the growth of the outer layer are smaller in BWR conditions could be tentatively related to the formation of a hematite phase in this layer which growth mechanism is expected to be different from that of the spinel trevorite phase usually found in PWR water.

Influence of Zn on the Incorporation of Co in Simulated BWR Conditions

This chapter focuses on the effect of ionic additions to the BWR coolant on the distribution of the main source of radiation field in BWR - radioactive cobalt - in the oxide films formed on primary loop recirculation piping [[49]-[52]]. GDOES profiles for Fe, Cr, Ni, Zn and Co in the oxide formed on AISI 304L after 1000 h of exposure to simulated BWR water containing 1 ppb of Co and different amounts of Zn [[52]] are shown in Figure 30. In these profiles, besides the usual inner and outer layers, a third layer probably formed by deposition of foreign cations is observed. Such three-layer structures have been observed in laboratory loop experiments [[53]] due to the fact that saturation is achieved more readily than in real reactor conditions as flow rates are typically much smaller. Co is incorporated in both layers, but mostly in the inner layer, and its incorporation is hindered by Zn addition in analogy to the PWR results described above. In the present context, the growth of the deposit layer has been treated in analogy to the outer layer, i.e. using the formalism described by equations (5) and (7) and defining a formal diffusion coefficient in the deposit layer. As it has been shown that also the deposit layer consists of loosely packed crystallites with electrolyte in between (see e.g. Ref. [43]), a pure diffusion mechanism has been used in complete analogy to the treatment for the outer layer. The results of the calculations with such an extended version of the model are presented in Figure 30 with solid lines and demonstrate the ability of the approach to account for this type of data as well. The kinetic and transport parameters in the three layers used to reproduce the experimental profiles are collected in Figure 31 (rate constants at the interfaces) and Figure 32 (diffusion coefficients in the three layers and the field strength in the inner layer).

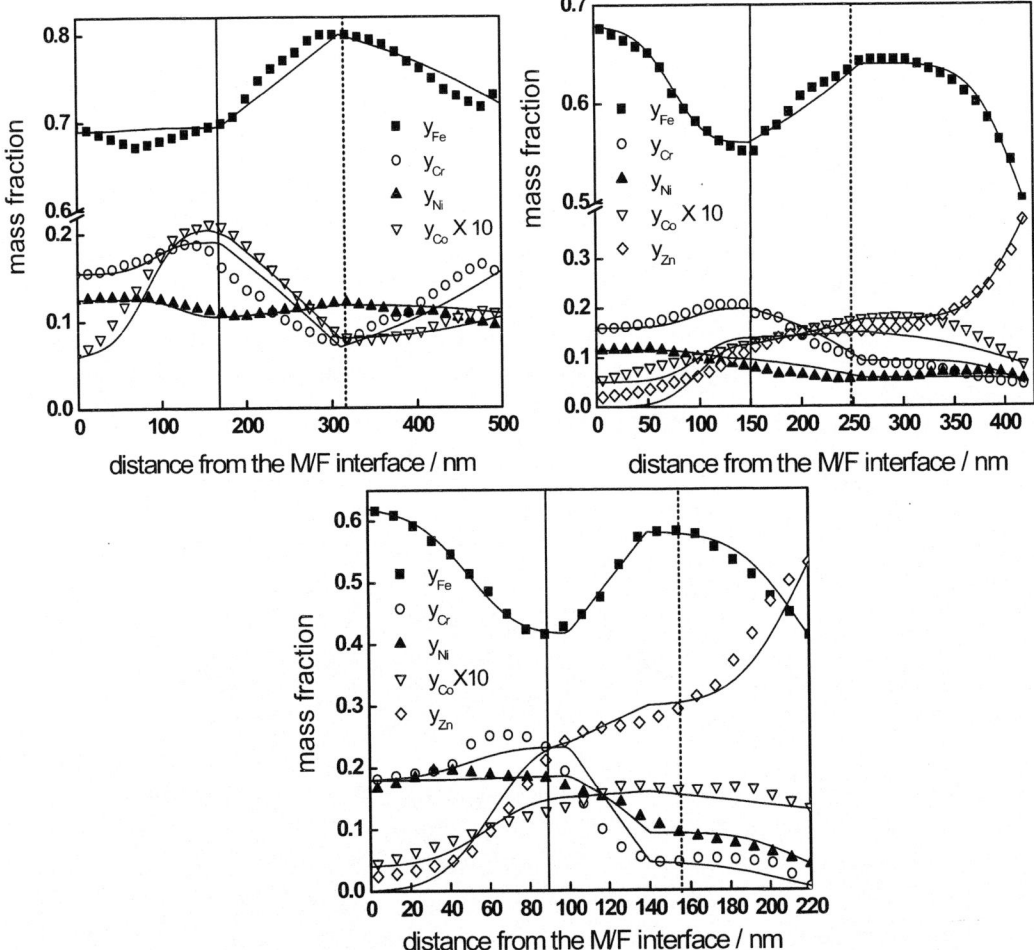

Figure 30. Experimental (points) and calculated (solid lines) depth profiles of the mass fractions of Fe,Cr, Ni,Co and Zn in the films formed on AISI 316 for 1000 h in simulated BWR water without Zn (above left), with 10 ppb Zn (above right) and 50 ppb Zn (below). The inner layer / outer layer and outer layer/deposited layer boundaries indicated with a vertical line. Data taken from Ref. [52].

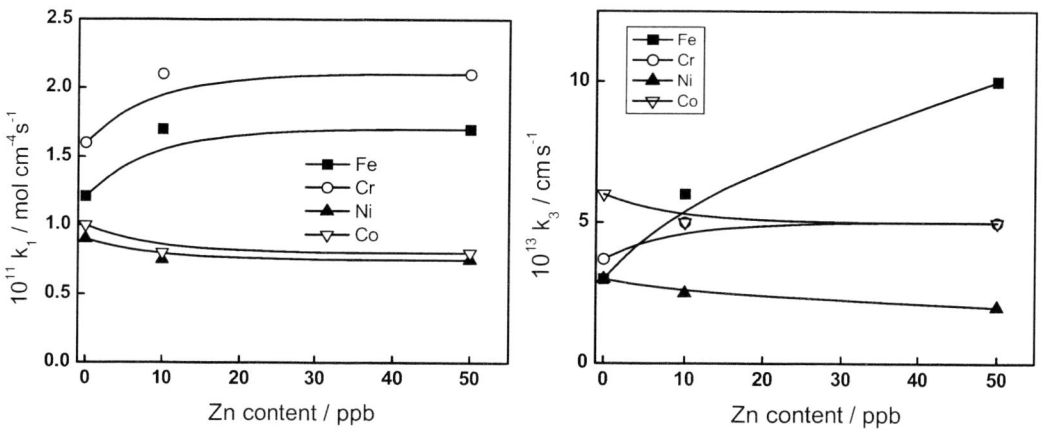

Figure 31. Dependence of the rate constants at the steel / inner layer interface (left) and the inner layer/electrolyte interface (right) on the Zn content in the simulated BWR water.

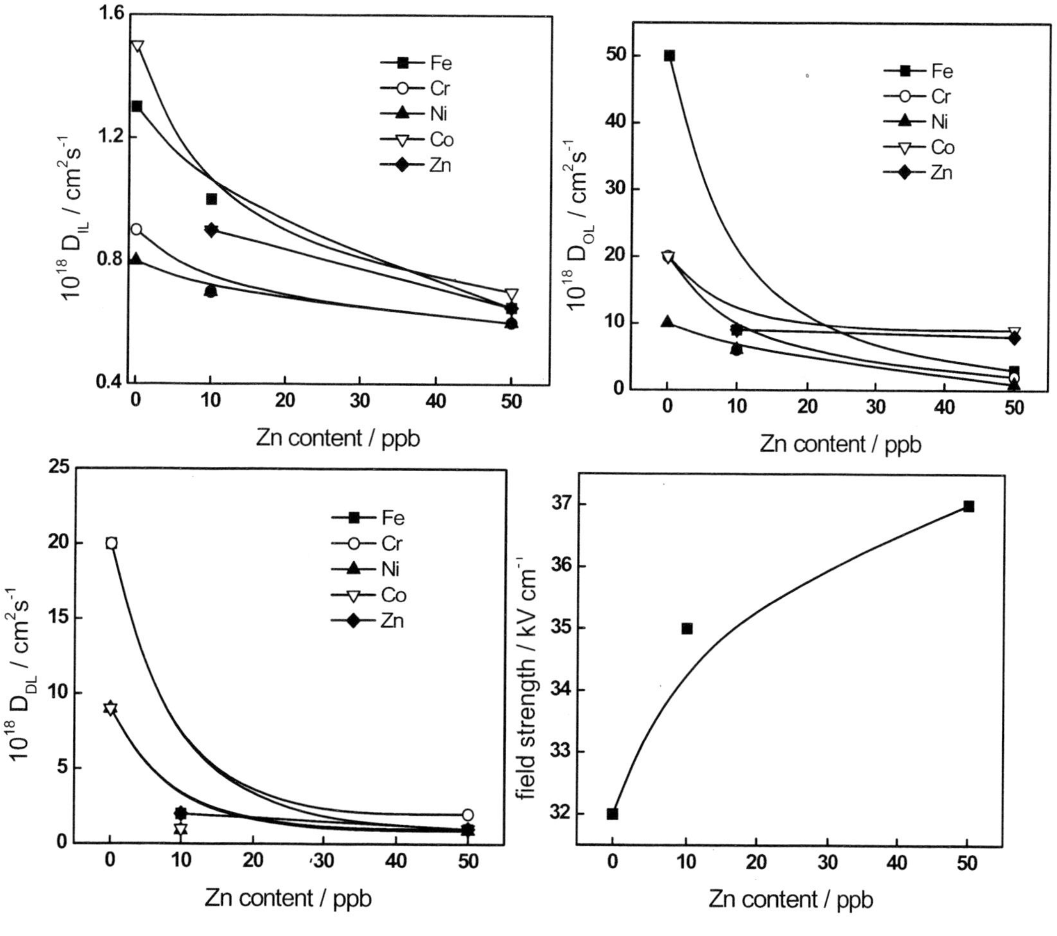

Figure 32. Dependence of the diffusion coefficients in the inner layer (above, left), the formal diffusion coefficients in the outer layer (above, right) and the deposited layer (below, left) and the field strength in the inner layer (below, right), on the Zn content in the simulated BWR water.

The following conclusions on the effect of Zn on the kinetic and transport parameters can be drawn on the basis of the calculational results:

- The effect of Zn on the rate constants at the alloy/inner layer interface is small, which can be explained by the fact that Zn incorporation does not in fact reach the inner interface
- In analogy to what has been observed in PWR conditions, the incorporation of Zn in the inner layer leads to an increase of the rate constant of dissolution of Fe at the inner layer/water interface. This once more lends support to the reaction scheme of incorporation described by equations (8)-(11)
- The diffusion coefficients of constituents in the inner layer, the formal diffusion coefficients of growth of the outer and deposit layer all decrease with the increase of the Zn content in the water, the decrease being the most pronounced for the outer and deposit layers. The significant concentrations of incorporated Zn in these layers could lead to the formation of new Zn containing phases, the growth mechanism of which is different and anyway the rate of growth much slower than that of the outer and deposit layers formed in the absence of Zn.

The equilibrium constant of the exchange reaction between Zn and Fe can also be estimated from the present results using the values of the mass fractions of Zn and Fe at the inner layer/electrolyte interface (Figure 30), the surface site concentration on hematite estimated earlier from titration experiments in simulated BWR water and the speciation of soluble Zn in the electrolyte calculated using literature data [[54]]. A value of $\log K = 2.8$ is obtained, which lies in between the values of 5.18 and -0.23 estimated in ref. [54] for surface complexation reactions of the type

$$2 \equiv FeOH^0_{1/2} + Zn^{2+}_{aq} + nH_2O \rightleftarrows (\equiv FeO)_2 Zn(OH)_n^{1-n} + (n+1)H^+ \qquad (18)$$

for n=2 and 5, respectively. Taking into account the fact that the experiments in Ref. [54] were performed with pure hematite in simulated BWR water, the correspondence between these two types of results is assumed to be reasonable.

Anyway, the equilibrium constant of Zn incorporation reaction into the oxide formed in BWR conditions is considerably (ca. 3-4 times) larger than that in PWR conditions, which could be traced to the larger equilibrium concentration of soluble Zn ions at the neutral pH of the BWR water, and also the smaller equilibrium concentration of surface sites for adsorption for the oxide formed in BWR conditions.

A compilation of layer thicknesses on AISI 304 and AISI 316 in simulated and in-reactor BWR coolant with or without Zn addition is shown in Figure 33 (data from Refs. [44]-[47],[52]) together with the thicknesses calculated by using the estimated model parameters and equation (14).

Once more a satisfactory agreement is obtained, allowing for the conclusion that the model is successful in predicting the kinetics of oxide growth also in these conditions.

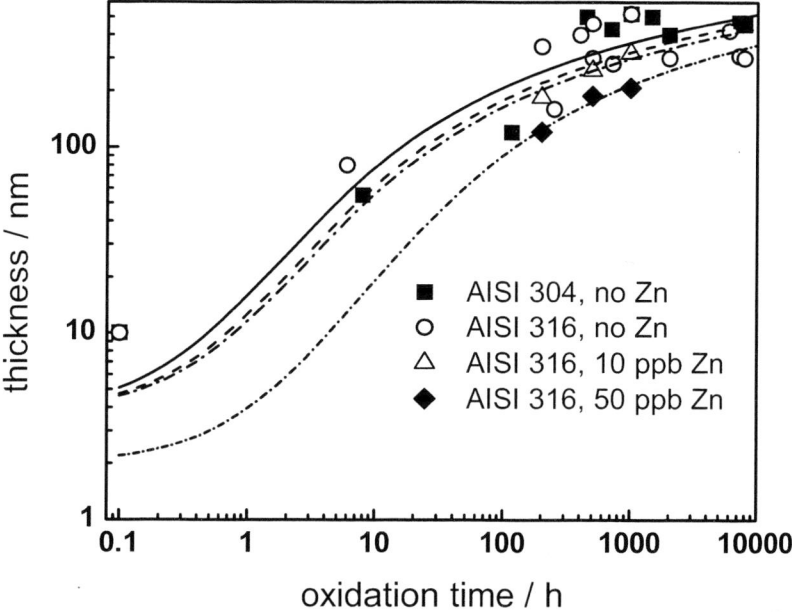

Figure 33. A compilation of thickness vs. time data for AISI 304 and AISI 316 as depending on the time of exposure and Zn content of simulated and in-reactor BWR water. Model predictions shown with lines.

CONCLUSION

Several aspects of the processes of growth and restructuring of oxide films formed on structural materials in LWRs, such as stainless steels and nickel-based alloys, have been reviewed in the present chapter. The inner layer growth and internal transport properties of charge and matter were shown to be quantitatively modelled by use of the MCM, which is developed from the solid state properties of a passive film and takes into account the coupling between transport and reaction fluxes at the interfaces, as well as ionic and electronic conduction. The interaction between the surface of the solid film and the aqueous environment has been also described in terms of incorporation of solution-originating cations and the net effects resulting from this interaction have been discussed.

The basic available theory presented in this chapter is focused on the inner layer of oxide having an uniform composition, but the first steps towards taking into account the evolution of the oxide microstructure (such as grain size and grain boundary width) on the transport of defects through the inner alyer are also described. Although there are some observations that absence of a pronounced outer oxide layer will enhance the overall alloy corrosion rate, it is assumed that the overall alloy corrosion rate is limited by the rates of the net processes in this layer and can thus be explained by the MCM. The importance of the oxide film on stainless steels and nickel base materials in LWRs is, however, not limited to the formation, growth, and stability of the inner layer. In most cases, an outer oxide layer is also present. The outer layer has generally a different composition as a result of the interaction with the environment and is not considered to be part of the passivating oxide film, although the outer layer may be important in order to prevent direct chemical attack of the protective chromium-rich layer in oxidising environments, i.e. BWRs or PWRs during shut-down. Hence, the outer layer(s) is not passivating, but can be protective. This means also that the outer layer is of fundamental importance in any effort to model the interaction between material and environment leading to activity build-up as well as to localised attack resulting in e.g. stress corrosion cracking.

The outer layer is a result of the interaction of the environment with the oxide film. This interaction is of utmost importance for the modelling of the activity build-up. The pick-up and release of radioactive nuclides (and other ions) will depend on the diffusion from the outer oxide surface into the inner oxide material. The surface concentration of such species will be a function of their concentration in the water, the extent of surface complexation, and the transport rate into the inner oxide material. As yet another effect of the interaction with the environment, deposits are formed onto the outer oxide. The deposits are especially well developed in BWRs. While the outer oxide is a result of the solid state reactions in the barrier layer and within the outer layer providing a local environment controlling its formation and restructuring, the deposits are formed mainly as a result of the local bulk water chemistry conditions. This notion is shown by the very well developed rather large single crystals found on stainless steel and nickel base materials in LWR environments.

In reality, the interfaces between the two oxide layers and between the inner oxide film and the metal interface, as well as between the water and the outer oxide film, is mostly found to be non-uniform. The importance of an uneven water-oxide interface will be small due to the very fast relative diffusion rates in water at elevated temperatures with regard to the dimensions of the oxide films, i.e. micron or sub-micron. For such distances, diffusion in water will allow transport on the millisecond scale. Similarly, the uneven oxide film appearance will have no importance for the modelling of the impact of the deposited crystals for the same reason. The importance of an uneven inner layer could be of more importance, but in reality a slightly thicker oxide layer would constitute a more pronounced transport barrier and a variation in this thickness would be self-regulating to result in a barrier layer oxide of similar transport limiting properties for all areas of similar local conditions on similar materials. An important step in going from an idealised oxide model of a uniform and even oxide film to the real oxide would be the introduction of porosity instead of using the formal approach to the growth and restructuring of the outer and deposit layers as outlined in the present chapter.

To summarise, the simple uniform two- or three-layer model, which is the basis for the MCM, will require an extension in order to propose a more realistic model according to the picture of the oxide on stainless steels or nickel base alloys in LWRs. It is, however, implied that the extensions can be made using basic principles and less complicated submodels than the MCM in itself.

ACKNOWLEDGMENT

The funding of this work by the FP6 Euratom programme of the European Commission via contract No. FI6O-036367 "A deterministic model for corrosion and activity incorporation in nuclear power plants (ANTIOXI)" is gratefully acknowledged.

REFERENCES

[1] Chao, C.Y.; Lin, L.F.; Macdonald D.D. *J. Electrochem. Soc.* 1981, *128*, 1187-1194.
[2] Macdonald, D.D. *J. Electrochem Soc.* 1992, *139*, 3434-3442.
[3] Macdonald, D.D. *Pure Appl. Chem.* 1999, *71*, 951-978.
[4] Macdonald, D.D.*J. Electrochem. Soc.* 2006, *153*, B213-B224.
[5] Fromhold Jr., A.T.; Cook, E. L. *J. Appl. Phys.* 1967, *38*, 1546-1553.
[6] Young, L.*Trans. Faraday Soc.* 1955, *51*, 1250-1255.
[7] Cahan, B.D.; Chen, C.-T. *J. Electrochem. Soc.* 1982, *129*, 921-925.
[8] Bojinov, M.; Fabricius, G.; Laitinen, T.; Mäkela, K.; Saario, T.; Sundholm, G. *J. Electrochem. Soc.* 1999, *146*, 3238-3247.
[9] Bojinov, M.; Fabricius, G.; Laitinen, T.; Mäkela, K.; Saario, T.; Sundholm, G. *Electrochim. Acta* 2000, *45*, 2029-2048.
[10] Bojinov, M.; Fabricius, G.; Laitinen, T.; Mäkela, K.; Saario, T.; Sundholm, G. *Electrochim. Acta* 2001, *46*, 1339-1358.
[11] Bojinov, M.; Fabricius, G.; Kinnunen, P.Laitinen, T.; Mäkela, K.; Saario, T.; Sundholm, G. *J. Electroanal. Chem.* 2001, *504*, 29-44.
[12] Beverskog, B.; Bojinov, M.; Kinnunen, P.; Laitinen, T.; Mäkelä, K.; Saario, T.*Corros. Sci.* 2002, *44*, 1923-1940.
[13] Bojinov, M.; Kinnunen, P.; Sundholm, G. *Corrosion* 2003, *59*, 91-103.
[14] Betova, I.; Bojinov, M.; Kinnunen, P.; Mäkelä, K.; Saario, T. *J. Electroanal. Chem.* *2004*, *572*, 211-223.
[15] Bojinov, M.; Kinnunen, P.; Lundgren, K.; Wikmark, G. *J. Electrochem. Soc.* 2005, *152*, B250-B261.
[16] Bojinov, M.; Galtayries, A.; Kinnunen, P.; Machet, A.; Marcus, P. *Electrochim. Acta* 2007, *52*, 7475-7483.
[17] Betova, I.; Bojinov, M.; Kinnunen, P.; Lundgren, K.; Saario, T. *J. Electrochem. Soc.* 2008, *155*, C81-C92.
[18] Zhang L.; Macdonald, D.D. *Electrochim. Acta* 1998, *43*, 2673-2685.
[19] Tiedemann, W.; Newman, J. *J. Electrochem. Soc.* 1972, *119*, 186-188.
[20] Lundgren, K.; Kelen, T.; Gunnarsson, M.; Ahlberg, E. *Proceedings of CHIMIE 2002, La chimie de l'eau dans les réacteurs nucléaires -Optimisation de l'exploitation et développements nouveaux*, SFEN, Avignon, France, 2002.
[21] Beverskog, B.; Puigdomenech, I. *Corrosion,* 1999, *55*, 1077-1087.
[22] Ziemniak, S.E.; Hanson, M. *Corros. Sci.* 2002, *44*, 2209-2230.
[23] Ziemniak, S.E.; Hanson, M. *Corros. Sci.* 2006, *48*, 2525-2546.
[24] Lister, D.H.; Venkateswaran, G. *Nucl. Technol.* 1999, *125*, 316-331.
[25] Beverskog, B. *Proc. International Conference Water Chemistry of Nuclear Reactor Systems*, American Nuclear Society, San Francisco, CA, 2004, paper 2.7.
[26] Betova, I.; Bojinov, M.; Kinnunen, P.; Lehikoinen, J.; Lundgren, K.; Saario, T. *Proceedings of ICAPP'07*, SFEN, Nice, France, 2007, paper 7122.
[27] Bojinov, M.; Kinnunen, P.; Olin, M. VTT-R-10851-07, VTT Technical Research Centre of Finland, Espoo, 2008.

[28] Bennett, P.J.: Gunnerud, P.; Petersen, J.K.; Harper, A. *Water Chemistry of Nuclear Reactor Systems 7*, British Nuclear Energy Society, London, 1996; Vol.1, pp. 189-193.
[29] Bennett, P.J.: Gunnerud, P.; Loner, J.; Petersen, J.K.; Harper, A. *Water Chemistry of Nuclear Reactor Systems 7*, British Nuclear Energy Society, London, 1996; Vol.1, pp. 293-296.
[30] Beverskog, B.; Mäkelä, K. *Proc.of the JAIF International Conference on Water Chemistry in Nuclear Power Plants*, Japanese Nuclear Society, Kashiwazaki, Japan, 1998, p. 89-98.
[31] Szklarska-Smialowska, Z.; Chou, K.; Xia, Z. *Corros. Sci.* 1992, *32*, 609-619.
[32] Tapping, R.L.; Davidson, R.D.; McAlpine, E.; Lister, D.H. *Corros. Sci.* 1986, *26*, 563-576.
[33] Lister, D.H.; McAlpine, E.; Davidson, R.D. *Corros. Sci.* 1987, *27*, 113-140.
[34] Machet, A. *Etude des premiers stades d'oxydation d'alliages inoxydables dans l'eau à haute température*, PhD Thesis, Université Pierre et Marie Curie (Paris VI), Paris, 2004.
[35] Machet, A.; Galtayries, A.; Marcus, P.; Combrade, P.; Jolivet. P.; Scott, P. *Surf. Interf. Anal.* 2002, *34*, 197-200.
[36] Machet, A.; Galtayries, A.; Zanna, S.; Klein, L.; Maurice, V.; Jolivet. P.; Foucault, M.; Combrade, P.; Scott, P.; Marcus, P. *Electrochim. Acta* 2004, *49*, 3957-3964.
[37] Ziemniak, S.E.; Hanson, M. *Corros. Sci.* 2006, *48*, 498-521.
[38] Ziemniak, S.E.; Hanson, M. *Corros. Sci.* 2006, *48*, 3330-3348.
[39] Chevalier, S.; Desserey, F.; Larpin, J.P. *Oxid. Met.* 2005, **64**, 219-234.
[40] Carette, F.; Lafont, M.C.; Chatainier, G.; Guinard, L.; Pieraggi, B. *Surf. Interf. Anal.* 2002, *34*, 135-138.
[41] Delabrouille, F.; Viguier, B.; Legras, L.; Andrieu, E. *Mater. High Temp.* 2005, *22*, 287-292.
[42] Bojinov, M.; Kinnunen, P.; Laitinen, T.; Mäkela, K.; Saario, T.; Sirkiä, P. *Plant Life Management. Progress for structural integrity,* Solin, J.; Ed.; VTT Symposium 227, VTT Technical Research Centre of Finland, Espoo, 2003, pp. 121-131.
[43] Kim, Y.-J. *Corrosion* 1995, *51*, 849-860.
[44] Degueldre, C.; O'Prey, S.; Francioni, W. *Corros. Sci.* 1996, *38*, 1763-1782.
[45] Degueldre, C.; Buckley, D.; Dran, J.C.; Schenker, E. *J. Nucl. Mater.* 1998, *252*, 22-27.
[46] Hemmi, Y.; Uruma, Y.; Ichikawa, N. *J. Nucl. Sci. Technol.* 1994, *31*, 443-455.
[47] Wambach, J.; Wokaun, A.; Hiltpold, A. *Surf. Interface Anal.* 2002, *34*, 164-170.
[48] Bojinov, M.; Kinnunen, P.; Laitinen, T.; Mäkela, K.; Mäkela, M.; Saario, T.; Sirkiä, P. *Corrosion* 2001, *57*, 387-393.
[49] Ono, S.; Naginuma, M.; Kumagai, M.; Kitamura, M.; Tachibana, K.; Ishigure, K. *J. Nucl. Sci. Technol.* 1995, *32*, 125-132.
[50] Honda, T.; Ohashi, K.; Furutani, Y.; Minato, A. *Corrosion* 1987, *43*, 564-570.
[51] Hosokawa, H.; Uetake, N.; Ushida, K.; Nakamura, M.; Mochizuki, K.; Nagata, T.; Ogawa, N.; Baba, T.; Ono, S.; Ishigure, K. *Proceedings of CHIMIE 2002, La chimie de l'eau dans les réacteurs nucléaires -Optimisation de l'exploitation et développements nouveaux*, SFEN, Avignon, France, 2002, paper 136.

[52] Inagaki, H.; Nishikawa, A.; Sugita, Y.; Tsuji, T. *J. Nucl. Sci. Technol.* 2003, *40*, 143-152.

[53] Betova, I.; Bojinov, M.; Helin, M.; Laitinen, T.; Muttilainen, E.; Mäkelä, K.; Reinvall, A.; Sirkiä, P. *Proceedings of the International Conference Water Chemistry of Nuclear Reactor Systems*, San Francisco, CA, October 10-14, 2004, pp. 282-288.

[54] Kinnunen, P.; Bojinov, M.; Lehikoinen, J.; Lundgren, K. *SAFIR, The Finnish Research Programme on Nuclear Power Plant Safety 2003– 2006, Final Report,* Karita-Puska, E.; Ed.; VTT Research Notes 2363, V TT Technical Research Centre of Finland, Espoo, 2006, pp.115-121.

Chapter 4

THE FLEXURAL MODULUS OF POLYMER MATRIX COMPOSITES

C.J.R. Verbeek[1], P.D. Ewart and D. Murtagh
The Department of Engineering.
The University of Waikato, Private Bag 3105, Hamilton, New Zealand.

ABSTRACT

The flexural and Young's modulus of materials are widely used material properties in design. Young's modulus is often tested either in tensile or in bending, assuming that those values are the same. Bending is often preferred due to the simplicity of the test specimens. However, when testing composite materials, the results obtained using these methods vary significantly.

Planar randomly oriented glass and wood fibre reinforced polyethylene composites as well as talc filled polyethylene were used to evaluate and compare the tensile, compression, three and four point bending modulus. It was found that the tensile modulus was consistently higher than the flexural modulus. It was concluded that because a beam is under a mixed stress state of tension and compression, the measured flexural modulus values were intermediate of the tensile and compressive modulus. The differences were small at low volume fractions reinforcement and independent of the degree of interfacial adhesion, at any volume fraction fibre. At high fibre volume fractions fibre, it was assumed that non-uniform stress distribution lead to shear deformation, causing greater differences between the various modulus values, exacerbated by debonding and inefficient fibre packing.

It was found that the glass and wood fibre reinforced composites can be successfully modelled assuming a layered micro-structure. It was concluded that the number of plies used in the model strongly affects the predicted flexural modulus and that using five plies in the model element is most representative of the material. The geometry of the reinforcement became irrelevant at a large number of plies, leading to predictions similar to the isostrain model, scaled with the adhesion coefficient.

It was found that low aspect ratio reinforcements will lead to the flexural modulus being higher than Young's modulus. It was concluded that in bending, stiffness is

---

[1] Corresponding author. (Tel) +64 7 383 4947 (fax) +64 7 838 4835, Email: jverbeek@waikato.ac.nz

proportional to the arrangement or geometry of the reinforcement phase and an individual fibre's contribution to the flexural modulus is a function of the relative distance to the neutral axis. The orientation of low aspect ratio fillers is therefore not important with respect to the flexural modulus. Furthermore, the difference between the various flexural tests was found to be consistent regardless reinforcement geometry and it was concluded that shear deformation may be the most important factor contributing to the difference between these.

Keywords: Flexural Modulus, Modelling, Interfacial Bonding

ABBREVIATIONS

Δf	change in deflection, m
ΔP	change in load, N
A_m	cross sectional area of a matrix layer, m^2
a_m	thickness of a matrix layer
A_r	cross sectional area of the fibre layer, m^2
a_r	thickness of a reinforcement layer, m
b	width, m
d	depth, m
E	Young's modulus, Pa
E_C	Young's modulus of the composite material, Pa
E_{cr}	compression modulus, Pa
E_{f3}	flexural modulus tested under three point bending, Pa
$E_{f3,shear}$	flexural modulus tested under three point bending adjusted for shear deformation, Pa
E_{f4}	flexural modulus tested under 4 point bending, Pa
E_{pd}	Young's modulus, partial adhesion, Pa
E_r	modulus of the reinforcement, Pa
E_t	tensile modulus, Pa
f	fractional shear stress representing interfacial shear strength
G	shear modulus, Pa
h_{mj}	distance from an individual matrix layer's central axis to the neutral axis of the beam, m
h_{ri}	distance from an individual fibre layer's central axis to the neutral axis of the beam, m
I	second moment of area, m^4
I_C	second moment of area of the composite beam, m^4
I_{db}	second moment of area of a debonded of debonded layers in the layered beam, m^4
I_{dC}	second moment of area of a homogeneous beam with dimensions b x y_d, m^4
I_{fibre}	second moment of area for all fibre layers in the layered composite, m^4
I_{k_lam}	fully bonded second moment of area, based on k-plies, m^4
I_{LC}	layered composite second moment of area under partial adhesion, m^4

I_{matrix}	the total second moment of area of all the matrix layers in the layered material, m^4
$I_{removed}$	second moment of area of a beam based on k$_d$ layers with full adhesion, m^4
I_{xxi}	second moment of area around axis XX for the ith fibre layer, m^4
I_{xxj}	second moment of area around axis xx for the jth layer of matrix material, m^4
k	number of plies
k$_d$	number of debonded layers
L	support span, m
n	number of layers
n$_f$	number of fibre layers
n$_r$	number of matrix layers
Q	first moment of area, m^3
R	modulus ratio
S	one minus fractional shear stress
V	shear force, N
v$_m$	volume fraction matrix
v$_r$	volume fraction reinforcement
y$_m$	position where debonding occurs, m
σ	stress, Pa
ε	strain
δ	deflection, m
τ	shear stress, Pa
τ$_{int}$	interfacial shear strength, Pa
τ$_{max}$	maximum shear stress in a beam, Pa

1. INTRODUCTION

The purpose of modelling mechanical properties of composite materials is not only an attempt to better understand and explain material behaviour, but also to enable the designer enhanced tools during the design process. Modern day computer assisted design relies on accurate material properties, or the ability to accurately model the system. Composite materials require a lot of computing power to accurately account for the microstructure of the material in addition to complex geometries of design. The ability to predict mechanical properties from the constituent phases is therefore considered a valuable tool.

In the design process, care must be taken to use the appropriate modulus for design. Materials for which the compressive and tensile moduli are different would imply that the flexural modulus would be different [1, 2]. Alternatively, it could be said that if the tensile and compressive modulus are the same, measuring the flexural modulus is simply a different method to determine either of the aforementioned moduli.

The tensile and compressive modulus for fibre reinforced composites are rarely the same and hence the flexural modulus is not equal to either of those [2-4]. In earlier work, attempts have been made to determine the flexural modulus based on the knowledge of the tensile and compressive modulus, using classical beam theory [1, 2].

In this paper, the relationship between the flexural and tensile modulus is further explored in light of micro-structure and geometry. Glass, wood and talc reinforced polyethylene were tested for flexural, tensile and compressive modulus to serve as model systems to explore the discrepancies between the various modulus values.

Furthermore, a model is presented that predicts the flexural modulus of reinforced thermoplastics where the geometry of the reinforcement represents a planar random orientation. Assumptions made to simplify the model are discussed and the model was compared to experimental data.

2. THE FLEXURAL MODULUS

Simply put, the elastic modulus of a material is its resistance to deformation. It is often accepted, without further consideration, that the elastic modulus in tension and under flexure should be the same. This if often exacerbated by the misconception that the modulus calculated using the bending of beams (equations 1 and 2) is the same as the Young's modulus of the material ($E_t = \dfrac{\sigma}{\varepsilon}$).

2.1. Calculating the Flexural Modulus

Three and four point bending tests are commonly used for testing the flexural modulus of several types of materials, including composites. There are some obvious differences between the two tests, most importantly that in the four point bending test, the beam section between the two loading points is in pure bending, i.e. no shear stresses are present. In Figure 1, the shear stress along the beam axis is shown for a beam under three and four point loading, respectively.

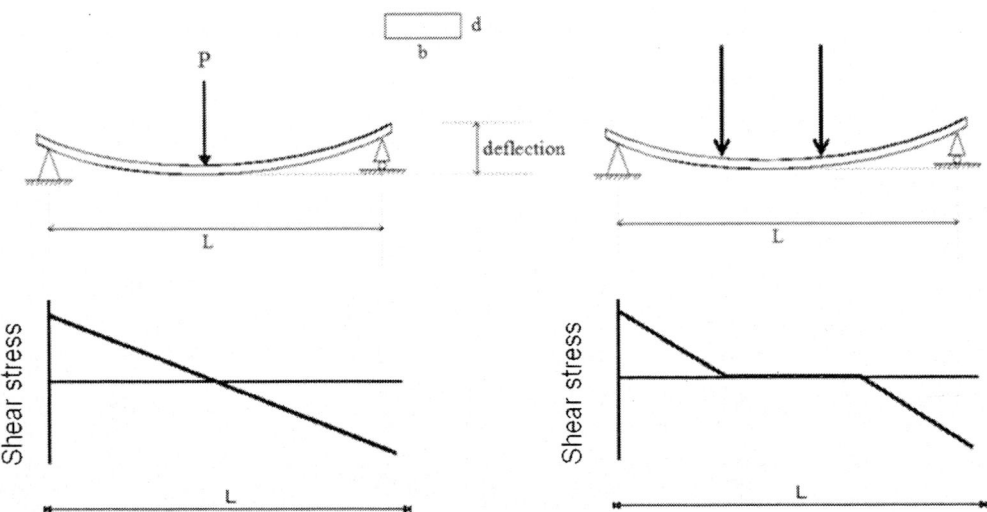

Figure 1. Shear stresses in beams under 3-point and 4-point loading, respectively.

The flexural modulus can then be determined using classical beam theory, where $\frac{\Delta P}{\Delta f}$ is the slope of the load deflection curve (other symbols are defined in Figure 1):

Three point bending:

$$E_{f3} = \frac{L^3}{4bd^3} \frac{\Delta P}{\Delta \delta} \tag{1}$$

Four point bending:

$$E_{f4} = \frac{0.17 L^3}{bd^3} \frac{\Delta P}{\Delta \delta} \tag{2}$$

Using these expressions, a number of implicit assumptions are made:

- the beam axis is the principle axis of orthotropy
- the beam material is linear elastic
- the length of the beam is significantly longer than the greatest dimension of the cross section
- the beam's cross section remains plane during bending
- the beam is straight before loading
- all bending is in the plane of the applied load
- the cross-section of the beam is symmetrical about the plane of bending
- stresses are uniform across the width of the beam
- transverse shear deformations are minimal

Most importantly, however, it is assumed that the modulus of elasticity in tension and compression are the same. This implies that the flexural modulus is equal to the Young's modulus.

Another important assumption is that there is no shear deformation during bending. The consequence is that the previously assumed plane cross section now becomes curved [5]. However, it has been shown that the effect of shear deformation can be offset by using a large span-to-depth ratio [3, 5].

Using an insufficient span-to-depth ratio gives rise to large shear deformation invalidating equations 1 and 2. Accounting for shear deformation has been described elsewhere [3, 5] and is repeated here as equation 3, for a simply supported beam subjected to a point load:

$$E_{f3,shear} = E_{f3}\left[1 + \frac{E_{f3,shear}}{\tfrac{5}{6}G}\left(\frac{d}{L}\right)^2\right] \tag{3}$$

It can now be seen that the ratio $\frac{d}{l}$ strongly influences the shear deformation, but that the ratio of the tensile modulus to shear modulus (G) is equally important. It has been shown that the error introduced by not using equation 3 is as much as 20% when $\frac{E_{f3}}{G}$ is 50 at a span-to-depth ratio of 16. When $\frac{E_{f3}}{G}$ was equal to 2.5 the error became negligible. When $\frac{E_{f3}}{G}$ is large, a large span-to-depth ratio is required, often as much as 60 for unidirectional fibre reinforced composites.

Other factors may also compromise the accuracy of equations 1 and 2, such as the variation of the support span during testing [6] and support indentation on the sample [7]. However, for composite materials where the tensile and compressive moduli are not the same, some additional complications exist and are described below.

2.2. Multi-Modulus Materials

Under certain conditions the compressive and tensile modulus of materials may be equal, for example materials that are homogeneous and isotropic, however, for composites this is rarely the case. As early as 1976, Jones [1] has shown that composite materials are characterised by an apparent flexural modulus that is intermediate of the tensile and compressive modulus. This was attributed to the difference in stress states in the material under bending. In Figure 2, a simplified stress state is shown for a multi-modulus beam subjected to a moment. It has been shown that strain varies linearly across the depth of the beam, while stress varies in a bilinear way. It is important to note that the neutral surface is not at the midpoint of the beam and consequently, because of the mix tension and compression state, the flexural modulus is neither E_{cr}, nor E_t. Jones derived the following expression for the apparent flexural modulus (E_f), based on the compressive and tensile modulus: $E_f = \frac{4 E_t E_{cr}}{\left(E_t^{1/2} + E_{cr}^{1/2}\right)^2}$.

This equation assumes that the beam is subjected to a pure moment and the axial strain is linear across the section depth [1].

It has been shown that in unidirectional carbon/epoxy composites, elastic instabilities and shear deformation cause a reduction in the flexural modulus compared to the tensile modulus [8]. In the same study it was also found that:

- the flexural modulus varied with strain, due to the variation of the tensile (E_t) and compressive (E_{cr}) modulus with strain.
- the neutral axis was displaced from the centre axis because of the difference between E_t and E_{cr} and that this caused a reduction in flexural modulus.

Even though the above considerations were for specific composites materials, the findings are generally valid. Lastly, the geometry of the reinforcement fraction in the composites may also influence the flexural behaviour of the material.

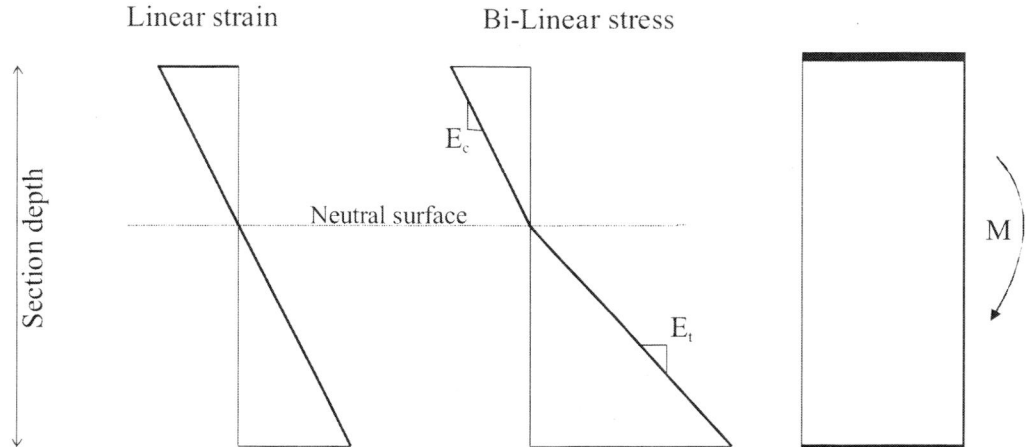

Figure 2. Linear strain and bi-linear stress variation for multi-modulus beam, subjected to a bending moment.

2.3. Geometry of Reinforcement

When an object is subjected to a load that produces a bending moment, it is common knowledge that the shape of that object determines its stiffness. This is different than the material's elastic modulus, which is effectively normalised with respect to area. Under bending, the second moment of area is what characterises the object's stiffness. For a rectangular beam, the second moment of area is: $I = bd^3/12$. If one then considers a composite material, it may be expected that the arrangement of the reinforcement fraction in the polymer matrix may have an effect on the stiffness of the material. It is well known that different orientations of reinforcement will lead to different properties, such as increased stiffness in a particular direction.

The micro-structure of composites is considered heterogeneous and therefore, the second moment of area of a section of the beam could be evaluated based on the particular arrangement of the constituents in the composite material.

As an example, consider the composite micro-structures as simplified in Figure 3. When any of these sections is subjected to a flexural load, the difference between the elastic modulus of the reinforcement and matrix should be considered.

It can be appreciated that in each case presented above, the influence of the stiffer reinforcement (shown as shaded) would be different. If one is to consider the second moment of area of the structure around axis XX, the geometry of the reinforcement phase therefore becomes important as well as the relative stiffness of each phase.

This concept was used in the development of a micro-mechanical model for flexural modulus of reinforced polymers and is discussed in the following section.

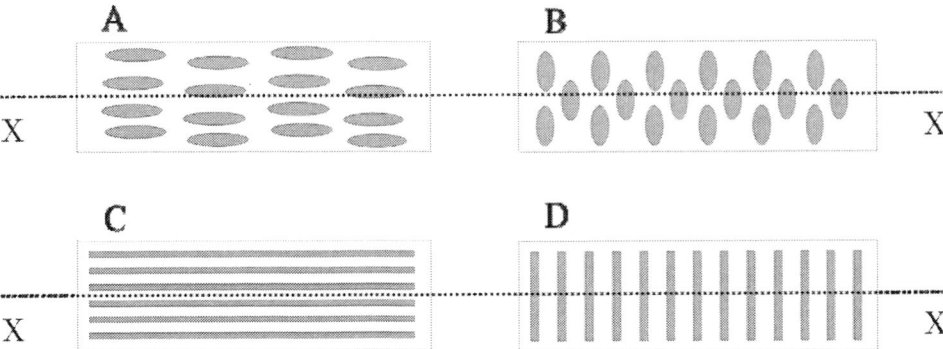

Figure 3. Examples of various simplified micro-structures that would lead to differences in the second moment of area of the section considered around axis XX.

3. MODELLING

In the model, short fibre reinforced composite materials were analysed by considering a small section (Figure 4A) of the composite and analysing it as a multilayer composite beam. By selecting the element as shown, it is assumed that:

- the effect of fibre ends on the modulus is negligible
- the fibres are randomly oriented in the x-z plane
- the composite can be presented as a layered structure, in which each layer acts as a monolithic material, with isotropic properties.
- the layer thickness of each material will be proportional to the volume fraction of that component

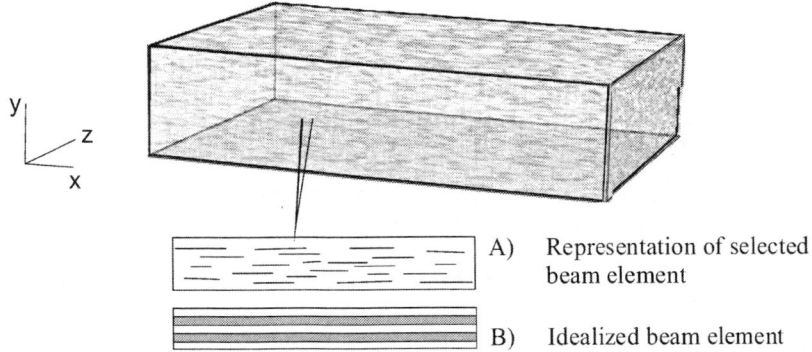

A) Representation of selected beam element
B) Idealized beam element

Figure 4. Selection of model element from the composite structure.

The selected element is then analysed as a simply supported beam subject to a point load, P, where the following relationship is valid for the deflection of the beam: $\delta = \dfrac{PL^3}{48EI}$. If it is assumed that flexural stiffness for the composite is equal to that of the layered model element, the layered model element and composite material will have the same deflection and hence:

$$E_c I_c = E_L I_L \quad (4)$$

The composite beam elements' second moment of area is defined as: $I_c = bd^3/12$, which leaves the layered model element's second moment of area to be determined in terms of the number of layers and the reinforcement volume fraction.

By defining the model element such that it will always be symmetrical around a central polymer layer (Figure 5), the neutral axis will be fixed in the central polymer layer, regardless of the amount of layers stipulated in the model. In the model element, k is defined as the number of plies where one ply consists of 4 layers. A ply is defined as one fibre and one polymer layer on each side of the neutral axis. It can therefore be seen that the first ply will consist of five layers, with a central polymer layer and each subsequent ply will consist of four layers. The total number of layers, n, is then defined as: $n = 4k + 1$.

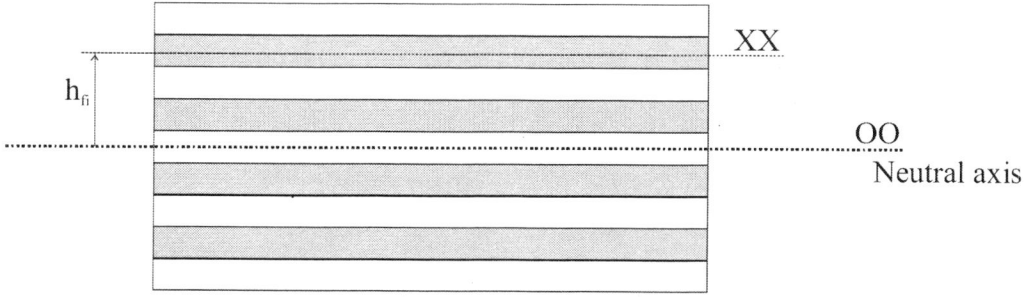

Figure 5. Individual layers and the neutral axis.

The number of fibre layers and matrix layers are now determined in terms of k as: $n_r = 2k$ and $n_m = 2k + 1$. The thicknesses of individual layers are defined in terms of the volume fraction of each component, the section depth, d, and the number of layers for each component: $a_r = \dfrac{v_r}{n_r}$ and $a_m = \dfrac{v_m d}{n_m}$.

It is a well know fact that in composite materials the interface between reinforcement and matrix material is of paramount importance. Insufficient bonding between reinforcement and matrix leads to inadequate load transfer and therefore weak mechanical properties [9, 10].

Experimentally, interfacial adhesion can be improved by, for example, including a surface active material in the formulation [9, 11]. For polypropylene, maleic acid grafted polypropylene is often added to compatibilise the fibre and matrix. This leads to improved composite properties by means of more efficient load transfer [10].

In modelling mechanical properties of composites, accounting for interfacial adhesion, or rather the lack thereof, is accounted for in numerous ways. Some of the earliest and simplest models assumed perfect adhesion in their derivation [9]. Alternatively, in many models no adhesion is assumed but neither of these situations accurately describes the real situation.

Consequently a lot of effort has been focussed on accounting for the level of interfacial adhesion in modelling the mechanical properties of polymer matrix composites [10].

In this paper three possible scenarios are modelled:

- perfect adhesion between reinforcement and matrix
- no adhesion
- partial adhesion.

Perfect Adhesion

Based on the above assumptions and the proposed model element, a model for the flexural modulus can be derived. To determine the overall second moment of area for the composite, the fibre layers and matrix layers are considered separately. When determining the second moment of area of each individual layer the parallel axis theorem is used to transform each layers second moment around its centroid (XX) to that of the beams' neutral axis (OO), as shown in Figure 5. The total second moment of area for all fibre or matrix layers would be the sum of the individual layers second moment about the beams neutral axis.

If the second moment of area for the ith fibre layer around its centroid is I_{xxi} and the distance from its centroid to the beams' neutral axis is h_{ri} then the total second moment of area for the fibre fraction is:

$$I_{fibre} = \sum_{i=1}^{k}\left(I_{xxi} + A_r h_{ri}^2\right) \tag{5}$$

The distance between the ith layers' centroid and the beams neutral axis is expressed in terms of the fibre layer thickness, a_r, and the polymer layer thickness, a_m.

These are: $h_{ri} = \frac{1}{2}(2k-1)(a_m + a_r)$ and $h_{mj} = k(a_m + a_r)$, where $1 \leq i, j \leq k$.

Equation 5 can also be rewritten as: $I_{fibre} = \sum_{i=1}^{k} I_{xxi} + \sum_{i=1}^{k} A_r h_{ri}^2$,

where $\sum_{i=1}^{k} I_{xxi} = \frac{bd^3}{12}\frac{v_r^3}{4k^2} = I_C \Gamma_r$,

and

$$\sum_{i=1}^{k}\left(A_r h_{ri}^2\right) = \frac{bd^3}{12}\left[v_r(2k+v_r)^2 \frac{(2k-1)}{(2k)^2(2k+1)}\right] = I_C \Lambda_r$$

The total second moment of area for the fibre fraction now becomes:

$$I_{fibre} = \sum_{i-1}^{k} I_{xxi} + \sum_{i-1}^{k} A_r h_{ri}^2 = I_C(\Gamma_r + \Lambda_r) \qquad (6)$$

A similar analysis is followed for the matrix layer with the total second moment of area for the matrix layers shown as:

$$I_{matrix} = \sum_{j-1}^{k} \left(I_{xxj} + A_m h_{mj}^2 \right) \qquad (7)$$

After manipulation the total second moment of area for the matrix fraction becomes:

$$I_{matrix} = \sum_{j-1}^{k} I_{xxj} + \sum_{j-1}^{k} A_m h_{mj}^2 = I_C(\Gamma_m + \Lambda_m) \qquad (8)$$

The above analysis of matrix and fibre layers did not account for differences in Young's modulus for the components. The equivalent section method uses a scaled Young's modulus for the matrix component based on the ratio $R = E_m / E_r$. The beam elements' second moment of area is thus given by Equation 9.

$$I_L = \frac{I_{matrix}}{R} + I_{fibre} \qquad (9)$$

From Equation 4, the composite flexural modulus, $E_C = \frac{E_L I_L}{I_C}$, is defined in terms of the laminate beam element's second moment of area, I_L, the composite beam second moment of area, $I_C = bd^3/12$ and the Young's modulus, $E_L = E_r$ (the fibre modulus is now used after applying the equivalent section method). Using Equations 6, 8, 9 and the above equation for E_C, the composite flexural modulus is given by Equation 10.

$$E_C = E_m(\Gamma_m + \Lambda_m) + E_r(\Gamma_r + \Lambda_r) \qquad (10)$$

$$\Gamma_m = \frac{(1-v_f)^3}{(2k+1)^2} \qquad \Lambda_m = \left[(1-v_r)(2k+v_r)^2 \frac{(k+1)}{4k(2k+1)^2} \right]$$

$$\Gamma_r = \frac{v_r^3}{4k^2} \qquad \Lambda_f = \left[v_r(2k+v_r)^2 \frac{(2k-1)}{2k^2(2k+1)} \right]$$

A graphical representation of Equation 10 is shown in Figure 6A. It is interesting to note that as the number of plies is increased, the equation quickly approaches the isostrain model or the linear rule of mixture $\left(E_C = v_r E_r + (1-v_r)E_m \right)$. This is because increasing the number of layers effectively homogenises the composite, thereby making the linear mixing rule more applicable.

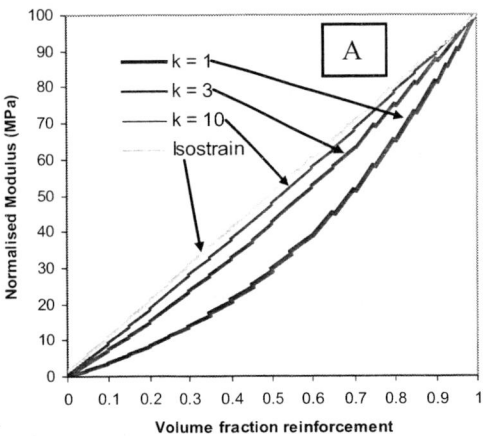 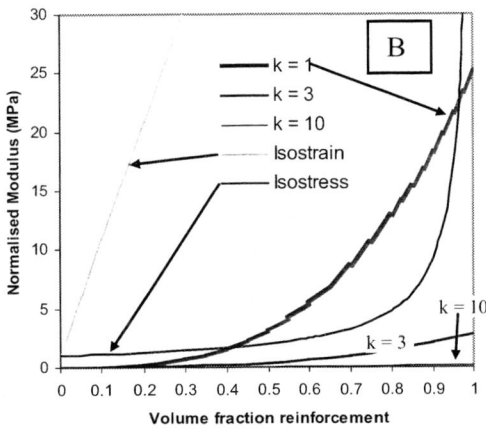

Figure 6. Graphical representation of the variation of flexural modulus for A) Perfect adhesion and B) No adhesion with respect to the number of layers (k).

No Adhesion

When there is absolutely no adhesion between layers, the second moment of area for the layered model element becomes the sum of the second moment for all the individual layers around their own neutral axis. The modulus of the composite beam is then determined similarly to the reasoning in the previous section and is presented in Equation 11:

$$E_C = E_m(\Gamma_m) + E_r(\Gamma_r) \tag{11}$$

From Figure 6B it can be seen that assuming no adhesion between layers quickly leads to a model that predicts a composite modulus far too low. This is understandable from the nature of the calculation. When the number of plies is large, the individual layer thicknesses will be very small. Even when stacked to the same section depth as the fully bonded beam, the overall modulus will be much lower because of the absence of shear between each layer. As a comparison, the isostress and isostrain models are also shown in the figure. It can be seen that this model even predicts values below the lower limit typically set by the isostress model $\left(\dfrac{1}{E_C} = \dfrac{v_r}{E_r} + \dfrac{(1-v_r)}{E_m} \right)$.

In order to account for partial adhesion, some degree of interaction is necessary between individual layers in the model element. In the next section a model is derived that accounts for partial adhesion between layers by considering the interfacial shear strength between the matrix and reinforcement.

Partial Adhesion

Insufficient adhesion typically leads to poor stress transfer between the load bearing reinforcement and matrix material. It has previously been shown that this stress transfer happens through a shear mechanism [9, 13]. If the shear stress in the beam therefore exceeds the interfacial shear strength, localized loss of adhesion occurs, lowering the mechanical properties of the material. It has to be acknowledged, however, that the field of composite failure is a complex field and various mechanisms might be at play e.g. distinguishing between fibre pull out or fibre breaking in fibre reinforced composites, but is not considered here.

In this model, a simplistic approach is taken whereby it is assumed that if the shear stress in the beam exceeds a certain value, based on interfacial shear strength of the composite, some of the layers will act as if they were debonded from the remaining layers in the model. This will effectively result in an averaging technique between the full adhesion and no adhesion models previously discussed.

The shear stress in homogenous beam is: $\tau = VQ/Ib$. In this expression V is the shear force, Q is the first moment of area of the area above the point where the shear stress is calculated, I is the second moment of area of the complete section and b is the width of the beam.

If debonding is assumed to occur at τ_{int} or $\tau_{int} = f\tau_{max}$. The shear stress at any position, y_m, is $\tau = \dfrac{3Py_m(d-y_m)}{bd^3}$. With symbols their usual meaning. The maximum stress occurs at the neutral axis of the beam and is: $\tau_{max} = \dfrac{3P}{4bd}$.

If it is then assumed that debonding will occur at a distance, y_d, from the neutral axis and the shear stress at that point exceeds the interfacial shear strength, y_d can be calculated from: $\tau_{int} = \dfrac{\tau_{max} y_m (d - y_d)}{d^2}$. Solving for y_d and using $\tau_{int} = f\tau_{max}$ gives: $y_d = \dfrac{d\sqrt{1-f}}{2}$ or $y_d = \dfrac{dS}{2}$.

The number of debonded plies, k_d, is then determined from the total thickness of the debonded, y_d, and the sum of the individual component's thicknesses: $2y_d = 2k_d(a_m + a_r) + a_m$. Solving for k_d leads to Equation 12:

$$k_d = \frac{(d\sqrt{1-f}) - a_m}{2(a_m + a_r)} \qquad (12)$$

Equation 12 is truncated to use only integer values in subsequent calculations since it is considered unrealistic to have fractional plies.

To determine the modulus of the composite under partial adhesion, a new second moment of area is defined based on a combination of the full and no adhesion cases:

$$I_{LC} = I_{k\_lam} - I_{removed} + I_{db}$$

I_{LC} = layered composite second moment of area, partial adhesion
I_{k_lam} = fully bonded second moment of area, based on k plies

$I_{removed}$ is the second moment of area of a beam based on k_d layers with full adhesion. This is subtracted from I_{LC} in order to be replaced by I_{db}, which is the second moment of area of the same number of plies, but calculated based on no adhesion between layers.

The second moment of area of the area to be removed, $I_{removed}$, can be determined from the fraction of the total cross sectional area occupied by the area to be removed: $I_{removed} = \frac{y_d}{d/2} I_{k\_lam}$. The derivation of I_{k_lam} is part of Equation 6, and was shown in the previous section. With $y_d = \frac{dS}{2}$ the second moment of area of the removed section becomes:

$$I_{removed} = S I_{k\_lam}.$$

The second moment of area for the section with no adhesion between layers is determined from the second moment of area of each individual layer: $I_{db} = I_{dC}\left(\frac{\Gamma_{dbm}}{R} + \Gamma_{dbr}\right)$. I_{dC} is the second moment of area of a homogeneous beam section with dimensions: b = width and y_d depth. I_{dC} can also be expressed as: $I_{dC} = S I_C$ with I_C the second moment of area of a homogeneous beam with dimensions as defined before. The laminate composite modulus, E is then determined from $E_C I_C = E_L I_{LC}$, using $E_c = E_r$ and a modulus ratio, $R = \frac{E_m}{E_r}$ (transformed section method for composite beams [5]):

$$E_{pd} = (1-S)[E_m(\Gamma_m + \Lambda_m) + E_r(\Gamma_r + \Lambda_r)] + S[E_m \Gamma_{dbm} + E_r \Gamma_{dbr}] \qquad (13)$$

Based on the magnitude of interfacial adhesion, Equation 13 will then predict values intermediate of the full and no adhesion model. If the adhesion factor, f, is increased the model approaches the perfect adhesion case, as expected. It is worth noting that the model does not predict the composite modulus to be equal to that of pure reinforcement at $v_f = 1$. This is expected since there will only be partial adhesion between the reinforcement layers, i.e. the material is not monolithic.

4. EXPERIMENTAL

The composite test samples were produced using a Dr Boy 15kN injection moulder with digital controller and a PA20 die set, machined to produce suitable test specimen pieces with dimensions of 12.5 x 3.2 x 118mm. The feed stock materials for the lab based test specimens was manually batched from separate components, extruded in a Prism 24:1 twin screw extruder and then granulated using a Castin tri-blade. The composite components (Table 3) were linear low density polyethylene (LLDPE) or meleic anhydride modified LLDPE matrix and various volume fractions of glass fibre, wood fibre or talc. Five specimens of each sample were tested using a Lloyd LR30N universal testing instrument. A 500N load cell was used and a load rate of 5mm/min.

Table 3. Material properties

Material	Young's Modulus	Aspect ratio	Source
Glass fibre shorts	78 GPa	~460	Dow Corning
Radiata wood fibre	15.0 GPa	~20 to 100	Kawarau pulp mill (NZ)
Talc	43 GPa	< 20	
LLDPE	0.19 GPa	n/a	Du Pont
Functional LLDPE	0.4 GPa	n/a	Du Pont

Specimens were tested according to the appropriate ASTM standards for:

- 3-point bending (ASTM D790), using a variable span three point bend apparatus, at a span-to-depth ratio of either 16 or 32.
- 4-point bending (ASTM D6272), using a third of the span length for the load span, and a span-to-depth ratio of ether 16 or 32.
- Young's modulus (ASTM D638), with a gauge length of 50mm.
- Compression modulus (ASTM D695)

5. RESULTS AND DISCUSSION

The mechanical properties of the composites studied highlighted four main aspects, and will be discussed accordingly:

- the difference between the various modulus values,
- the effect of interfacial adhesion,
- the correlation with the partial adhesion model and
- the geometry of the reinforcement phase.

The Flexural Modulus

The mechanical properties of the wood fibre and glass fibre composites are shown in

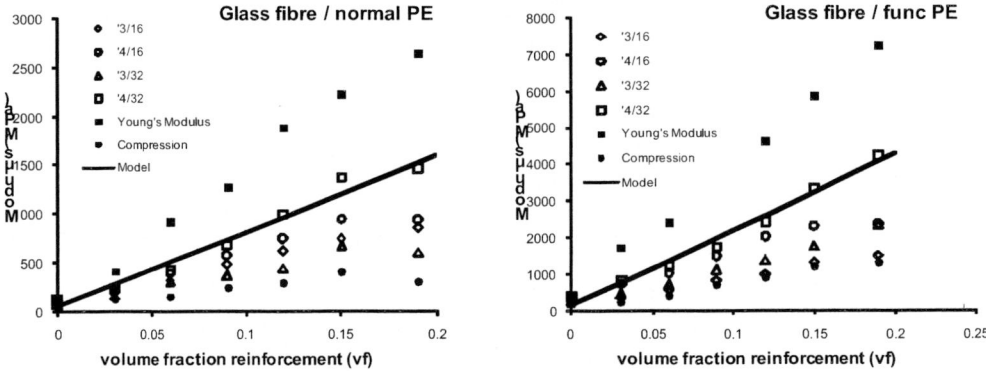

Figure 7 and Figure 8. If shear deformation were to influence the flexural modulus, one would expect 3-point bending to yield a lower flexural modulus than 4-point bending because the 4-point bending beam is partially under pure shear, as explained earlier. Additionally, due to shear deformation, a short span-to-depth ratio would further reduce the measured flexural modulus. From

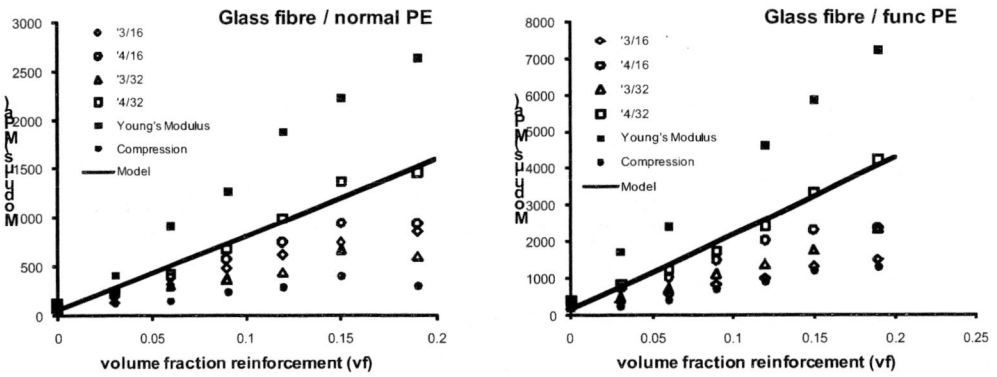

Figure 7 and Figure 8 it can be seen that every flexural test resulted in a different flexural modulus value, the order of which is consistent with what was expected:

4-point, L/D = 32 > 4-point, L/D = 16 > 3-point, L/D = 32 > 3-point, L/D = 16

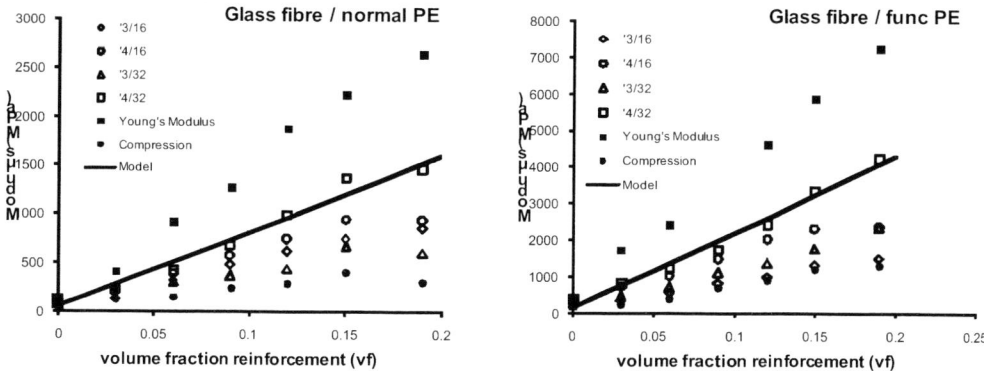

Figure 7. Elastic modulus for glass fibre reinforced LLDPE and functionalised LLDPE. Model curves shown for f = 0.22 and f = 0.53 for normal and functional PE, respectively using k = 5.

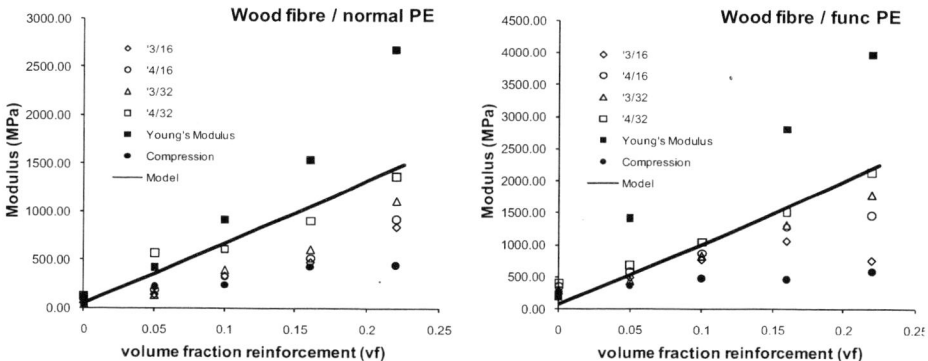

Figure 8. Elastic modulus for wood fibre reinforced LLDPE and functionalised LLDPE. Model curves shown for f = 0.73 and f = 0.92 for normal and functional PE, respectively using k = 5.

However, the difference is small at low volume fractions reinforcement. Increasing the fibre volume fraction amplifies the effect of non-uniform stress distribution leading to shear deformation. Larger differences between the various modulus values are therefore observed for wood and glass fibre composites at high volume fraction.

Also evident from the figures is that all the measured flexural modulus values were well below the tensile modulus, and were more pronounced when using functionalised polyethylene. In addition, the compressive modulus was consistently lower than any of the measured flexural modulus values, resulting in the flexural modulus being intermediate of those. This is consistent to what has previously been observed regarding composite materials, regarding the difference between compressive modulus and tensile modulus [1]. The same was observed for both functional and standard LLDPE indicating that this was independent of the degree of interfacial adhesion. It would therefore seem that the composites are more efficient at load sharing between matrix and reinforcement under tension than compression, leading to a lower compressive modulus.

Jones proposed a method to determine the flexural modulus, based on the compressive and tensile modulus [1]. In Figure 9, the calculated flexural modulus for the glass fibre reinforced system is shown. It can be seen that this model predicted values that were intermediate to that of the measured flexural modulus, but that it followed the same trend. It highlights the importance of the geometry of the reinforcement, especially for the fibrous materials. When aligned parallel to the applied force, fibres may not be able to resist as much of the compressive force as can be expected from a tensile load. The decrease in compressive modulus at higher volume fractions may also be a result of void space, possibly exacerbating the tendency for the fibres to buckle under the compressive load. Under a flexural load, the material experiences a combination of tensile and compressive forces, and hence the averaged result, as seen in Figure 9.

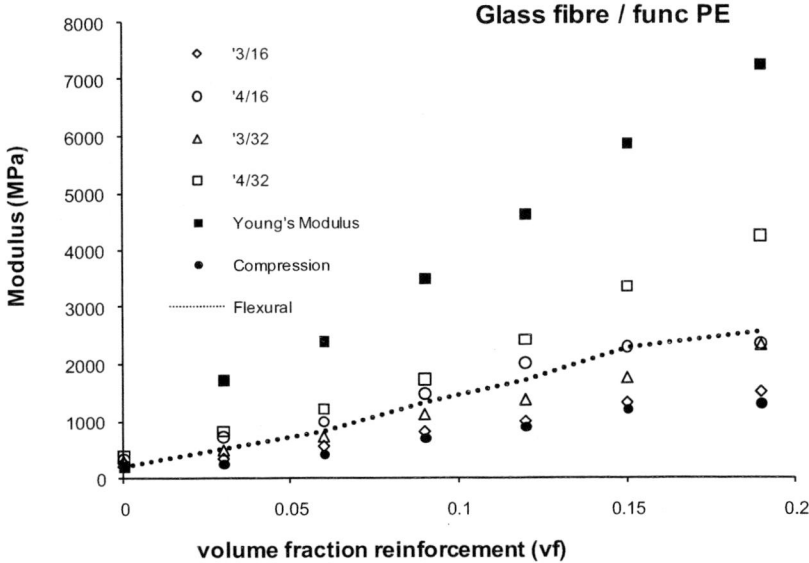

Figure 9. Calculated flexural modulus, based on the work of Jones [1].

It is noteworthy to point out that at relatively higher volume fractions fibre, the flexural modulus values all seem to flatten off with increasing volume fraction fibre. This was observed for glass and wood fibre composites, regardless the matrix type. It is most likely due to void formation at high volume fraction fibre. The flexural modulus is proportional to the second moment of area of the structure, therefore, void space between the fibres would result in a lower second moment of area because of the lack of adhesion. This would not be the case for perfect adhesion in which case the situation may be similar to having a stiff material with a flexible core or a sandwich beam. The effect of void formation is not as pronounced in the Young's modulus in this range of volume fractions, but also become important as the maximum packing fraction of the reinforcement is reached [14].

Interfacial Adhesion

Because of the inherent different modulus between standard LLDPE and functional LLDPE, it is not possible to say if the increase in modulus is only due to improved adhesion. However, for wood and glass fibre, the increase in modulus with increasing volume fraction fibre is very significant and the data would therefore suggest that improved adhesion is resulting in higher modulus values. This is not unexpected and corresponds well with general knowledge in the field.

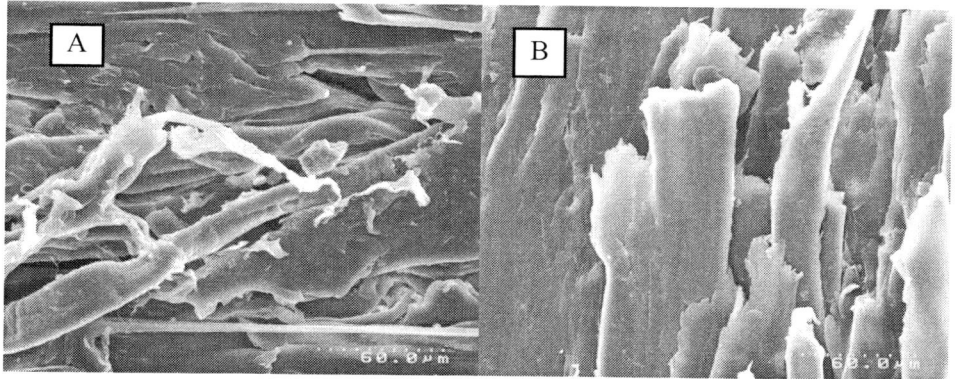

Figure 10. Scanning electron micrograph of 30 wt% wood fibre reinforced A) LLDPE and B) functional LLDPE, illustrating the degree of adhesion between the fibres and matrix in each case.

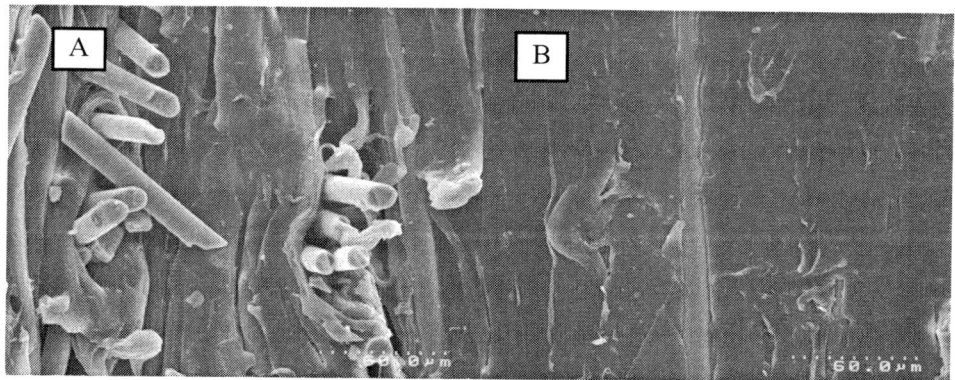

Figure 11. Scanning electron micrograph of 30 wt% glass fibre reinforced A) LLDPE and B) Functional LLDPE, illustrating the degree of adhesion between the fibres and matrix in each case.

A fracture surface of 30 wt% wood fibre reinforced LLDPE is shown in Figure 10. It was concluded that adhesion was improved using functional LLDPE, as evident from broken fibres and little separation between matrix and fibre. This is consistent to what is expected regarding the improved bonding between LLDPE and wood fibre when functional groups are introduced on the polymer backbone.

The difference in adhesion is also clearly visible from the SEM micrographs for glass fibre reinforced LDPE, as shown in Figure 11. Distinct debonded regions are visible for standard LLDPE, in contrast to much less pronounced fibre-matrix separation when functional LLDPE is used.

Also highlighted from these results is the effect of poor adhesion on the flexural modulus of the composites. Because of poor adhesion, there may be areas of debonding between the fibres and matrix. When considering the second moment of area of the fibres and matrix around the beams neutral axis, the lack of adhesion would result in a much lower modulus.

Poor adhesion that results in debonded regions can also be seen as a composite structure of air, fibre and matrix or in other words, a foam. The result is a reduction in flexural modulus, as seen from the experimental and modelled data. This was successfully modelled using the partial adhesion model, presented in this paper.

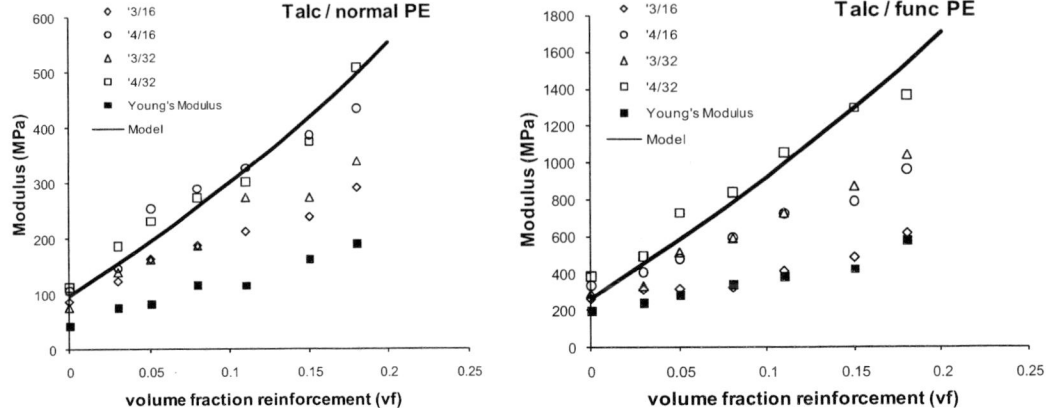

Figure 12. Elastic modulus for talc filled LLDPE and functionalised LLDPE.

Comparison with the Partial Adhesion Model

Flow during injection moulding leads to orientation of fibres in the direction of flow. A layered structure is formed and is clearly visible from the SEM micro-graph of wood fibre reinforced LLDPE (Figure 13) as well as for glass fibre reinforced LLDPE (Figure 14) and is consistent with the layered structure in the proposed model.

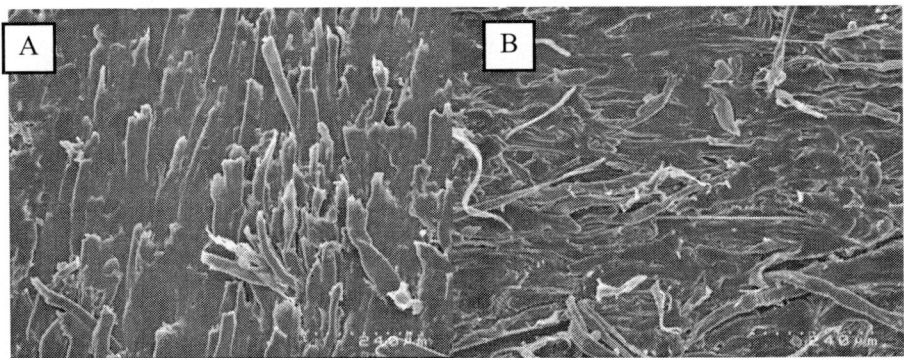

Figure 13. Scanning electron micrograph of 30 wt% wood fibre reinforced A) LLDPE and B) functional LLDPE, illustrating the fibre alignment in each case

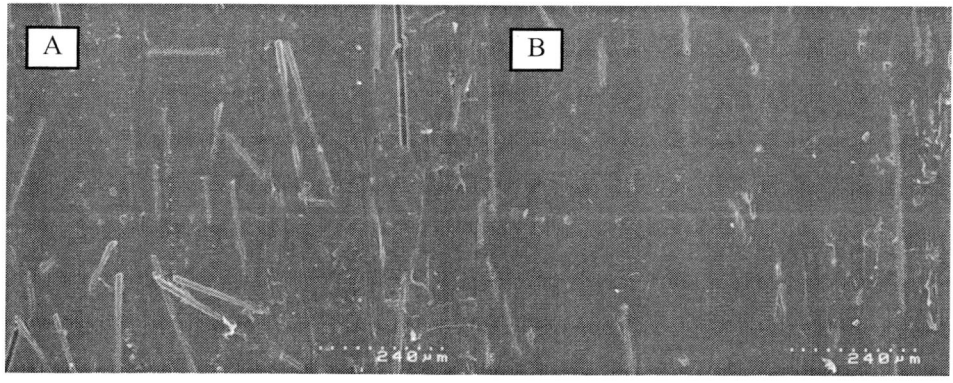

Figure 14. Scanning electron micrograph of 30 wt% glass fibre reinforced A) LLDPE and B) Functional LLDPE, illustrating the fibre alignment in each case

The partial adhesion model is in good agreement with the experimental data when a large span-to-depth ratio is used with 4-point bending

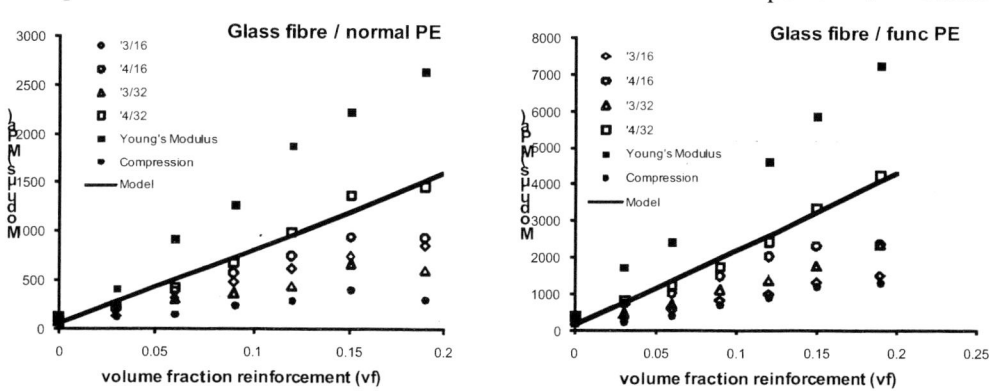

(Figure 7 and Figure 8), even though the model was derived assuming a point load on a simply supported beam and shear deformation wasn't considered. The 4-point bending results, at $L/D = 32$, most likely shows the least shear deformation and therefore are closer to the predicted modulus values.

For wood fibre reinforced composites, using 5 plies in the partial adhesion model (k = 5), good agreement is seen with the experimental results when the adhesion coefficient is set at 0.73 for standard LLDPE and 0.92 for the functionalised LLDPE. For glass fibre, best agreement is seen using 5 plies and a adhesion coefficient of 0.22 and 0.52 respectively. Poor adhesion between glass fibre and LLDPE is confirmed by this low adhesion coefficient as well as the distinct difference between the normal and functionalised LLDPE.

It was concluded from this work that fewer number of plies (k) in the model element is more representative than a large number of plies. The number of plies strongly affects the predicted flexural modulus. Increasing the number of plies essentially homogenises the modelled material and it can easily be shown from Equation 5 and Equation 8 that that the model becomes the same as the isostrain model or the linear mixing rule. The only difference would be that the flexural modulus is scaled for the adhesion coefficient in the partial adhesion model. At this point, the geometry of the reinforcement becomes irrelevant and the flexural modulus only depends on the volume fraction reinforcement.

Reinforcement Geometry

The situation changes when talc is used as reinforcement. Glass and wood have a large aspect ratio compared to talc. For glass fibre the aspect ratio is about 460, for wood fibre about 100 and for talc it is would be less than 20. The result is that the relationship between flexural and Young's modulus is reversed compared to that of glass and wood fibre composites. In this case, the flexural modulus is higher than the tensile modulus at all volume fractions reinforcement, as shown in Figure 12. However, the relationship between 3-point and 4-point bending at different span-to-depth ratios are the same. This suggests that shear deformation may be the most important factor contributing to the differences between the various testing techniques in bending.

One of the most significant differences between talc and fibre composites would be the aspect ratio of the reinforcement phase. In bending, stiffness is proportional to the arrangement or geometry of the reinforcement phase (Figure 3). For the fibre composites, the fibres are preferentially aligned perpendicular to the applied load. In that direction, an individual fibre's contribution to the flexural modulus becomes a function of the relative distance to the neutral axis. Because of the large aspect ratio, the orientation of the fibre in the matrix is crucial.

For Young's modulus, it can therefore be expected that the stiffness be the highest if the fibres are aligned in the direction of the applied load. From the experimental data

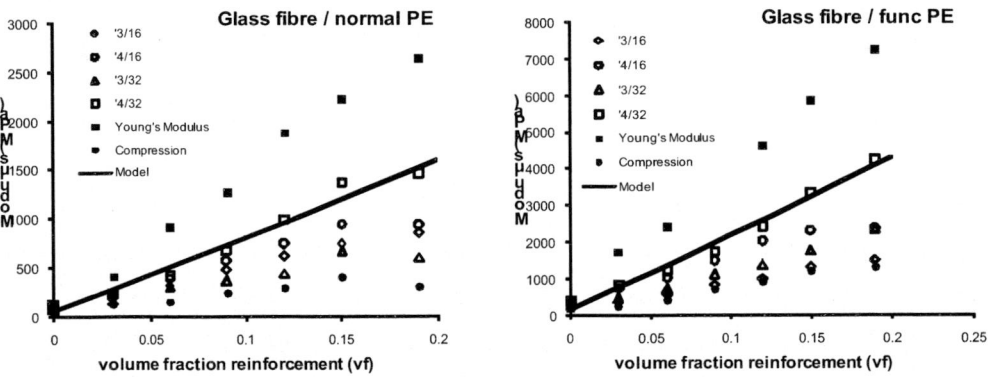

(Figure 7 and Figure 8) it was clearly seen that the Young's modulus was consistently higher than the flexural modulus. However, for talc (low aspect ratio), the directionality associated with the fibres no longer plays a role and therefore the Young's modulus was seen to be lower than the flexural modulus (Figure 12). This result is expected in light of the second moment of area. In this case, because particles are used as reinforcement, the stiffness would be similar irrespective of loading direction.

In Figure 15, SEM images are shown for 22.5 wt% talc in standard and functionalised LLDPE. Judging by the absence of any debonded areas surrounding the particles it can be concluded that the adhesion between the two phases must be reasonably good. However, when the partial adhesion model is applied to this system, using a modulus ratio of 55, the adhesion coefficient must be set as low as 0.15 for the standard LLDPE and 0.44 for the functionalised system. These values seem excessively low for a system that is displaying good adhesion and is in contradiction to the glass and wood fibre systems. This is most likely due to the huge difference between the aspect ratios of the reinforcing materials, making the model inapplicable to the talc system. The more isotropic nature of particulate reinforced systems is therefore makes them inappropriate to be modelled in a similar manner than the fibrous systems.

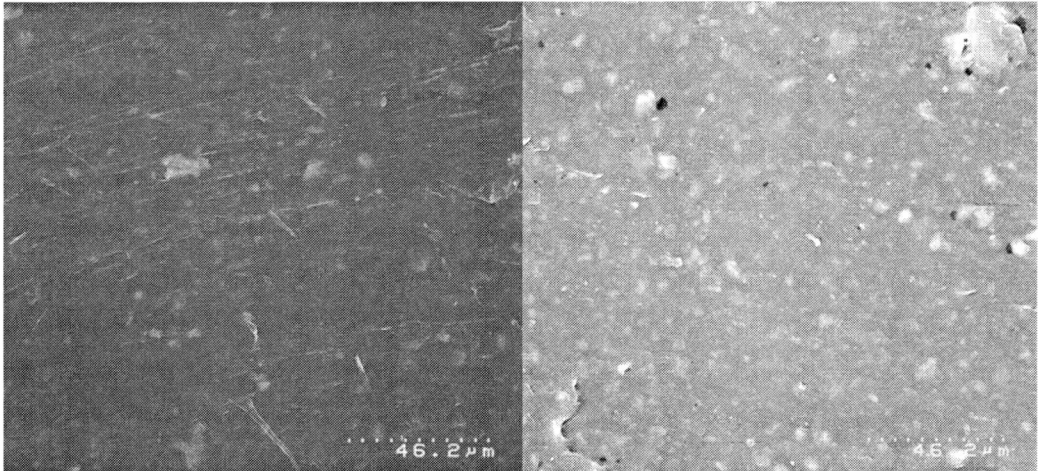

Figure 15. Scanning electron micrograph of 22.5 wt% talc filled A) LLDPE and B) Functional LLDPE, showing good adhesion in both cases

6. CONCLUSIONS

It was concluded that because a loaded beam is under a mixed stress state of tension and compression, the measured flexural modulus values will be intermediate of the tensile and compressive modulus. It was found that for wood and glass reinforced systems, the differences were small at low volume fractions reinforcement and independent of the degree of interfacial adhesion, at any volume fraction fibre. At high fibre volume fractions, it was assumed that non-uniform stress distribution lead to shear deformation, causing greater differences between the various modulus values, exacerbated by debonding and inefficient fibre packing.

It was found that the partial adhesion model was in good agreement with the experimental data when a large span-to-depth ratio was used during four-point bending in order to minimise the effect of shear deformation. It was concluded that the assumed layered structure was in good agreement with experimentally observed micro-structures, caused by flow induced alignment during injection moulding. It was concluded that the number of layers strongly affects the predicted flexural modulus and that fewer number of layers in the model element is more representative of the material. The geometry of the reinforcement became irrelevant at a larger number of layers, leading to predictions similar to the isostrain model, scaled with the adhesion coefficient.

It was concluded that the opposite relationship between flexural and Young's modulus for talc-filled composites was mainly due to the large difference between aspect ratio of the materials. In bending, stiffness is proportional to the arrangement or geometry of the reinforcement phase and an individual fibre's contribution to the flexural modulus is a function of the relative distance to the neutral axis. The orientation of low aspect ratio fillers is therefore not important with respect to the flexural modulus, however, the difference between the various flexural tests was found to be consistent regardless reinforcement geometry. It was concluded that shear deformation may be the most important factor contributing to the difference between these.

Lastly, because of the lack of orientation and appropriate geometry, the partial adhesion model was found to be ineffective for predicting the flexural modulus of a talc-filled composite. It was concluded that the more isotropic nature of particulate reinforced systems made it inappropriate to be modelled in a similar manner than the fibrous systems.

REFERENCES

[1] Jones, R., *Apparent Flexural Modulus and Strength of Multimodulus Materials.* Journal of Composite Materials, 1976. **10**: p. 342-354.

[2] Mujika, F., et al., *Determination of tensile and compressive moduli by flexural tests.* Polymer testing, 2006. **25**: p. 766-771.

[3] Tolf, G. and P. Clarin, *Comparison between flexural and tensile modulus of fibre composites.* Fibre Science and Technology, 1984. **21**(4): p. 319.

[4] Bos, H.L., J. Mussig, and M.J.A. van den Oever, *Mechanical properties of short-flax-fibre reinforced compounds.* Composites Part A-Applied Science And Manufacturing, 2006. **37**(10): p. 1591-1604.

[5] Timoshenko, S., *Strength of Materials, Part 1: Elementary Theory and Problems.* 3rd edition ed. 2004: S.K. Jain for CBS Publishers and Distributors.

[6] Mujika, F., *On the difference between flexural moduli obtained by three-point and four-point bending tests.* Polymer testing, 2006. **25**: p. 214-220.

[7] Brancheriau, L., H. Bailleres, and D. Guitard, *Camparison between modulus of elasticity values calculated using 3 and 4 point bending tests on wooden samples.* Wood Science and Technology, 2002. **36**: p. 367-383.

[8] Stevanovic, M. and D. Sekulic, *Macromechanical characteristics deduced from three-point flexure tests on unidirectional carbon/epoxy composites.* Mechanics of Composite Materials, 2003. **39**(5): p. 387-392.

[9] Cahn, R., P. Haasen, and E. Kramer, *Materials science and technology : a comprehensive treatment.* Vol. 13. 1986: Weinheim ; Cambridge : VCH.

[10] 1Thomason, J.L., *Micromechanical parameters from macromechanical measurements on glass reinforced polypropylene.* Composites Science And Technology, 2002. **62**(10-11): p. 1455-1468.

[11] Rozman, H.D., et al., *Polypropylene-oil palm empty fruit bunch-glass fibre hybrid composites: a preliminary study on the flexural and tensile properties.* European Polymer Journal, 2001. **37**(6): p. 1283.

[12] Ewart, P.D. and C.J.R. Verbeek. *Prediction of the Flexural Modulus of Composite Materials for Sporting Equipment.* in *Asia-Pacific Conference on Sports Technology.* 2005. Tokyo, Japan: Australian Sports Technology Alliance Pty Ltd.

[13] Cox, H., *The Elasticity and Strength of Paper and other Fibrous Materials.* British Journal of Applied Physics, 1952. **3**: p. 72-79.

[14] Verbeek, C.J.R. and W.W. Focke, *Modelling the Young's modulus of platelet reinforced thermoplastic sheet composites.* Composites Part A-Applied Science and Manufacturing, 2002. **33**: p. 1697–1704.

In: Structural Materials and Engineering
Editor: Ference H. Hagy

ISBN 978-1-60692-927-8
© 2009 Nova Science Publishers, Inc.

Chapter 5

PROBABILISTIC MODELING OF CLEAVAGE FRACTURE IN THE DUCTILE-TO-BRITTLE TRANSITION REGION

Xiaosheng Gao

Department of Mechanical Engineering, University of Akron, Akron, OH 44325

ABSTRACT

Transgranular cleavage fracture in the ductile-to-brittle (DBT) region is of a statistics nature and follows a weakest link mechanism. The Weibull stress model, derived from the failure mechanism, provides a framework to quantify the relationship between the macro and micro-scale driving forces for cleavage fracture. It has been successfully applied to predict the constraint effect on cleavage fracture and the scatter of macroscopic fracture toughness values. This chapter summarizes the research work conducted by the author and his co-workers in recent years to extend the engineering applicability of the Weibull stress model to predict cleavage fracture in structural steels. These recent efforts have introduced a threshold parameter in the Weibull stress model, developed more robust calibration methods for determination of model parameters, predicted experimentally observed constraint effect and toughness scatter, demonstrated temperature and loading rate effects on model parameters, and expanded the Weibull stress model to include the effects of plastic strain and microcrack nucleation.

INTRODUCTION

Cleavage fracture in structural steels can be simplified as a two-step process: cracking of grain boundary carbides to form microscopic cracks followed by unstable propagation of these microcracks into the surrounding ferrite matrix [1, 2]. It has been recognized that formation of carbide cracks is associated with slip or twinning dislocation pile-ups [3], indicating cleavage fracture has to be preceded by plastic flow. These carbide cracks, under sufficient applied stresses, lead to grain size microcracks in the ferrite [1] or complete failure of the structure. Fractographical examinations reveal that cleavage fracture in the ductile-to-brittle transition (DBT) region can often be traced back to the propagation of a single microcrack [4, 5], suggesting a weakest link mechanism. Due to the stochastic effect of microcrack distribution and the nonlinear response of plastic deformation, the cleavage fracture toughness exhibits a large amount of scatter, a dependence on the crack front length, and a strong sensitivity to the local stress and deformation fields [4, 6]. These issues present major obstacles in application of elastic-plastic fracture mechanics principles to assess the integrity of structural components, which in turn have stimulated an increasing amount of research in the past two decades. These research efforts have led to a quantitative understanding of the scatter and temperature dependence of macroscopic fracture toughness (in terms of J_c or K_{Jc}) under high constraint, small scale yielding (SSY) conditions, where K denotes the stress intensity factor, J denotes the J-integral, and K_J is the value of K converted from J based on the relationship between them under the elastic, plane strain condition. Scatter of the SSY toughness data can be described by a three-parameter Weibull distribution, where the Weibull modulus for the K_{Jc} distribution is 4 (or 2 for J_c distribution) and the minimum fracture toughness for common ferritic steels is $K_{min} \approx 20 \text{MPa}\sqrt{m}$ [4, 7]. This three-parameter Weibull distribution has been adopted in ASTM standard E1921 [8]. E1921 also adopts a so-called "Master Curve", empirically derived by Wallin and others [9, 10], to describe the dependence of the median fracture toughness on temperature for ferritic steels in the DBT region.

The approach described in E1921 [8] characterizes the cleavage fracture toughness under the high constraint, SSY conditions. But in engineering applications the crack front often experiences constraint loss. This motivates the development of micromechanics-based models to address the transferability of cleavage fracture toughness across varying levels of crack-front constraint. The Weibull stress model, originally proposed by the Beremin group [11], provides a framework for quantifying the relationship between macro-scale and micro-scale driving forces for cleavage fracture. The introduction of the so-called Weibull stress (σ_w) provides the basis for generalizing the concept of a probabilistic fracture parameter and supports the development of procedures that adjust toughness values across different crack configurations and loading modes (tension vs. bending). This model does not limit the extent of plastic deformation and has been successfully applied to predict the constraint effect on cleavage fracture and the scatter of the macroscopic fracture toughness values [12 - 15]. Combined with the ductile crack growth models, it can also predict the effect of prior ductile tearing on cleavage fracture in the upper transition region [16 - 18].

This chapter is not intended to provide an extensive literature review on the probabilistic models for cleavage fracture (interested readers are referred to a recent review article by Pineau [19] for more references on this subject). The focus here is to provide a brief summary of the research work conducted by the author and his co-workers in recent years to extend the engineering applicability of the Weibull stress model. Following a brief description of the original Beremin model [11], issues discussed in this chapter include development of the Weibull stress based toughness scaling model, introduction of a threshold parameter in the Weibull stress model, development of more robust calibration methods for determination of model parameters, prediction of the experimentally observed constraint effect and scatter of fracture toughness data, effects of temperature and loading rate on model parameters, and expansion of the Weibull stress model to include the effects of plastic strain and microcrack nucleation.

THE WEIBULL STRESS MODEL

Cleavage fracture initiation depends on the probability of finding a critical microcrack in the stressed volume. Assuming (i) carbides fracture to form microcracks when the surrounding ferrite matrix undergoes plastic deformation, (ii) the probability density function for microcracks with size l is proportional to $(1/l)^\gamma$, where γ is a material constant, and (iii) propagation of microcracks is governed by the Griffith criterion [20], Beremin [11] derived a two-parameter Weibull distribution to describe the cumulative failure probability based on the weakest link statistics

$$P_f(\sigma_w) = 1 - \exp\left[-\left(\frac{\sigma_w}{\sigma_u}\right)^m\right] \qquad (1)$$

In the above expression, σ_w represents the Weibull stress defined as the integral of a weighted value of the maximum principal (tensile) stress (σ_1) over the process zone for cleavage fracture (i.e., the plastic zone)

$$\sigma_w = \left[\frac{1}{V_0}\int_{V_p}\sigma_1^m dV\right]^{1/m} \qquad (2)$$

where V_p represents the volume of the fracture process zone, V_0 is a reference volume and m denotes the Weibull modulus which relates to the size distribution of the microcracks, $m = 2\gamma - 2$. In Eq. (1), σ_u represents the scale parameter of the Weibull distribution which defines the micro scale material toughness corresponding to 63.2% cumulative failure probability.

In the context of probabilistic fracture mechanics, the Weibull stress, σ_w, thus emerges as a near-tip fracture parameter to describe the coupling of remote loading with a micromechanics model which incorporates the statistics of microcracks. The Weibull stress enables construction of a toughness scaling model between different crack configurations based on equal failure probability. Under increased remote loading described by J or K_J, differences in evolution of the Weibull stress, $\sigma_w(J)$, reflect the potentially strong variations in crack-front stress fields due to the effects of constraint loss and volume sampling. The inherently 3D formulation for σ_w defined by Eq. (2) readily accommodates variations in J or K_J along the crack front. Constraint *corrections* then become possible between different crack geometries and/or applied loading conditions at identical σ_w-values. Figure 1 illustrates schematically the toughness scaling between a test specimen (or structural component) and the plane strain, SSY (*T*-stress = 0) configuration. To reach the same failure probability (i.e., σ_w-value), the specimen requires loading to $J_{specimen} > J_{SSY}$, reflecting the effect of constraint loss, Fig. 1(a). Toughness scaling diagrams is constructed as shown in Fig. 1(b), where each pair of *J*-values ($J_{specimen}$ and J_{SSY}) produces the same σ_w-value. Given a set of measured toughness values, the constraint-corrected toughness values can be found from the toughness scaling diagram.

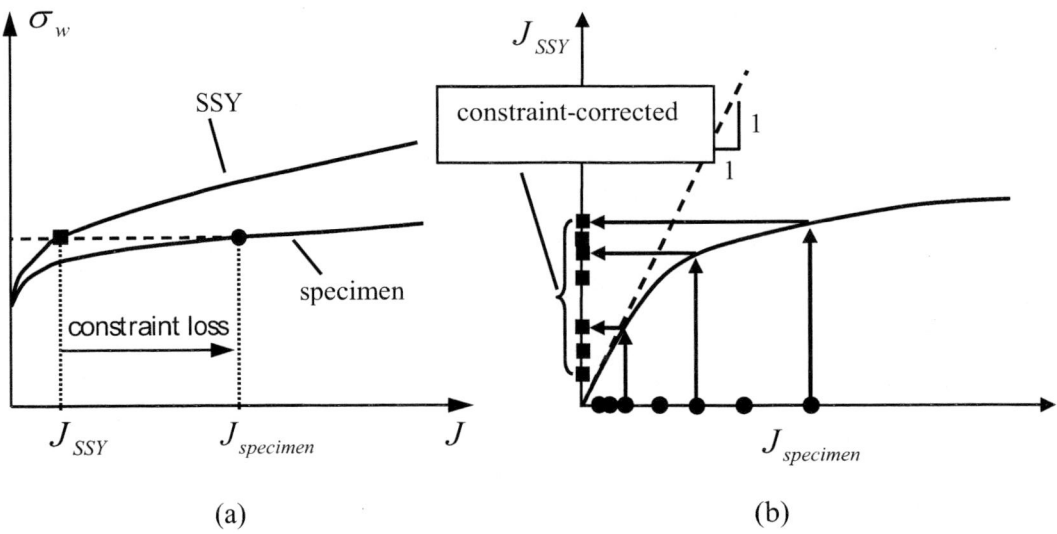

Figure 1. Illustration of toughness scaling between a test specimen and the plane strain, small scale yielding (SSY) configuration. (a) σ_w versus J curves showing the evolution of σ_w as J increases; (b) toughness scaling diagram.

Under plane-strain, SSY conditions with T-stress = 0, the relationship between the Weibull stress and J follows as

$$\sigma_w^m = \phi BJ^2 \qquad (3)$$

where ϕ denotes a material constant and B the crack-front length. This relationship holds due to the existence of self-similar fields with amplitudes depending only on J [14]. Constraint loss in specimens leads to the breakdown of the relationship in Eq. (3). Gao and Dodds [21] quantified the effects of constraint loss by introducing a non-dimensional constraint function, $g(M)$, and expanding Eq. (3) to

$$\sigma_w^m = \phi BJ^2 g(M) \qquad (4)$$

where M defines the non-dimensional deformation, $M = b\sigma_0/J$, σ_0 is the yield stress and b is the length of the remaining ligament. Figure 2 shows an example of $g(M)$ vs. M relationship for SE(B) specimens with $B = W$, $S = 4W$ and $a/W = 0.5$ and having material properties of $E/\sigma_0 = 500$, $v = 0.3$, $n = 10$ and $m = 20$, where E represents the Young's modulus, v the Poisson's ratio and n the strain hardening exponent. The value of $g(M)$ exceeds unity for $M > 200$ because of the positive T-stress in the deep-notch SE(B) specimen early in the loading. The value of $g(M)$ decreases as deformation increases, reflecting constraint loss in the SE(B) specimen.

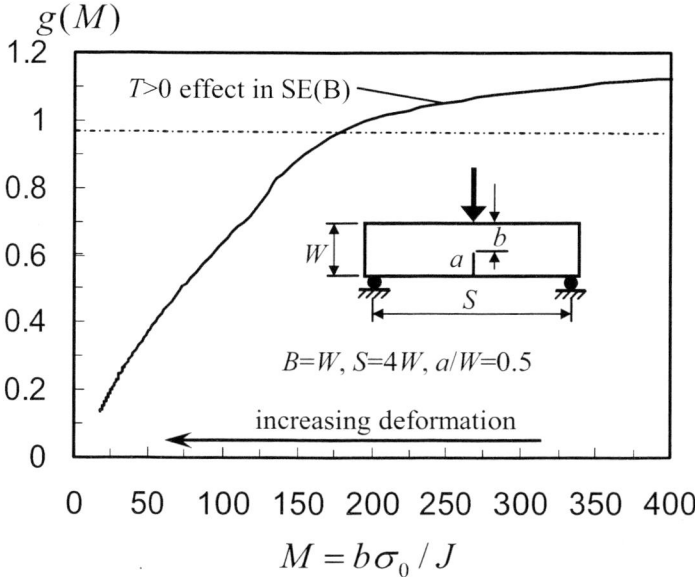

Figure 2. The non-dimensional constraint function, $g(M)$, for SE(B) specimens with $B = W$, $S = 4W$ and $a/W = 0.5$ and having material properties of $E/\sigma_0 = 500$, $v = 0.3$, $n = 10$ and $m = 20$ [21].

The g-function quantifies the constraint loss relative to the plane-strain, SSY (T-stress=0) configuration. It can also be used to compare constraint conditions between different types of specimens. Petti and Dodds [22, 23] analyzed the standard C(T) specimens, the square cross-section (BxB) SE(B) specimens and the rectangular cross-section (Bx$2B$) SE(B) specimens. The results show a clear constraint ranking. The C(T) specimen exhibits the highest constraint behavior, followed by the rectangular SE(B), with the square SE(B) being the lowest in constraint, consistent with experimental observations.

Under plane strain, SSY conditions, the non-sigular T-stress characterizes the effect of geometry (specimen size, crack configuration, etc.) and loading mode (bending vs. tension) on crack-tip fields. Negative values of T-stress strongly affect fracture toughness by lowering the crack-front stress triaxiality. Using the Weibull stress based toughness scaling model, Gao and Dodds [24] presented a simplified approach to parameterize the T-stress effect on the fracture toughness in the DBT region. The ratios of toughness values between $T \neq 0$ configurations and the reference $T = 0$ configuration were fitted into closed form functions of T and m for typical ferritic steels. These functions provide a practical tool to assess constraint effect on fracture toughness for ferritic steels in the DBT region under conditions of contained yielding.

CALIBRATION OF THE WEIBULL STRESS MODEL

The applicability of the Weibull stress model to predict cleavage fracture in structural components relies heavily on the calibrated model parameters. Minami et al. [13] first performed the calibration of Weibull stress parameters (m, σ_u) from measured fracture toughness values, although Mudry [25] had earlier suggested such an approach. In this approach, the value of Weibull stress at cleavage fracture is computed at each measured J_c-value with an assumed Weibull modulus (m). An iterative procedure then finds values of m and σ_u that minimize differences between rank probabilities of the test data and the cleavage probabilities given by Eq. (1). The least square and maximum likelihood methods represent two commonly used approaches within the iterative scheme to determine m and σ_u. Gao et al. [14] demonstrated analytically and numerically that a non-uniqueness arises in the calibrated pair (m, σ_u) when the calibration process uses only fracture toughness data (J_c or K_{Jc}) measured under the SSY conditions. This result explains the often broad inconsistencies in reported m-values calibrated using high constraint fracture specimens, such as deep-notch SE(B) or C(T) specimens, which fail virtually under SSY conditions.

Numerous experimental studies have shown that the cleavage fracture toughness data follow a Weibull distribution with a Weibull modulus of 2 in terms of J (or 4 in terms of K_J) under the plane strain, SSY conditions [4, 8]. The non-uniqueness of the calibrated Weibull stress parameters stems from the fact that regardless of the m-value, Eq. (1) is always equivalent to $P_f = 1 - \exp[-(J/\beta)^2]$ under the plane strain, SSY conditions [14], where β represent the fracture toughness value at the 63.2% failure probability. To overcome this inherent non-uniqueness issue, Gao et al. [14] and Ruggieri el al. [26] introduced a new procedure to calibrate m and σ_u using measured fracture toughness data. This calibration procedure utilizes the constraint difference between two sets of measured toughness values and the toughness scaling model discussed in the previous Section. One set of toughness data is obtained from high constraint specimens (e.g., deep-notch SE(B) or C(T) specimens) and the other set from low constraint specimens (e.g., shallow-notch SE(B) specimens). The calibrated Weibull stress parameters (m, σ_u) enable the two measured sets of toughness distributions to collapse to the same distribution of cumulative fracture probability when scaled to the reference SSY configuration.

The Gao-Ruggieri-Dodds calibration procedure has received increasing acceptance over the last few years. Wasiluk et al. [27] refined the Gao-Ruggieri-Dodds procedure by assigning a weight factor to each measured toughness value in defining the error function used in the calibration process. The parameters calibrated using the new error function lead to improved predictions of toughness values at the lower- and upper-tails of the toughness distribution. The following outlines the calibration procedure:

1) Test two sets of specimens that exhibit markedly different constraint levels at the same loading rate ($\dot{K}_J$) and a fixed temperature in the DBT region. Denote the high constraint dataset as HC and low constraint dataset as LC. The specimen geometries and test temperature should be selected to ensure different evolutions of constraint levels with no ductile tearing prior to cleavage.

2) Perform detailed, 3D finite element analyses for the tested specimens and the plane strain, SSY boundary layer model (T-stress = 0). The mesh refinement must be sufficient to ensure converged σ_w vs. J histories over the range of loading levels.

3) Assume an m-value and compute the σ_w vs. J history for the HC, LC and SSY configurations respectively. Constraint-correct the HC and LC toughness distributions to the SSY configuration.

4) Define an error function to characterize the difference between the two sets of constraint-corrected toughness data and evaluate this error function.
5) Repeat steps 3) and 4) for additional values of m. The calibrated m-value minimizes the error function.
6) After m is determined, σ_u takes the σ_w-value at $J = \beta$ in the SSY configuration, where evaluation of β uses the constraint-corrected HC (or LC) toughness data and follows the procedures described in E1921 [8].

THE THREE-PARAMETER WEIBULL STRESS MODEL

The two-parameter Weibull stress model, Eqs (1) and (2), describes an unconditional fracture probability, implying that any applied J (or K_J) yields a finite failure probability. However, the newly formed microcracks cannot propagate in polycrystalline metals unless sufficient energy exists to break the atomic bonds, to drive the crack across grain boundaries and to perform additional plastic work. Consequently, there exists a minimum toughness value (K_{min}) below which the microcracks arrest. A simplified, three-parameter Weibull distribution describing the macroscopic fracture toughness under SSY conditions has the form [7, 8]

$$P_f(K_I) = 1 - \exp\left[-\left(\frac{K_I - K_{min}}{K_0 - K_{min}}\right)^4\right] \quad (5)$$

where K_0 defines the fracture toughness value at the 63.2% failure probability and K_{min} represents the threshold toughness for the material. For common ferritic steels, K_{min} has an empirical value of $20 \text{MPa}\sqrt{m}$ (often assumed to be independent of temperature and loading rate) [8].

Therefore, Eqs (1) and (2) should be modified to include a threshold parameter. Some investigators [12, 16] have introduced a threshold stress directly into the Weibull stress computation so that the integrand of Eq. (2) becomes $(\sigma_1 - \sigma_{th})^m$. But determination of a physically meaningful value for σ_{th} remains an open issue. Moreover, the introduction of σ_{th} does not lead to a non-zero K_{min}.

Gao et al. [18] found the introduction of σ_{th} reduces the m-value and raises the σ_u-value but the predicted failure probability is not sensitive to the choice of σ_{th}. Figure 3 shows the variation of m and σ_u with σ_{th} for a low-strength, high-hardening pressure vessel steel (2.25Cr1Mo), where both σ_{th} and σ_u are normalized by the yield stress (σ_0). To check the effect of σ_{th} on the predicted failure probability, consider three σ_{th}-values, 0, $2\sigma_0$ and $4\sigma_0$. The corresponding values of (m, σ_u) are (11.08, 11.86σ_0), (6.5, 12.88σ_0) and (1.95, 109.24σ_0) respectively. Figure 4 compares the predicted P_f vs. J curves calculated using the three σ_{th}-values and their corresponding (m, σ_u) values for a 1T-C(T) specimen. The symbols in Fig. 4 represent the experimental data expressed in terms of measured fracture toughness (J_c) versus rank probabilities. It is clear different choice of σ_{th} does not result in noticeable difference in the predicted failure probability.

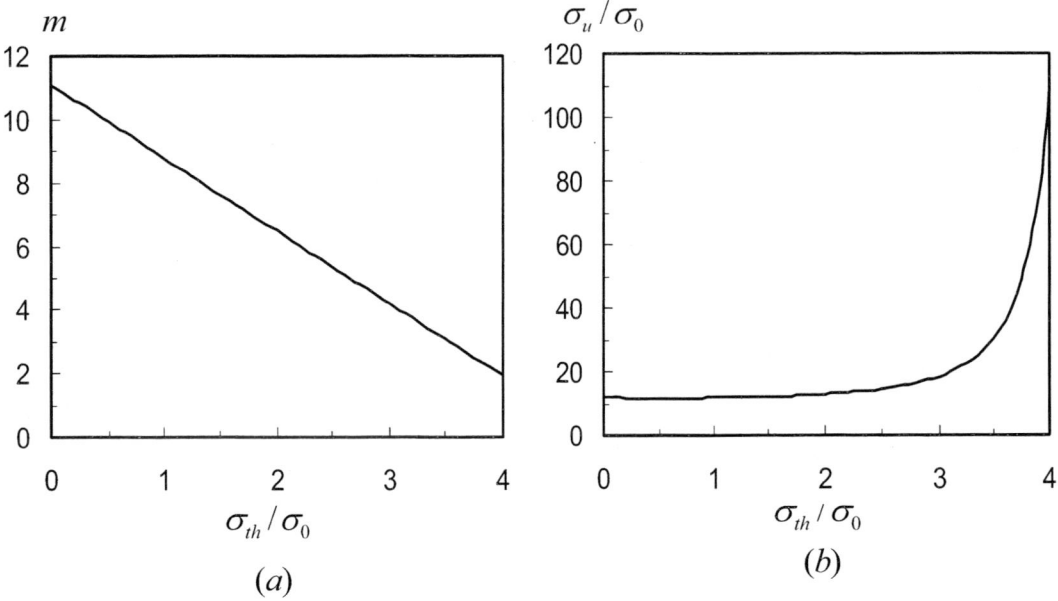

Figure 3. Variation of m and σ_u with σ_{th} for a low-strength, high-hardening pressure vessel steel (2.25Cr1Mo) [18].

Recognition of the shortcomings of the two-parameter Weibull stress model and the issues related to the introduction of σ_{th} led to Gao et al. [14, 21] to propose an alternative way to introduce a threshold parameter into the Weibull stress model. Similar to Eq. (5), Eqs (6) and (7) define two of the possible forms to include a threshold Weibull stress value, σ_{w-min}, in the failure probability expression

$$P_f(\sigma_w) = 1 - \exp\left[-\left(\frac{\sigma_w - \sigma_{w-min}}{\sigma_u - \sigma_{w-min}}\right)^m\right] \qquad (6)$$

or

$$P_f(\sigma_w) = 1 - \exp\left[-\left(\frac{\sigma_w^{m/4} - \sigma_{w-min}^{m/4}}{\sigma_u^{m/4} - \sigma_{w-min}^{m/4}}\right)^4\right] \qquad (7)$$

where σ_{w-min} represents the minimum σ_w-value at which macroscopic cleavage fracture becomes possible and is defined as the value of σ_w when $K_J = K_{min}$. Cleavage fracture cannot occur when $\sigma_w < \sigma_{w-min}$. Eq. (6) has a simpler form and predicts less toughness scatter than Eq. (7) does, but only Eq. (7) is consistent with Eq. (5) under plane strain, SSY conditions.

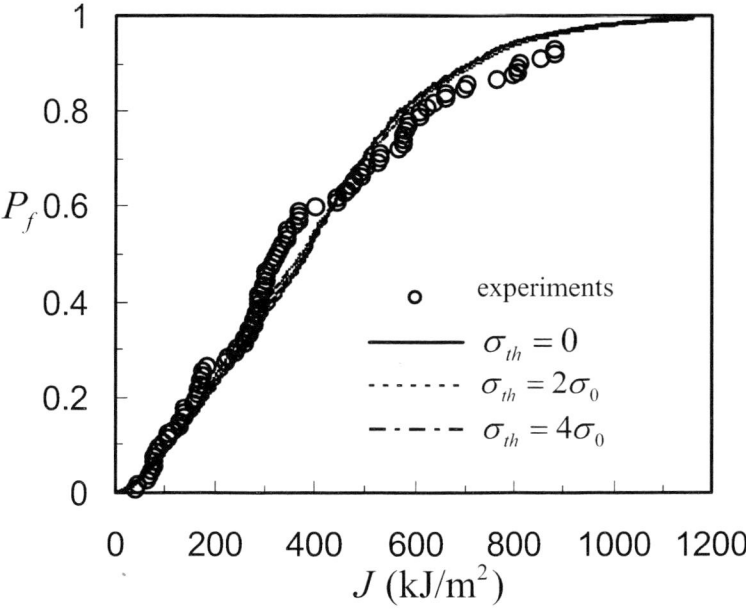

Figure 4. Comparison of the predicted P_f vs. J curves calculated using different σ_{th}-values and their corresponding (m, σ_u) values for a 1T-C(T) specimen, where the symbols represent the measured fracture toughness and the rank probabilities [18].

PREDICTION OF CLEAVAGE FRACTURE IN A PRESSURE VESSEL STEEL

The three-parameter Weibull stress model described above has been successfully applied to predict cleavage fracture in a wide range of fracture specimens of various ferritic steels. This Section describes an example of such applications. Tregoning, Joyce and Link conducted extensive fracture tests on an A515-70 pressure vessel steel in the DBT region [15, 28]. Their test program with quasi-static loading includes 12 plane-sided C(T) specimens ($a/W = 0.6$, $B = 25$ mm, $W = 50$ mm) tested at -28°C, 12 plane-sided SE(B) specimens ($a/W = 0.2$, $B = 25$ mm, $W = 50$ mm, $S = 4W$) tested at -7°C, 7 pin-loaded and 7 bolt-loaded surface crack specimens [SC(T)] tested at -7°C. Figure 5 shows the geometries of the surface crack specimens, where the cracks are semi-elliptical with a surface length of 19.05 mm and depth of 6.35 mm. The pin-loaded specimen subjects to a higher bending moment whereas the bolt-loaded specimen subjects to predominantly tensile loading. All specimens failed by cleavage without prior ductile tearing.

Following the procedure proposed by Gao et al. [14], toughness data of the C(T) and SE(B) specimens are used to calibrate the Weibull stress parameters. Since these two sets of fracture tests are conducted at different temperatures, the Master Curve approach of ASTM E-1921 [4] is employed to adjust the high-constraint C(T) toughness values for the temperature change. The calibrated m-value at -7°C is 11.2. Extracted from [15], Figure 6 compares the predicted failure probabilities using the calibrated three-parameter Weibull stress model with

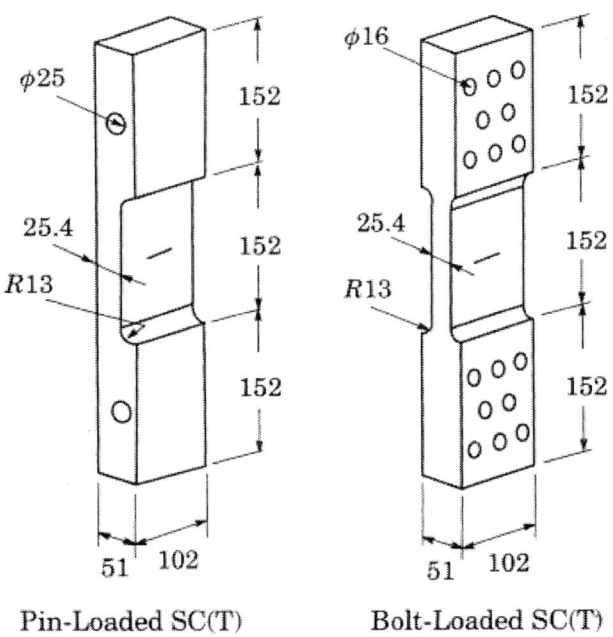

Figure 5. Geometries of the SC(T) specimens, where all dimensions are in mm.

the rank probabilities for measured J_c-values. For C(T) and SE(B) specimens, the through-thickness average J-values are used to characterize the intensity of far field loading on the crack front, while for SC(T) specimens, the J-values are computed at the center-plane (the deepest point on the crack front). In Fig.6, the solid lines represent model predictions, the symbols indicate median rank probabilities computed as $P_i = (i - 0.3)/(N + 0.4)$, where i denotes the rank number and N represents the total number of fracture tests [8], and the dashed lines denote 90% confidence limits for the estimates of rank probability of the experimental data. To compute these confidence limits, it is assumed that the (continuous) P_f-values from Eq. (6) provide the expected median rank probabilities for an experimental dataset containing the number of measured toughness values. Details of the procedures to compute these 90% confidence limits for the rank probabilities can be found in [15, 29]. Model predictions agree very well with experimental data and capture the strong constraint effect on fracture toughness due to specimen geometry, crack configuration and loading mode (bending vs. tension).

LOADING RATE EFFECT ON THE WEIBULL STRESS MODEL PARAMETERS

Defect assessments often must consider dynamic loading due to collision, sudden pressure rise, thermal shock, etc. It is, therefore, necessary to investigate the applicability of the Weibull stress model under dynamic loading. Joyce conducted an extensive fracture test program on a 22NiMoCr37 steel [30]. This material is similar to the A508 pressure vessel steel widely used in North American commercial nuclear power plants. In Joyce's program,

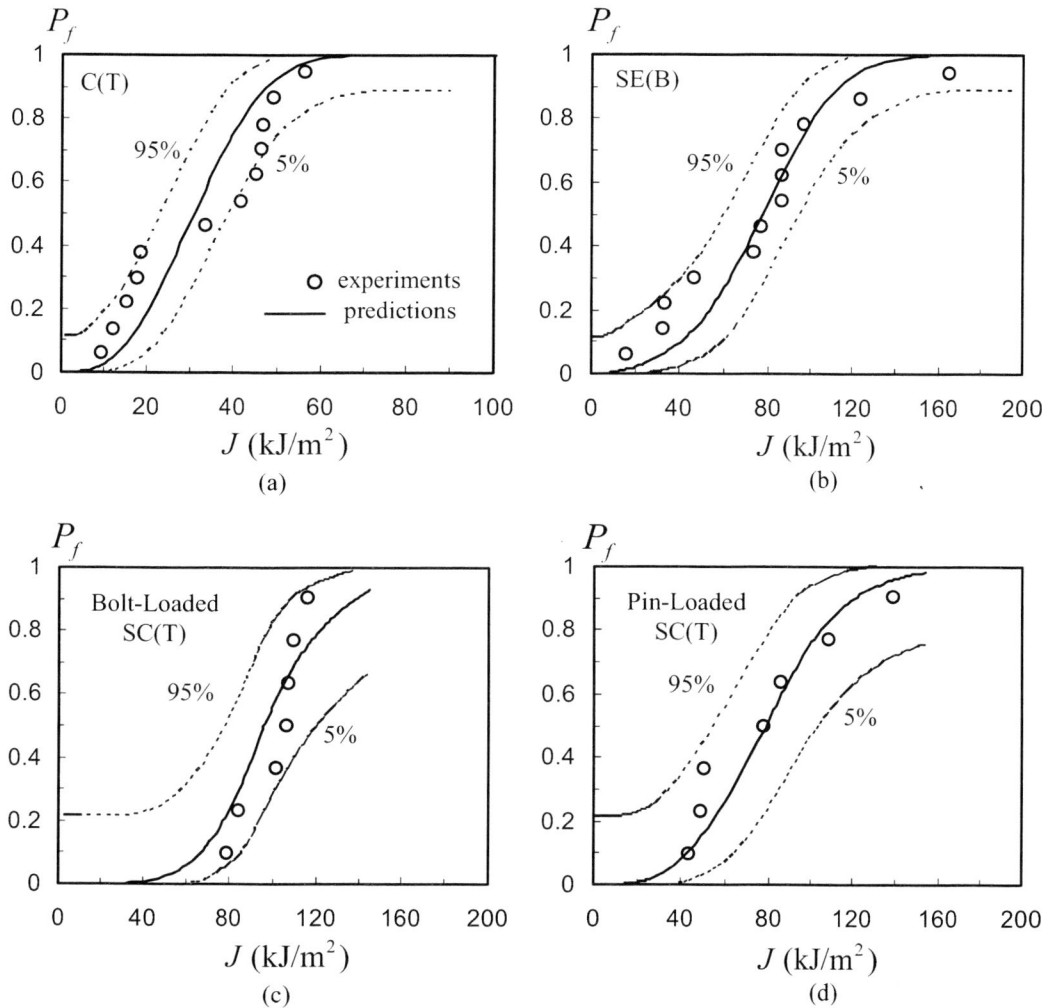

Figure 6. Comparison of the predicted failure probabilities with the rank probabilities for measured J_c-values [15].

all fracture toughness tests were conducted at -100°C using SE(B) specimens with the cracks oriented in the LS orientation. The shallow-cracked specimens have an a/W ratio of 0.1 – 0.15 and the deep-cracked specimens have an a/W ratio of 0.5 – 0.55. In addition to the quasi-static fracture tests, three dynamic loading rates (in terms of $\dot{K}_J$) in the low-to-moderate range are considered. The load-line velocities for the dynamic tests are chosen so that they lead to almost the same $\dot{K}_J$ value for the deep- and shallow-cracked cracked specimens at each given loading rate. To limit the possible effects due to material variability, all datasets were checked for material homogeneity using the procedure and the criteria proposed by Wallin et al. [31] and the material was found to be homogeneous for all cases except for the intermediate loading rate, shallow-cracked specimens.

Independent calibration at each of the four loading rates (quasi-static plus three dynamic rates) results in a very small variation in the m-value among the four loading rates. This small variation is due to the material variability and the uncertainties in the calibration process, e.g., the limited number of toughness values in the datasets. Therefore, it is concluded that m for this material is independent of loading rate and the calibrated m-value is 7.1. Figure 7 displays the variations of σ_{w-min} and σ_u with loading rate, where the values are normalized by the quasi-static yield stress (σ_0) for simplicity of presentation. Fig. 7 indicates that both σ_{w-min} and σ_u vary with loading rate: σ_{w-min} increases slightly with $\dot{K}_J$, which follows simply from the rate-sensitive material flow properties, while σ_u decreases with $\dot{K}_J$, reflecting the combined effects of loading rate on material flow properties and toughness.

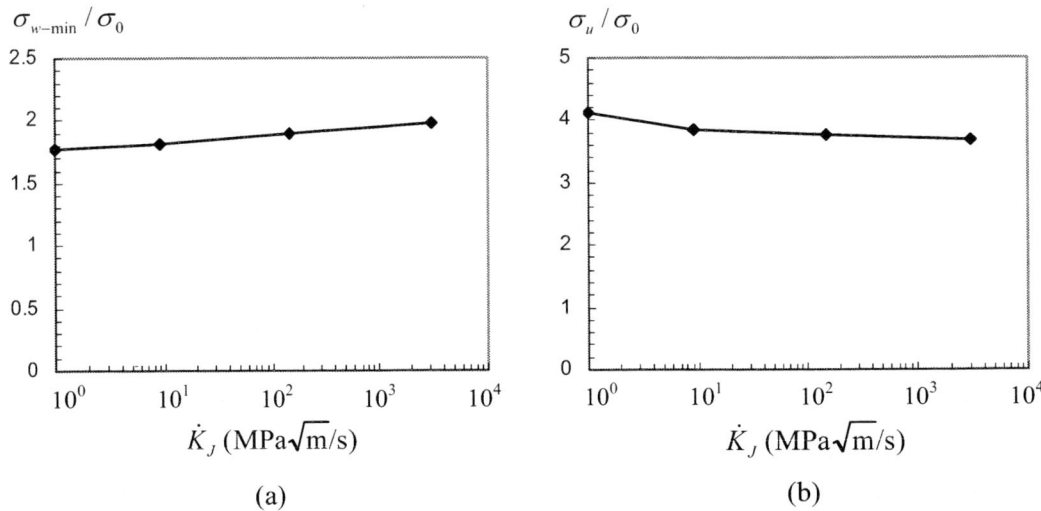

Figure 7. Variations of σ_{w-min} and σ_u with loading rate for the 22NiMoCr37 steel [30].

Figure 8 compares the predicted cumulative failure probabilities and the experimental rank probabilities for both deep- and shallow-cracked specimens at the intermediate loading rate. Results at other loading rates show similar comparisons. In general, the model predictions agree very well with the experimental data and accurately capture the effects of loading rate and constraint ($a/W = 0.5$ vs. $a/W = 0.1$; square cross-section vs. rectangular cross-section) on fracture toughness.

Based on the above results and the studies of Gao et al. [32, 33], we can argue that m is invariant of loading rate for typical ferritic steels in the low-to-moderate loading rate range ($\dot{K}_J < 3 \times 10^3$ MPa$\sqrt{m}/s$). In addition, studies of [30, 32, 33] also show that dynamic loading reduces constraint loss, i.e., it drives the response towards the SSY configuration. Two major factors are believed to be responsible for this response: i) the yield stress increases with increasing strain rate; ii) under higher loading rates, fracture occurs at lower toughness values. This rate effect tends to saturate as loading rate increases.

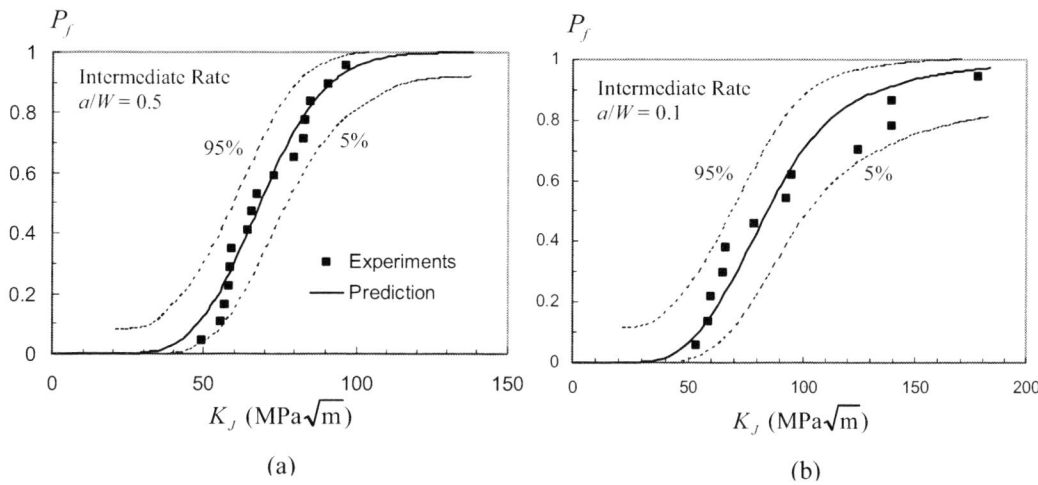

Figure 8. Comparison of the predicted cumulative cleavage probabilities (solid lines) with rank probabilities of the measured K_{Jc}-values (symbols) for the intermediate dynamic loading case [30].

Although m can be taken as loading rate invariant, σ_u requires calibration for the loading rates of interest because the model prediction depends very sensitively on σ_u [32]. With m taken as rate invariant and σ_{w-min} computable from the SSY boundary layer model, the σ_u parameter becomes key to predict the failure probability and its value must be known at each loading rate of interest in an application. Petti and Dodds [34] recently described an approach to calibrate the temperature dependence of σ_u using the Master Curve approach with the reference temperature (T_0) estimated by testing standard fracture specimens at one temperature. They assume m is independent of temperature and select σ_u values that force the Weibull stress model to predict the Master Curve temperature dependence of K_{Jc} values for the material. Through experimental studies, Wallin [35] has shown that the Master Curve and reference temperature (T_0) framework applies as well for dynamic loading. Since m has now been shown to be independent of loading rate, the Petti-Dodds approach may also be applied to estimate the loading rate dependence of σ_u.

FURTHER MODIFICATION OF THE WEIBULL STRESS MODEL AND THE EFFECT OF TEMPERATURE ON MODEL PARAMETERS

The original Weibull stress model assumes microcracks nucleate at carbide particles once plasticity develops, i.e., all carbides become microcracks in the plastic zone. However, experimental observations suggest that the number of microscopic cracks initiated at carbide particles increases with an increase in plastic strain [36, 37]. Thus, the probability density function for microcrack size should be modified to reflect the increase of the microcrack numbers with the increasing plastic strain, i.e., $f(l) = h(\varepsilon_p)l^{-\gamma}$, where $h(\varepsilon_p)$ represents a function of the local plastic strain. With this argument, Gao et al. [38, 39] modified the Weibull stress expression as follows

$$\sigma_w = \left[\frac{1}{V_0} \int_{V_p} h(\varepsilon_p) \sigma_1^m dV \right]^{1/m} \tag{8}$$

Inspired by the experimental studies of Lindley [36] and Gurland [37], Gao et al. [38, 39] approximated $h(\varepsilon_p)$ to be proportional to ε_p. Gao et al. [38, 39] also proposed to exclude a small region closest to the crack front, where material experiences large deformation and low stress triaxiality, from the fracture process zone. The reason to exclude this small region derives from the observation that microcracks in this region tend to blunt and become incapable of causing cleavage fracture. Detailed descriptions of the modified Weibull stress model can be found in [38, 39]. The modified model has been applied to predict cleavage fracture in several ferritic steels. Compared to the original Weibull stress model, improved predictions were found with the modified model, especially for specimens exhibiting very low constraint levels [38].

As an example, consider the experimental program conducted by Faleskog et al. [40], who tested a large number of SE(B) specimens of a modified A508 steel at three different temperatures, -30, 25 and 55°C. The specimens are all plane-sided with B = 20mm, W = 40 mm and $S = 4W$. All specimens fail by cleavage fracture with no evidence of ductile tearing. The dataset contains 20 specimens for a/W = 0.5, and 12 specimens each for a/W = 0.1 and 0.25 at each temperature. The modified Weibull stress model is calibrated at each temperature using the fracture toughness data of the a/W=0.5 and a/W=0.1 specimens according to the calibration procedure described in the previous Section. The toughness data obtained from the a/W=0.25 specimens are used to verify the calibrated model. The results suggest that m is independent of temperature and the calibrated m-value is 11.8 [39]. Figure 9 shows the variations of σ_u and σ_{w-min} on temperature: σ_{w-min} decreases with increasing temperature because of the change of material flow properties with temperature, while σ_u increases with temperature, reflecting the combined effects of loading rate on material flow properties and toughness. Figure 10 compares the predicted cumulative cleavage probabilities with the median rank probabilities for the three sets of measured J_c-values at 25°C. Virtually all experimental data fall within the 90% confidence bounds. Results at the other two temperatures show similar comparisons.

A recent study by Wasiluk et al. [27] also found that the Weibull modulus, m, is invariant of temperature, where the material considered was a 22NiMoCr37 pressure vessel steel. These results support the argument of Petti and Dodds [34] that m is independent of temperature for common ferritic steels and the Master Curve can be used to estimate the σ_u dependence on temperature once the m-value is determined. This will dramatically reduce the amount of experimental data needed to characterize the cleavage fracture behavior with varying temperature.

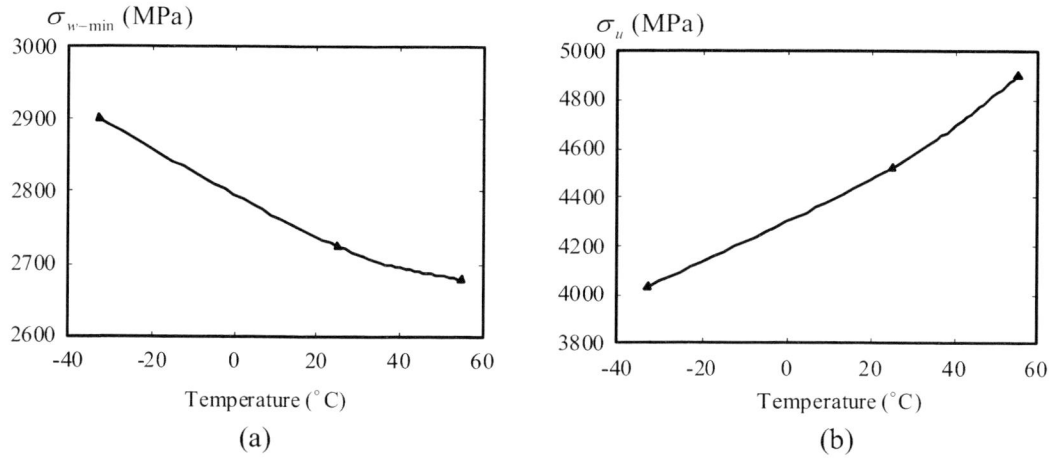

Figure 9. Variations of $\sigma_{w-\min}$ and σ_u with temperature for the modified Weibull stress model, where the material under consideration is a modified A508 steel [39].

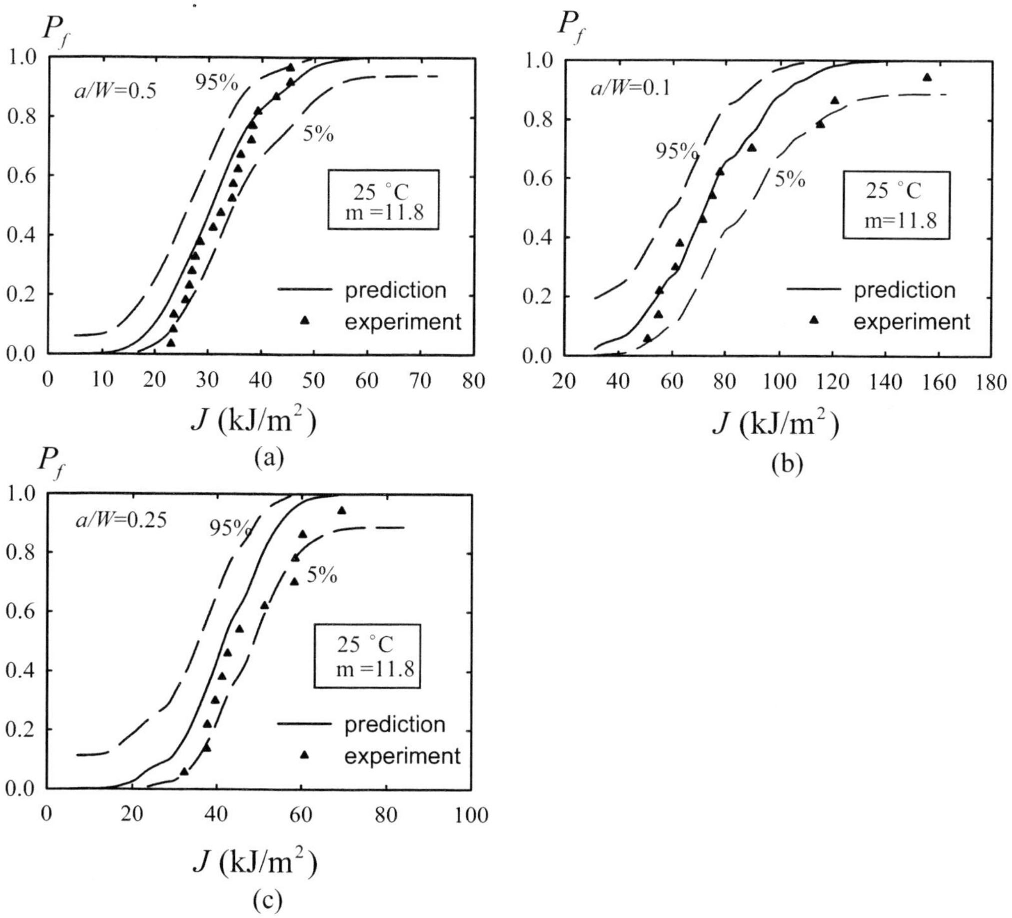

Figure 10. Comparison of the predicted cumulative cleavage probabilities (solid lines) with rank probabilities of the measured J_c-values (symbols) at 25°C [39].

CONCLUSION

Transgranular cleavage fracture in the DBT region of ferritic steels often leads to spectacular and catastrophic failures of engineering structures. The research conducted over the past two decades by several groups worldwide has extended the engineering applicability of the Weibull stress model to predict cleavage fracture. This chapter summarizes recent research conducted by the author and his co-workers on this subject. These new developments have introduced a threshold parameter into the model, developed more robust calibration methods, predicted experimentally observed constraint effect and toughness scatter, demonstrated temperature and loading rate effects on model parameters, and expanded the Weibull stress model to include the effects of plastic strain and microcrack nucleation. Although continued research is needed to further develop this model, the work summarized in this chapter, when taken together with those published by other investigators, provides a convincing case for the predictive power of the Weibull stress model.

ACKNOWLEDGMENTS

The author would like to thank Prof. R.H. Dodds of the University of Illinois and Prof. J.A. Joyce of the US Naval Academy for their contributions. The work presented in this chapter summarizes our collaborative efforts in this area in recent years.

REFERENCES

[1] McMahon, C. J. Jr.; Cohen, M. *Acta Metall.* 1965, *vol.* 13, 591–604.
[2] Smith, E. In *Physical Basis of Yield and Fracture: Conference Proceedings*; Strickland, A. C.; Ed.; Physical Society of Oxford; 1966; 35-46.
[3] Barnby, J. T. *Acta Met.* 1967, *vol.* 15, 903.
[4] Wallin, K. *Eng. Fract. Mech.* 1984, *vol.* 19, 1085–1093.
[5] Wallin, K.; Saario, T.; Torronen, K. *Metal Sci.* 1984, *vol.* 18, 13-16.
[6] Sorem, W. A.; Dodds, R. H.; Rolfe, S. T. *Int. J. Fract.* 1991, *vol.* 47, 105–126.
[7] Anderson, T. L.; Stienstra, D.; Dodds, R. H. A In *Fracture Mechanics, ASTM STP 1207*; Landes, J. D.; McCabe, D. E.; Boulet, J. A.; Eds.; American Society for Testing and Materials, Philadelphia, 1994; *vol.* 24, 186–214.
[8] ASTM E1921-06; *Test Method for the Determination of Reference Temperature, T_0, for Ferritic Steels in the Transition Range*; American Society for Testing and Materials, West Conshohocken, 2006.
[9] Wallin, K. *Int. J. Mater. Product. Technol.* 1999, *vol.* 14, 342-354.
[10] McCabe, D. E.; Merkle, J. G.; Wallin, K. In *Fatigue and Fracture Mechnics, ASTM STP 1360*; Paris, P. C.; Jerina, K. L.; Eds.; American Society for Testing and Materials, Philadelphia, 1999; *vol.* 30, 21-33.
[11] Beramin, F. M. *Metall. Trans.* 1983, *vol.* 14A, 2277–2287.

[12] Bakker, A.; Koers, R. W. J. In *Defect Assessment in Components–Fundamentals and Applications, ESIS/EG9;* Blauel, J. G.; Schwalbe, K. H.; Eds.; Mechanical Engineering Publications, London, 1991, 613–632.
[13] Minami, F.; Brückner-Foit, A.; Munz, D.; Trolldenier, B. *Int. J. Fract.* 1992, *vol.* 54, 197–210.
[14] Gao, X.; Ruggieri, C.; Dodds, R. H. *Int. J. Fract.* 1998, *vol.* 92, 175-200.
[15] Gao, X.; Dodds, R. H.; Tregoning; R. L.; Joyce, J. A.; Link, R. E. *Fatigue Fract. Engng Mater. Struct.* 1999, *vol.* 22, 481-493.
[16] Xia, L.; Shih, C. F. *J. Mech. Phys. Solids* 1996, *vol.* 44, 603-639.
[17] Ruggieri, C.; Dodds, R. H. *Int. J. Fract.* 1996, *vol.* 79, 309-340.
[18] Gao, X.; Faleskog, J.; Shih, C. F. *Fatigue Fract. Engng Mater. Struct.* 1999, *vol.* 22, 239-250.
[19] Pineau, A. *Int. J. Fract.* 2006, *vol.* 138, 139-166.
[20] Griffith, A. A. *Phil. Trans. Roy. Soc.* 1920, *vol.* A221, 163.
[21] Gao, X.; Dodds, R. H. *Int. J. Fract.* 2000, *vol.* 102: 43-69.
[22] Petti, J. P.; Dodds, R. H. *Engng Fract. Mech.* 2004, *vol.* 71, 2079-2103.
[23] Petti, J. P.; Dodds, R. H., *Engng Fract. Mech.* 2004, *vol.* 71, 2677-2683.
[24] Gao, X.; Dodds, R. H. *Engng Fract. Mech.* 2001, *vol.* 68, 263-283.
[25] Mudry, F. *Nucl. Engng Design* 1987, *vol.* 105, 65–76.
[26] Ruggieri, C.; Gao, X.; Dodds, R. H. *Engng Fract Mech* 2000, *vol.* 67, 101-117.
[27] Wasiluk, B.; Petti, J.; Dodds, R. H. *Engng Fract Mech* 2006, *vol.* 73, 1046-1069.
[28] Joyce, J. A.; Link, R. E. In *Fracture Mechanics, ASTM STP 1321*; Underwood, J. H.; Macdonald, B. D.; Mitchell, M. R.; Eds.; American Society for Testing and Materials, Philadelphia, 1996; *vol.* 28, 243–262.
[29] Wallin, K. Research Report 604, Technical Research Center of Finland, Espoo, Finland, 1989.
[30] Gao, X.; Joyce, J. A.; Roe, C. *Engng Fract. Mech.* 2008, *vol.* 75, 1451–1467.
[31] Wallin, K.; Nevasmaa, P.; Laukkanen, A.; Planman, T. *Engng Fract Mech.* 2004, *vol.* 71, 2329-2346.
[32] Gao, X.; Dodds, R. H.; Tregoning, R. L.; Joyce, J. *Fatigue Fract. Engng Mater. Struct.* 2001, *vol.* 24, 551-564.
[33] Gao, X; Dodds, R. H. *Engng Fract. Mech.* 2005, *vol.* 72, 2416-2425.
[34] Petti, J. P.; Dodds, R. H. *Engng Fract. Mech.* 2005, *vol.* 72, 91-120.
[35] Wallin, K. In *Recent Advances in Fracture*; Mahidhara, R. K.; Ed.; The Minerals, Metals and Materials Society 1997, 171-182.
[36] Lindley, T. G.; Oates, G.; Richards, C. E. *Acta Met.* 1970, *vol.* 18, 1127-1136.
[37] Gurland, J. *Acta Met.* 1972, *vol.* 20, 735-741.
[38] Gao, X.; Zhang, G.; Srivatsan, T. S. *Mater. Sci. Eng. A* 2005, *vol.* 394, 210-219.
[39] Gao, X.; Zhang, G.; Srivatsan, T. S. *Mater. Sci. Eng. A* 2006, *vol.* 415, 264-272.
[40] Faleskog, J.; Kroon, M.; Oberg, H. *Engng Fract. Mech.* 2004, *vol.* 71, 57-79.

In: Stuctural Materials and Engineering
Editor: Ference H. Hagy

ISBN 978-1-60692-927-8
© 2009 Nova Science Publishers, Inc.

Chapter 6

TOWARDS PERFORMANCE-BASED SEISMIC DESIGN FOR BUILDINGS IN TAIWAN

Qiang Xue and Cheng-Chung Chen
Civil and Hydraulic Engineering Research Center,
Sinotech Engineering Consultants Incorporated, Taipei, Taiwan

ABSTRACT

Lessons learned from the devastating Chi-Chi earthquake has wakened the scientific community in Taiwan to consider new seismic design methodologies. Performance-based seismic design has been recognized as an ideal method for use in the future practice of seismic design. In this chapter, lessons learned from the Chi-Chi earthquake have been summarized. The state-of-the-art and the state-of-the-practice in performance-based seismic engineering have been reviewed. Possible misleading in engineering practice has been pointed out. The outline, design flowchart and key issues of performance-based seismic design (PBSD) methodology implemented in the performance-based seismic design draft code for buildings in Taiwan have been described. A direct displacement-based design method has been demonstrated through examples. Performance criteria regarding displacement to be applied in a direct displacement-based design method for non-ductile or less ductile structures are discussed. Suggestions on seismic performance criteria and seismic evaluation of existing buildings have been made. Feedback and difficulties encountered during the development of the draft code have been discussed. Finally, future research topics related to performance-based seismic design have been summarized. It is believed that performance-based earthquake engineering will bring a new era to engineering practices with increased confidence in, and reliability on, seismic performance and safety.

1.1. INTRODUCTION

Observations on the performance or damage of structures after strong earthquake ground motions have always served as an effective means to evaluate the current seismic regulations and guidelines and make further improvements afterwards. Loss from earthquakes such as the

1989 Loma Prieta, 1994 Northridge, 1995 Kobe and 1999 Chi-Chi earthquakes forces the scientific communities in many countries to consider new seismic design methodologies to control structural performance within an acceptable level. Performance-based seismic design [ATC-34 1995, Vision 2000 1995, Bertero 1996b, Hamburger 1997, Fajfar and Krawinkler 1997, Krawinkler 1999a] means that plan, design, evaluation, construction and maintenance of engineered facilities will result in their performance to meet the diverse needs and objectives of the owners, users and society reliably with accepted confidence under different levels of earthquake ground motion. This chapter first provides an overview of the lessons learned from the 1999 Chi-Chi earthquake to highlight the need for performance-based seismic design in Taiwan. Then, the development of the seismic design draft code for buildings in Taiwan using performance-based seismic design methodology is described.

The local seismic design code issued in 1999 [ABRI-CPAMI, 1999] was based on a research study [Tsai et al., 1994] in 1994 in Taiwan. According to international research development and practice, the need for performance-based earthquake engineering has been recognized in Taiwan. Based on other research studies [Yeh, 2001], concepts of multi-level seismic hazard and design response spectra of seismic microzonation have been introduced in the updated seismic design code issued in 2005 [ABRI-CPAMI, 2005]. During this period, research studies on performance-based seismic evaluation of existing structures and performance-based seismic design of new structures have been carried out in research organizations in Taiwan. Since 2003, PBSD methodologies have been requested for use in more and more projects regarding seismic evaluation and retrofitting of existing buildings. Recognizing an increasing need for performance-based seismic design guidelines or provisions for local engineers to follow, in 2003, the Architecture and Building Research Institute (ABRI), Ministry of the Interior, sponsored Sinotech Engineering Consultants Incorporated to propose a framework for a performance-based seismic design code applicable to current engineering practices. Based on the state-of-the-art and the state-of-the-practice, the current seismic design code provisions have been examined according to the theoretical basis of PBSD to identify which methodologies of PBSD need to be incorporated into the current seismic design code. Following the framework, the ABRI sponsored another project for the same research group to propose a seismic design draft code for buildings by introducing performance-based design methodologies. In this chapter, a PBSD flowchart, key issues and outline of the draft code are presented. Transparent seismic design objectives for buildings of different use groups are established qualitatively and interpreted quantitatively as performance criteria, including drift limits. Site feasibility requirements, conceptual design scopes and basic rules are proposed. Performance-objective–oriented procedures for preliminary design and contents for detailed seismic performance evaluation are explained. Particularly, an efficient direct-displacement–based design method is presented with demonstration examples. Suggestions on seismic performance criteria and the evaluation of existing buildings are made. Feedback and difficulties encountered are discussed. There is still a lot of work to do in order to achieve the most economic solution with reliable safety. Future research activities related to performance-based seismic design in Taiwan are summarized. We believe that all of these efforts will bring a new era to engineering practices with increased confidence in, and reliability on, seismic performance and safety.

1.2. LESSONS LEARNED FROM THE 1999 CHI-CHI EARTHQUAKE

Taiwan is mainly located in an unstable tectonic zone between the Philippine Sea and Eurasian Plates (Figure 16). It is situated at the junction of the Manila and Ryukyu Trench in the Western Philippine Sea where the Philippine plate is being forced under the Eurasia plate. The Philippine plate appears to be propagating northward, causing a significant strike-slip component along the northern portion of the Manila Trench and eventually popped up the island of Taiwan. Taiwan has a rich seismic history. Seismic safety is the main concern.

On 1:47:12.6 a.m., September 21, 1999, local time, a violent earthquake—with Richter scale local magnitude (M_L) 7.3 by the Central Weather Bureau (CWB), Taiwan—occurred near Chi-Chi in Nantou county, Taiwan. Thousands of aftershocks, several of which with magnitudes greater than 6.0 (M_L), were detected. In addition to a great amount of injury and loss of life, collapsed buildings and destroyed bridges, roads, communication, gas, electric power and water supply systems were observed. This series of earthquakes has affected Taiwan's economy. Fortunately, strong-motion records have been able to be retrieved because the seismic center of the Central Weather Bureau has completed an extensive seismic instrumentation program. This is the reason that the intensity scale in Taiwan corresponds to an instrument-based intensity scale proposed by CWB.

The need for performance-based earthquake engineering is addressed in this section according to the damage evidence observed after the Chi-Chi earthquake.

1.2.1. Need of Multi-Level and Transparent Performance Objectives

The picture in Figure 17 shows a 32-level building in Fengyuan. Although the whole building did not collapse due to the earthquake, severe damage of local structural elements and the outside wall was found. For a lot of buildings damaged in this event, the mainframe structural systems were capable of resisting the earthquake, but the exterior and/or interior walls or stairs, etc. were badly damaged. Although there was no life lost, offices in these buildings had to be closed or relocated until further evaluation and rehabilitation are completed.

The purpose of establishing an individual performance objective is to provide a platform for information exchange between the owners and the designers so that they can achieve a common view about the structural performance under different earthquake levels. Although performance objectives associated with small, medium and large earthquake levels have been defined qualitatively in the Taiwan seismic design code, traditional earthquake-resistant design of new buildings focuses on a single performance goal – life safety quantitatively.

Seismic capacity of an existing building has been quantified by a so-called "peak ground acceleration to cause building collapse" of the designated earthquake. It is estimated approximately without detail analysis before associated with the intensity scale. Life safety

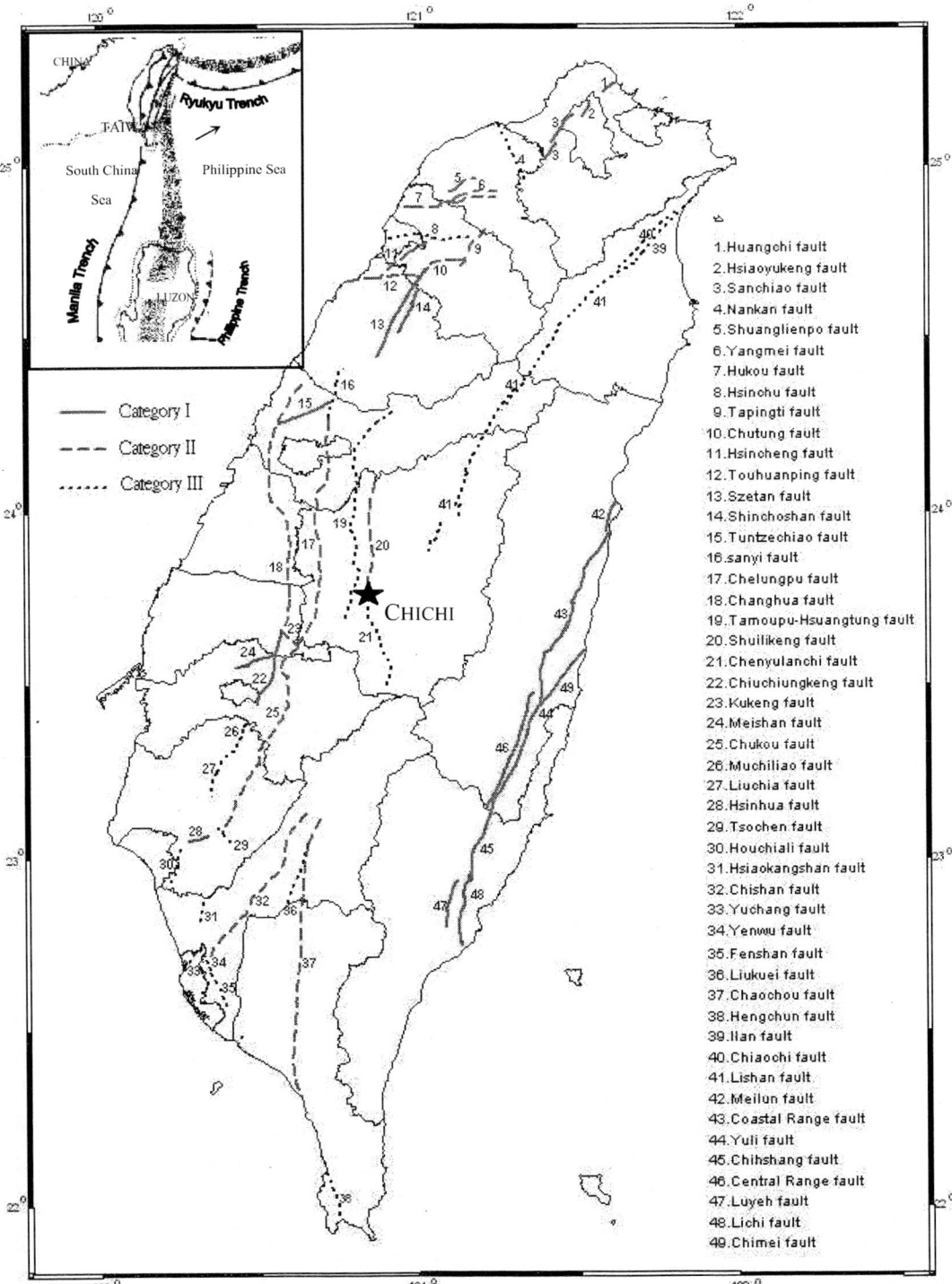

Figure 16. Taiwan and its faults

and collapse prevention are indeed very important. However, a structure, which is safe against collapse, may deflect or vibrate excessively, or the non-structural elements are badly damaged so as to interfere with further use. Therefore, additional performance goals such as damage control are necessary. The so-called "peak ground acceleration to cause building collapse" estimated according to traditional seismic design methodology is neither accurate nor suitable to indicate seismic performance under different earthquake level.

Figure 17. Function loss due to damage

1.2.2. Need of Site Suitability Study and Site-Specific Seismic Hazard Analysis

A large number of structures within the fault zone were severely damaged during the Chi-Chi earthquake. Figure 18(a) catches the foundation damage of a surface faulting passing through a building in FengShi road, Fengyuan city. In Figure 18(b), the fault rupture is clearly seen in the sports ground of Kuangfu junior high school after the earthquake. A transmission tower in Minjian village built on the fault rupture is tilted (Figure 18(c)). Figure 18(d) illustrates the effects of liquefaction in Yuanlin, Changhua county. Fault rupture is seen especially along the bank of Dajia river as shown in Figure 18(e) and 3(f). The fault passing through the Dajia river lifted up the riverbed, forming a waterfall and also damaged the south part of the Pifeng bridge (Figure 18(e)). Figure 18(f) presents the damaged Hsigang dam. This dam was believed not in the fault zone according to the 1977 version geologic map when its construction was commenced. However, the Chi-Chi earthquake proves that it is located about only 3Km away from the Chelungpu fault. A triggered fault from Chelungpu thrust fault passing through the dam foundation lifted the left side of the dam body for about 2.1m, and the right for about 9.8m, thus partially destroying this dam.

Figure 18. Damage in fault zone or by liquefaction

An important step in PBEE before the preliminary design of a structure is to make the site suitability study and site-specific seismic analysis. From the damage observed in the above figures, it is generally undesirable for structures, especially important structures to be built on or very close to the fault rupture unless the structural performance can be guaranteed. For the areas where it is probable that the fault rupture will reach the ground surface, that is where the probability for severe ground failure is very high, construction needs to be prohibited. Taiwan is a small island with a relatively large population. Site suitability study usually focused on liquefaction potential only. It is hard to reach the same conclusion about the back off distance from the fault zone. However, the severely damaged facilities within the fault zone during the Chi-Chi earthquake demonstrate the need to consider this factor in a proper manner in future design projects.

It is often necessary to select a design strong-motion record or a response spectrum that represents the anticipated earthquake shaking at the project site. Site-specific seismic hazard analysis using the probabilistic approach [McGuire 1995] or the deterministic approach must be performed on a basis of geologic, seismologic and soil characteristics of the site. For instance, for buildings within a few kilometers of a possible major event, near-fault effects must be considered. Traditional seismic hazard analysis is deterministic. Although the deterministic approach is useful and maybe applicable when locations of all affective seismic faults, that may affect the site, and focal mechanisms are clear, the probabilistic approach is preferable. According to the Chi-Chi earthquake records, response spectra obtained at many sites exceed the design values. Site-specific design response spectra to be built by employing the probabilistic approach with special soil or near fault characteristics considered are needed.

1.2.3. Need for Proper Selection of the Configuration or Structural System

A lot of wood structures, brick structures and reinforced brick structures are severely damaged or collapsed in this event. In addition to the weak and non-ductile or less ductile lateral resisting systems, inadequate lateral bracing and/or eccentricity due to improper wall layout for wood structures, Π type plan layout, soft or weak story for brick and reinforced brick structures are some of the main reasons.

Figure 19 presents partial collapse of an unsymmetrical building in Taichung. Figure 20 shows a type of typical buildings with a pedestrian corridor and open front at the ground floor. Significant lateral displacement and the disappearance of the ground story are clearly seen.

Figure 21 depicts a building with several columns damaged in a specific area of the basement. This localized or concentrated damage is due to the stop of the nearby shear wall at the ground story, which leads to the sudden reduction of lateral stiffness. Also, the lateral reinforcement seems inadequate. Figure 22 illustrates local damage evidence due to improper structural system with beam to beam and beam to column centerline offsets. Figure 23 presents buildings constructed at different times but tied to each other. Pounding effect damaged the structures particularly at the interface.

From the damage described above, selection of a proper configuration and structural system before the preliminary design is necessary. However, only a few guidelines are mentioned in the local seismic regulations because the scientific community thinks that a design engineer should have had such basic knowledge. A structural engineer may understand that regularity and symmetry in plan and elevation are desirable because the behavior of such structures is relatively simple and so are the analysis and design. Sudden changes in stiffness, strength, or mass in either vertical or horizontal planes of a building can result in distributions

Figure 19. Failure due to irregularity

Figure 20. Failure due to soft stories

Figure 21. Damage due to discontinued vertical element

Figure 22. Damage due to improper structural system

Figure 23. Damage due to inadequate separation between buildings

of lateral loads and deformations different from those of uniform structures. However, they always have to compromise with a creative architecture that is usually irregular. Although such drawbacks could be overcome by proper design based on suitable yet complicated analysis tools, the associated uncertainties are better avoided whenever possible. Bertero [1982] suggested that in order to overcome or decrease the uncertainties, it is necessary to pay more attention to conceptual design, which is defined as the avoidance or minimization of seismic effects by applying an understanding of the behavior rather than using numerical computations.

1.2.4. Need for Proper Selection of the Analysis Tools and Design Detailing

Figure 24 illustrates the column failure due to incorrect detailing. Figure 24(a) shows the most common column failure due to insufficient transverse reinforcement. The ninety-degree hooks do not provide adequate anchorage. These have limited the development of the yield capacity of the plan ties provided. Furthermore, the thick concrete cover with a small core results in a reduced effective cross-section area, which leads to a weak column. Figure 24(b) shows spalling of unconfined thin concrete cover. Figure 24(c) shows the damage evidence due to column centerline offset between the basement and the upper storey.

(a)

(b)

(c)

Figure 24. Column failure due to improper reinforcement or concrete detailing.

Design detailing should base on a suitable analysis. In order to achieve an economic solution, PBEE accepts damage in seismic events and thus requires the performance of the structural system be accurately characterized and realistically predicted. A reasonable model and a suitable analysis method must be chosen, particularly for the purpose to capture the inelastic behavior of a structural system. In the traditional seismic regulations of Taiwan, guidelines for suitable modeling of different structural components are not provided in detail. A structure is analyzed elastically under a reduced level of equivalent lateral force based on equal energy or equal displacement rule according to its fundamental period (stiffness). It is then designed by the working stress or the ultimate limit state design method to provide sufficient strength and with special ductile detailing to withstand the earthquake loading. Nonlinear structural behavior is not explicitly evaluated. Therefore, a lot of engineers cannot understand why the expected strong-column-weak-beam mechanism disappears, particularly when a structure deforms further under a high level earthquake.

There is a need to develop the guidelines for simulation and detailed analysis methods for different structural systems. For example, torsional effects or three-dimensional analysis must be considered for unsymmetrical structures or they should be avoided. Soil-foundation-structure interaction should be modeled wherever possible. Effects of non-structural systems such as infilled walls and stairs on the seismic resistance of the structural system need to be simulated. Non-linear analysis must be used to predict the inelastic behavior of structures under severe earthquake ground motions. Nonlinear three-dimensional time-history analysis is necessary for important structures with irregularity and significant higher-mode effects. Duration of the ground shaking must be addressed in the design method to account for the cumulative damage. Deterioration in strength and stiffness are important issues. In PBEE, displacement-based [Moehle 1992, Priestley 1995] rather than the force-based design approaches are desirable to control the drift of a structural system for architectural integration, limiting structural damage, avoiding structural instability and avoiding human discomfort. On the other hand, energy-based approaches [Bertero et al. 1996] are appropriate for structures that might be subjected to high cumulative damage or whose response is controlled through energy dissipation devices.

1.2.5. Need for Construction Quality Assurance and Maintenance of Building Integrity throughout Its Life Cycle

Some of the Chi-Chi earthquake damage is caused by poor construction quality. For example, stirrups presented in Figure 25 were cut improperly. The reinforcement bars shown in Figure 26 are not placed at the right positions as designed. Also, the lateral reinforcement seems inadequate. Shear failure and stress concentration of the beam in Figure 27 have been clearly shown due to improper cutting through.

Figure 25. Unexpected cutting of the stirrups

Figure 26. Incorrectly-positioned reinforcement bars

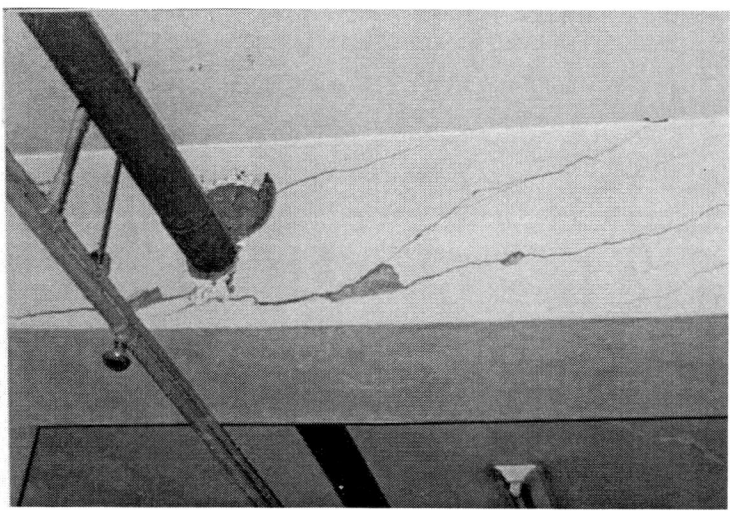

Figure 27. Damage due to improper cutting through

Design and construction of a structure are intimately related. Designers provide drawings and instructions to the builder. The quality of the product relies on design and supervision. The construction quality of a properly designed structure is crucial in order to achieve its expected performance. The performance of the system also depends on its state when the earthquake strikes. Therefore, construction and maintenance, which include surveillance, inspection, monitoring, retrofitting, upgrading and rehabilitation, are important. Change of the non-structural elements without approval by the designers or other competent structural engineers should be prohibited. Before the Chi-Chi earthquake, it was commonly seen in Taiwan that occupants rearrange the interior or exterior walls and/or windows by their own will without knowing the consequence of the change of stiffness of the building. Owners who own two units separate in two adjacent stories often link them by destroying part of the floor slab to make a town house. Particularly, for multi-story buildings, owners of the top-floor unit own the roof area above and often add an additional penthouse for their own use. As a result, additional mass is added on the top floor, which is not initially considered in the original design. This increases the possibility of damage during an earthquake. Thus, it is crucial to issue the guidelines for construction quality control and maintenance in Taiwan.

1.2.6. Need for Comprehensive Loss and Cost Estimation

As mentioned earlier, PBEE implies the acceptance of damage in seismic events for economic concern. Apart from the conventional cost estimation of design and construction as in Taiwan, cost of maintenance, direct and indirect property damage or loss of life after an earthquake, loss due to interfere with intended usage or proper operation of properties after the earthquake, cost of retrofit, rehabilitation or demolition should be considered. Cost and loss estimation will provide the owner/client with important information whether it is economic to use advanced materials such as Fibre Reinforced Polymer (FRP), composite material, etc. or to use advanced technique such as using any energy dissipative device or base isolation, and whether it is better to build a stronger building with negligible damage under strong earthquakes, or to build a relatively weaker building with moderate damage and retrofitting, or simply a much weaker building without collapse but requiring demolition after the event, and whether the economic loss can be paid back by insurance. In addition to property and life loss, social cost and environmental cost are also important, particularly for seismic hazard mitigation planning. Earthquake loss estimation must be done with collaboration among engineers, economists and social scientists [National Research Council 1999].

1.2.7. Need for Taking Account of Uncertainties and Risk

Uncertainties and variability exist in defining the earthquake design values, modeling the structures for analysis, employing analytical method and design tools, involving engineer experience, design and construction. For instance, the eastern coast of Taiwan is considered most seismically active but the recent earthquake occurred in the central part. According to the available historical data, Taichung has been classified as the seismic zone with design PGA of 0.23g ~ 0.28g either in the "regulations for seismic design of buildings" [Tsai et al. 1994] or the "regulations for seismic design of roads and bridge". Compared with the PGA values that could be 3 times the design value, from the stations in this area, the design PGA values appear not large enough for an adequate seismic resistant design. Therefore, buildings in the Taichung area were severely damaged in the Chi-Chi earthquake. This fact implies that a building, whose design has satisfied design codes or regulations and constructed with high level quality, may still expose to seismic risk. Such risk may be reduced through maturity of regulation system regarding the whole life cycle of buildings, establishment of risk management guideline, enhancement of risk assessment methodologies, paying more attention to building maintenance and management, as well as emergency preparation and practicing.

In addition to paying more attention to the conceptual design phase as discussed in section 2.3 to deal with the uncertainties, a probabilistic basis with the probability of exceeding a certain desired performance and the confidence in characterizing this probability, must be implemented. That is, the structure is protected for a specified performance level with a given annual probability of exceedance. This is different from the traditionally using single performance objective design, which is expressed in a deterministic manner. The probabilistic reliability format is a useful way to design and build better and more economical facilities [Collins 1995].

1.2.8. Need for Collaboration among Seismologists, Geologists, Architects, Structural and Geotechnical Engineers, Socioeconomic Scientists and Contractors

As discussed in section 2.2, site suitability study requires geological information. Identification of seismic zones needs the knowledge of seismology and geology. As given in section 2.3, architects, engineers and contractors will have to work together and reach the same idea. Balance between engineering and socioeconomic issues should be achieved. This multidiscipline system [Krawinkler 1999b] is indeed a major challenge in PBEE.

There are many collaboration technologies that enable people to meet and work together from different geographical locations through the Internet. Virtual Engineering is one of the above technologies that fill the gaps among people from different disciplinary. Traditional methods for earthquake design collaboration sometimes fall short of conveying the complicated design concept when the designers need to exchange their ideas with other groups. In particular, engineering drawings and documents grow exponentially when the complexity of a project increases. With the use of Virtual Reality (VR), better understanding and communication is achieved because the designers' concept is exemplified with multimedia, animation and interaction in a virtual reality environment. The interactive nature of VR made it a natural extension to the 3-D graphics that enable the designers, project managers, site workers and clients to visualize and sense real life structures before actually building them.

VR is a three-dimensional visualization technique using computers to simulate the real world and provide 3D-navigation for better immersion and analysis. On the other hand, the development of the World Wide Web (WWW) over the last decade has led to many applications constructed by Virtual Reality Modeling Language (VRML) [Hartman and Wernecke 1996] in access to information over the Internet. These VRML models offer many advantages including ease of use, quick access, low cost and cross platform compatibility. Moreover, VRML is available without the limitations of time or location and flexible in allowing users to examine every design details. The important advantages of the virtual reality environment over other computer based design tools are that it enables the user to interact with the simulation to conceptualize relations that are not apparent from a less dynamic representation, and to visualize models that are difficult to understand in other ways.

1.3. INTRODUCING PERFORMANCE-BASED SEISMIC DESIGN METHODOLOGY INTO TAIWAN

Since performance-based seismic design covers a wide range of fields, introducing performance-based design methodology into the current seismic design code for buildings must be done one step at a time. In order to identify which methodologies of PBSD need to be incorporated into the current seismic design code in Taiwan, the general concept, the state of the art and the state of the practice have been reviewed and summarized briefly in this section.

1.3.1. General Concepts

General concept of PBSD can be found in many papers as mentioned above. Performance of a building may generally cover serviceability, earthquake resistance, fire resistance, wind resistance and durability, etc. In the field of earthquake engineering as in this chapter, it is confined to seismic performance unless otherwise stated in this chapter.

Performance of a building means its behavior or response and the corresponding consequence including physical damage, function loss, down time, life safety and environmental, cultural and social impact. Regarding seismic design, this chapter focuses on structural response that is closely related to damage, serviceability, possibility to repair, life safety and collapse prevention.

Performance-based seismic design procedure consists of three phases in sequence, the conceptual design phase, the numerical design phase and the implementation phase. The conceptual phase includes selection of performance objective and criteria, site suitability study and the conceptual design. The numerical design phase includes preliminary analysis, sizing, detailing, performance check and final design. The implementation phase includes peer review, construction quality control, maintenance and management.

The purpose of PBSD is to ensure that buildings whose performance under different level of earthquake ground motions respond to the diverse needs and objectives of the owners, users and society. Therefore, it is essential that the design objective is clear and transparent for the owners or users and designers to understand the expected performance and the inherent risks of the structures under various levels of ground motions expected during their life cycles. The design objective is called performance objective, which is defined as the acceptable performance levels under the considered levels of seismic hazard. Theoretically, performance objective associated with continuous earthquake levels is preferable. For practical application, discrete ones are usually adopted for efficiency. The purpose of establishing multi-level performance objective is to provide a platform for information exchange between the owners and the designers so that they can achieve a common view about the structural performance under different earthquake levels. Performance objective must be transformed into quantitative performance criteria, including quantitative performance levels and quantitative seismic hazard levels, for engineers to use during engineering design.

Selection of performance objectives depends on the usage or importance of the buildings. A more important building corresponds to a higher performance objective. A higher performance objective means a higher performance level under the same earthquake level, or the same performance level under a higher earthquake level. Earthquake level is typically expressed by the probability of exceedance P_{ET} for a period of time T, or by the corresponding return period P_R. Their relationship is shown in Equation (1). Site-specific seismic hazard analysis employing probabilistic methodology is preferable. Through seismic hazard analysis, the peak ground acceleration (PGA) corresponding to P_{ET} is found for the construction site. The intensity scale is then obtained. In such a way, owners or users may understand how strong the considered earthquake is.

$$P_R = \frac{1}{1-e^{(1/T)\ln(1-P_{ET})}} \quad (1)$$

For building design, the earthquake level is expressed by the corresponding design response spectrum or time history of the design earthquake ground motion.

The performance level of a building is qualitatively described in terms of building function, performance of the structural system and components, non-structural system and components such as facilities, equipments and internal contents. Quantification of performance level involves selecting an appropriate performance indicator or a damage index with consideration of the design approaches to be used.

Site suitability study is to understand the environmental sensitivity, particularly geotechnical condition and potential risk factors. The purpose is to ensure that the selected performance objective can be met on the construction site. Otherwise, special engineering techniques, adjusted performance objective or a new project must be applied.

Conceptual design includes selection of structural material, configuration, layout, system type, foundation and energy dissipation mechanism. Engineers experience plays an important role at this stage. The conceptual design results govern the design spectra, preliminary design results and selection of analytic methods.

PBSD means design of a building is oriented by the expected seismic performance, that is, such a design started with the quantified performance levels under the design earthquake levels. Regarding structural performance, which is the focus of this chapter, performance indicator may be based on force, ductility, displacement (stiffness) or deformation, energy dissipating, their combination or other damage index. Therefore, force or strength-based approach, displacement-based approach, energy-based approach and damaged-based approach can be applied. With consideration of uncertainties or cost, reliability-based approach and cost-based approach are useful. Preliminary sizing and detailing can be determined as a result. Since performance of a building is governed by many factors, only some of them can be considered in the preliminary design stage for efficiency. Thus, acceptance of the designated performance is usually checked by employing an appropriate analysis method, through which the real structural behavior can be obtained rationally before final design.

The advantages of PBSD over the traditional design methodologies are summarized as the following six key issues:

1. Multi-level seismic hazards are considered with an emphasis on the transparency of performance objectives;
2. Building performance is guaranteed through limited inelastic displacement or damage in addition to strength and ductility;
3. Seismic design is oriented by performance objectives interpreted by engineering parameters, performance criteria;
4. An analytical method through which the structural behavior, particularly the nonlinear behavior is rationally obtained, is required;
5. The building will meet the prescribed performance objectives reliably with accepted confidence;
6. The design will ensure the minimum life-cycle cost.

1.3.2. Brief Summary on the State of the Art and the State of the Practice

Although there are many papers and reports worth referring to, in this paper, summaries of the state of the art and the state of practice are confined to the widely-known reports and the latest seismic design codes with more or less PBSD methodologies applied, respectively.

Starting from 1989, particularly after 1992, specific researches have been carried out by several universities, research institutes and international workshops to develop the PBSD guidelines and standards in the United States. Their efforts lead to some exciting achievements including the publication of *ATC-34* [1995] report sponsored by the National Center for Earthquake Engineering Research (NCEER, now MCEER), Vision 2000 [1995] and SEAOC blue book [SEAOC 1999] by the Structural Engineers Association of California (SEAOC), ATC-40 [1996] report sponsored by the California Seismic Safety Commission (CSSC), the NEHRP guidelines FEMA 273 [1997] sponsored by the Federal Emergency Management Association (FEMA) and prepared by Applied Technology Council (ATC) for the Building Seismic Safety Council (BSSC) and its updated version FEMA 356 [2000] and FEMA 440 [2004], FEMA 350 [2000] Prepared by the SAC Joint Venture for FEMA. These reports including FEMA 302 [1997], FEMA 368 [2000] and FEMA 450 [2004] form the basis of methodology, which has been implemented in the updated seismic design regulations [IBC 2000, 2003, 2006]. Related researches [ATC-58 2003, Comartinet et al. 2004] are still carried on.

The concept of PBSD has received a lot of attention and has been introduced in the latest seismic design codes of some countries such as the USA [IBC 2003, 2006], Canada [Heidebrecht 2004], Japan [JSCA 2000], Australia and New Zealand [King and Shelton 2004], etc. For example, after the 1995 Kobe earthquake, Seismic design in Japan has been focused on functionality and possibility to repair in addition to life safety. The 13th World Conference on Earthquake Engineering has made the conclusion that PBSD is the methodology to use in the future practice of seismic design.

1.3.2.1. Performance Objectives and Criteria

Establishment of performance objectives means selecting a performance level and a corresponding earthquake level according to the importance of a building. For seismic design in practice, quantified performance criteria must be established. A design level earthquake for a given P_{ET} or P_R is quantitatively represented either by a response design spectrum or by an acceleration time history. A traditional response spectrum is in spectral acceleration (S_a) vs. period (T) format because the design approach is strength-based. As displacement-based design approaches receive more attention, spectral displacement (S_d) vs. period (T) format and $S_a \sim S_d$ format are widely used. For other design approaches such as energy-based approach and damage-based approach, an energy spectrum [Bertero 1997, Riddell and Garcia 2001] and a damage spectrum [Bozorgnia, Y, Bertero, 2004] may be needed.

A response spectrum can be categorized into one of the three types, elastic, equivalent elastic [Chopra 1995, ATC-40 1996, Kowalsky et al. 1994, Priestley et al. 1996, Reinhorn 1997, Kunnath et al. 1996, Iwan and Gates 1979, Freeman 1998] and inelastic [Newmark-Hall 1982, Krawinkler and Nassar 1992, Vidic et al.1994, Chopra 1995, Reinhorn 1997, Fajfar 1999, Chopra and Goel 1999]. Design response spectra originally provided in codes are elastic. For elastic spectra, the relationship in equation (2) exists.

$$S_a = \omega_n S_v = \omega_n \times (\omega_n S_d) = \left(\frac{2\pi}{T_n}\right)^2 S_d \tag{2}$$

Theoretically, for long period buildings, equal displacement rule is applied. However, in the traditionally seismic design code such as the current seismic design code in Taiwan, $S_a \sim T$ *format* elastic design spectra provided in code are limited to a certain level in the long period range in order to prevent that the design strength is too low.

Design spectra accounting for the nonlinear inelastic behavior of a structural system can be represented either by an equivalent elastic design spectrum or an inelastic design spectrum. Both can be obtained through reduction from the elastic design spectra. For an equivalent elastic spectrum, reduction factors based on the Newmark-Hall [1982] amplification factors in Equation (3) can be used.

$$SR_A = \frac{3.21 - 0.68 \ln(\zeta_{eff})}{2.12} \tag{3}$$

$$SR_V = \frac{2.31 - 0.41 \ln(\zeta_{eff})}{1.65} \tag{4}$$

These reduction factors are associated with effective viscous damping ζ_{eff}, which is composed of the equivalent viscous damping ζ_{eq} and the inherented viscous damping ζ_0 (e.g. 5% for RC structures). Several damping models have been developed. Examples are described in the following:

- The ATC-40 [1996] damping model is based on that the energy dissipated by the inelastic structure is equal to that dissipated by an equivalent viscous system in a single cycle of motion. The effective damping ζ_{eff} is estimated by equation (5).

$$\zeta_{eff} = \zeta_0 + \zeta_{eq} = \left[5 + \kappa \frac{2(\mu - 1)(1 - r)}{\pi\mu(1 + r\mu - r)}\right]/100 \tag{5}$$

where μ and r are the corresponding ductility ratio and the post-yielding stiffness ratio, respectively. κ is a damping modification factor associated with the hysteresis behavior type [ATC-40 1996] A, B and C of the structure.

- The damping model used by Kowalsky et al. [1994] is based on the laboratory test results and curve fitting.

$$\zeta_{eff} = \zeta_0 + \zeta_{eq} = 0.05 + 0.39372\left[1 - \frac{1}{\sqrt{\mu}}\right] \tag{6}$$

- The damping model used by Preistley et al. [1996] is based on Takeda hysteresis model.

$$\zeta_{eff} = \zeta_0 + \zeta_{eq} = 0.05 + \frac{1 - \mu^n(\frac{1-r}{\mu} + r)}{\pi} \tag{7}$$

where the stiffness degradation factor n equals to 0 for steel structures and 0.5 for RC structures.

- The damping model used by Reinhorn [1997] and Kunnath et al. [1996] is based on the average stiffness and energy method presented by Iwan and Gates [11].

$$\zeta_{eff} = \left(\frac{3}{2\pi\mu^2}\right) \frac{\pi\zeta_0\left[(1-r)\left(\mu^2 - \frac{1}{3}\right) + \frac{2}{3}r\mu^3\right] + 2(1-r)(\mu-1)^2}{(1-r)(1+\ln\mu) + r\mu} \tag{8}$$

Inelastic response spectra have been recognized preferable comparing with equivalent elastic spectra. Different type of inelastic response spectra as mentioned above can be found in the references [Newmark-Hall 1982, Krawinkler and Nassar 1992, Vidic et al.1994, Chopra 1995, Reinhorn 1997, Fajfar 1999, Chopra and Goel 1999]. In Taiwan, Newmark-Hall [1982] reduction factors have been employed with minor modifications according to the local seismic hazard analysis.

Based on a time history record on the construction site or nearby site with similar geology, a series of artificial earthquake ground motions that are compatible with the design response spectrum are usually needed.

Qualification of performance levels can be found in the above-mentioned documents. Although the considered performance levels may be different from one document to the other, the idea is identical. According to these documents, typical performance levels include Operational, Immediate Occupancy, Damage Control, Life Safety and Collapse Prevention. Each performance level corresponds to a certain state along the force-displacement capacity curve (Figure 28) of the structure. Only two or three of these performance levels together with their corresponding earthquake levels may be used to establish the performance objectives of a certain building, whose usage or importance is determined.

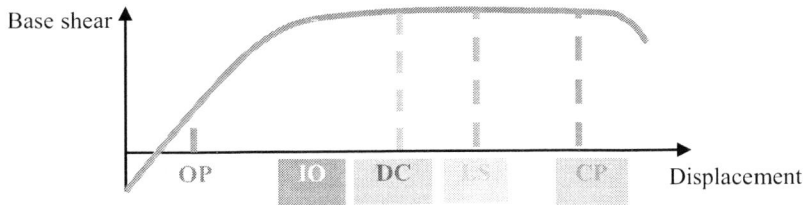

Figure 28. Performance levels along the force-displacement capacity curve

Selection of performance indicator depends on the factors that may affect or control the seismic performance of the building. Seismic performance of a structure is closely associated with its strength, stiffness and ductility. In a conventional seismic design, structural strength and ductility [Park 1986, Wallace and Moehle 1992] are mainly considered as the performance indicator. However, under large earthquake loading, when input energy is dissipated through inelastic deformation, a small change in strength may cause a large jump in displacement. Up to 50% error may be found in the estimated yielding displacements through traditional design method [Sozen et al. 1992]. Using strength and ductility as the only performance indicators may not always be rational [Priestley 1993, Moehle 1996]. In PBSD, performance indicator must be associated with structural system and components, non-structural systems and components and contents. For structural systems, in addition to strength and ductility, additional performance indicators may be structural stiffness or displacement [Mahin & Bertero 1976, Sozen 1981, Priestley 1993, Miranda 1997, Seneviratna and Krawinkler 1994, FEMA 273 1997, ATC-40 1996], energy [Bertero et al. 1996C, Riddell and Garcia 2001] and damage index [Park-Ang 1985, Krawinkler and Nassar 1992, Reinhorn and Valles 1996, Cosenza et al. 1993] which may be defined by one of the former indicators or their combinations. With consideration of uncertainties, reliability index or confidence level [FEMA 350 2000, Collins 1995] associated with the mentioned indicators may be used. For structural components or elements, performance indicators may be strength or deformation [FEMA 273 1997, FEMA 356 2000]. Regarding non-structural performance, floor acceleration or economic loss may be used as the performance indicator.

As mentioned above, selection of appropriate performance indicator is closely related to the design approach to be used. According to the current state of the practice, performance-based preliminary design is usually oriented by the performance of the structural system. Performance of structural components will be checked through detail analysis. Non-structural systems are designed to accommodate floor acceleration or drift after the structural system design is completed. For seismic design, economic loss, as well as downtime, replacement cost and number of casualties are not employed as a quantified design criterion, but could be assessed through loss or risk evaluation for the purpose of hazard mitigation [Kircher et al. 1997].

The performance objectives in IBC 2000 [2000] and AS/NZS 1170 [King and Shelton 2004] are of particular interest. In IBC, performance objectives for buildings with three use groups I, II and III, respectively, are established. At the preliminary design stage with strength being the performance indicator, the performance level, life safety is of the only concern for all building groups. Performance goals for buildings of different use groups are achieved by selecting different design earthquake levels through multiplying an importance factor, 1, 1.25 and 1.5 for use group I, II and III, respectively. The maximum considered earthquake level

MCE is for use group III only. After preliminary design, performance check is performed with focus on the structural displacement. Only the design level earthquake is considered, Performance goals for buildings of different use groups are achieved by selecting different displacement limit under the same earthquake level. The more important the building is, the smaller the limit.

In AS/NZS 1170, performance objective for each of the four groups can be found in Table 4. In general, two performance goals associated with the serviceability limit state 1 (SLS1) under the minimum considered earthquake level and the ultimate limit state (ULS) under the maximum considered earthquake level are considered. The minimum considered earthquake level corresponds to a return period of 25 years. The maximum considered earthquake level depends on the importance of the building. For buildings with low risk, only one performance goal associated with ULS is considered. For buildings with very high risk, additional performance goal associated with serviceability limit state 2 (SLS1), which means that the building is functional or immediate repair is not needed, under a moderated level earthquake is considered

Table 4. Performance objectives for buildings of different importance in AS/NZS 1170

Annual probability of exceedence	Building importance			
	Low risk	General risk	High risk	Particularly important or very high risk
1/2500				ULS
1/1000			ULS	
1/500		ULS		SLS2
1/100	ULS			
1/25		SLS1	SLS1	SLS1

1.3.2.2. Site Suitability Study and Conceptual Design

Site suitability study usually focuses on potential of soil liquefaction, land sliding, submerge, etc. A lower potential or a larger offset distance from the active fault zone [SEAOC 1999, IBC 2000~2006] is needed for a more important building with a higher performance objective. Importance of conceptual design and the commonly suggested design rules can be found in references by Bertero and Bertero [2002] and by Paulay and Preistley [1992]. Building height and type of the lateral structure system are limited according to the established performance objective [SEAOC 1999, IBC 2000~2006]. For important buildings, limitations on system irregularity, discontinuity, etc. are provided [SEAOC 1999].

1.3.2.3. Preliminary Design Approaches

Performance-based seismic design means that the design is oriented by the performance objective. With the performance indicators selected, preliminary design approaches can be categorized as strength-based or force-based, displacement-based [Sozen 1981,Qi and Moehle 1991, Moehle 1992 & 1996, Priestley 1993, Thomsen and Wallace 1995, Court and Kowalsky 1998, Bachmann and Dazio 1997, Fajfar 1999, Xue 2001a], energy-based [Teran-Gilmore 1996, Mander and Dutta 1996, Bertero and Uang 1992], damaged-based [Park et al. 1987, Cosenza and Manfredi 1997], reliability-based [Wen et al. 1994, Collins et al. 1995], cost-based [pires et al. 1996] approaches and comprehensive design approach [Bertero et al. 1996C].

The concept of displacement-based seismic design has received a lot of attention in recent years because such design focuses on displacement instead of force as the direct performance or damage indicator. Although the displacement quantifier is not able to reflect both the deformation and loading path and accumulates over time as the energy quantifier, it is simple and transparent. Besides, by employing energy parameters, the displacement-based approach can be applied to account for the accumulative damage due to strong motion duration [Cosenza and Manfredi 1997, Fajfar 1992].

Displacement-based design approach can be further classified as indirect and direct. An indirect displacement-based design approach means that a force-based approach is employed together with check on the displacement criteria through rational analysis. Compared with the indirect approach, a direct displacement-based design method is believed to be more efficient because it is oriented by a target displacement. As shown in Figure 29, all direct displacement-based design methods are theoretically similar. Seismic demand is estimated according to the target displacement associated with the performance point which is the intersection of a demand curve represented by a response spectrum and a capacity curve usually traced by nonlinear static pushover analysis. If the building performs elasticly, an elastic design spectrum is applicable. Otherwise, either an equivalent elastic design spectrum or an inelastic design spectrum must be used. In this chapter, existing direct displacement-based design approaches are classified as the substitute structure approach and the inelastic structure approach.

The substitute structure approach includes the so-called DBD (Direct-Displacement Based Design) method [Kowalsky et al 1994, Priestley and Kowalsky 2000, SEAOC 1999] and EBD (Equal-Displacement Based Design) method [Court and Kowalsky 1998, SEAOC 1999]. As shown in Figure 29, the substitute structure DBD method utilizes an equivalent elastic structural system with a higher damping ratio ζ_{eff}. By using an equivalent elastic response spectrum, strength demand and stiffness demand are obtained by the secant stiffness K_{eff} associated with the performance point at the target displacement. The EBD method is a simplified version of DBD by applying the equal displacement rule for long-period structures.

The so-called capacity spectrum method (CSM) [Fajfar 1999, Chopra and Goel 1999] is typically an inelastic structure approach. Seismic demand is estimated according to the tangent stiffness K_0. The yielding spectrum method (YSM) [Aschheim and Black 2000, Aschheim 2002] and the generalized numerical method as will be discussed later in this chapter adopt similar principle to CSM and yields the same results if all related conditions are the same. The yielding spectrum method is an alternative method for the same purpose because the yielding spectrum is obtained through reduction of the demand curve by the ductility ratio R_μ, which is also used to reduce the target displacement (Figure 30). The generalized numerical method, in which both an equivalent elastic response spectrum and an inelastic response spectrum can be used, is completely numerical without any graphical assistance. Feasibility and simplicity of some of the direct and indirect displacement-based design methods have been concluded according to example studies by FIB [2003].

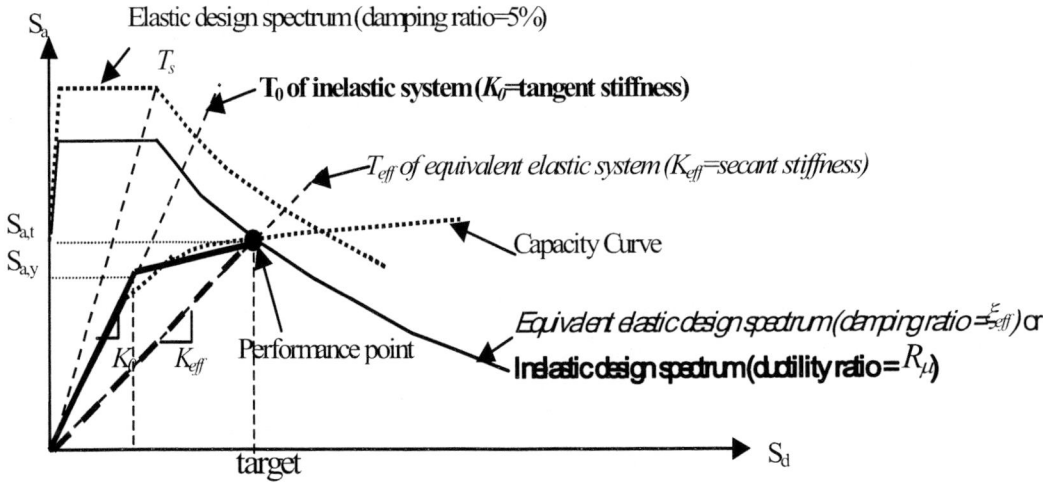

Figure 29. The substitute structure DBD method and the capacity spectrum method

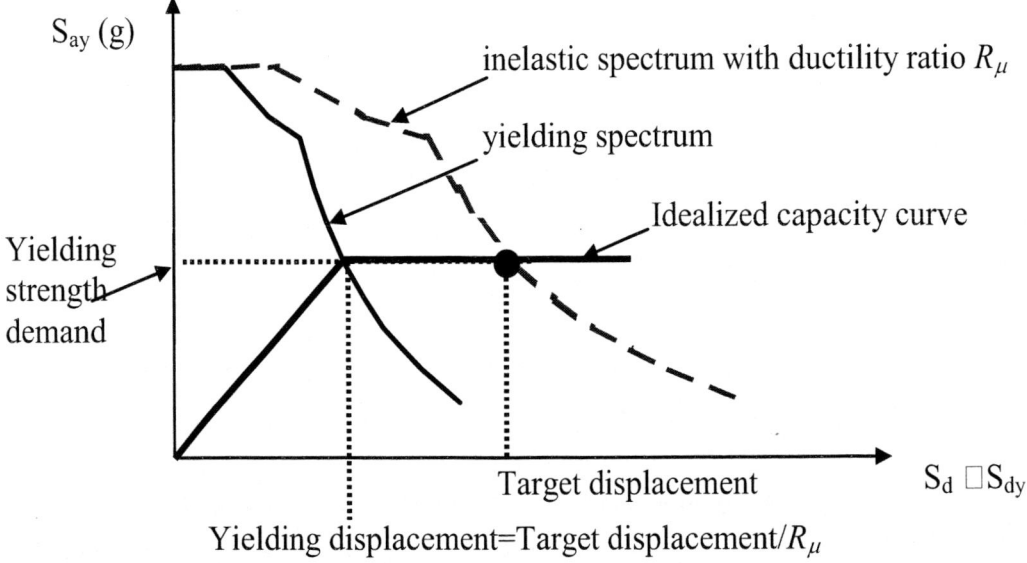

Figure 30. The yielding spectrum method

Although guidelines [SEAOC 1999] for direct displacement-based seismic design have been published, it is the indirect displacement-based design approach that has been widely implemented in the current state of the practice.

1.3.2.4. Performance Evaluation

An analytical method through which the structural performance can be rationally obtained must be chosen based on the building height, regularity and earthquake level. Linear or non-linear, static or dynamic methods and the application scope in the references have been briefly summarized in Table 5.

Table 5. Analytical methods and applicable scope

References	Analytical methods	Application scope
SEAOC[1999] Appendix I-A	Linear static	Regular buildings with floor number <=2
	Linear dynamic	Other buildings
SEAOC[1999] Appendix I-B	Linear static, Linear dynamic	Regular buildings with uniform ductility demand and appropriate detailing; low reliability requirement
	Nonlinear static pushover	Other buildings
	Nonlinear dynamic	
	Other nonlinear analysis	
FEMA 350	Linear static	Low story buildings under small or moderate earthquakes; Low story regular buildings under large earthquakes
	Linear dynamic	High story buildings under small or moderate earthquakes; High story regular buildings under large earthquakes
	Nonlinear static	Low story irregular buildings under large earthquakes
	Nonlinear dynamic	High story irregular buildings under large earthquakes
FEMA 356 (FEMA 273) ATC-40	Linear static	Regular buildings without either significant nonlinearity or significant higher mode effect
	Linear dynamic	Regular or (torsion, vertical mass or stiffness) irregular buildings with insignificant nonlinearity and higher mode effect
	Nonlinear static	Buildings with insignificant (torsion, vertical mass or stiffness) irregularity and higher mode effect, but with significant
	Nonlinear dynamic	Irregular buildings with significant nonlinearity and higher mode effect

In the current state of the practice, nonlinear static pushover analysis has been widely used to understand the expected behavior of the structure under large earthquake loading that may cause the structure to behave nonlinearly. However, under certain circumstances, nonlinear dynamic analysis must be carried out [Elnashai, 2002]. When employing existingcommercial packages, structural simulation is very important to obtain a correct solution. In addition to modeling of structural geometry, material, mass, loadings and boundary conditions, for RC structures, selecting an appropriate hysteresis curve to simulate the component behavior including cracking [FEMA 356 2000], strength degradation, stiffness deterioration and pinching effect [Krawinkler 1999a] must be concerned.

According to the analytical results, the multi-level performance objective will be checked. Regarding the performance goals under large earthquakes, when using non-linear static pushover analysis, structural performance indicators associated with structural system and structural components must then be compared with the acceptance performance criteria this is through either the capacity spectrum methods, or [ATC-40 1996, FEMA 440 2004] or the displacement coefficient method [FEMA 273 1997, FEMA 356 2000, FEMA 440 2004].

1.3.3. Deficiencies in the Existing Seismic Design Code in Taiwan

Traditional seismic design codes are partially performance-based. For example, design objective has been qualified by the following three requirements:

- A building is able to resist a low level earthquake without damage.
- A building is able to restore its function after repair under a moderate level earthquake.
- A building is able to resist a major level earthquake without collapse.

Besides, drift limit under low level of earthquake needs to be controlled in addition to ensuring life safety under a strong earthquake level. Structural strength and ductility have also been employed as the performance indicator. However, according to the theoretical basis of PBSD. Several deficiencies have been found in the existing seismic design codes in Taiwan. Conventional seismic design objectives lack transparency. For example, the performance objective of an important building is adjusted by increasing the considered earthquake level through an importance factor (I) while keeping the expected performance level the same as that of a general building. This is the case even for the maximum considered earthquake level with a return period of 2500 years. Engineers do not know the corresponding intensity scale of such an amplified earthquake because an earthquake with a return period of 2500 years corresponds to the maximum intensity scale already for most places in Taiwan [Loh and Hwang 2002]. Therefore, the design criteria and expected performance of the building are not understood well by the public. Building stiffness or displacement is not as important as strength and ductility. The current seismic design code emphasizes structural strength and ductility. Drift limits, associated with structural stiffness, are only provided for earthquakes that occur frequently. The expected performance of buildings has not been evaluated using rational analysis methods. Structural responses to high-level earthquakes may not be predicted accurately. Even engineers may not be sure why the criteria specified in the current seismic design code would allow the building to meet the performance goals. Therefore, introducing performance-based seismic design methodology in the current seismic design code will focus on the first 4 key issues listed in section 0.

1.3.4. Misleading in Engineering Practice

The idea of performance-based design or performance evaluation has been widely accepted in Taiwan. However, the concept of "performance-based" has not been understood well by a lot of engineers or even some of the researchers. As mentioned earlier, only the peak ground acceleration of the designated earthquake, which may cause collapse of the structure, serves as a performance index of an existing building. It is obvious that structural behavior and response may be quite different under ground motions with the same PGA but different strong motion duration or frequency content, etc. Performance-based earthquake engineering implies that performance of a structure is evaluated according to its response and behavior under the designated earthquake. Even if the performance level of "collapse" is defined according to structural behavior and response, such estimation may mislead engineers to concern only about building collapse. The so-called "collapse PGA" may serve as an index of seismic resisting capacity. Estimation of this index is different from seismic performance evaluation regarding various levels of earthquakes.

Difference of performance evaluation between new buildings and existing buildings has not been clearly understood by engineers. This issue is considered when developing the draft code and will be explained later in this chapter.

Performance objectives in code only provide the minimum protection to buildings. Increase of performance objective, which means increase of cost, is allowed upon the owner's acceptance. Engineers should be aware of seismic risk because of uncertainties.

1.4. DEVELOPMENT OF THE PERFORMANCE-BASED SEISMIC DESIGN DRAFT CODE FOR BUILDINGS IN TAIWAN

1.4.1. The Framework and the Design Flowchart

According to the-state-of-the-art and the-state-of-the-practice review and the deficiency found in the existing seismic design code, it is decided that introducing PBSD methodology into the current seismic design codes focuses on the first 4 issues listed in section 0. That is, to establish transparent seismic design objectives for buildings of different use groups qualitatively and interpreted quantitatively as performance criteria including drift limits under all hazard level, to address site feasibility requirements and conceptual design basic rules considering different performance objectives, to employ performance-based approach for preliminary design, to select a rational analysis method and to provide the contents for seismic performance evaluation. For the completeness, guidelines for direct-displacement-based design methods and suggestions on both performance criteria and performance evaluation of existing buildings are also provided.

Seismic performance of a building is associated with both structural and non-structural systems and elements. In this research, a non-structural system is designed to accommodate acceleration or drift after the structural system design is completed as in the current seismic design code. Economic loss is not employed as a quantified design criterion. Therefore, this research focuses on the structural performance. Issues regarding non-structural systems, buildings including base isolation systems and energy dissipative devices are outside the scope of this paper.

Figure 31 illustrates the flowchart for the performance-based seismic design of new buildings. Topics in Figure 31 except those illustrated with dotted lines will be discussed in the following sections.

Performance criteria are the acceptance criteria necessary to meet the performance objective. The purpose of establishing performance criteria is for engineering design. The criteria are composed of both quantified hazard levels and quantified performance levels. Each seismic hazard level is quantified by an elastic design spectrum corresponding to the construction site. Unlike in the current seismic design code, where the spectral acceleration associated with a structural period larger than 2.5 times the characteristic period T_s is limited to 0.4 times the peak spectral acceleration (PSA) to prevent a very low level design strength demand, the elastic design spectrum obtained according to probabilistic seismic hazard.

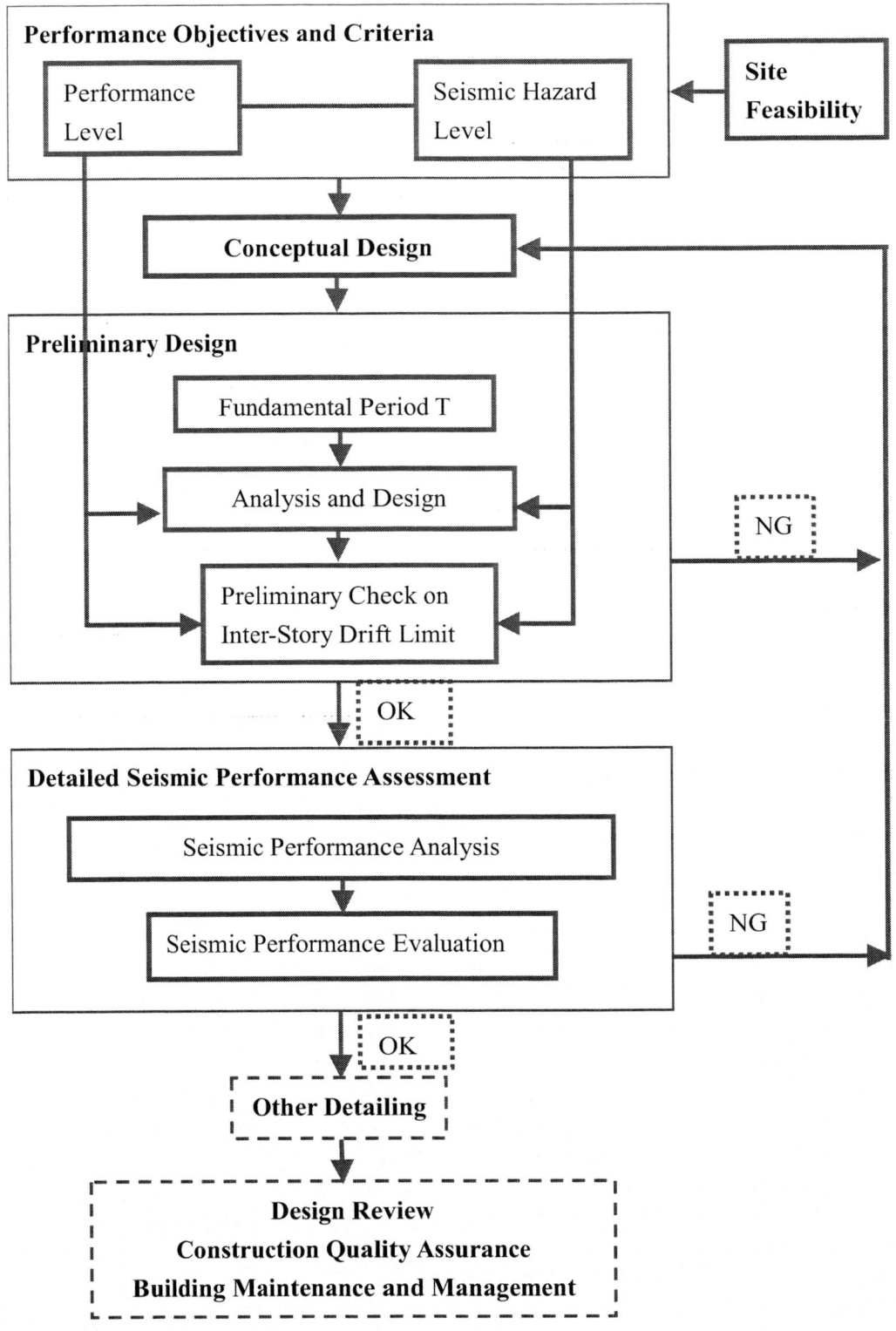

Figure 31. Performance-based seismic design flowchart for new buildings

Table 6 shows the table of contents of the draft codes. Chapters are mainly organized according to the design sequence as in the flowchart (Figure 31).

Table 6. Table of contents — draft code of performance-based seismic design for buildings

Chapter 1 General Principle	4.4.6 Lateral design force for underground part
1.1 Application Scope	4.4.7 Vertical design force
1.2 Nomenclature	4.5 Member design
1.3 Fundamental principle for seismic design	4.6 Preliminary check on the maximum inter-story drift ratio
1.4 definition of "base-level"	Chapter 5 Seismic performance assessment
1.5 Necessity of ductile design	5.1 General principle
1.6 Requirements on analysis methods	5.2 Seismic performance estimation under all considered hazard levels
1.7 Notation	5.2.1 Selection of analytical method for seismic performance estimation
Chapter 2 Seismic performance objectives and criteria	5.2.2 Structural simulation and modeling
2.1 Seismic use group classification for buildings	5.2.3 Linear time-history analysis
2.2 Seismic performance objectives and criteria for buildings of different use groups	5.2.4 Nonlinear static pushover analysis
2.3 Seismic hazard	5.2.5 Nonlinear dynamic analysis
2.4 Regional elastic design spectral acceleration (S_a) for short periods and at 1 second in the horizontal direction for general regions	5.2.5.1 Nonlinear time-history analysis
2.5 Site specific elastic design spectral acceleration for short periods and at 1 second in the horizontal direction for general regions	5.2.5.2 Nonlinear dynamic pushover analysis
2.6 Site specific elastic design spectral acceleration for short periods and at 1 second in the horizontal direction for near fault regions	5.3 Seismic performance evaluation
2.7 Site specific elastic design spectral acceleration for short periods and at the characteristic period in horizontal direction for Taipei Basin	5.3.1 The maximum inter-story drift ratio and allowable ductility ratio
2.8 Site specific elastic design spectra with 5% damping ratio in S_a-T format in the horizontal direction	5.3.1.1 Hazard level EQ1
2.9 Site specific elastic design spectra with 5% damping ratio in S_d-T format and S_a-S_d format in the horizontal direction.	5.3.1.2 Hazard level EQ2 and EQ3
2.10 Site specific elastic design spectra with damping ratio other than 5%	5.3.2 Damage and weak story mechanism
2.11 Constant ductility inelastic design spectra	5.3.3 Extremely soft story and extreme torsion irregularity
2.12 Earthquake time history in horizontal direction	5.3.4 Building separation distance
2.13 Elastic design spectrum in vertical direction	Chapter 6 Special requirement on structural detailing
Chapter 3 Structural system conceptual design	6.1 General principle
3.1 General principle	6.2 Special requirement on structural detailing
3.2 Structural systems	6.2.1 Buildings with multiple structural systems
3.3 Building height limitation	6.2.2 Connection
3.4 Structural system irregularity	6.2.3 Deformation conformity
3.5 Yielding or damage mechanism	6.2.4 Link element and continuity
Chapter 4 Preliminary design	6.2.5 Force transmission element
4.1 General principle	6.2.6 RC wall or brick wall anchorage
4.2 Selection of analytical method for preliminary design	6.2.7 Diaphragm
4.3 Linear static analysis	6.2.8 Cantilever element
4.3.1 Fundamental period T	6.2.9 Reinforcement of elements when force-transmission path is not continuous.

Table 6. (Continued)

4.3.2 Lateral design force	Chapter 7 Seismic performance assessment and rehabilitation of existing buildings
4.3.3 Reduction factor of lateral design force	7.1 General principle
4.3.4 System yielding level amplification factor from the first yielding point.	7.2 Seismic performance assessment and rehabilitation principle
4.3.5 Modification of the lateral design force	7.3 Rehabilitation construction
4.3.6 Distribution of the lateral design force	7.4 Confirmation of rehabilitation effect
4.3.7 Lateral design force for underground part	Chapter 8 Others
4.3.8 Structural simulation and modeling	8.1 Site feasibility
4.3.9 Incidental torsion	8.1.1 General principle
4.3.10 Overturning moment	8.2 Considering earthquake during construction
4.3.11 Vertical design force	8.3 Seismic instrumentation
4.4 Modal spectrum analysis	References
4.4.1 Reduced design spectral acceleration in the horizontal direction	Appendix A Flowchart for seismic hazard quantification
4.4.2 Adjustment of the total lateral design force	Appendix B Guidelines for displacement-based preliminary design (for Review Purpose)
4.4.3 Structural simulation and modeling	Appendix C Flowchart to calculate lateral design force
4.4.4 Multi-mode superposition	Appendix D Construction quality control
4.4.5 Dynamic torsion	

1.4.2. Performance Objectives and Criteria

The purpose to establish individual performance objective for buildings of different seismic use group is to provide a platform for information exchange between the owners and the designers so that they can achieve a common view regarding structural performance under different level of earthquake excitation. Therefore, transparency of the design objectives is very important.

In the draft code, a building is classified as being in one of three use groups I, II and III with increasing importance or damage consequences. In the current seismic design code, multi-level performance objectives have also been considered. However, as pointed above, performance objectives lack transparency. Intensity scale of the amplified major level earthquake of an important building is not well understood. In the draft code, a good alternative to adjust the performance objective of an important building is to decrease the expected performance level while keeping the considered earthquake level. As shown in Figure 32, three seismic hazard levels are considered and can be distinguished by return period, probability of exceedence, or corresponding site intensity scale. The intensity scale in Taiwan corresponds to an instrument-based intensity scale proposed by the Central Weather Bureau. Performance of a building has been classified into 5 levels, Operational (OP), Immediate Occupancy (IO), Damage Control (DC), Life Safety (LS) and Collapse Prevention (CP). For each building, a performance objective is composed of three performance goals, each of which is constructed by one considered seismic hazard level and an expected performance level. Performance objectives for seismic use groups I, II and III are depicted in Figure 32.

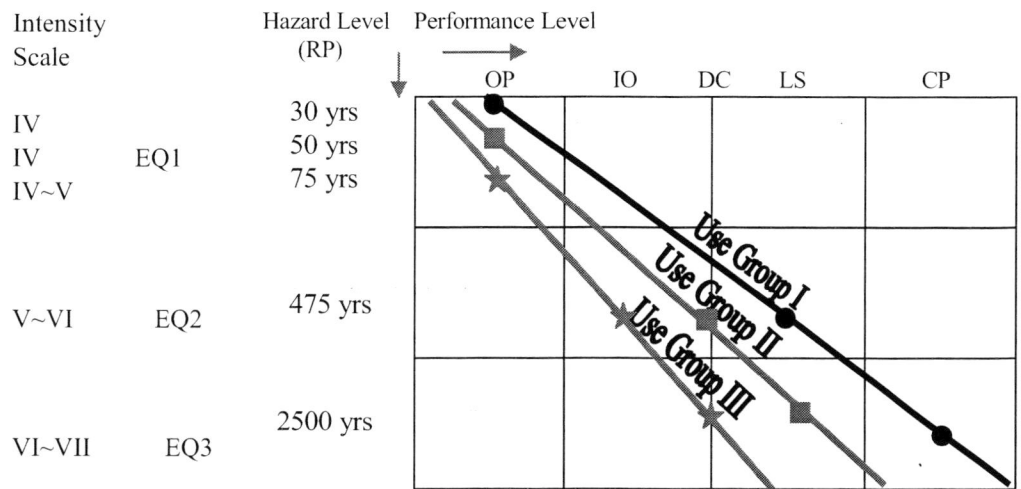

Figure 32. Performance objectives

elastic design spectrum or an inelastic design spectrum will be adopted. Regarding the inelastic design spectrum, Newmark-Hall [1982] reduction factors have been employed with slight modifications according to the local seismic hazard analysis. Regarding the equivalent elastic design spectrum, the effective damping model in equation (8) proposed by Iwan and Gates [1979] is suggested to use so as to conform to the inelastic design spectrum [Xue 2001b].

Each performance level is quantified by parameters associated with strength, stiffness and ductility. Regarding strength, OP corresponds to an elastic behavior. Overstrength must be ensured for other performance levels and no large strength degradation can occur beyond the ductility limit. No weak story exists and the structure has enough vertical capacity.

Regarding ductility, the concept of inelastic displacement demand ratio (*IDDR*) [SEAOC 1999, Court and Kowalsky 1998] is employed. *IDDR* represents the ratio of inelastic displacement demand over the ultimate inelastic displacement capacity. Acceptable values $IDDR_a$ associated with structural system performance levels OP, IO, DC, LS, and CP are 0, 0.2, 0.4, 0.6 and 0.8, respectively. Subscript *a* stands for the allowable value. For structural members, no less than 80% of the elements need to meet the same criteria.

Regarding stiffness, the maximum inter-story drift ratio (*IDR*) is considered to limit building lateral displacement. It is not easy to reach an agreement on the drift limit. In this research, based on references such as the SEAOC blue book [1999], FEMA450 [2003], IBC 2003 [2003], JSCA [2000], AS/NZS 1170 [King and Shelton 2004] and NBCC 2005 [Heidebrecht 2004] and according to the local seismic hazard analysis and opinions from the advisory committee of this project, the *IDR* limits in Table 7 are suggested. Structural systems are mainly classified into four types, namely, the load-bearing walls, the frame systems, the moment resisting frames and the dual systems. It should be noted that the limit of the maximum inter-story drift ratio is used as a performance assessment criterion but may result in an unreasonable demand for some structural systems, particularly for less ductile structural systems if a direct displacement based design method is used for the preliminary design. A detailed discussion on this issue can be found through the comparison study of an example to be presented later in this chapter.

Table 7. The maximum allowable inter-story drift ratio

Structural System	Performance Level				
	CP	LS	DC	IO	OP
Systems with Masonry Shear Walls	0.009	0.007	0.007	0.007	0.005
Other Systems	0.025	0.020	0.015	0.010	0.005

An obvious change that will result from the implementation of the proposed performance criteria over the current seismic design code is an increase in structural stiffness demand for more important buildings under the performance goal with low level earthquakes.

It is important to point out that the draft code suggested performance objectives and criteria provide the minimum protection to the buildings. An increase in the performance objectives is allowed upon the owner's request.

1.4.3. Site Feasibility

The site feasibility study is done to ensure that the established performance objectives can be met on the construction site. Similar to the current seismic design code, this research focuses on the determination of very weak soil layers and the liquefaction potential of sandy soil so that the coefficient of subgrade reaction can be reduced by a reduction factor D_E. Under EQ1 level earthquakes, liquefaction is not acceptable. For EQ2 and EQ3 level earthquakes, soil liquefaction may occur but is limited to an allowable degree through the maximum allowable value of D_E, which is set to be 0, 1/3 and 2/3 for buildings of seismic use groups I, II and III, respectively. If this limit is exceeded, soil improvement is necessary.

For buildings of seismic use groups II and III, the back off distances from type 1 active faults, which are identified active during the past 10,000 years, and other hazards such as landslides must be taken into consideration.

1.4.4. Conceptual Design

Basic conceptual design rules are given with an emphasis on the redundancy and uniform continuity of strength, stiffness and ductility, respectively.

The ductile capacity of each system is represented by a given ultimate ductility ratio μ according to its relative capacity to dissipate energy after yielding. Building height limits, horizontal and vertical irregularities, and expected energy dissipation and yielding mechanisms have been defined. Flexibility is given to allow design of special structural systems based on reliable technology.

1.4.5. Preliminary Design

For the purpose of application in engineering practice, at the current stage, the code should be used together with the existing strength-based material design codes for structural component detailing. Repeated nonlinear static analysis during the design procedure is not preferable. Therefore, indirect displacement-based design approach is suggested to use in the draft code. Guidelines on direct displacement-based design are provided in the appendix.

Similar to traditional force-based design method, fundamental period of the structure system is given by empirical formula. Regarding EQ1, the lateral design force is $V_d = \dfrac{S_a}{\alpha_y} \times W$, where S_a is the spectral acceleration corresponding to the fundamental period reading from the elastic design spectra of the considered hazard level. Because traditional strength-based design approach has been used, for structural period larger than 2.5 times Ts, S_a is limited to 0.4 times PSA. W is the total weight. α_y is the system yielding level amplification factor by which the first significant yielding level is amplified. For examples, for RC structures designed by the ultimate strength design method, $\alpha_y = 1.5$.

Regarding EQ2 and EQ3, the lateral design force is $V_d = \dfrac{1}{1.4\alpha_y} \times \left(\dfrac{S_a}{F_u}\right) \times W$, where 1.4 stands for the overstrength from the system yielding level to the ultimate strength. F_u is the ductility reduction factor defined in Equation (9).

$$F_u = \begin{cases} \mu_a & ; \text{ for } T \geq T_S \\ \sqrt{2\mu_a - 1} + \left(\mu_a - \sqrt{2\mu_a - 1}\right) \times \dfrac{T - 0.6T_S}{0.4T_S} & ; \text{ for } 0.6T_S \leq T \leq T_S \\ \sqrt{2\mu_a - 1} & ; \text{ for } 0.2T_S \leq T \leq 0.6T_S \\ \sqrt{2\mu_a - 1} + \left(\sqrt{2\mu_a - 1} - 1\right) \times \dfrac{T - 0.2T_S}{0.2T_S} & ; \text{ for } T \leq 0.2T_S \end{cases} \quad (9)$$

T_S is the characteristic period at the intersection of the constant acceleration and constant velocity regions of the response spectra. The target allowable ductility ratio μ_a is related to $IDDR_a$ by Equation (10).

$$\mu_a = 1 + IDDR_a(\mu - 1) \quad (10)$$

where μ and $IDDR_a$ have been defined earlier in this paper. Since μ has been derived by including margin of safety, it corresponds to 80-90% of the ultimate inelastic displacement capacity. Accordingly, during preliminary design, acceptable values $IDDR_a$ associated with structural system performance levels OP, IO, DC, LS and CP are revised to be 0, 2/9, 4/9, 2/3 and 1, respectively. However, for sites within the Taipei Basin, these values are revised to be 0, 1/6, 1/3, 1/2 and 1, respectively, considering a large number of cyclic excitations.

Equation (9) means that for structural period T> Ts and 0.2Ts>T>0.6Ts, equal displacement rule and equal energy rule apply, respectively. Linear interpolation applies for other periods.

1.4.5.1. Analysis and Design

Just as in the current seismic design code, elastic analysis can be used in this step. Static analysis or modal spectrum analysis can be selected according to building height and system irregularity. Element design is conducted in the usual way according to the existing design specifications and tools.

1.4.5.2. Preliminary Check on the Inter-Story Drift Ratio

Regarding concerns arising from the complexity of inelastic analysis during detailed seismic performance assessments in the next step, a simple method for conducting the preliminary check of the inter-story drift ratio is employed so that any deficiency associated with stiffness can be discovered at an earlier stage.

Since modal pushover analysis [FEMA 440 2004□Chopra and Goel 2002] has been demonstrated to estimate the maximum inelastic displacement with acceptable accuracy, this concept is employed in this research. The number of modes N with 90% mass participating may be considered. Alternatively, N=3 is allowed. Under earthquake levels EQ2 or EQ3, for each mode i with period of T_i, inelastic spectral displacement S_d^i associated with the corresponding performance level is calculated through Equation (11).

$$S_d^i = \frac{\mu_a^i}{F_u} \times \left(\frac{T_i}{2\pi}\right)^2 \times S_{ae}^i \tag{11}$$

where S_{ae}^i is the elastic design spectral acceleration corresponding to T_i. For the first mode, μ_a^1 is equal to μ_a estimated by Eq. 2 with a revised $IDDR_a$. For other modes, $\mu_a^{i \neq 1} = 1$ by assuming an elastic response or other reasonable value may be assigned. S_d^i represents the inelastic target displacement of the equivalent single degree of freedom (ESDOF) system under mode i. Modal superposition is applied by either the SRSS or CQC rule to obtain the target displacement of the ESDOF system x^*. Finally, the maximum inter-story drift ratio is calculated by Equation (12) [SEAOC 1999]:

$$IDR = x^* / (h_n \times k_1 \times k_2) \tag{12}$$

where h_n, k_1 and k_2 are the building height at the roof level, factor of effective height, factor of displacement shape function, respectively as given in SEAOC blue book. It should be mentioned that this approximate method is most suitable for a structure whose behavior is dominated by its first mode.

To maintain the flexibility of the code, a preliminary design according to the current force-based design method is allowed as long as the performance criteria are acceptable after a detailed performance assessment. It is suggested that the current design method should be used together with the preliminary check of the inter-story drift ratio.

1.4.6. Detailed Seismic Performance Assessment

Although the preliminary design procedure is somewhat objective oriented, the assumed seismic ductility capacity may not be developed in the designed structure due to consideration of sizing standards and construction convenience. Besides, nonlinear behavior is not explicitly considered in the preliminary design phase. The real overstrength factor may be different from the prescribed one as used in the preliminary design. Therefore, a detailed seismic performance assessment is needed to ensure, although not with 100% reliability, that the structural seismic performance meets the prescribed performance criteria and objective.

1.4.6.1. Seismic Performance Analysis
Analysis method to be employed for performance assessment is suggested in Table 8. The method is selected according to the level of seismic hazard and the method used in the preliminary design based on the building height and system irregularity.

Table 8. The minimum allowable analysis method for performance assessment

Seismic hazard level	Static analysis is allowed in the preliminary design	Dynamic analysis is used in the preliminary design	
		$T \leq 3.5 T_s$	$T > 3.5 T_s$
EQ1	Linear Static	Linear Dynamic	Linear Dynamic
EQ2	Nonlinear Static	Nonlinear Static	Nonlinear Dynamic
EQ3	Nonlinear Static	Nonlinear Static	Nonlinear Dynamic

Note: material nonlinearity is the only focus in this table.

Methods in Table 8 are used to evaluate seismic performance under all considered levels of the design earthquake. In the case when nonlinear time-history analysis is employed, nonlinear dynamic pushover or incremental dynamic analysis [FEMA 350 2000, Vamvatsikos and Cornell., 2002, Xue and Meek, 2001] may be carried out to evaluate structural system ductility. Concerning the complexity of nonlinear dynamic analysis, the advisory committee suggested using a simplified model such as a stick model. At least three time history records that are compatible with the design spectrum are required. Strong motion records considering earthquake magnitude, near fault, site and source effect are recommended.

The following sections focus on nonlinear static analysis.

1.4.6.2. Structural Simulation and Modeling
Structural simulation and modeling are as important as selection of the analysis method. In addition to geometry and basic material property, plastic hinge properties of different elements can be adopted from reliable experiments or research reports in consideration of their failure mechanisms. Cracking of a RC element is considered in the same way as

suggested in FEMA 356 [2000]. Influence, particularly adverse influence of non-structural walls (such as the infill walls) on the structural damage mechanism and deformation capacity must be considered.

1.4.6.3. Nonlinear Static Pushover Analysis

Based on a constant load pattern, nonlinear static pushover analysis is an incremental iterative method to obtain the base shear versus roof displacement relationship. It is suggested to employ either of two types of load patterns [FEMA 356 2000, Fajfar 2000]. One is proportional to the fundamental mode. The other is the uniformly distributed load pattern proportional to the story mass. Other adaptive load patterns based on reliable studies are also allowed. Besides, a 100% dead load plus no less than a 25% live load are applied priory to the lateral load in the nonlinear pushover analysis. The monitoring point of pushover analysis is the mass center of the roof story including the penthouse. The end point of the capacity curve corresponds to an ultimate or failure state which is identified when the structure loses stability, or 80% of the elements reach their deformation limit, or more than 20% (maximum or effective) of the base shear strength was reduced after yielding, whichever is critical.

1.4.6.4. Seismic Performance Evaluation

The performance point of the structure that responds to the considered hazard level is evaluated through the seismic coefficient method [FEMA 356 200, FEMA 440 2004] or the capacity-spectrum method [ATC-40 1996, FEMA 440 2004, Fajfar, 2000, Xue 2001a, Chopra and Geol 1999]. The structure is pushed again to the target displacement associated with the performance point to access the behavior of both the structural system and the elements if such a hazard level occurs. Performance criteria regarding all performance goals are examined regarding structural systems and elements.

Structural systems

- Regarding vertical load carrying capacity, the structure should not collapse due to loss of any element [ATC-40, 1996].
- Regarding lateral strength, damage and weak story mechanisms can be clearly understood through hinge location and hinge formation sequences. Strength deterioration is controlled by the definition of failure state during nonlinear pushover analysis.
- Regarding lateral deformation capacity, the maximum inter-story drift ratio *IDR* and inelastic displacement demand ratio *IDDR* are calculated and compared with the acceptable performance criteria. Extremely soft story and extreme torsion irregularities must be avoided [FEMA 450 2003]. That is, each story's stiffness should not be less than 60% of that of the story directly above it or less than 70% of the average of the 3 stories above it. The maximum displacement along the horizontal direction that the earthquake applies should not exceed 1.4 times the average displacement along both horizontal directions.
- Building separation distance must not be less than either the maximum story displacement under EQ2 or 70% of that under EQ3 to avoid pounding [FEMA 450 2003].

Structural elements

- Elastic behavior examination is conducted for elements that need to remain elastic according to the damage mechanism.
- Deformation capacity examination is conducted for deformation-controlled behavior based on the moment-rotation or force-displacement relationship employed in the nonlinear static analysis. According to the performance criteria, only no less than 80% of the elements need to meet the same allowable IDDRa as that of the structural system.
- Strength capacity examination is conducted for force-controlled behavior so that the lowest bound of strength will not be less than the estimated force induced by the considered loading.

In this research, element performance examination is not applied to the performance goal associated with the CP performance level where global system stability is the main concern.

1.4.7. Seismic Performance Evaluation of Existing Buildings

For an existing building, detailed seismic performance evaluation is conducted in the same way as for a new building. The major difference is that structural modeling should be based on site investigation of the real structure and the performance objective and acceptance criteria may be not as restrict as that for new buildings.

In addition to collecting existing data associated with design, monitoring and earthquake history, site investigation is very important in order to understand the real material properties, noticeable element damage mechanisms and existing rehabilitation methods and effects if any. In consideration of uncertainties, hinge properties given in seismic performance evaluation of a new building may be relatively more conservative than an existing building. For example, in the seismic evaluation competition of in-situ pushover tests in Kouhu Elementary School at Yunlin county, shear yielding strength of a column based on experiment results is adopted instead of the code suggested value to obtain an actual seismic capacity instead of a conservative one. Preliminary screening may be conducted through a preliminary seismic evaluation form if a group of buildings are subjected to evaluation under limited budget.

The seismic use groups, seismic hazard levels, and performance levels are considered in the same way for new buildings. However, looser restrictions on seismic performance goals and criteria except for those associated with life safety and collapse prevention are allowed for existing buildings in comparison with new buildings. There are two approaches to loosen restrictions. One is to keep the performance level the same while considering a lower level of earthquake associated with the buildings remaining life span, which is not less than the minimum requested. The other is to keep the earthquake level the same while using a lower performance level. However, it should not be lower than the highest among all the lower levels for new buildings. Similarly, an increase in the performance objective is allowed upon the owner's requirement.

1.5. CASE STUDIES

The presented example is a regular RC building (Figure 33) being in use group I. Step-by-step procedures using the draft code provisions are described except site feasibility study that suggested there is no liquefaction potential. A direct displacement-based design procedure is also presented to address that performance criteria regarding different structural systems are critical.

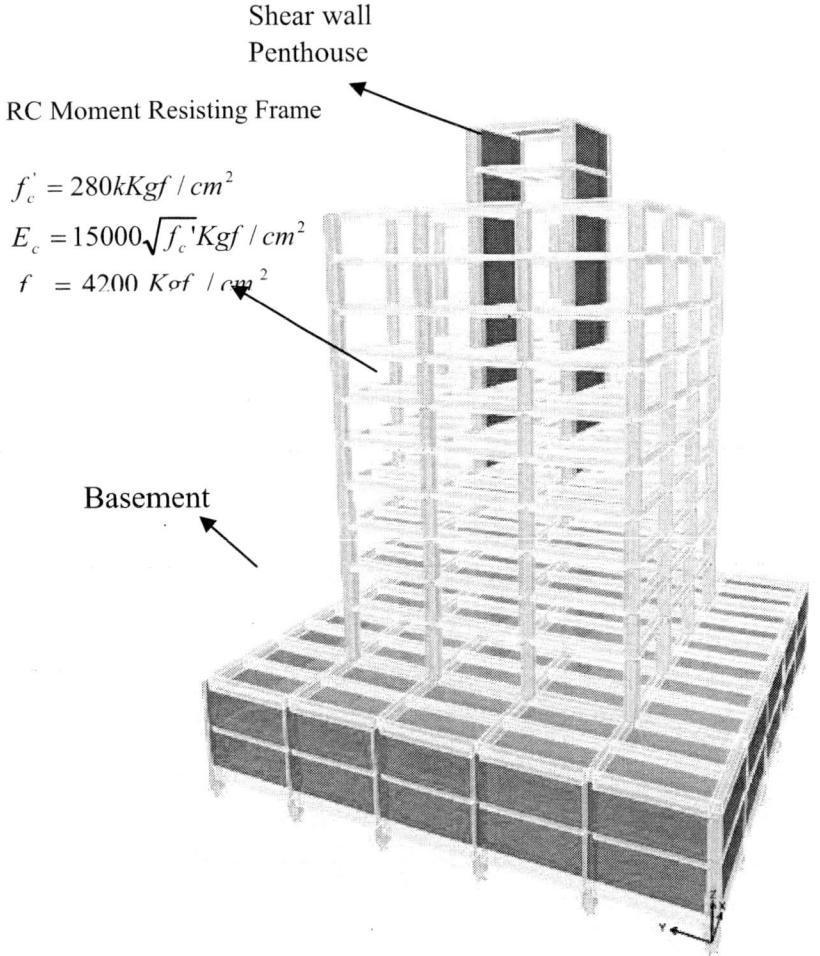

Figure 33. Layout of Example 1

1.5.1. An Example by Using the Draft Code Provisions

Performance Objectives and Criteria

Performance objective in Figure 34 is established with performance levels quantified by $IDDR_a$ and IDR_a in Sec. 4.1. Elastic design spectrum with 5% damping ratio regarding EQ1 and inelastic design spectra regarding EQ2 and EQ3 can be obtained according to the draft code.

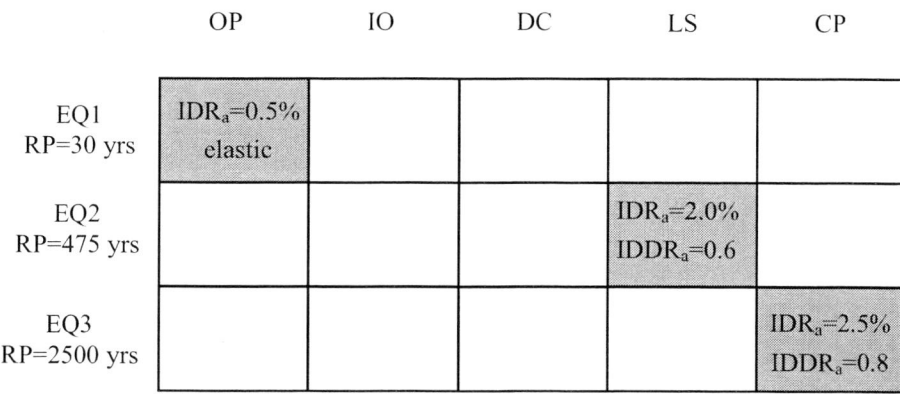

Figure 34. Performance objectives of the Example

Conceptual Design

Structural configuration, layout, material and foundation type have been decided. The building is a three-bay regular building equally spanned at 8(m) in both directions. It has a uniform story height of 3.2(m) except for 4.5(m) for the first story and 3(m) for both stories of the penthouse. A mat foundation has been used. A moment resisting frame system is in the Y direction. A dual system including additional shear walls is in the X direction. Structural system and preliminary sizing (Figure 35) are determined according to engineering experience. Considering non-structural brick infill walls, the system ductility capacity ratio is $\mu = 4$ in both directions. The structural height is within the code specified limit.

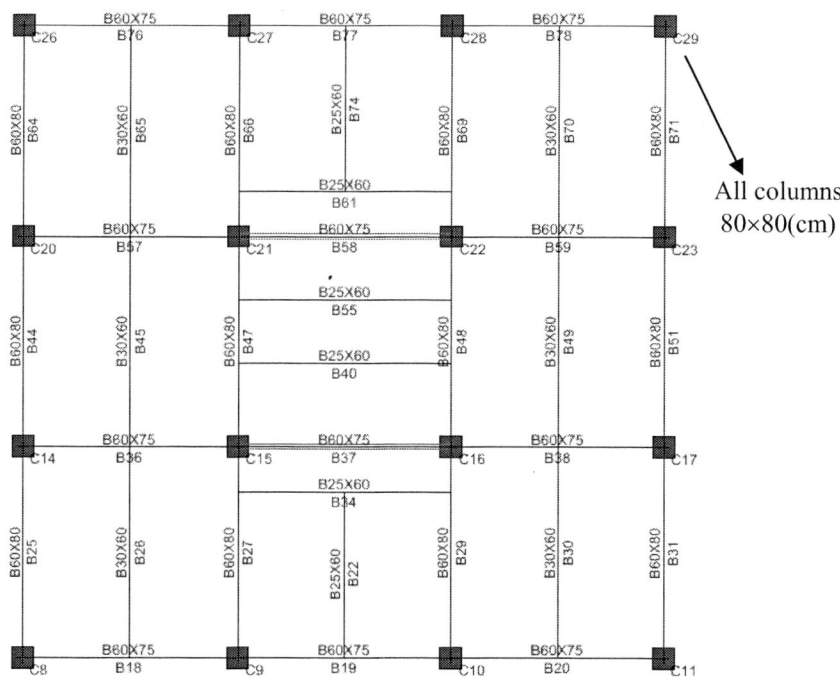

Figure 35. Initial sizing according to engineering experience (6th FL)

Preliminary Design

The building is preliminarily designed by the force-based method together with check on the displacement limit. A fundamental period of 0.69(s) is estimated according to the code suggested empirical formula. An upper limit of 0.97(s) is allowed to estimate the lateral design force. An eigenvalue analysis indicates that the 1st mode corresponds to a period of 1.39(s) in the Y direction. The dominant mode in the X direction is the 3rd mode, which corresponds to a period of 0.84(s). Therefore, the fundamental periods T used in estimating the design lateral force in the X and Y directions are 0.84(s) and 0.97(s), respectively. Since the structure is regular, a linear static analysis method is applicable at this stage. The equivalent lateral loading is applied individually in each of the major axes.

Regarding the performance level LS, $IDDR_a$=2/3 after revision and $\mu_a = 3$. Regarding CP, $IDDR_a$=1 and $\mu_a = 4$. A lateral design force is calculated for each performance goal. The maximum value corresponding to EQ2 controls the design. The design base shear in the X and Y directions are 0.086W and 0.071W, respectively.

The design base shear force has been distributed vertically. Member design is then conducted conventionally through elastic static analysis and by using the existing material design code [CPAMI 2002].

A preliminary check of IDR_a indicated that no stiffness deficiency has been found at this stage in this example.

Performance Evaluation

An elastic static analysis method is employed for EQ1. Regarding EQ2 and EQ3, a nonlinear static pushover analysis is used.

Structural Simulation and Modeling

Each shear wall element is modeled as an equivalent column element with a shear hinge assigned at the midpoint. A moment hinge is assigned at both ends of each main girder. The coupled axial force and biaxial bending moment hinge is assigned at both ends of each column element. Each hinge property is presented by a multi-linear hysteresis curve. Except for the shear hinges whose properties are estimated based on the local engineering practice, the properties of the moment related hinges suggested by ATC-40 [1996] have been adopted. The strain hardening ratio is taken as 5%. Effective stiffness is used when considering concrete cracking based on FEMA356 [2000]. For the girder type and column type elements, flexural rigidity is taken as half and 70% of the full section, respectively. Shear rigidity is reduced to 40%. $P-\Delta$ effect is considered.

Seismic Performance Analysis

Nonlinear static pushover analysis is carried out in each principal direction with a preload including 100% dead load plus 25% live load. Both a dominant mode pattern and a uniform load pattern proportional to the story mass are employed as pushover load patterns to trace the structural base shear versus the roof displacement capacity curve. The effective period (2.36s) associated with the effective stiffness is very close to that found by using a displacement-based design method as will be described later in this chapter.

The seismic performance point under the considered hazard level is estimated by the capacity-spectrum method. The performance point under EQ3 corresponds to

$S_a = 0.185(g)$ and $S_d = 0.31(m)$ in the Y direction and $S_a = 0.33(g)$ and $S_d = 0.22(m)$ in the X direction. The Performance point under EQ2 corresponds to $S_a = 0.18(g)$ and $S_d = 0.257(m)$ in the Y direction and $S_a = 0.25(g)$ and $S_d = 0.176(m)$ in the X direction. As an example, based on the dominant mode in the Y direction, the performance points under hazard level EQ3 is illustrated in Figure 36. According to the idealized bilinear capacity curve, the system yielding displacement and the ultimate displacement are known. *IDDR* is then estimated. Because member sizing standard and construction convenience are usually considered during design, it is found that the system overstrength factor is higher than that assumed in the preliminary design model. The ductility capacity ratio from the pushover analysis is also less than that provided in the code because the structural ductility is not uniformly distributed.

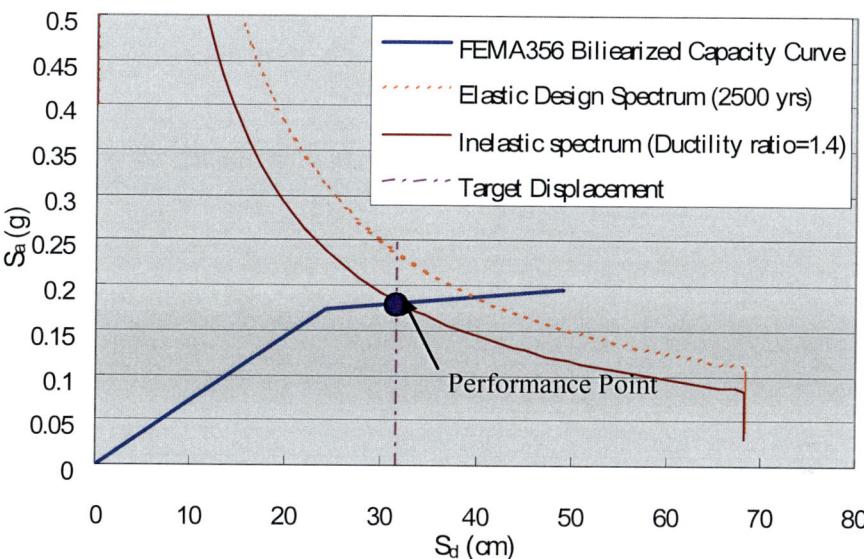

Figure 36. Performance points under hazard level EQ3 in the Y direction

Seismic Performance Evaluation of the Structural System

Local seismic demand is estimated by pushing the structure again to the state associated with the performance point. Under seismic hazard level EQ1, the structure remains elastic. The maximum inter-story drift ratio is *IDR*=0.14% and 0.24% in the X and Y directions, respectively. Such performance meets the prescribed performance criteria. Under seismic hazard levels EQ2 and EQ3, the system performance evaluations regarding *IDDR* and *IDR* are shown in Tables 6 and 7 in the X and Y directions, respectively.

It is worth mentioning that the seismic capacity curve (strength and stiffness) obtained from the nonlinear static pushover analysis is different from the preliminary design model. This is because, as mentioned earlier, the member sizing standard and construction convenience are usually considered during design. In addition, concrete cracking is considered in the pushover analysis.

Table 9. System performance evaluation regarding *IDDR* and *IDR* in the Y direction

Pushover Load Pattern	Dominant Mode		Story Mass Proportional	
Seismic Hazard Level	475yr RP	2500yr RP	475yr RP	2500yr RP
IDR_a	2.0	2.5	2.0	2.5
IDR (%)	1.4	1.6	1.35	1.6
Check if $IDR \leq IDR_a$	Yes	Yes	Yes	Yes
$IDDR_a$	0.6	1.0	0.6	1.0
IDDR	<0 (Elastic)	0.29	<0 (Elastic)	0.06
Check if $IDDR \leq IDDR_a$	Yes	Yes	Yes	Yes

Table 10. System performance evaluation regarding *IDDR* and *ID* in the X direction

Pushover Load Pattern	Dominant Mode		Story Mass Proportional	
Seismic Hazard Level	475yr RP	2500yr RP	475yr RP	2500yr RP
IDR_a	2.0	2.5	2.0	2.5
IDR (%)	0.92	1.1	0.78	0.9
Check if $IDR \leq IDR_a$	Yes	Yes	Yes	Yes
$IDDR_a$	0.6	1.0	0.6	1.0
IDDR	<0 (Elastic)	0.11	<0 (Elastic)	<0 (Elastic)
Check if $IDDR \leq IDDR_a$	Yes	Yes	Yes	Yes

The damage mechanism is clearly understood from the hinge formation sequence. As depicted in Figure 37, no single or partially weak story mechanism is found. No soft story mechanism is found. The ratio of the maximum story displacement to the estimated average story displacement is not larger than 1.2 for all stories. No torsion irregularity is found.

The maximum roof inelastic displacements under hazard levels EQ2 and EQ3 are estimated. The building separation distance is controlled by EQ2. At least 24.5(cm) and 33.3 (cm) must be provided in the X and Y directions, respectively.

Performance Evaluation of the Structural Members

Member strength capacity has been ensured during the member design phase. All members are elastic under hazard level EQ1. Under hazard level EQ2, the maximum number of yielding elements in the X and Y directions are 72 (Table 11) and 42 (Table 12)

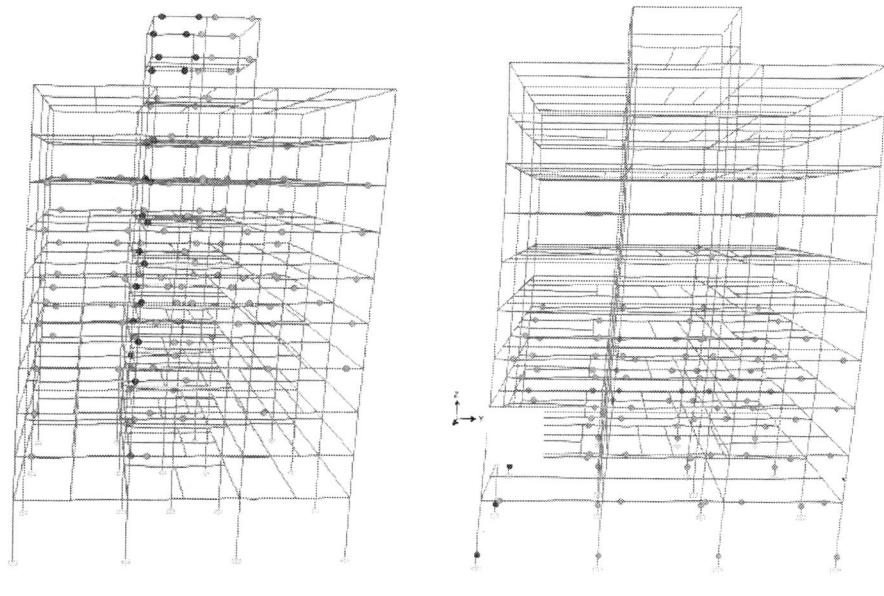

a) pushover in the X direction   b) pushover in the Y direction

Figure 37. Yielding mechanism of the structural system

respectively. *IDDR* values of all members are within acceptable limits. Although element performance evaluation for the performance goal associated with CP performance level in not necessary, the corresponding results for hazard level EQ3 are also presented in Tables 8 and 9 for reference purposes.

Table 11. Performance evaluation of the members in the X direction

Pushover Load Pattern	Dominant Mode		Story Mass Proportional	
Seismic Hazard Level	475yr RP	2500yr RP	475yr RP	2500yr RP
Allowable Number of Yielding Elements	904	904	904	904
Number of Yielded Elements	72	135	46	74
Number of Yielded Elements with $IDDR>IDDR_a$	0	0	0	0
Percentage of Satisfactory Elements	100%	100%	100%	100%
Meet 80% Requirement	Yes	Yes	Yes	Yes

It is worth mentioning that for RC structures, the structural stiffness is mainly controlled by the member size, whereas the structural strength is mainly controlled by the reinforcement ratio. By considering member sizing standards and construction convenience, structural ductility is not uniformly distributed. Once a plastic hinge is formed at the column base, it

Table 12. Performance evaluation of the members in the Y direction

Pushover Load Pattern	Dominant Mode		Story Mass Proportional	
Seismic Hazard Level	475yr RP	2500yr RP	475yr RP	2500yr RP
Allowable Number of Yielding Elements	904	904	904	904
Number of Yielded Elements	42	100	36	78
Number of yielded elements with $IDDR>IDDR_a$	0	0	0	0
Percentage of Satisfactory Elements	100%	100%	100%	100%
Meet 80% requirement	Yes	Yes	Yes	Yes

may develop very quickly and cause local damage. Based on the same column size for the first several floors, a higher reinforcement ratio assigned to the first 2 stories is helpful for uniform distribution of system ductility. It is better to modify the element performance criterion as follows.

- All columns must meet the same $IDDR_a$ as that of the structural system. Only 80% of the other members need to meet the same allowable value $IDDR_a$ as that of the structural system. Regarding unsatisfactory members, their performance level should not exceed one level higher than that of the structural system.

1.5.2. An Example by Using a Direct Displacement-Based Design Approach

Considering performance criteria associated with structural displacement, IDR_a of example 1, target displacement of the equivalent single degree of freedom (ESDOF) system x^* regarding each considered performance level is estimated. In this paper, x^* is estimated based on the normalized inter-story drift vector δ_Φ of the considered mode $\{\Phi\}$. Only the dominant mode has been considered in either direction. The ith story target displacement δ_i is calculated by equation $\delta_i = \delta_{\Phi i} \times IDR_a \times h_i$, where IDR_a stands for the drift criteria associated with the considered performance level and h_i is the height of the ith story above the building base. Finally, the target displacement is $x^* = \dfrac{\sum_{i=1}^{n}\left(m_i \times \delta_i^2\right)}{\sum_{i=1}^{n}\left(m_i \times \delta_i\right)}$. The values are 6.78(cm), 27.13(cm) and 33.91(cm) in the Y direction and 9.99(cm), 39.99(cm) and 49.99(cm) in the X direction under EQ1, EQ2 and EQ3, respectively. Story drift ratio and story displacement at the time when the structural system yields are assumed by using the empirical formula proposed by Priestley and Kowalsky [2000]. The corresponding yielding displacements of the ESDOF system are estimated in the same way as above and are 21.57(cm) and 20.2(cm) in the X and Y directions, respectively.

Several direct displacement-based design approaches were employed by considering both an idealized elasto-plastic hysteresis model and a bi-linear hysteresis model with a post yielding stiffness r=0.05. The former model gives relatively little conservative results. The substitute structure method that includes the so-called DBD approach [SEAOC 1999, Kowalsky et al. 1994, Priestley and Kowalsky 2000] and the EBD approach [SEAOC 1999, Court and Kowalsky, 1998] was adopted. Since the same equivalent elastic design spectrum in the draft code is employed, both approaches give the same design base shear. Using an inelastic design spectrum [Fajfar 1999], a yielding point spectrum [Aschheim 2002] and the simplified numerical procedure [Xue 2001a] in the capacity spectrum method resulted in the same design base shear because the equal displacement rule applies in this example. If the procedure is conducted with graphical assistance, all approaches need similar computational effort.

Without losing its accuracy, the simplified numerical procedure (Table 13) is easier because the procedure is completely numerical. No graphical assistance is needed. The overstrength factor is taken as $\Omega = 1.4\alpha_y = 2.1$ according to the code. It is worth noticing that the displacement-based method is based on the capacity curve obtained from a pushover analysis where it is necessary to consider the cracking of the RC elements. Therefore, stiffness demand in terms of fundamental period is also associated with reduced flexural and shear rigidity. That is, T_0 in Table 13 is an effective period associated with the effective stiffness K_0 of the RC structure. Such a period (2.43s in the Y direction) is usually larger than that (1.39s) obtained by an elastic dynamic analysis but is similar to that (2.36s) when taking into consideration the element cracking in the force-based preliminary design procedure. Besides, in Table 13, after S_{ay} has been estimated, equations $T_0 = 2\pi \sqrt{\dfrac{x^*}{S_{ay}}}$,

$$K_0 = \frac{4\pi^2 \times M^*}{9.8 \times T_0^2}$$ and $V_y = K_0 \times PF \times x^*_y$ have been used for calculation under EQ1

when the structure is expected to be elastic. The value, 9.8, used in the equations in Table 10 is for the purpose of unit transformation.

By using the simplified method [Xue 2001a], T_0 and V_d in Table 13 are equal to 2.43(s) and 0.072W, respectively. The design base shear is almost the same as that in the force-based procedure.

The seismic assessment above has also been conducted by considering the displacement-based design result in the Y direction. By replacing 50×80 (cm) girders and 60×80 (cm) girders with 50×75 (cm) girders and 60×75 (cm) girders, respectively in the Y direction, the effective period is 2.5(s). The maximum inter-story drift ratios are 1.5% and 1.76% under EQ2 and EQ3, respectively. System and element performance criteria are met as well.

The same approach in Table 13 has also been used in the X direction by adopting the draft code criteria that have been suggested without consideration of the difference between a moment resisting frame and a shear wall system. Stiffness and strength demands in terms of

Table 13. Displacement-Based Design by the Simplified Method in the Y Direction

Parameter	EQ1	EQ2	EQ3	Remark
x^* (m)	0.068	0.271	0.339	
x^*_y (m)	0.202	0.202	0.202	Subscript y means yielding Priestley and Kowalsky [2000]
μ_a	1.000	1.341	1.677	$\mu_a = \dfrac{x^*}{x^*_y} \geq 1$
T_S (s)	0.563	0.563	0.550	Parameters of elastic design spectrum
PSA (g)	1.867	7.840	9.800	PSA=Peak Spectral Acceleration
PSV (m/s)	0.167	0.702	0.858	PSV=Peak Spectral Velocity
$S_{ay,V}$ (m/s^2)	0.412	1.354	1.294	$S_{ay,V} = \dfrac{PSV^2}{\mu_a \times x^*}$
T_P (s)	2.550	2.813	3.216	$T_P = \dfrac{\sqrt{\mu} \times 2\pi \times PSV}{\mu \times A_{yv}}$
T_S^* (s)	0.563	0.630	0.652	$T_S^* = \sqrt{\dfrac{2\mu - 1}{\mu}} \times T_S$
S_{ay} (m/s^2)	0.412	1.354	1.294	$S_{ay} = S_{ay,V}$ for $T_P > T_S^*$
V_y (tonf)	879.183	969.299	926.698	For EQ2 and EQ3, $V_y = PF \times M^* \times S_{ay}/9.8$ $PF = \dfrac{\{\Phi\}^T[M]\{1\}}{\{\Phi\}^T[M]\{\Phi\}} = \dfrac{\sum m_i \Phi_i}{\sum m_i \Phi_i^2} = 1.304$ $M^* = k_3 \times M = 0.84 \times M = 5381.802(ton)$
K_0 (tonf/m)	3333.654	3675.354	3513.820	For EQ2 and EQ3, $K_0 = \dfrac{V_y}{PF \times x^*_y}$
T_0 (s)	2.551	2.43	2.48	For EQ2 and EQ3, $T_0 = 2\pi \sqrt{\dfrac{M^*}{9.8 \times K}}$
V_d/W	0.065	0.072	0.069	$V_d = V_y/\Omega$

T_0 and V_d/W are 3.58(s) and 0.036, respectively. The design base shear is much lower than that using the force-based approach. This is because the draft code criteria IDR_a is particularly good for a moment resisting frame but may be too large for some other systems. Although the deformation pattern of the shear wall in the X direction has been considered, target displacement in the X direction has been underestimated and is even less than that in the Y direction. For this dual system or other less ductile systems, such criteria can only be used as double check criteria after force-based design. They cannot be used as criteria either for an optimized design or in a direct displacement-based design procedure.

The displacement limit of a shear wall system suggested in the Appendix I of SEAOC [1999] has been considered. IDRa with respect to performance levels OP, LS and CP are 0.5%, 1.6% and 2.4%, respectively. By applying the same procedure in Table 13, stiffness and strength demands in terms of T_0 and V_d/W are 2.86(s) and 0.056, respectively. A nonlinear static pushover analysis has shown that a structure designed according to the demands failed to meet the prescribed performance criteria. All evidence indicated that the performance criteria are critical. Considering that the target displacement of the dual system is not larger than that of the moment resisting frame, based on the target displacement in the Y direction, IDR_a with respect to performance levels OP, LS and CP are 0.34%, 1.35% and 1.69%, respectively in the X direction. The inter-storey drift ratio of 1.69% associated with CP corresponds well to the experimental results. The same procedure is conducted again. The stiffness and strength demands in terms of T_0 and V_d/W are 2.416(s) and 0.078, respectively. Based on this design base shear, the shear wall thickness can be reduced by 5cm. Comparing the new capacity curve with that of the force-based design results (Figure 38), not much difference has been found for the initial stiffness. However, structural strength and ductility capacity have been reduced significantly. The maximum inter-storey drift ratios corresponding to the performance points under EQ2 and EQ3 are 1% and 1.2%, respectively. All elements are elastic or within IO performance level. Performance criteria are satisfied.

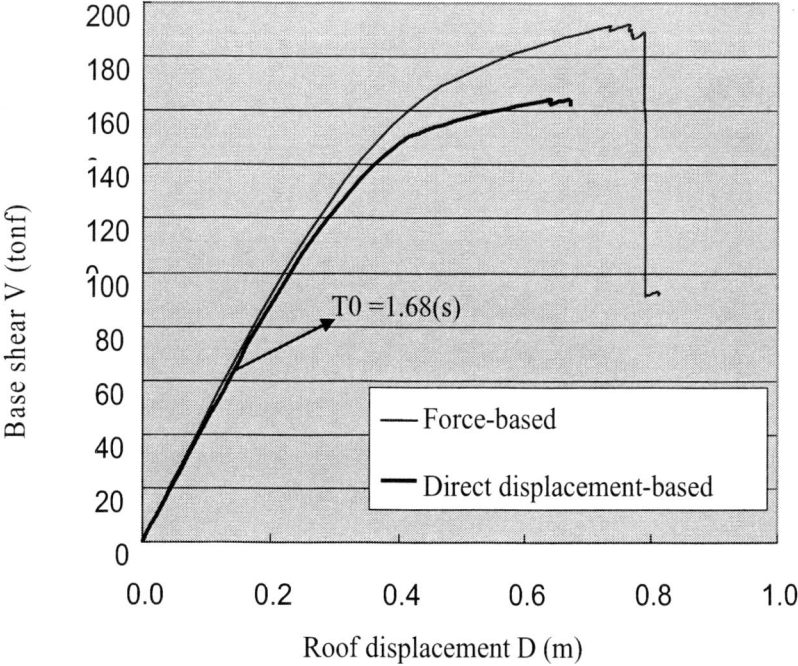

Figure 38. Capacity curves by the force-based and direct displacement-based design methods

1.6. FEEDBACK AND DIFFICULTIES ENCOUNTERED DURING THE DEVELOPMENT OF THE DRAFT CODE

Code development is not only a matter of technique, but also related to policy and implementation. Engineers, particularly senior engineers, who have been familiar with the history of code updating, are keen on the old and simple design methods. They have complained about frequent change of the code, increasing complexity of the design procedure and analysis methods and very fine seismic zone. Because engineering projects in this region are mainly design-build projects, increasing effort in the design phase does not mean reciprocally increasing design payment.

Necessary flexibility is considered when developing the code. For example, any rational method from reliable researches or provided by organizations that are credible or worth to be trusted can be used to simulate time history records that are compatible with the design spectrum. Again, engineers have to face new challenges to simulate time history records or they have to obtain the results through sub-contract with some credible organizations so that their design project will surely pass peer view. Again, this causes additional effort and cost. Therefore, a series of time history records that are compatible with the design spectrum of each seismic zone are needed.

1.7. MULTIDISCIPLINE COLLABORATION THROUGH VIRTUAL ENGINEERING

A construction project usually consists of several stages, which include feasibility study, planning, preliminary design, detailed design and construction, according to its task contents. At different stages, engineers need to produce related engineering drawings, documents and budgets to describe the functions, settings, geometry, dimensions, construction methods and costs of the design target. These design drawings and documents are very important communication means for the participants of the project, including clients, designers, site engineers, contractors, construction workers and government agencies. The amount of documents and drawings grows with the complexity of the project at different stages. This situation is even worse for large scale turnkey projects, which are very popular today. Collaboration between the design and construction team becomes frequent and important issue. To ensure the consistency between design and built structures is a challenge that project managers have to cope with everyday. Let the design team foresee construction difficulties of its designs and let the construction team fully understand the reasoning behind designs to achieve close and effective communication between all the participants is a demanding goal to reach.

1.7.1. VRML-Based Design Collaboration Model

The VRML models provide the convenience of design information search and exchange during the design and construction stages. The fact that VRML models are described by objects rather than geometrical data formats enable the simulation of real life conditions such as unable to go across a wall, gravity, collision detection etc. Herewith, we propose a idea of using VRML models, the so-called VRML-based Design Collaboration Model (VRMLDCM), to enhance the communication among the design and construction groups as illustrated in Figure 39. In Figure 39, the visualization models that are described by VRML are served as a universal outer communication interface for many construction work groups to understand the design intents within the center of circles. The products of design activities, e.g. drawings, guidelines, design documents, estimation reports, analysis data and other related data are located beyond the outer universal interface. Users navigate the VRML models through standard web browser and access all related design information by selecting arbitrary elements intuitively within the outer communication interface. Further, with the use of 3D VR models, engineers are able to feel, examine, simulate and evaluate their design works and identify inconsistent problems between design and construction stages to ensure high quality construction. As the process of building 3D VR models is normally time consuming and labor intensive, it is key to the success of VRMLDCM. Hence, it is important to develop an effective architecture to relax the aforementioned limitation for rapid prototyping VR models.

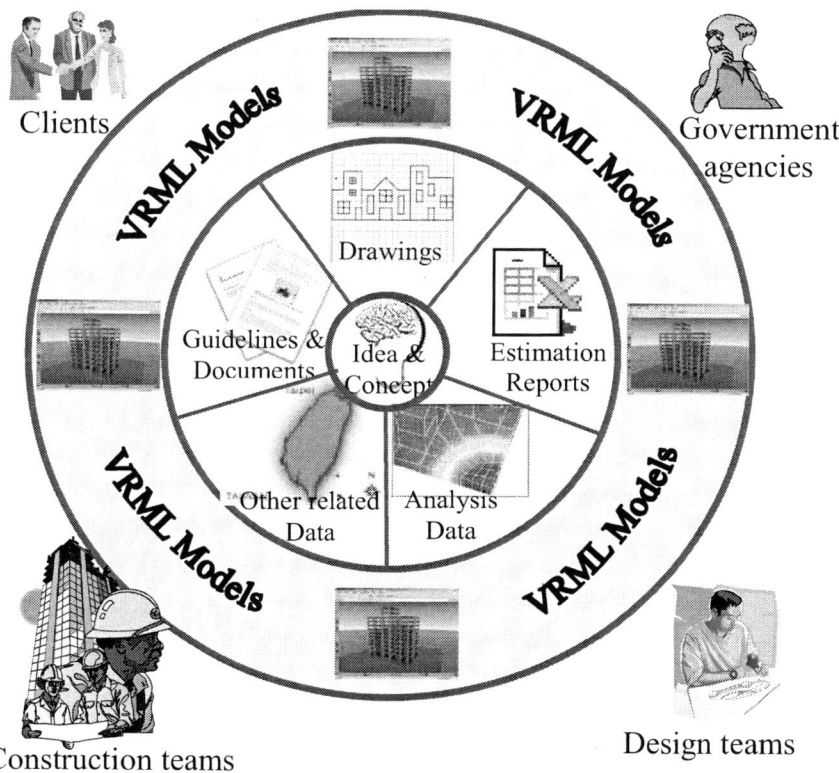

Figure 39. VRML-based design collaboration model diagram

The advantages of VRMLDCM require to the implementation of VRML models into the generic design activities of designers, who may not understand VRML. On the other hand, most designers are capable of writing high level programming languages and the data for building the VRML models actually are under fully controlled by the designers. Hence, we propose a VRML Implementation Architecture (VRMLIA) to bridge the gaps among generic analysis programs, post processors, VRML graphic kernel, world wide web and the users. In the proposed VRMLIA, several VRML models are generated automatically after the ending of programs and post processors as shown in Figure 40.

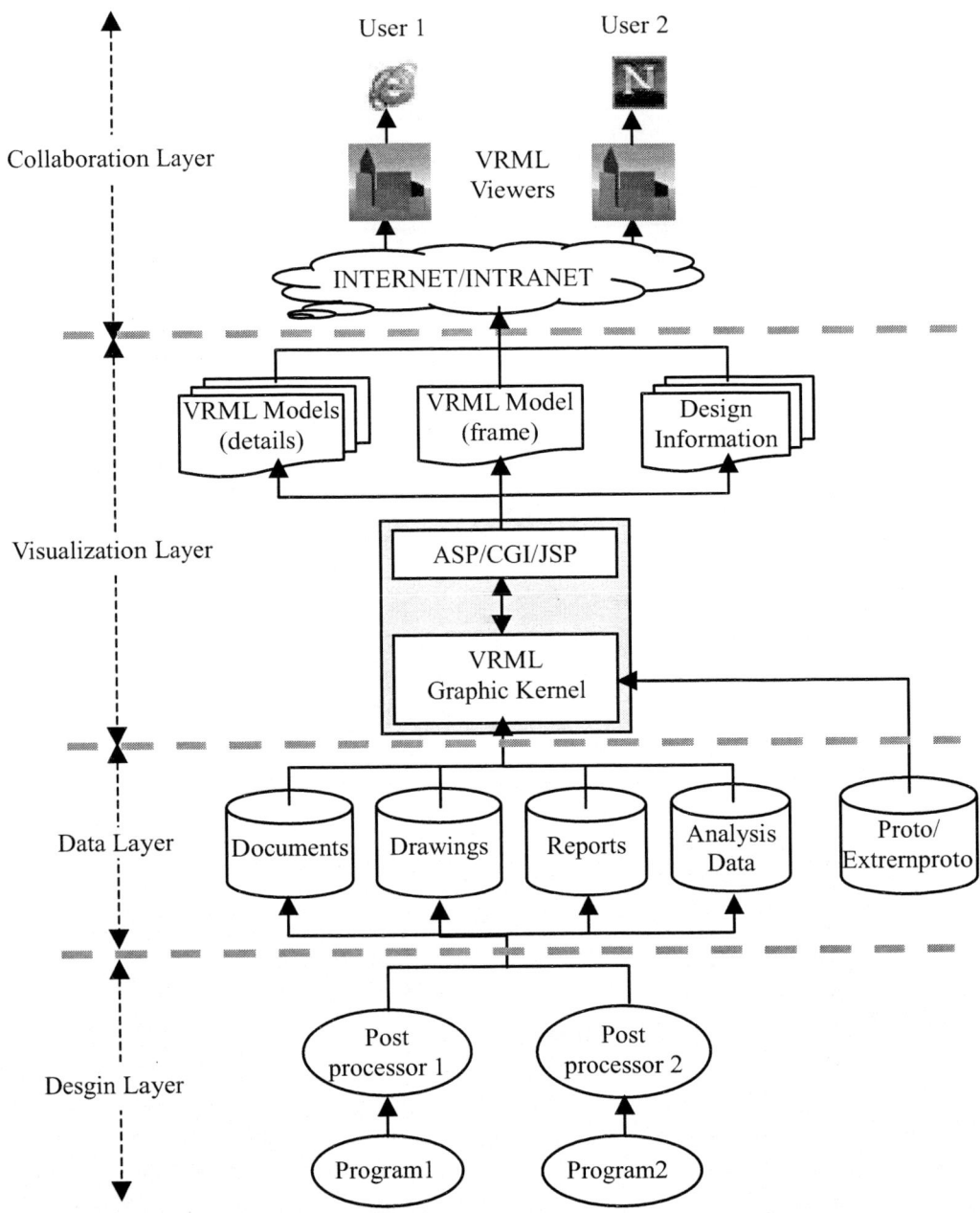

Figure 40. The VRML implementation architecture

The VRMLIA consists of four distinct layers. Starting from the designer side, the first layer is the so-called "Design Layer" where commercial software, self developed programs and post processors for constructing output data, documents, reports and drawings for specific engineering applications are located. The post processors are developed by engineers who are familiar with the application and enterprise based graphic libraries. The enterprise based graphic libraries provide graphic functions to display, import, export, convert and print CAD, bitmap images and VRML models for all enterprise users. In the graphic libraries, calling functions to generate VRML models are very similar to those for CAD drawing output.

The second layer is the "Data Layer" which includes a collection of output data from the "Design Layer," e.g., documents, drawings, reports and analysis data. A standard PROTO/EXTERPROTO library is also included in this layer. PROTO and EXTERNPROTO are two constructs in VRML to allow new nodes to be created by prototyping, which is a more powerful mechanism than instancing. A PROTO gives both the definition of the node and its implementation, whilst an EXTERNPROTO gives a node definition and a reference to a PROTO in another file. This mechanism allows libraries to be created and reused between files.

"Visualization Layer" is the third layer that consists of VRML graphic kernel, ASP/CGI/JSP server side scripts, VRML models and design information requested by the last layer, "Collaboration Layer." Using data in "Data Layer," VRML models for the entire frame and components of RC structures are created dynamically by the VRML graphic kernel and the user request collecting by sever-side scripts. Related design information is also generated and correlated with relevant objects in VRML models automatically. The visualization layer is located at designated Web site where information is opened to all network users. Finally, in the collaboration layer, design and construction teams are able to watch, examine and walk into the VRML models through the VRML viewer. These VRML models are also served as a universal interface for the retrieval of related design information.

1.7.2. Examples and Discussions

The design of a 12-story RC building is considered. In the design layer, a commercial software ETABS [Computers and Structures 2003] is used for the structural analysis of example structure while a post processor called ETABSD [Chen et al.], developed by Sinotech Engineering Consultants, Inc, is applied for the ductility design of high-rise RC structures. The data layer consists of four types of data that are the output of ETABS and ETABSD. The prototyping schemes provide 48 types of bar hooks listed on ACI [ACI Committee 318, 1995] and local design code and guidelines [CICHE 1997].

The VRML graphic kernel in the visualization layer is a component object extension of Sinotech's existing graphic libraries. There are two types of VRML models exporting from VRML graphic kernel in visualization layer. The first is a frame model for the whole frame of the RC building and is created in original scale based on the outcome of ETABS program. All related documents, drawings, calculation reports, analysis data and other related data are automatically linked to relevant VRML objects within the frame model. The second type of VRML models reveals the reinforcement details of the structural components in the frame VRML model, for example the placement of bars and stirrups in the beams, columns and joints. All VRML models and related design information are published on one appointed web site and opened to design and construction groups. Users are required to obtain valid accounts to login relevant project in the web site in the collaboration layer.

A screen shot of frame model is given in Figure 41 with a user control panel on the upper right side of the screen. Design information linked to the frame model includes beam calculation reports (Figure 42), column calculation reports (Figure 43), bar drawings (Figure 44), a summary report of beam and column quantities (Figure 45), design check detail (Figure 46), user manual and other details of arbitrary floor.

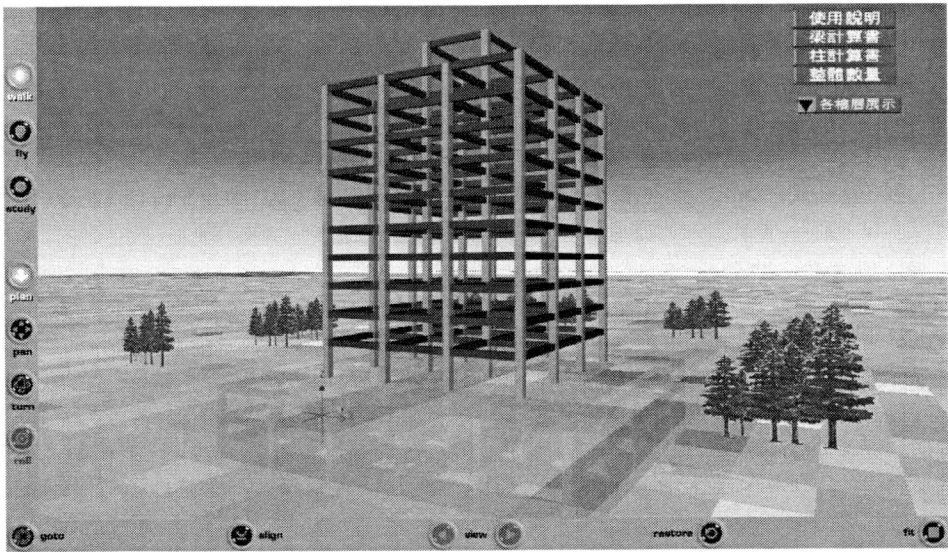

Figure 41. The frame model of example structure

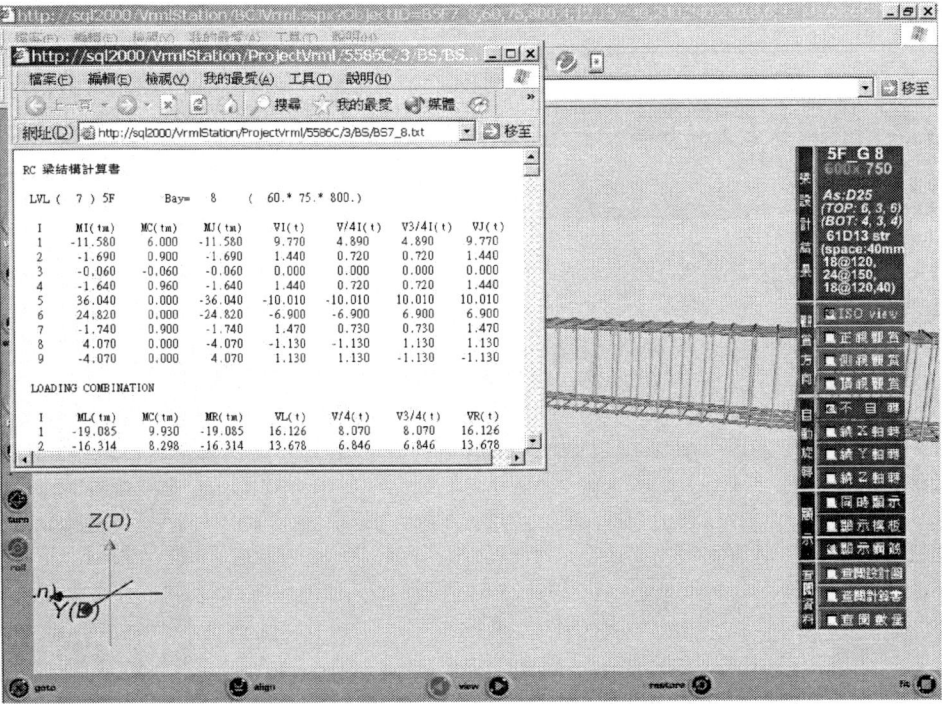

Figure 42. Calculation report and the reinforcement details of Beam 5F_G8

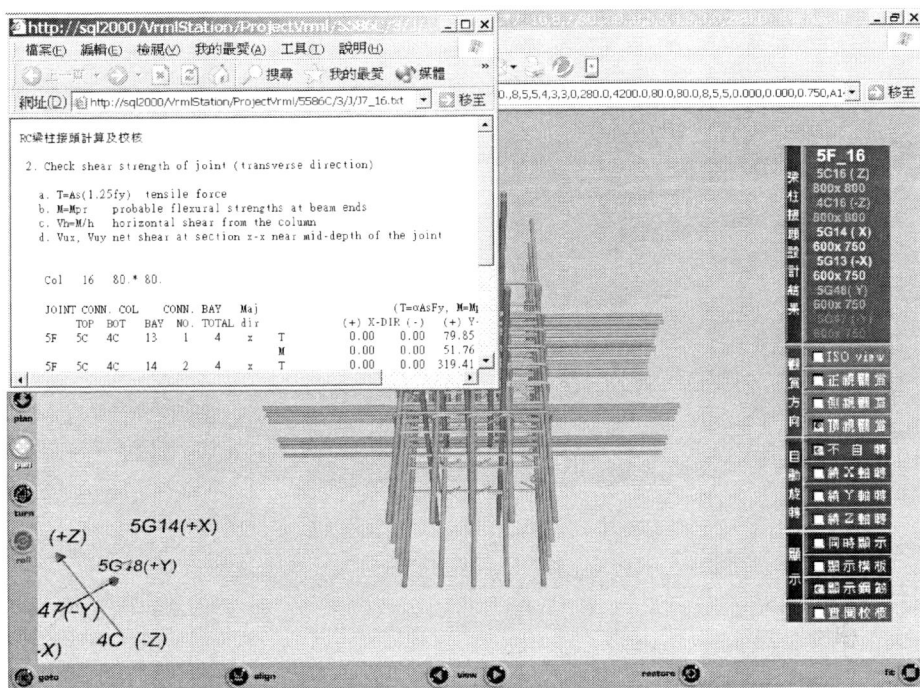

Figure 43. Calculation report and the reinforcement details of Column 5F_16

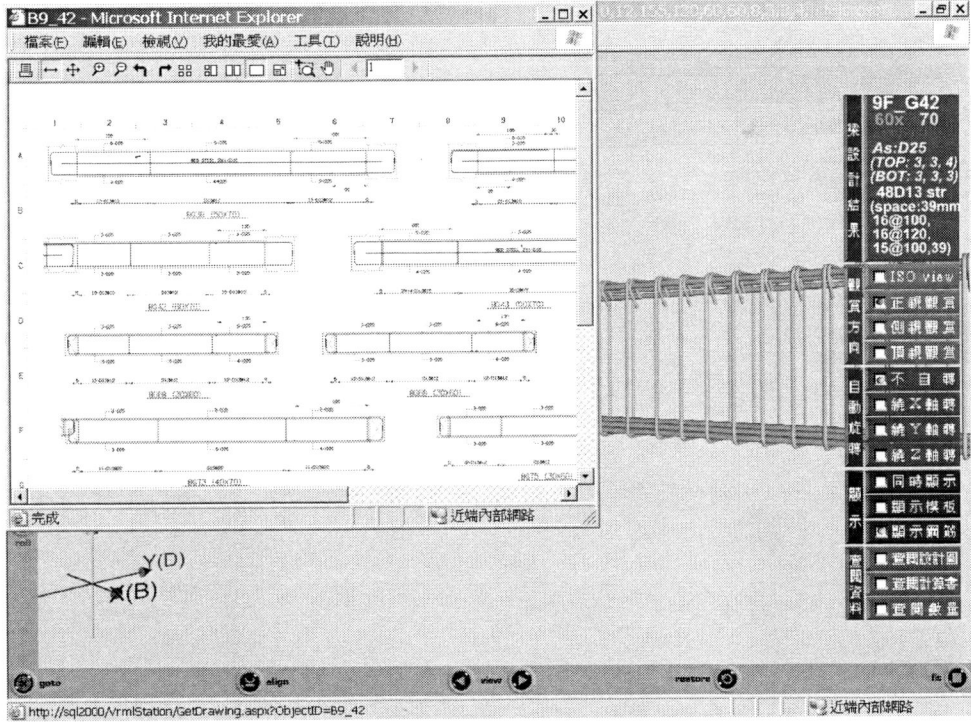

Figure 44. Bar drawings of the reinforcement details of Beam G42

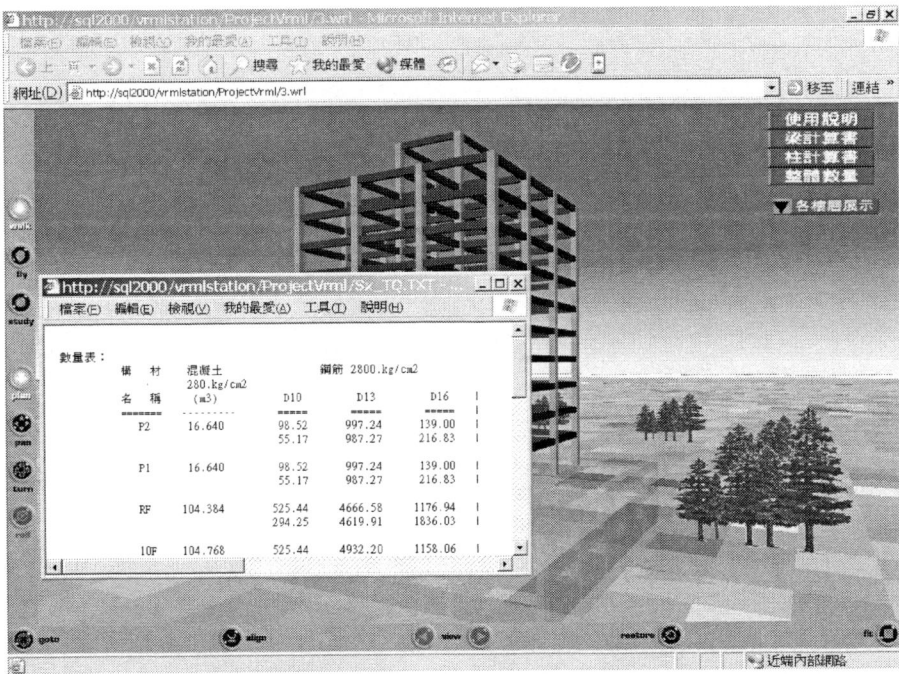

Figure 45. A summary report of beams and columns

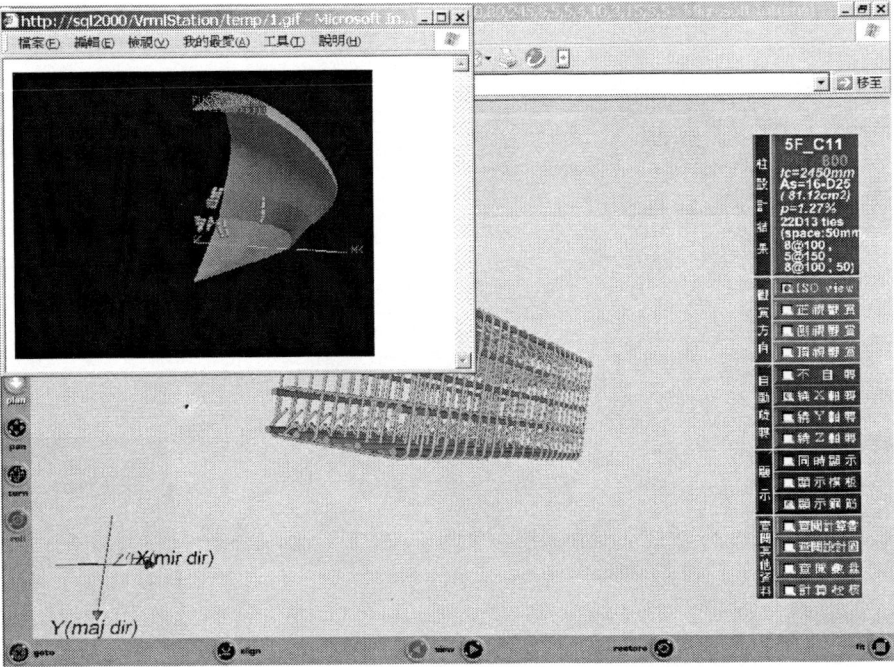

Figure 46. A design check diagram for column 5F_C11

Apart from the advantages of VRML models discussed previously, another significant benefit brought by VRMLDCM is that all related design products and information together with the automatically generated VRML models can be stored in a tiny CD as shown in

Figure 47. Such that, the users of the same construction group without internet/intranet access are still able to reach the VRML models and related design information.

Figure 47. A CD storing all related design information of the example structure

With VRML models, engineers are possible to conduct cursory examination on the design since view point and direction are able to be controlled by the viewers in the VRML browser, e.g. Figure 48and Figure 49. Further, the models can be setting up for automatically rotating to help the practitioners conduct design check.

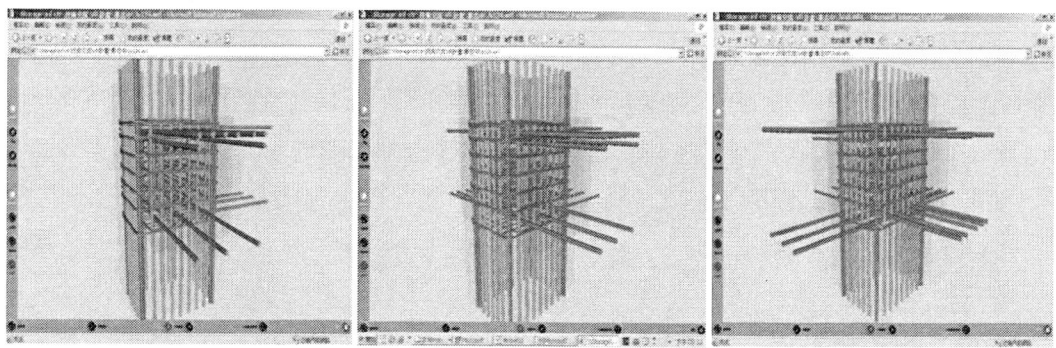

Figure 48. Automatically rotated reinforcement details at joints

Figure 49. A magnified view of column reinforcement details

Placement of bars, stirrups and bar hooks and bar continuation have been important concerns of the construction team. The VRML models of reinforcement details of beams, columns and joints provide a vivid demonstration for site workers before actually placing bars into the structures. Users can click on any structural components to view further design information when they navigate in the model as given in Figure 50.

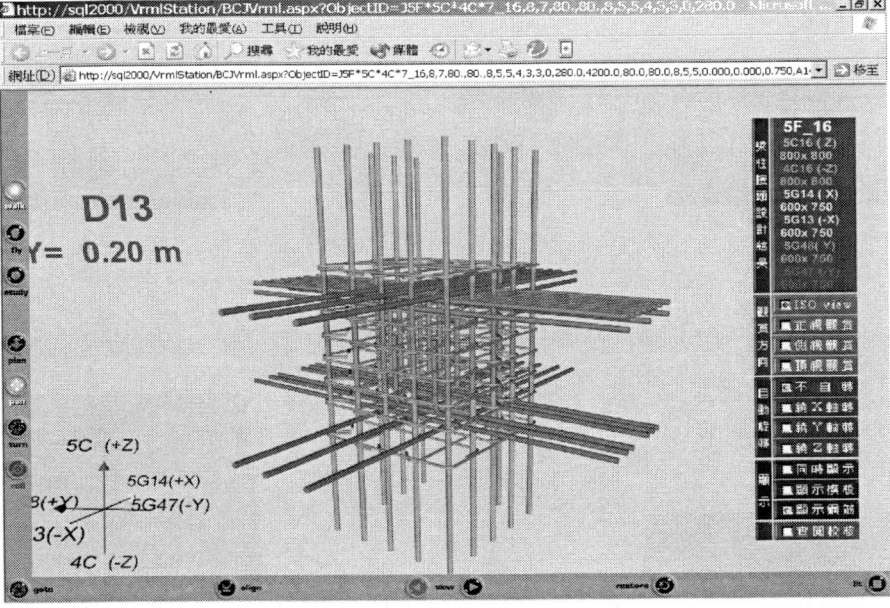

Figure 50. Bar placement details at Joint 5F_16

Finally, constructability issues such as the sequence of concrete placing, bars, stirrups and formworks that can be demonstrated by VRML models are very important to the construction group as shown in Figure 51. Figure 52 presents two animated screen shots showing the sequence and location of stirrup and bar placement on the column whenever viewer clicks on arbitrary bars or stirrups on the left or right hand side.

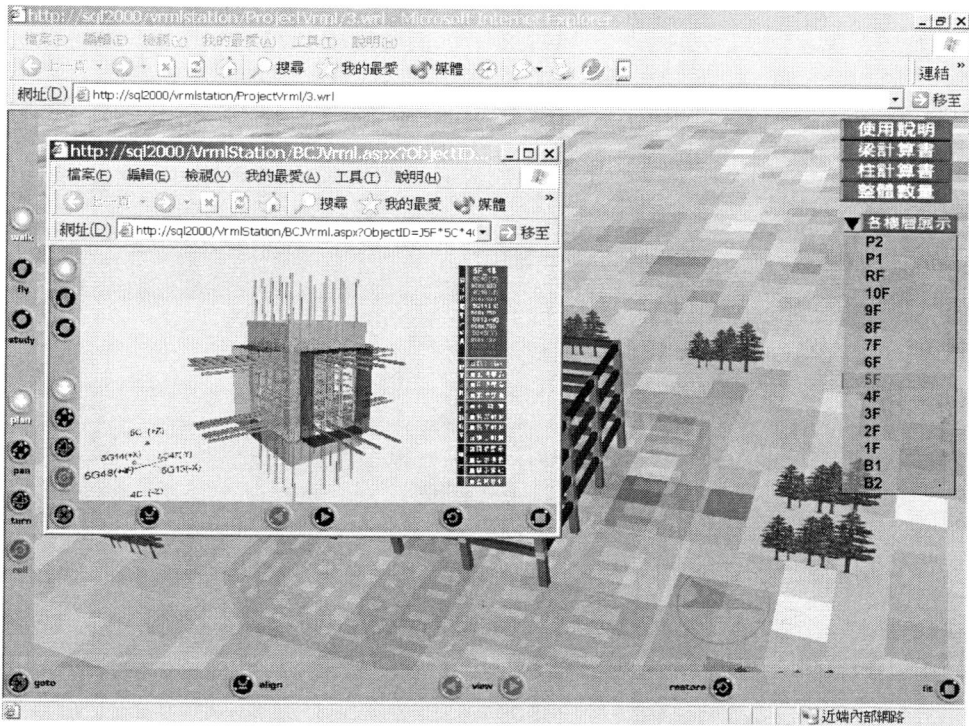

Figure 51. Displaying construction sequence with the joint reinforcement details

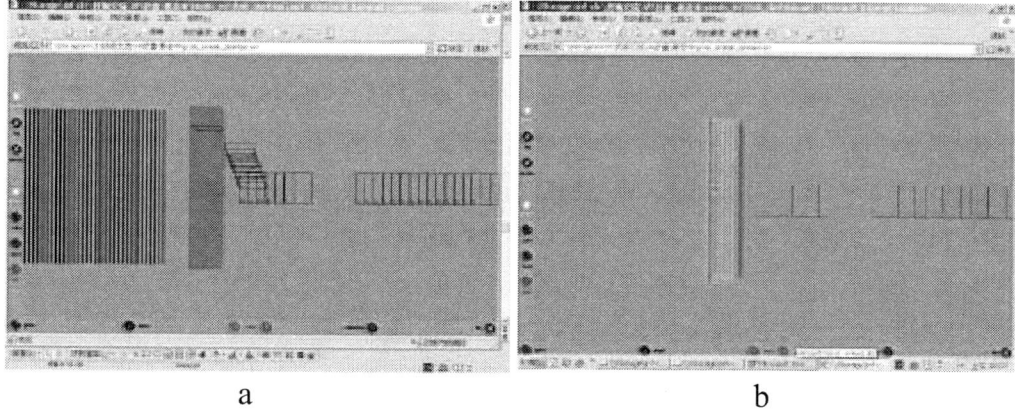

a   b

Figure 52. Bars and stirrups fly to the location of placement after clicking

1.8. SUMMARY AND FUTURE ACTIVITIES

The performance-based seismic design draft code introduces a transparent platform in which the owners and designers can exchange their views on the expected seismic performance of the buildings under different levels of earthquakes. For buildings of different seismic use groups, specific performance goals are established without employing an importance factor. Performance levels are quantified through parameters associated with structural strength, stiffness and ductility. Limitations on the degree of liquefaction potential of the construction site are given. Conceptual design rules with focuses on redundancy and uniform continuity of strength, stiffness and ductility are specified. A performance-objective–oriented preliminary design procedure is presented with consideration of flexibility. Preliminary checks on the inter-story drift limit may help in finding the stiffness deficiencies earlier in the preliminary design stage and save some computational effort, particularly for steel structures. Instead of ensuring a system possessing the prescribed ductility capacity through ductile design, the expected behavior or performance of the preliminarily designed structure is analyzed and evaluated through an analysis method, which may reasonably capture the real behavior. Engineers may have more confidence in the seismic performance of the designed structure under different levels of earthquake excitation. For performance evaluation of existing buildings, structural modeling should be based on site investigation of the real structure and the performance objective and acceptance criteria may be not as restrictive as that for new buildings.

By adopting the performance criteria in the draft code, direct displacement-based design procedures have been applied successfully for a moment resisting frame without iteration. The performance criteria associated with stiffness or displacement as suggested in the draft code should not be used either as optimized design criteria or in a direct displacement-based design procedure for structural systems other than moment resisting frames. For the dual structural system as illustrated in this chapter, alternative performance criteria can be obtained by considering that the target displacement of the dual system is not larger than that of the moment resisting frame.

The performance-based seismic design code is not, at present, intended to replace the current seismic design code. After authorization of the performance-based seismic design code in the future, it is supposed to exist with the current seismic design code for a period of time, during which possible amendments may be introduced into the performance-based seismic design code. Finally, it will replace the current seismic design code.

It is obvious that performance-based design increases design effort for engineers. Under the condition of design-build projects, systematic policies must be proposed to encourage engineers to use complicated but more advanced technique. At present, ABRI has established a policy system to give accreditation examination for both seismic design and construction of buildings. A certificate is awarded only to buildings passing the examination. The Construction and Planning Agency, Ministry of Interior has also modified the building law to prohibit adding a penthouse based on the owner's will only.

The code provisions only provide the minimum protection to the buildings. People have to face potential seismic risk due to uncertainties. Although the current version performance-based design code would allow engineers to judge if the building is functional with or without repair, or may threaten life safety under a certain level of earthquake, no one can gauge the engineer's confidence in such results. Reliability-based design method is also an important topic. Even if the confidence level is known, seismic risk, which may be very small, still exists. The public must understand this well and search for other tools such as insurance to transfer such risk. Further research studies on a risk-based design method, oriented to the probability of failure and a corresponding consequence including casualties, down time, loss and repair cost, are worth carrying out.

In this draft code, the design of non-structural components is done as usual to accommodate either acceleration or displacement. Non-structural damage is limited by structural drift limit. No specific criterion regarding economic loss of non-structural components or contents is provided. Life cycle cost including both structural and non-structural cost in plan, design, construction and maintenance phase is a valuable topic for future research. Both structural vulnerability and durability must be considered to estimate maintenance cost including repair.

Regular and intermittent inspections and performance monitoring are helpful in early warning to mitigate disaster. If individually monitored parameters can be related to structural performance, economic loss and potential risk, the results may serve as the basis for risk management.

With the increase in the complexity of modern construction projects, the need for information exchange is also increased because various programs are adopted in different stages. The technology of VRML provides a robust means to improve the communication among the members of construction projects. The presented collaboration model VRMLDCM is further implemented by VRMLIA to correlate engineering design process and World Wide Web technology to dynamically generate VRML models of RC structures and reinforcement details of beams, columns and joints. This architecture also helps engineers to connect design information with VRML objects. The practicability and robustness of VRMLIA have been demonstrated by applying it to the design of an RC building. It is possible to extend this versatile architecture to other engineering applications such as plumbing and piping design of building or others if proper programs are adopted in the architecture. Possible directions for future studies include the extension to X3D for VRML graphic kernel and its integration with assorted XML-based technology and the IFC/STEP information exchange models.

Although performance-based engineering covers a wide range of fields and the methodology sounds ideal, it is believed that the methodology will lead to a safer and more economic society if the performance can be clearly related to damage, life cycle cost, reliability and risk.

ACKNOWLEDGMENTS

The contents of this chapter are the result of several studies sponsored by Sinotech Engineering Consultants, Incorporated; Sinotech Engineering Consultants Company, Ltd.; and by the Architecture and Building Research Institute, Ministry of the Interior. Their financial support as well as the reviewers' comments are much appreciated. Special thanks are given to our colleagues: Mr. Li-Ming Chen, CEO, for his consistent inspiration and serving as a wonderful example as a research scientist; to the Vice CEO, Mr. Chuen-Cheng Lin, who has always been wise and considerate; to the center manager, Mr. Kuo-Chin Chen, and the vice manager, Mr. Li-ming Hsu, for their trust and friendship; to Mr. Chia-Wei Wu, who has done his best working with us for the past several years; to Mr. Weng-Yang Hzou and Mr. Chen-Hsun Wang of Sinotech Engineering Consultants Company, Ltd; Prof C. T. Lee of National Central University; and Prof. C. H. Loh on behalf of NCREE for sharing some of the photos.

REFERENCES

ABRI-CPAMI, [1999]. *Seismic Design Provisions and Commentary for Buildings*, Architecture and Building Research Institute, Construction and Planning Agency, Ministry of the Interior, CPRMI publisher, Taiwan. (in Chinese)

ABRI-CPAMI, [2005]. *Seismic Design Provisions and Commentary for Buildings*, Architecture and Building Research Institute, Construction and Planning Agency, Ministry of the Interior, CPRMI publisher, Taiwan. (in Chinese)

ACI Committee 318, [1995], *ACI 318-95 & Commentary (ACI 318R-95)*, Building Code Requirements for Structural Concrete, USA.

Aschheim, M. and Black, E. F., [2000], "Yielding Point Spectra for Seismic Design and Rehabilitation", *Earthquake Spectra*, Vol. 16(2): 317-335.

Aschheim, M., [2002]. "Seismic design based on the yield displacement", *Earthquake Spectra*, 18: 581–600.

ATC-34, [1995], *A Critical Review of Current Approaches to Earthquake-Resistant Design*, Applied Technology Council, Redwood City, California.

ATC-40, [1996], *Seismic Evaluation and Retrofit of Concrete Buildings*, Vol.1, Applied Technology Council, Redwood City, California, USA.

ATC-58, [2003], *Development of Next-Generation Performance-based Seismic Design Procedures For New and Existing Buildings*, The ATC-58 Project in Brief. http://www.atcouncil.org/atc-58.shtml.

Bachmann, H. and Dazio, A., [1997], "A Deformation-based Seismic Design Procedure for Structural Wall Buildings", *Proceedings of the International Workshop on Seismic Design Methodologies for the Next Generation of Codes*, Fajfar & Krawinkler (eds): 159-170.

Bertero, R. D., Bertero, V. V. and Teran-Gilmore, A. , [1996] "Performance-Based Earthquake-Resistant Design Based on Comprehensive Design Philosophy and Energy Concepts", Report No. UCB/EERC-96/01, *Earthquake Engineering Research at Berkeley*, Earthquake Engineering Research Center, College of Engineering, University of California at Berkeley, 1-7, 1996.

Bertero, R. D., Bertero V. V. and Teran-Gilmore, A., [1996C], *Performance-Based Earthquake-Resistant Design Based on Comprehensive Design Philosophy and Energy concepts*, Earthquake Engineering Research at Berkeley –1996, Report No. UCB/EERC-96/01: 1-7, May 1996. Earthquake Engineering Research Center, UC Berkeley.

Bertero, R.D. and Bertero, V.V., [2002], "Performance-based seismic engineering: The need for a reliable conceptual comprehensive approach." *Earthquake Engineering and Structural Dynamics*, 31, pp627-652.

Bertero, V. V. and Uang, C. M., [1992], "Issues and Future Direction in the Use of Energy Approach for Seismic-resistant Design of Structures", in P. Fajfar & H. Krawinkler (eds), *Nonlinear Seismic Analysis of RC Buildings*: 3-22. London: Elsevier.

Bertero, V. V., [1996b], *The Need for Multi-Level Seismic Design Criteria*, Earthquake Engineering Research at Berkeley –1996, Report No. UCB/EERC-96/01: 25-32, May 1996. Earthquake Engineering Research Center, UC Berkeley.

Bertero, V. V., [1997], "Performance-Based Seismic Engineering: A Critical Review of Proposed Guidelines", *Seismic Design Methodologies for the Next Generation of Codes*: 1-31, Bled/Slovenia/24-27 June 1997.

Bertero, V. V., [1982], "State of the art in seismic resistant construction of structures", *Proceedings of The Third International Earthquake Microzonation Conference*, University of Washington, Seattle, Washington, Volume II, pp.767-805.

Bozorgnia, Y, Bertero, V. V., [2004] "Damage spectrum and its applications to performance-based earthquake engineering", *13WCEE*, Vancouver, Canada.

Chopra, A. K., [1995], *Dynamics Of Structures, Theory And Application To Earthquake Engineering*, Prentice Hall, Englewood Cliffs, New Jersey.

Chopra, A. K., and Goel, R. K., [1999], *Capacity-Demand-Diagram Methods for Estimating Seismic Deformation of Inelastic Structures: SDF Systems*, Report No. PEER-1999/02, Pacific Earthquake Engineering Research Center, College of Engineering, University of California, Berkeley, April.

Chopra, A.K. and Goel, R.K., [2002], "A modal pushover analysis procedure for estimating seismic demands for buildings", *Earthquake Engineering and Structural Dynamics*; 31: 561-582.

CICHE, 1997, *Reinforced Concrete Engineering Research Report 14- Reinforced Concrete Design Guidelines and Explanations (Civil 401-86)*, Chinese Institute of Civil and Hydraulic Engineerings.

Collins, K. R., [1995], "A Reliability-Based dual Level Seismic Design procedure for building structures", *Earthquake Spectra*, 11(3), 417-429, California.

Comartin, C.D., Aschheim, M., Guyader, A., Hamburger, R., Hanson, R., Holmes, W., Iwan, W., Mahoney, M., Miranda, E., Moehle, J., Rojahn, C., Stewart, J., [2004], "A Summary of FEMA 440: Improvement of Nonlinear Static Seismic Analysis Procedures," *13WCEE*, Paper No.1476, Vancouver, B.C., Canada.

Computers and Structures, [2003], *ETABS—Integrated Building Design Software*, Berkeley.

Cosenza, E. and Manfredi, G., [1997], "The Improvement of the Seismic-Resistant Design for Existing and New Structures using Damage Concept", *Seismic Design Methodologies for the Next Generation of Codes*, Fajfar & Krawinkler (eds): 119-130, Balkema, Rotterdam.

Cosenza, E., Manfredi, G. and Ramasco, R., [1993], "The Use of Damage Functionals in Earthquake-Resistant Design: A Comparison among Different Procedures", *Earthquake engineering & Structural Dynamics*, Vol.22: 855-868.

Court, A. B. and Kowalsky, M. J., [1998], "Performance-Based Engineering of Buildings – A Displacement Design Approach", T109-1, *SEWC'98*, Edit by N.K. Srivastava, July 19-23, 1998, San Francisco, USA.

CPAMI, [2002]. *Code Requirements for Structural Concrete*, Construction and Planning Agency, Ministry of the Interior, CPRMI publisher, Taiwan. (in Chinese)

Elnashai, A. S., [2002], "Do We Really Need Inelastic Dynamics Analysis", *Journal of Earthquake Engineering*, Vol.6, Special Issue 1:123-130.

Fajfar P., [2000]. "A nonlinear analysis method for performance-based seismic design", *Earthquake Spectra*, 16(3): 573-592.

Fajfar, P. and Krawinkler, H. (eds.), [1997], *Proceedings of the International Workshop on Seismic Design Methodologies for the Next Generation of Codes*, Balkema, Rotterdam.

Fajfar, P., [1992], "Equivalent Ductility Factors, Taking into Account Low-Cycle Fatigue", *Earthquake Engineering and Structural Dynamics*, Vol.21: 837-848.

Fajfar, P., [1999], "Capacity Spectrum Method Based on Inelastic Demand Spectra", *Earthquake Engineering and Structural Dynamics*, Vol. 28: 979-993.

FEMA 273 (ATC-33), [1997], *Second Ballot Version "NEHRP Guidelines for Seismic Rehabilitation of Buildings"*, BSSC, Washington, DC, USA.

FEMA 302, [1997], *NEHRP Recommended Provisions for Seismic Regulations for New Buildings and Other Structures, Part 1: Provisions*. Prepared by Building Seismic Safety Council, National Institute of Building Sciences for Federal Emergency Management Agency, Washington D. C.

FEMA 350, [2000], *Recommended Seismic Design Criteria for New Steel Moment-Frame Buildings*, prepared by the SAC Joint Venture, a partnership of the Structural Engineers Association of California, the Applied Technology Council, and universities for Research in Earthquake Engineering; published by the Federal Emergency Management Agency, Washington, DC.

FEMA 356, [2000], *Prestandard and Commentary for The Seismic Rehabilitation of Buildings*, prepared by ASCE, published by the Federal Emergency Management Agency, Washington, DC. USA.

FEMA 368, [2000], *NEHRP Recommended Provisions for Seismic Regulations for New Buildings and Other Structures, Part 1: Provisions*. BSSC, Washington, DC, USA.

FEMA 440 (camera ready draft), [2004], *Improvement of Nonlinear Static Seismic Analysis Procedures*. Prepared by Applied Technology Council (ATC-55 Project) for Federal Emergency Management Agency, Washington, DC.

FEMA 450, [2003], *NEHRP Recommended Provisions for Seismic Regulations for New Buildings and Other Structures, Part 1: Provisions*. BSSC, Washington, DC, USA.

FIB, [2003], *Displacement-based seismic design of reinforced concrete buildings*, The International Federation for Structural Concrete, Switzerland.

Freeman, S. A., [1998], "Development and Use of Capacity Spectrum Method", *Proceedings of 6th U.S. National Conference on Earthquake Engineering*, Seattle, EERI, Oakland, California.

Hamburger, R. O., [1997], "A Framework for Performance-Based Earthquake Resistive Design", *EERC-CUREE Symposium in Honor of Vitelmo V. Bertero*, Jan. 31-Feb. 1, 1997, Berkeley, California.

Hartman, J. and Wernecke, J., [1996], *The VRML 2.0 Handbook*, Addison-Wesley.

Heidebrecht, A. [2004] "Code Development Issues Arising from The Preparation of The Seismic Provisions of The National Building Code of Canada", *13WCEE*, Vancouver, Canada, Aug.1~6.

IBC 2000, [2000], *International Building Code 2000*, International Code Council, CA, USA.

IBC 2003, [2003], *International Building Code 2003*, International Code Council, CA, USA.

IBC 2006, [2006], *International Building Code 2006*, International Code Council, CA, USA.

Iwan, W. D. and Gates, N. C. (1979). "Estimating Earthquake Response of Simple Hysteretic Structures," *Journal of The Engineering Mechanics Division*, Vol. 105, No. 3, pp. 391-405.

JSCA, [2000], *Design Methodologies for Response Controlled Structures*, Japan Structural Consultants Association (Specifications), Tokyo, JAPAN. (In Japanese)

K.C. Chen, T.W. Lin, F.W. Wang, G.C. Chiang, C.C. Chin, [1998], *Ductility Design and Graphic Programs for High Rise RC Structures*, Research Report SEC/R-CS-98-01, Sinotech Engineering Consultants, Inc. (In Chinese)

King, A. and Shelton, R. [2004], "New Zealand Advances in Performance-Based Seismic Design", *13WCEE*, Vancouver, Canada, Aug.1~6.

Kircher, C., Nassar, A., Kutsu, O., and Holmes, W., (1997), "Development of building Damage Functions for Earthquake Loss Estimation", *Earthquake Spectra*, 13, 663-682.

Kowalsky, M.J., Priestley, M.J.N. and MacRae, G. A. (1994). "Displacement-based design, a methodology for seismic design applied to single degree of freedom reinforced concrete structures," Report No. *SSRP-94/16*, Structural Systems Research, University of California, San Diego, La Jolla, California.

Krawinkler, H. and Nassar, A. A., [1992], "Seismic Design Based on Ductility and Cumulative Damage Demands and Capacities", *Nonlinear Seismic Analysis and Design of Reinforced Concrete Buildings*, eds (P. Fajfar and H. Krawinkler):23-40, Elsevier Applied Science.

Krawinkler, H., [1999a], "Challenges and Progress in Performance-Based Earthquake Engineering", *Proceedings of Seminar on Seismic Design for the Next Century*, NCREE-99-034. Taiwan, Dec. 1999.

Krawinkler, H., [1999b] *Advancing Performance-Based Earthquake Engineering*, http://peer.berkeley.edu/news/1999jan/advance.html.

Kunnath S. K., Valles-Mattox R.E. and Reinhorn A. M. (1996). "Evaluation of Seismic Damageability of Typical R/C Building in Midwest United States," 11th *World Conference on Earthquake Engineering*, Paper No. 1300.

Loh, C.H. and Hwang, Y.R., [2002]. *Seismic Hazard Analysis of Taiwan Area — considering multiple ground motion parameters*, Report No. NCREE-02-032, National Center for Research on Earthquake Engineering, Taipei, Taiwan. (in Chinese)

Mahin, S.A. and Bertero, V. V., [1976], "Problems in Establishment and Prediction of Ductility in Aseismic Design," *Proc. Int. Symp. EQ Struct. Engrg*. University of Missouri, Rolla, Missouri, U.S. I: 613-628.

Mander, J. B. and Dutta, A., [1996], " A Practical Energy-Based Design Methodolodgy for Performance Based Seismic Engineering", *Proceedings, 65th Annual Convention*: 319-338 SEAOC.

McGuire, R. K., [1995], "Probabilistic Seismic Hazard Analysis and Design Earthquakes: Closing the Loop", *Bull. Seism. Soc. Am*. Vol. 85(5): 1275-1284.

Miranda, E. [1997], "Estimation of Maximum Inter-story Drift Demands in Displacement-Based Design", *Proceedings of the International Workshop on Seismic Design Methodologies for the Next Generation of Codes*, Fajfar & Krawinkler (eds).

Moehle, J. P. [1992], "Displacement-Based Design of RC Structures Subjected to Earthquakes", *Earthquake Spectra*, 8(3), 403-428.

Moehle, J. P., [1992], "Displacement-Based Design of RC Structures Subjected to Earthquakes", *Earthquake Spectra*, EERI, Vol 8: 403-428.

Moehle, J. P., [1996], *Displacement-Based Seismic Design Criteria*, Earthquake Engineering Research at Berkeley, Report No. UCB/EERC-96/01:139-146, May 1996. Earthquake Engineering Research Center, UC Berkeley.

Moehle, J. P., [1996], *Displacement-Based Seismic Design Criteria*, Earthquake Engineering Research at Berkeley, Report No. UCB/EERC-96/01:139-146, May 1996. Earthquake Engineering Research Center, UC Berkeley.

National Research Council, [1999], *The Impacts of Natural Disasters: A Framework for Loss Estimation*, National Academy Press, Washington, D.C., 1999.

Newmark, N. M. and Hall, W. J., [1982], *Earthquake Spectra and Design*, Earthquake Engineering Research Institute, Berkeley.

Park Y. J., Ang A. H-S., Wen, Y. K., [1987], "Damage-Limiting Aseismic Design of Buildings", *Earthquake Spectra*, Vol.3(1): 1-26.

Park, R., [1986], "Ductile Design Approach for Reinforced Concrete Frames," *Earthquake Spectra*, EERI, Vol.2(3): 565-619.

Park, Y. J. and Ang, M., [1985], "Mechanistic Seismic Damage Model for Reinforced Concrete", *J. Struc. Eng.* ASCE, Vol. 111(4): 722-739.

Pires, J. A., Ang, A. H-S. and Lee, J. C., [1996], "Target Reliabilities for Minimum Life Cycle Cost Design: Application to A Class of R. C. Frame Wall Buildings", Paper No. 1062, *Eleventh World Conference on Earthquake Engineering*, Acapulco, Mexico.

Priestley, M. J. N. [1995], "Displacement-Based Seismic Assessment of Existing Reinforced Concrete Buildings", *Pacific Conference on Earthquake Engineering*, Australia, 20-22 Nov., 225-244.

Priestley, M.J.N. and Kowalsky, M.J., [2000], "Direct Displacement-Based Seismic Design of Concrete Buildings", *Bulletin of The New Zealand Society for Earthquake Engineering*, Vol. 33(4).

Priestley, M.J.N., [1993], "Myths and Fallacies in Earthquake Engineering – Conflicts between Design and Reality," *ACI*, SP 157: 231-254.

Priestley, M.J.N., Kowalsky, M. J., Ranzo, G. and Benzoni, G. (1996). "Preliminary development of direct displacement-based design for multi-degree of freedom systems," *Proceedings 65th Annual Convention*, SEAOC, Maui, Hawaii.

Qi, X. and Moehle, J. P., [1991], *Displacement Design Approach for Reinforced Concrete Structures Subjected to Earthquakes*, Report No. UCB/EERC-91/02. Earthquake Engineering Research Center, University of California at Berkeley, Berkeley, California.

Reinhorn, A. M. [1997]. "Inelastic analysis techniques in seismic evaluations," *Seismic Design Methodologies for the Next Generation of Codes, Proc. The International workshop on Seismic Design Methodologies for the Next Generation of Codes,* Balkema, Rotterdam, pp. 277-287.

Reinhorn, A. M., and Valles, R. E., [1996], D*amage Evaluation in Inelastic Response of Structures: A Deterministic Approach*, Report No. NCEER-95, National Center for Earthquake Engineering Research, State University of New York at Buffalo.

Riddell, R. and Garcia, J. E., [2001], "Hysteretic Energy Spectrum and Damage Control ", *Earthquake Engineering and Structural Dynamics*, Vol.30:1791-1816.

SEAOC, [1999], *Recommended Lateral Force Requirements and commentary* (SEAOC Blue Book), Structural Engineers Association of California-Seismology Committee.

Seneviratna, G. D., and H. Krawinkler, [1994], "Strength and Displacement Demands for Seismic Design of Structural Walls", *Proceedings of the Fifth U. S. National Conference on Earthquake Engineering*, Chicago, Illinois.

Sozen, M.A., [1981], "Review of Earthquake Response of Reinforced Concrete Buildings with a View to Drift Control", *State of the Art in Earthquake Engineering*: 383-418, Ankara.

Sozen, M.A., Monterio, P., Moehle, J. P., and Tang, H. T., [1992], "Effects of Cracking and Age on Stiffness of Reinforced Concrete Walls Resisting In-Plane Shear", *Proceedings, 4th Symp. On Current Issues Related to Nuclear Power Plant Structures, Equipment, and Piping*, Orlando, Florida.

Teran-Gilmore, A., [1996], *Performance-Based Earthquake-Resistant Design of Frame Buildings Using Energy Concept*, PhD. Thesis, University of California, Berkeley, CA.

Thomsen, J. H. IV, and J. W. Wallace, [1995], *Displacement-Based Design of RC Structural Walls: An Experimental Investigation of Walls with Rectangular and T-Shaped Cross-Sections*, CU/CEE-95/06, Clarkson University, Department of Civil and Environmental Engineering, Structural Engineering Research Lab, Potsdam, New York.

Tsai, I. C., Hsiang, W. B., Tsai, K. C., and Chang, K. C., [1994], *Regulations and Provisions, Commentary and Examples of Seismic Design for Buildings*, Structural Engineering Association, ROC. 7/1994 (in Chinese).

Vamvatsikos D and Cornell C.A., [2002]. "Incremental dynamic analysis", *Earthquake Engineering and Structural Dynamics*; 31(3): 491–514.

Vidic, T., Fajfar, P., and Fischinger, M., [1994], "Consistent Inelastic Design Spectra: Strength and Displacement", *Earthquake Engineering and Structural Dynamics*, Vol. 23: 507-521.

Vision 2000, [1995], *Conceptual Framework for Performance Based Seismic Engineering of Buildings*, SEAOC, Sacramento, CA, USA.

Wallace, J. W., and Moehle, J. P., [1992], "Ductility and Detailing Requirements for Bearing Wall Buildings", *Journal of Structural Engineering*, ASCE, Vol.118, No. 6, PP. 1525-1644.

Wang, M. T., Tserng, H. P. and Hsieh, S. H., [1998] 'A Benefit-Cost Analysis Model for Implementing Construction CALS', *Proceedings of the 15th International Symposium on Automation and Robotics in Construction*, Munich, Germany, March 31 – April 1, 173-182, 1998.

Wen Y. K., Hwang, H., and Shinozuka, M., [1994], *Development of Reliability-Based Design Criteria for Buildings under Seismic Load*, Technical Report 94-0023, National Center for Earthquake Engineering Research, State University of New York at Buffalo.

Xue, Q and Meek, J.L., [2001]. "Dynamic Response and Instability of Frame Structures", *Computer Methods in Applied Mechanics and Engineering*, 190(40-41): 5233-5242.

Xue, Q., [2001a], "A Direct Displacement-Based Seismic Design Procedure of Inelastic Structures," *Engineering Structures*, Vol. 23/11:1453-1460, Sep. 2001 (SCI)

Xue, Q., [2001b]. "Assessing the Accuracy of the Damping Model Used in Displacement-Based Seismic Demand Evaluation and Design of Inelastic Structures", *Earthquake Engineering and Engineering Seismology*, 3(2): 37-45.

Yeh, C. S, [2001], *Seismic Microzonation and the Associated Design Response Spectrum in Taipei Basin*, ABRI, Ministry of Interior. (in Chinese)

In: Structural Materials and Engineering
Editor: Ference H. Hagy

ISBN 978-1-60692-927-8
© 2009 Nova Science Publishers, Inc.

Chapter 7

RECENT ADVANCES IN THE DESIGN OF INDUSTRIAL GROUND-FLOOR SLABS WITH SPECIAL EMPHASIS ON PERMISSIBLE DEFORMATIONS

Ali A. Abbas[1,1], *Milija N. Pavlović*[2] *and Michael D. Kotsovos*[3]

[1]Department of Civil and Environmental Engineering,
Imperial College London, London, UK
[2]Department of Civil and Environmental Engineering,
Imperial College London, London, UK
[3]Department of Civil Engineering, National Technical University of Athens,
Athens, Greece

ABSTRACT

This chapter begins by briefly summarizing an extensive programme of research aimed at producing more accurate and economical guidelines for the analysis and design of ground-floor slabs (GFS), thus improving long-standing formulae that have been in use by engineers for many decades. This research addressed primarily the question of stress analysis and strength, as summarized in five recent publications. Current guidelines on GFS place emphasis on initial construction flatness and stresses (or strength) due to subsequent structural loading but, surprisingly, give very little guidance on displacements (and no guidance on slopes) under such loading, despite the fact that increasing heights of storage and forklift machinery require that serious consideration be given to such deformational matters. Most of the present chapter is, therefore, devoted to the question of serviceability under operating conditions. This is done by means of a parametric linear finite-element analysis (LFEA) of GFS resting on a Winkler subgrade. Square-patch loads with widths ranging from zero to 300mm were applied at the centre, edge and corner of the slab unit. The ensuing LFEA-based design equations and charts enable maximum deflections and slopes to be estimated for the full range of slab thicknesses and subgrade qualities found in practice. The work presently reported complements a similar study (the first of the five articles mentioned earlier) based on similar LFEA which proposes a more accurate permissible-stress analysis of

---

1 Corresponding author

GFS and offers the potential for reducing currently used thicknesses. This is illustrated by a fully worked out example in the appendix to the present chapter, where the reduced-thickness slab (obtained from the LFEA-based stress analysis) is now further checked for permissible deflection and slope criteria to ascertain the effect of thickness reduction on serviceability.

Keywords: concrete ground-floor slabs; finite-element analysis; maximum deflection; maximum slope; Winkler foundation

ABBREVIATIONS

A	constant in expression for δ
a_1	distance between corner and load centroid measured along corner bisector
B	constant in expression for δ
b	parameter used in previous elastic solutions and related to r
C	constant in expression for δ
c	perpendicular distance between load centroid and corner sides
D	flexural stiffness of the slab ($= \dfrac{E_c h^3}{12(1-\mu^2)}$)
d	width of square-patch load area
E_c	Young's modulus for concrete
f_{ct}	modulus of rupture of concrete
h	slab thickness
k	modulus of subgrade reaction
l	radius of relative stiffness ($= \left(\dfrac{D}{k}\right)^{\frac{1}{4}}$)
P	applied (patch or point) load
r	radius of circular loaded area
VNA	Very Narrow Aisle fork-lift truck
γ	unit weight of concrete
β	correction factor for arbitrary loaded-area width (d) obtained by reference to "standard" case $d = 100$ mm or arbitrary location of corner load obtained by reference to "standard" case $c = 50$ mm
μ	Poisson's ratio for concrete
δ	Maximum deflection

Subscripts

i, e, c denote interior (central), edge and corner locations respectively

50   refers to corner-load location of $c = 50$ mm

100, 200   refers to patch-load widths $d = 100$ mm, 200 mm respectively

1. INTRODUCTION

Concrete ground-floor slabs (henceforth GFS) form a vital infrastructure component as they feature in road and airfield pavements as well as industrial and commercial buildings where they act as a platform for manufacturing, storage, distribution and retail. The GFS is a concrete foundation that is required to sustain highly-concentrated loads (e.g. from vehicle and plane wheels in the case of pavements) and transfer them to the supporting soil without any structural failures or unaccepted settlements. The structural elements of a typical concrete GFS are depicted in Fig. 1. Industrial concrete GFS are another example of this type of slab, with the concentrated load being transmitted through the wheels of fork-lift trucks and through stanchions of racking-system frames (which are often heavily loaded over heights exceeding 10 m). Due to this similarity, current design guidelines used for industrial GFS are largely based on research originally carried out for airfield pavements.

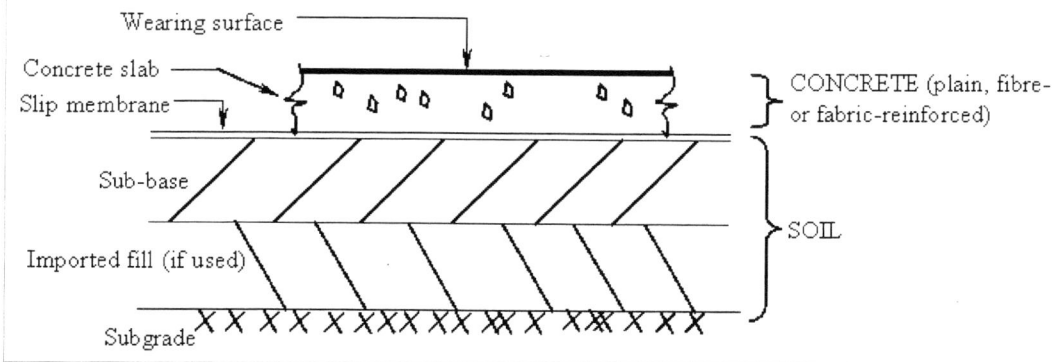

Figure 1. Structural elements of a typical concrete GFS.

Ground-floor slabs (GFS) are subject to loads acting on localized areas, and hence it is mandatory to study the stresses and deflections caused by such loads. Contrary to mat foundations of buildings (where columns are usually uniformly spaced over the entire surface and thus the soil pressure is assumed to be uniform) the local effects of concentrated loads on GFS should be accounted for. In order to study these effects, extensive numerical investigations of GFS were carried out [1] using linear and non-linear finite element analyses (FEA). These studies have been described in a series of five publications, encompassing a full appraisal of: elastic stresses [2], multiple loading [3], aisle-width considerations [4], non-linear FEA of plain- and fabric-reinforced concrete GFS [5, 6]. The proposed design equations and charts ensuing from these recent research publications by the authors have proven to be more economical than currently available guidelines and the potential savings are significant [1]. This is important since the concrete industrial-flooring sector is one of the largest users of concrete (e.g. in the UK an estimated 1.5 million cubic metres of floor are being laid every year). Clearly, addressing conservatisms inherent in current design guidelines is beneficial for both the economy and the environment since it helps reduce material waste.

As a companion to the aforementioned study of permissible stresses [2], elastic deflections are studied in the present chapter: both studies are required if potential savings in slab thickness (of which many experienced designers are aware) are to be justified and, eventually, realized.

2. HISTORICAL BACKGROUND

Despite their obvious significance, GFS design guidelines currently used in the UK [7, 8], USA [9] and Australia [10] date back to research work carried out 50 ~ 80 years ago, notably by Westergaard [11] and Meyerhof [12], and thus the need for their re-examination is evident. Among the first attempts to design GFS one should mention the works of Goldbeck [13] and Older [14], who studied the corner regions of concrete pavements. Soon afterwards – and based on the pioneering work of Winkler (1867), Zimmermann (1883), Hertz (1884) and Föppl (1922) [1] – Westergaard [11] presented a mathematical analysis to compute elastic stresses and deflections in concrete airfield pavements. His work has dominated GFS design ever since, and has been cited in almost every succeeding research publication. The solutions are available for the load situated at three different locations that are deemed to be critical (namely corner, interior and edge). Other researchers – e.g. Kelley [15], Teller and Sutherland [16] – attempted to improve these solutions, largely by means of experimental work. Westergaard's work forms the basis of elastic design guidelines in the UK [8] and elsewhere [9, 10].

Plastic solutions were also proposed for GFS, mainly by Meyerhof [6] and Losberg [17], which represent the basis of current design guidelines in the UK [7]. However, these studies have focused on the load-carrying capacity (i.e. strength) and have not investigated the deflection/slope (i.e. serviceability). This is due to the fact that they employ plastic-based solutions, which are inherently incapable of providing any information on deflections. This highlights the need for a method which does enable predictions of deflections and slopes. One such methodology is presented in this chapter for the elastic range that is the governing criterion in practice. (As for post-cracking deflections and slopes, these are beyond the scope of the present elastic study and have been discussed elsewhere [5, 6].) A more detailed survey of the state-of-the-art of GFS analysis and design can be found in Abbas [1].

3. PROBLEM DEFINITION

Although serviceability is a key issue in the design of GFS, it has not yet received the same degree of attention that has been devoted to stresses. The two key serviceability parameters are flatness and levelness, also known together as "surface regularity" [7]. As indicated schematically in Fig. 2, levelness can be defined as finishing the floor top surface so that this surface is kept normal to the gravity-force direction, while flatness can be defined as keeping every point on the top surface within a permitted variation from a datum plane.

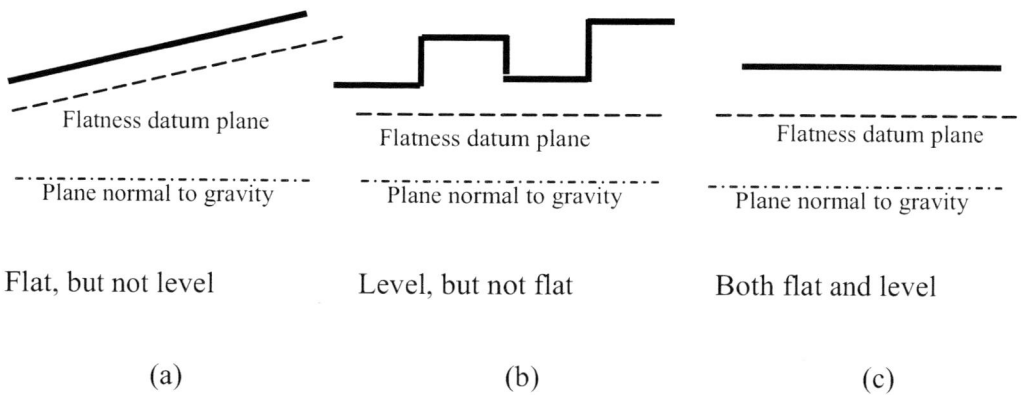

Figure 2. Schematic representation of levelness and flatness: (a) flat, but not level; (b) level, but not flat; (c) both flat and level.

The focus of current practice [7] is on surveying the slab after its construction, in order to check that adequate levels of flatness and levelness have been achieved. Consequently, there is a strong emphasis on improving construction techniques in order to achieve acceptable "surface regularity" levels prior to the application of the structural loading. On the other hand, there seems to be very little guidance on how to take into account deflections and slopes caused by structural loading itself. Paradoxically, it would appear that slab deflections are seen to be exclusively the result of an inadequate construction operation, rather than also being the result of poorly estimated structural loading. It is also puzzling that, whereas Westergaard's equations for estimating the maximum stresses [11] have been used extensively, his complementary equations for estimating maximum deflections [11] are omitted from most of the leading current design guidelines worldwide [8, 9, 10], without providing reasons for such an obvious omission (an exception is the latest set of UK guidelines [7] where a check for serviceability is recommended using Westergaard's deflection formulae, but even here no method for appraisal of slopes is provided). With such meagre guidance on deflections, and in the absence of information on resulting slopes, the designer can hardly anticipate the effect of applied loads on the targeted flatness and levelness standards. Consequently, predictions will be lacking on differential settlements within one panel of the slab or between two adjacent panels and associated slopes departing from levelness, which are critical for the performance of fork-lift trucks moving within one panel and/or across the edges of two adjacent panels. Therefore, design guidelines should, explicitly, contain equations for estimating maximum deflections and slopes due to structural loading, as well as criteria for defining acceptable deflections and slopes (in a similar fashion to those currently-available for suspended slabs). Until such deformations are properly estimated, hardly any rational recommendations can be made on the serviceability or performance of the slab. Construction techniques on their own do not guarantee a serviceable slab nor a level surface, since such techniques precede the application of loading.

Cracking is also a serious serviceability concern often voiced by industrial floor owners (purely for cosmetic and hygienic reasons or sometimes for more practical reasons such as controlling further damage to the floor caused by fork-lift steel wheels and durability). Although cracking considerations are beyond the scope of the present elastic investigation, they have been discussed in Abbas *et. al.* [5, 6].

4. METHODOLOGY USED

Maximum deflections and slopes were estimated using linear finite-element analysis (LFEA) of GFS resting on a Winkler subgrade. Single square-patch loads were applied in turn at the centre, edge and corner of the slab (for the case of multiple loads, influence lines for calculating maximum deflections are presented in Abbas *et. al.* [3]). The numerical investigations covered the full range of slab thicknesses and subgrade qualities found in practice. LFEA-based parametric studies were carried out which established relationships between the maximum deflections and the relevant variables. These studies were preceded by sensitivity analyses [1] which helped identify the key parameters and paved the way for the subsequent detailed parametric studies.

The studies were performed primarily to re-examine the accuracy of Westergaard solutions [11], which represent the current state-of-the-art of maximum deflection estimation. However, these are elastic solutions obtained for an uncracked section. Therefore, the LFEA-based investigations were also carried out for an uncracked section (the deflections at the post-cracking stage were studied by means of a non-linear concrete model and are described in Abbas *et. al.* [5, 6]).

As the design assumes the material to be linearly elastic, the deflection is directly proportional to the applied load and thus the entire LFEA was carried out for a unit load. Three-dimensional brick finite elements were used for both slab and subgrade and, as explained elsewhere [1, 2], the results are not affected by the slab dimensions beyond the *practical* range of 6.3 m by 6.3 m, which was adopted in the analyses. The variables considered in the parametric studies (carried out with an optimised FE mesh which, for all practical purposes, yields "exact" results [1, 2]) are listed below.

a. Thickness of the slab (h): the values considered are 150 mm, 175 mm, 200 mm, 225 mm, 250 mm, 275 mm and 300 mm, which cover the range of slab depths encountered in practice.

b. Modulus of subgrade reaction (k): the values considered are 14 MN/m^3, 27 MN/m^3, 54 MN/m^3 and 82 MN/m^3, which represent, respectively, orders of magnitude associated with very poor, poor, good and enhanced (very good) subgrades [8].

c. Width of the square-patch loaded area (d): the values considered are zero, 50 mm, 100 mm, 150 mm, 200 mm, 250 mm and 300 mm, which encompass all likely loading situations.

d. Location of single loads: in accordance with long-established practice, the critical locations of applied loads to be checked for are the centre, the edge and the corner of the slab (see Figs. 3).

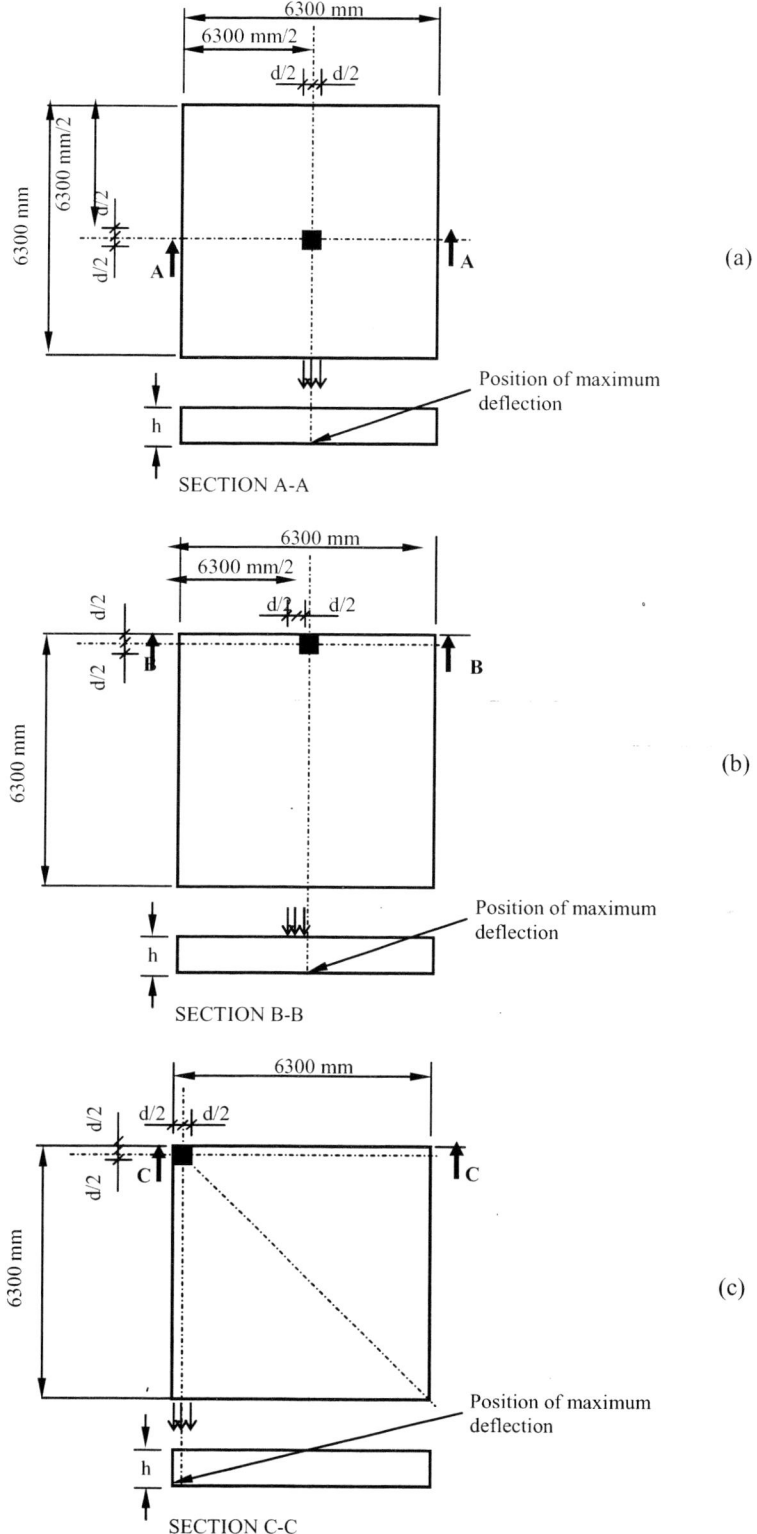

Figure 3. Layout of the single patch load acting on the slab at: (a) centre; (b) edge; (c) corner.

Table 1. Maximum deflections (in mm) due to a square-patch load of 1 kN only (i.e. no slab self-weight)

h (mm)	k (MN/m³)	Stiffness of Slab (MNm) $D = \dfrac{E_c h^3}{12(1-\mu^2)}$	Radius of relative stiffness (m) $l = \sqrt[4]{\dfrac{D}{k}}$	Loading position on slab							
				Corner		Edge				Central	
				Width of square-patch load (mm)								
				100	200	100	200	100	200	100	200
150	14	9.78	0.914	0.0893	0.084	0.0342	0.0327	0.0106	0.0105		
	27		0.776	0.0619	0.0572	0.0238	0.0225	0.00732	0.00729		
	54		0.652	0.0417	0.0379	0.0162	0.0152	0.00502	0.00499		
	82		0.588	0.0327	0.0294	0.0128	0.0119	0.00400	0.00397		
175	14	15.53	1.026	0.0729	0.0686	0.0281	0.0270	0.00866	0.00863		
	27		0.871	0.0508	0.0473	0.0194	0.0185	0.00598	0.00596		
	54		0.732	0.0344	0.0316	0.0132	0.0125	0.00408	0.00406		
	82		0.660	0.0271	0.0247	0.0105	0.00985	0.00326	0.00324		
200	14	23.19	1.134	0.0609	0.0576	0.0237	0.0228	0.00729	0.00727		
	27		0.963	0.0427	0.0400	0.0164	0.0157	0.00504	0.00502		
	54		0.810	0.0291	0.0269	0.0111	0.0106	0.00342	0.00341		
	82		0.729	0.0230	0.0211	0.00883	0.00833	0.00272	0.00271		
225	14	33.02	1.239	0.0520	0.0494	0.0204	0.0197	0.00626	0.00625		
	27		1.052	0.0365	0.0344	0.0141	0.0135	0.00433	0.00432		
	54		0.884	0.0250	0.0233	0.00954	0.00910	0.00294	0.00293		
	82		0.797	0.0199	0.0183	0.007577	0.00718	0.002335	0.00232		
250	14	45.29	1.341	0.0451	0.0430	0.0179	0.0173	0.00545	0.00544		
	27		1.138	0.0318	0.0300	0.0123	0.0118	0.00379	0.00378		
	54		0.957	0.0219	0.0204	0.00833	0.00797	0.00257	0.00256		
	82		0.862	0.0174	0.0161	0.00660	0.00629	0.002036	0.00202		
275	14	60.28	1.440	0.0396	0.0379	0.0159	0.0155	0.00481	0.00480		
	27		1.222	0.0280	0.0265	0.0109	0.0105	0.00335	0.00334		
	54		1.028	0.0193	0.0181	0.00737	0.00708	0.00228	0.00227		
	82		0.926	0.0154	0.0143	0.00584	0.00558	0.001801	0.00179		
300	14	78.26	1.538	0.0353	0.0338	0.0144	0.0140	0.00429	0.00429		
	27		1.305	0.0249	0.0236	0.00975	0.00945	0.00298	0.00298		
	54		1.097	0.0172	0.0162	0.00659	0.00635	0.00204	0.00203		
	82		0.988	0.0138	0.0129	0.005219	0.00500	0.001611	0.00160		

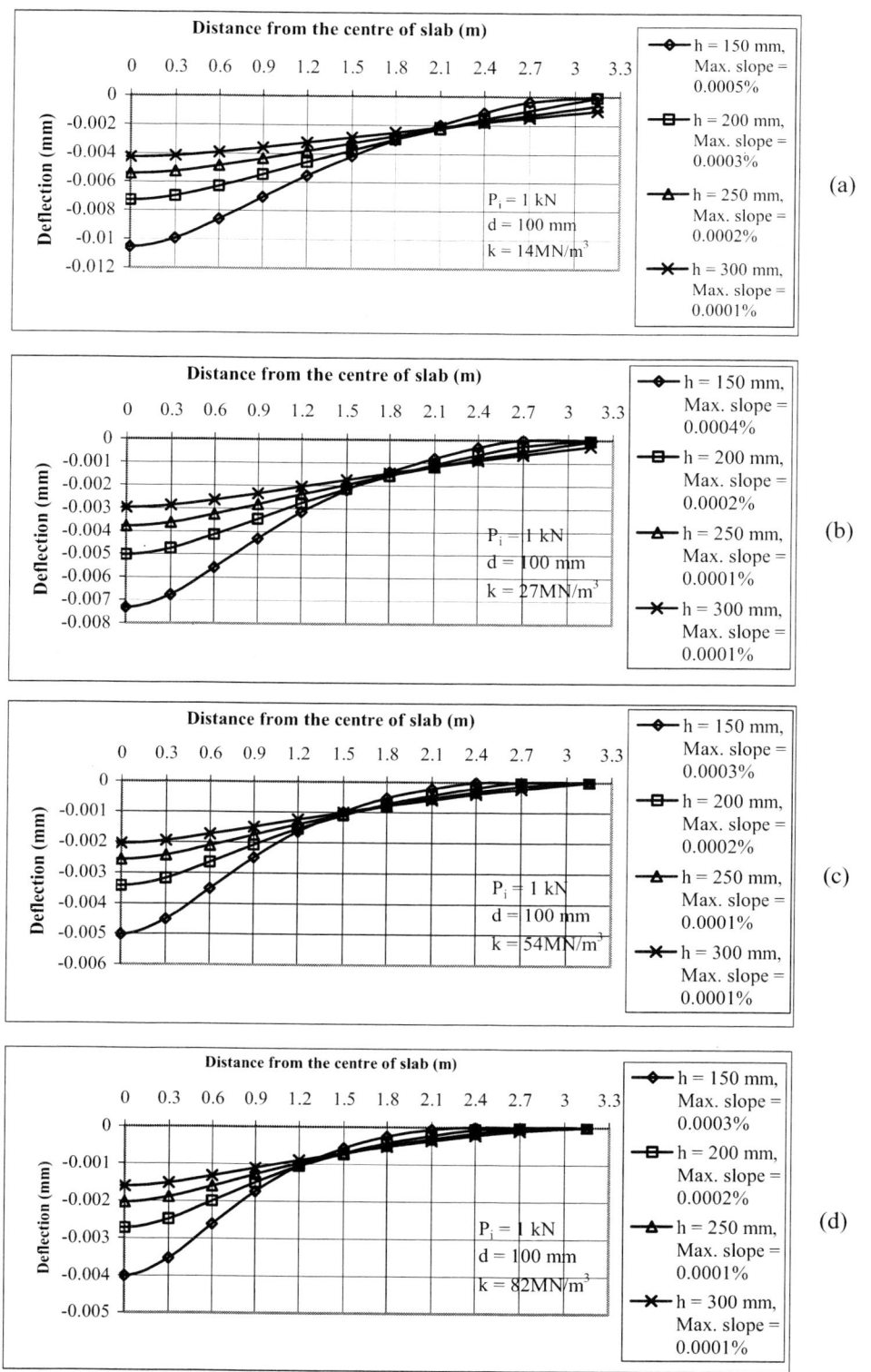

Figure 4. Deflected shape of the slab along the centreline due to a central load for different values of h and k. (Note: origin is at the centre of the slab.)

Table 1 shows the maximum deflections stemming from the present LFEA for a square-patch load of total value of 1 kN (and of widths d = 100 mm and 200 mm, respectively). Note that these results do not include the effect of the slab self-weight, which is considered subsequently.

In addition to estimating maximum deflections, it is important to calculate the maximum slopes because it helps estimate the effect of applied patch loads on the flatness and levelness of the slab, and enables the designer to check whether or not the prescribed "surface regularity" criteria are met.

Consideration of maximum slopes is particularly critical for Very Narrow Aisle (VNA) trucks, which are commonly used on industrial floors to store materials at heights exceeding 10 m. In order to calculate maximum slopes, plots of the deflected shapes were constructed for slabs under unit loads (e.g. such as those shown in Fig. 4 for the central-load case and various values of slab thickness and subgrade stiffness). Based on these graphs, all possible values of slopes were calculated at different sections along the span and then the maximum value was identified (again for every value of h and k). In this manner, the maximum-slope values were calculated as indicated in Fig. 4 for the unit central-load case. The same exercise was repeated for the edge- and corner-load cases and the results are also reported in this chapter. Clearly, the maximum-slope results are based on unit loads: thus, for serviceability checks, the currently reported values must be multiplied by the actual magnitude of the applied load.

5. MAXIMUM DEFLECTION FOR THE CENTRAL-LOAD CASE (δ_i)

Parametric studies were carried out to estimate the maximum deflection due to a central square-patch load acting on the slab. From the LFEA, it is clear that this maximum value occurs at the bottom of the slab and directly under the centre of the patch load, as depicted in Fig. 3a. The deflected shapes along the centreline of the slab, for a square-loaded area of width 100 mm and different values of h and k, are depicted in Figs 4a to 4d. These charts enable the designer to quickly assess the maximum deflection and slope (and their locations) caused by a central load.

Based on Westergaard's work [11], the following relationship was assumed between the maximum deflection δ_i and the total applied load P_i, modulus of subgrade reaction k and radius of relative stiffness l:

$$\delta_i \propto \frac{P_i}{kl^2} \qquad (1)$$

where l is defined by Westergaard as [11]:

$$l = \sqrt[4]{\frac{E_c h^3}{12(1-\mu^2)k}} \qquad (2)$$

where E_c is the modulus of elasticity for concrete and μ is its Poisson's ratio (taken as 0.15).

The validity of relationship (1) was confirmed by the FE runs as depicted in Figs 5a and 5b (which are based on Table 1) for $d = 100$ mm and 200 mm, respectively. It is clear that the difference between these relationships is small and, therefore, the two graphs can be combined in one (Fig. 5c). Thus, one average equation is sufficient to describe relationship (1) for the two widths of loaded area, namely

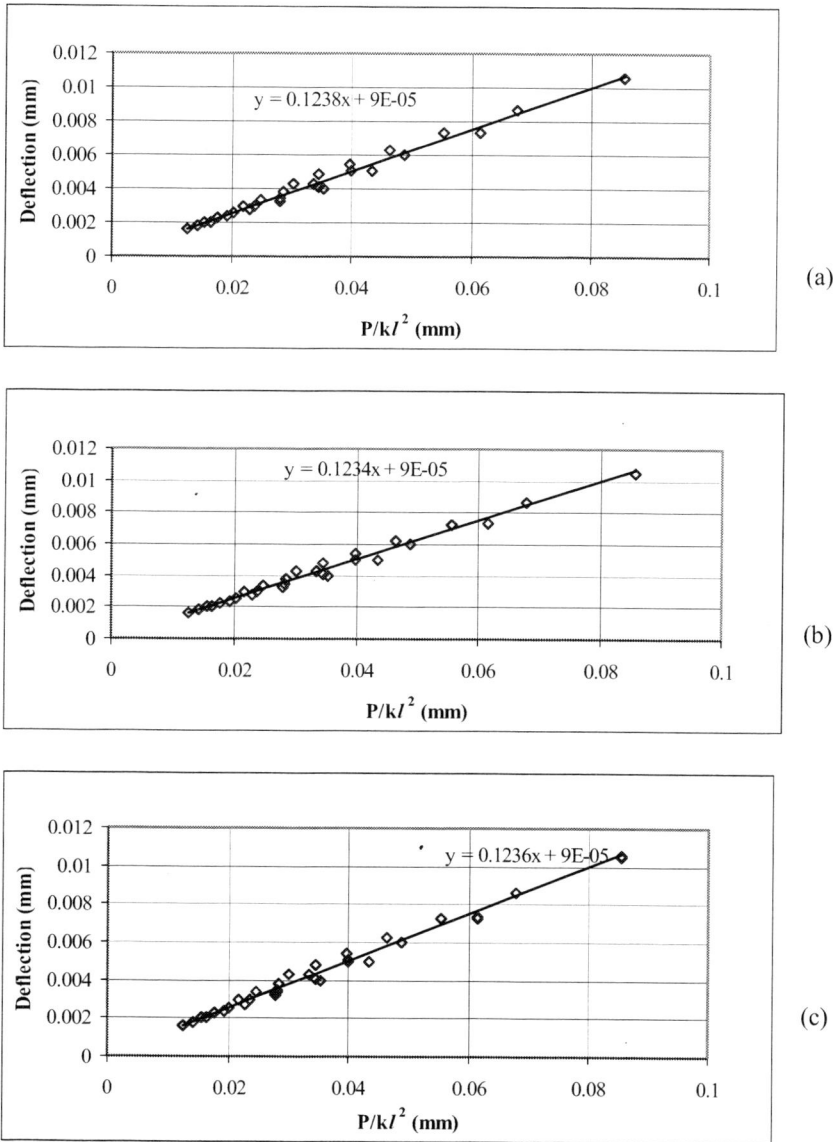

Figure 5. Linear relationship for estimating the maximum deflection due to a square patch load of width d acting at the centre: (a) $d = 100$ mm; (b) $d = 200$; (c) combined relationship for both $d = 100$ mm and $d = 200$ mm.

$$\delta_{i100,200} = 0.1236 \frac{P_i}{kl^2} + 0.00009 \qquad (3)$$

The value 0.00009 mm is small compared to the rest of the expression and, therefore, can be neglected. Hence, equation (3) reduces to

$$\delta_{i100,200} = 0.1236 \frac{P_i}{kl^2} \qquad (4)$$

Equation (4) represents the standard-width (i.e. $d = 100$ mm) design formula for estimating the maximum deflection due to a central patch load. In order to ascertain whether or not the equation is also applicable for other values of d likely to be encountered in practice, the investigations described in the next section were carried out.

5.1. Effect of Width of Loaded Area (d)

To extend the application of the standard-width equation (4) to other values of d, maximum deflections due to square-patch loads of widths zero, 50 mm, 150 mm, 250 mm, 300 mm and 900 mm were calculated, in turn. Only two FE runs were carried out for each width, as they are sufficient to obtain the corresponding constant of proportionality of relationship (1). In this manner, expressions similar to equation (4) were obtained for each value of d as depicted in Fig. 6. The generic form of these equations is

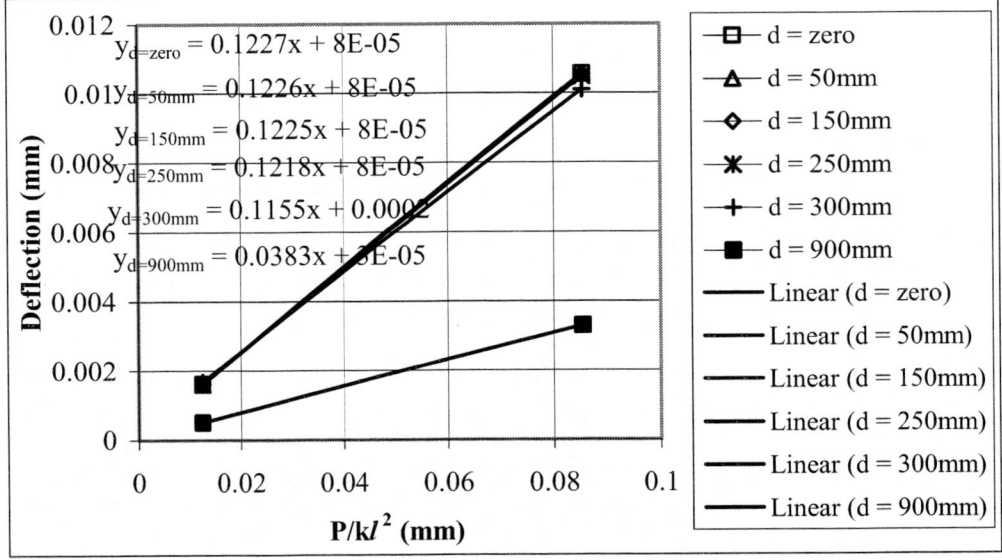

Figure 6. Deflection equations for a square patch load of different widths acting at the centre.

$$\delta_i = A_i \frac{P_i}{kl^2} \qquad (5)$$

which can be expressed in terms of the standard-width equation (4) by means of a correction factor for the width of loaded area (β_i) as follows:

$$\delta_i = \beta_i * 0.1236 \frac{P_i}{kl^2} \qquad (6)$$

From equations (5) and (6), β_i can be defined as

$$\beta_i = \frac{A_i}{0.1236} \qquad (7)$$

The values of β_i against the ratio d/d_{100} (where $d_{100} = 100$ mm) are depicted in Fig. 7, from which it is clear that no correction is required for $0 \le b \le 250$ mm (i.e. $\beta_i = 1.0$ in this range). On the other hand, the effect of d becomes significant for $d > 250$ mm. Therefore, the relationship between β_i and the ratio d/d_{100} is

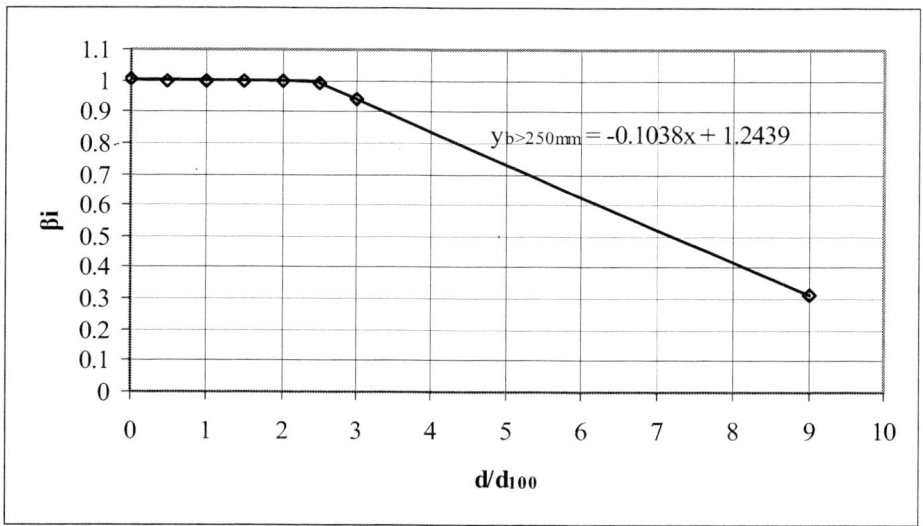

Figure 7. Effect of load width on central deflection.

$$\beta_i = 1.0 \qquad \text{for } 0 \le d \le 250 \text{ mm} \qquad (8a)$$

$$\beta_i = 1.2439 - 0.1038 \left(\frac{d}{d_{100}} \right) \qquad \text{for } d > 250 \text{ mm} \qquad (8b)$$

and, by substituting equations (8a) and (8b) into equation (6), the maximum deflection due to a central load can be expressed as

$$\delta_i = 0.1236 \frac{P_i}{kl^2} \qquad \text{for } 0 \leq d \leq 250 \text{ mm} \qquad (9a)$$

$$\delta_i = 0.1236 * \left[1.2439 - 0.1038\left(\frac{d}{d_{100}}\right)\right] * \frac{P_i}{kl^2} \qquad \text{for } d > 250 \text{ mm} \qquad (9b)$$

5.2. Effect of Concrete Grade

Based on the results of the sensitivity analysis for the effect of the concrete grade, which are described elsewhere [1], the value of the maximum deflection was found to be dependent on the value of E_c, as one would expect. In fact, the effect of E_c is embedded in the proposed equation for the maximum deflection within the term l^2, as is clear from expression (2). Since the results in Table 1 are based on $E_c = 34$ GN/m^2 (the value adopted for concrete grade 40), the generic equation (5) needs to be verified for other values of E_c, in order to check that the term l^2 fully encompasses the relationship between the deflection and the concrete grade. In order to do so, deflections due to $d = 100$ mm, but for different concrete grades, were plotted against P/kl^2 as depicted in Fig. 8 (only two FE runs were carried out for each grade, as they are sufficient to describe the linear relationship of equation (1), and suitable values of E_c were adopted to represent each grade [1]). From Fig. 8, it can be concluded that the generic equation (5) holds for all practical concrete grades.

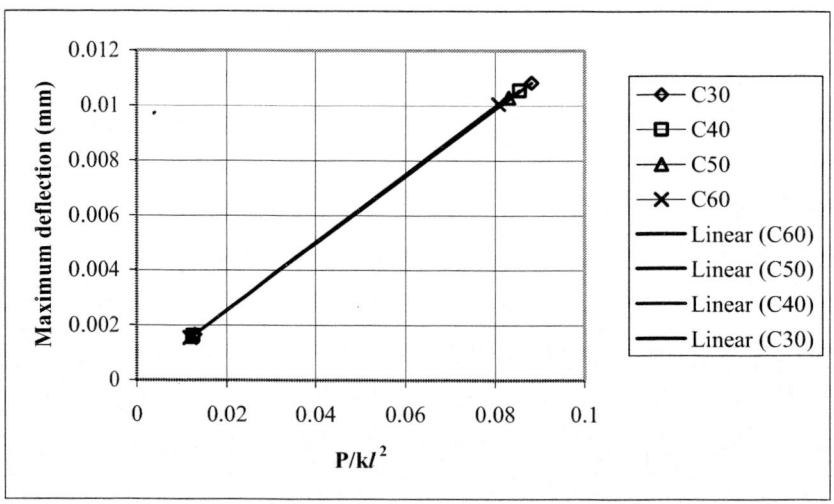

Figure 8. Maximum deflections due to a central load of 1 kN for different grades of concrete.

5.3. Comparison with Existing Experimental Results

The proposed generic design equation for the central-load case was verified in the light of experimental tests carried out by Childs et. al. [18], Childs and Kapernick [19] and Beckett [20]. The data input for the tests can be found in these three references, but the salient values are presently summarized in Table 2, which lists the corresponding theoretical values of δ_i based on the proposed LFEA-based equations (9), as well as the predictions obtained through the following equation due to Westergaard [11]:

Table 2. Comparison between δ_i based on the present LFEA, Westergaard's equation (1925) and some available experimental data

Experimental data reference	Table 12 in Childs et. al. (1957)	Fig. 16 in Childs and Kapernick (1958)		Beckett (1995)
E_c (GN/m^2)	41	31	41	34
h (mm)	152.4	203.2	152.4	150
k (MN/m^3)	163	20.37	27.17	35
d (mm)	212.5	270.12	212.5	100
P_i (kN)	26.69	44.48	26.69	90
l (m)	0.525	1.021	0.821	0.727
Westergaard δ_i (mm)	0.0743	0.262	0.182	0.608
LFEA δ_i (mm)	0.0735	0.259	0.180	0.601
Test δ_i (mm)	0.0838	0.254	0.203	0.800
(LFEA/Test)%	88%	102%	89%	75%
(Westergaard/Test)%	89%	103%	90%	76%

$$\delta_i = 0.125 \frac{P_i}{kl^2} \tag{10}$$

By reference to Table 2, it is clear that the maximum deflections predicted by the present LFEA-based expressions are in good agreement with the experimental results (as is Westergaard's prediction for this particular set of data – but see below).

5.4. Comparison of the LFEA-Based δ_i with The Westergaard Equations

In addition to his 1925 solution [11], in 1939 Westergaard [21] proposed the following modification to his original equation (see expression (10) above) in order to allow for larger sizes of loaded areas:

$$z_i = \frac{P_i}{8kl^2}\left[1 + \{0.3665 \log_{10}\left(\frac{a}{l}\right) - 0.2174\}\left\{\left(\frac{a}{l}\right)^2\right\}\right] \qquad (11)$$

where a is the radius of the circular loaded area. Equation (11) has a singularity at $a = 0$, but this solution was not intended for point loads. A comparison between the proposed LFEA-based expressions (9a) and (9b), and the Westergaard equations (10) and (11) [11, 21] has been carried out for $P_i = 1$ kN, $0 \le d \le 1000$ mm and different values of k and h as shown in Fig. (9). For comparison purposes, the radius a was transformed into the equivalent square width d by equating the two areas (thus $d = a\sqrt{\pi}$). It is clear from Fig. (9) that, for $d \le 250$ mm, the two sets of results practically coincide since the factor of 0.1236 of the LFEA-based equation (9a) is almost identical to the factor of 0.125 suggested by Westergaard (1925), i.e. equation (10). However, for $d > 250$ mm, it is evident that the LFEA-based equation (9b) is more economical (i.e. predicting smaller deflections) than the Westergaard (1939) solution (11) (which, in turn, is more economical than his original solution (10)). A further comparison between the LFEA-based equation and the Westergaard (1939) improved solution (adopted for all d except $d = 0$ [1]), shows that the former produces deflections typically 75% of the latter for $d = 500$ mm and as little as 25% for $d = 1000$ mm. As the LFEA-based equation consistently provides a more accurate and less conservative solution, it is suggested that it should replace Westergaard's (1925, 1939) equations for δ_i.

6. MAXIMUM DEFLECTION FOR THE EDGE-LOAD CASE (δ_e)

For this load configuration, the square-patch load was applied at the edge of the slab, with the load edge coinciding with the slab edge, and the ensuing maximum deflection was calculated. From the present LFEA results, it is clear that this maximum value occurs at the bottom of the slab at the very edge, as depicted in Fig. 2b. The deflected shapes of the slab along a line perpendicular to the edge-line due to a square loaded area of width 100 mm and different values of h and k are shown in Figs 10a to 10d. These curves are useful for the designer in order to check the slope caused by the edge load. (Similar plots, giving deflected shapes of the slab along the edge-line can be found in Abbas [1]: however, while the *maximum* deflection coincides (by definition, i.e. at the very edge) with that of plots along the line perpendicular to the edge, the latter set of results consistently give more critical (i.e. larger) slopes than the former; therefore, as the results perpendicular to the edge are critical, only they are shown in the present chapter.)

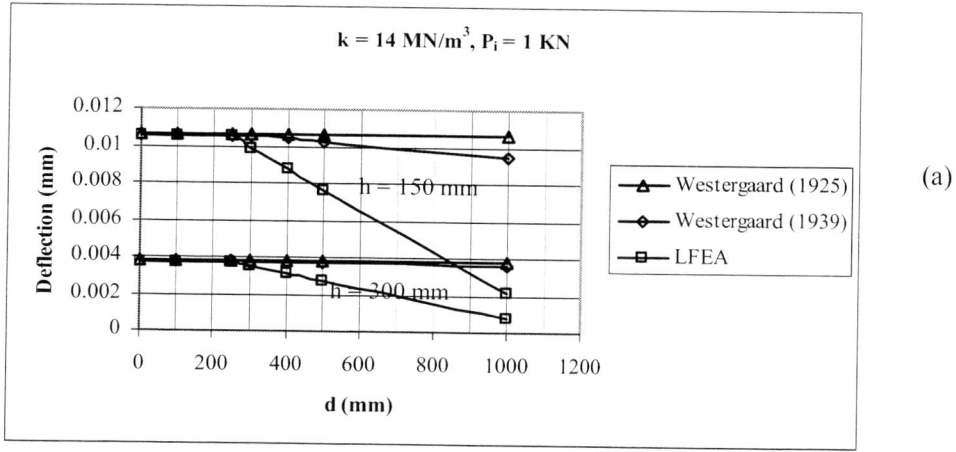

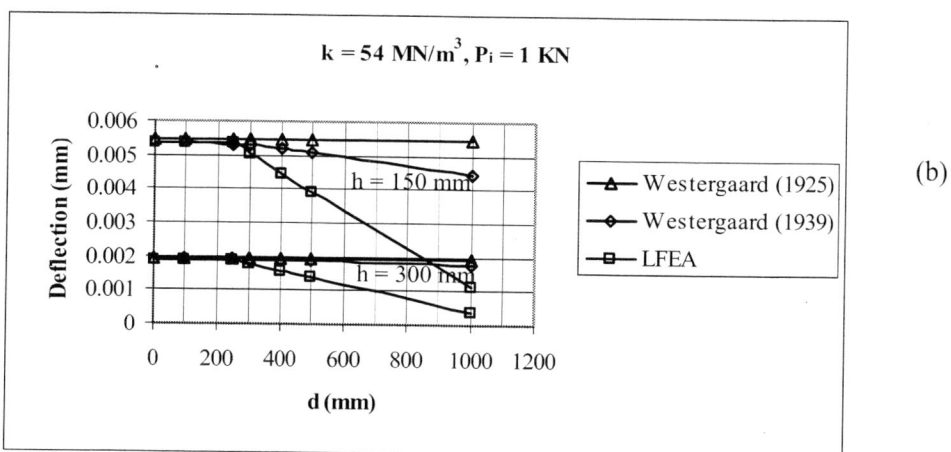

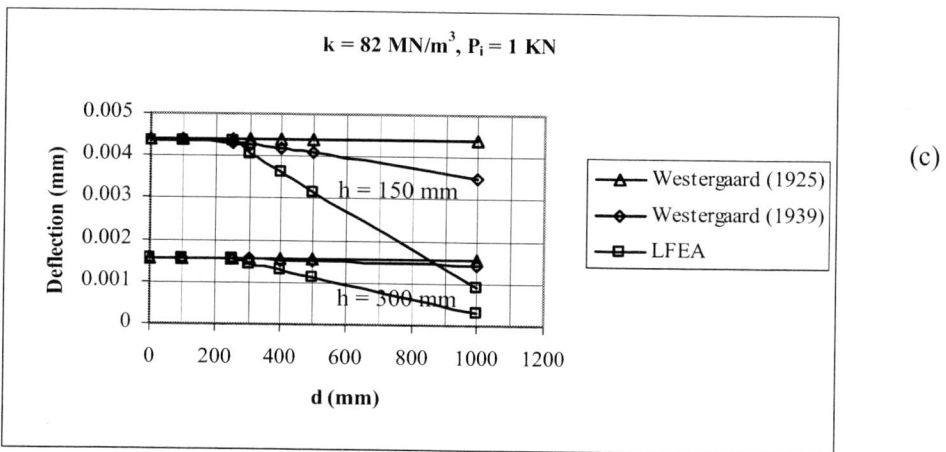

Figure 9. Maximum central deflections due to the proposed LFEA-based and Westergaard (1925, 1939) equations for various values of d, h and k. (Note: origin is at the centre of the slab.)

In 1925, Westergaard [11] suggested the following linear relationship between the maximum deflection δ_e, and the total applied load P_e, Poisson ratio for concrete μ, k and l:

$$\delta_e \propto (1+0.4\mu)\frac{P_e}{kl^2} \tag{12}$$

As a constant value, of 0.15, has been adopted for μ throughout the present LFEA, relationship (12) can be reduced to:

$$\delta_e \propto \frac{P_e}{kl^2} \tag{13}$$

Relationship (13) was assumed to hold for the present research on the edge load case. In order to check the validity of the assumed relationship, two square loaded areas were considered of widths 100 mm and 200 mm, respectively, and the ensuing results of δ_e are shown in Table 1. These results were then plotted against the right-hand term of relationship (13) [1], which yielded the following expressions confirming the validity of the assumed relationship (13):

$$\delta_{e100} = 0.4016\frac{P_e}{kl^2} \tag{14}$$

and

$$\delta_{e200} = 0.3838\frac{P_e}{kl^2} \tag{15}$$

Similarly to the central-load case, additional FE runs were performed to include square patch loads of width zero, 50 mm, 150 mm, 250 mm and 300 mm. Consequently, the following generic equation for the edge-load case was obtained:

$$\delta_e = 0.4016\left[1.0376 - 0.0434\left(\frac{d}{d_{100}}\right)\right]\frac{P_e}{kl^2} \tag{16}$$

where d_{100} = 100 mm and the edge of which coincides with the edge of the slab. Equation (16) was found to be valid for all practical grades of concrete [1].

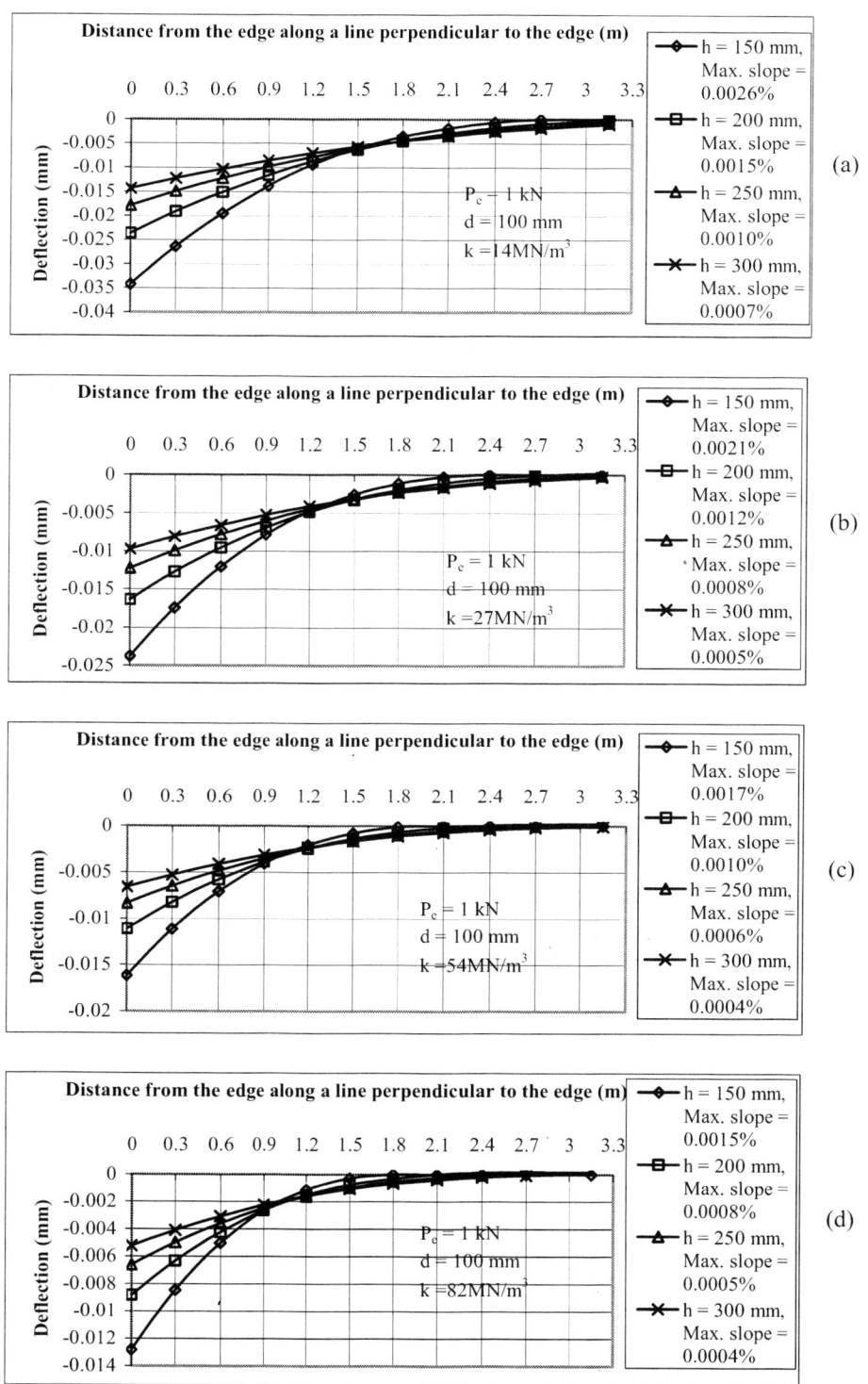

Figure 10. Deflected shape of the slab along a line perpendicular to the edge due to an edge load of width 100 mm for different values of h and k. (Note: origin is at the extreme edge of the slab.)

6.1. Comparison with Existing Experimental Results

The proposed design equation (16) was compared to the experimental data obtained by Childs *et. al.* [18] and Childs and Kapernick [19]. The tests' inputs and results are summarized in Table 3, along with the corresponding theoretical values of δ_e calculated based on the LFEA equation (16) and the Westergaard equation [11]. The latter can be expressed as follows:

Table 3. Comparison between δ_e based on the present LFEA, Westergaard's equation (1925) and some available experimental data

Experimental data reference	Table 9 in Childs et. al. (1957)		Table 12 in Childs et. al. (1957)	Fig. 16 in Childs and Kapernick (1958)	
E_c (GN/m^2)	41	41	41	31	41
h (mm)	152.4	152.4	152.4	203.2	152.4
k (MN/m^3)	163	27.17	163	20.37	27.17
d (mm)	212.50	212.50	212.50	270.12	212.50
P_i (kN)	31.136	20.016	26.69	44.48	26.69
l (m)	0.525	0.821	0.525	1.021	0.821
Westergaard δ_e (mm)	0.300	0.473	0.257	0.906	0.630
LFEA δ_e (mm)	0.263	0.415	0.226	0.774	0.553
Test δ_e (mm)	0.279	0.356	0.241	0.559	0.508
(LFEA/Test)%	94%	117%	94%	138%	109%
(Westergaard/Test)%	107%	133%	107%	162%	124%

$$\delta_e = 0.433 \frac{P_e}{kl^2} \tag{17}$$

where the constant 0.433 stems from using $\mu = 0.15$ as Westergaard suggested (as opposed to $\mu = 0.2$ in Concrete Society guide [7], which results in the slightly higher value of 0.442 for this constant). It is clear from Table 3 that both theoretical solutions (especially the LFEA one) agree well with Child's tests.

6.2. Comparison with the Westergaard Equation for δ_e

Unlike the Westergaard equation [11] which was derived for a point load only, the proposed LFEA-based equation (16) caters for the effect of d. A comparison between the two equations, for $P_e = 1$ kN, $0 \leq d \leq 1000$ mm and different values of h and k [1], shows that the proposed LFEA-based equation is more economical than the Westergaard equation, especially as d increases (e.g. the former produces deflections typically 95% of the former for d = zero, 85% for d = 300 mm and as little as 55% for d = 1000 mm). As the proposed LFEA-based equation provides an economically-sound solution, it seems sensible to replace Westergaard's equation for δ_e by this less restrictive estimate.

7. MAXIMUM DEFLECTION FOR THE CORNER-LOAD CASE (δ_c)

In the present case, the patch load is applied at the corner of the slab as illustrated in Fig. 2c. From the LFEA, it is clear that the maximum value of deflection (δ_c) occurs at the corner of the slab. The deflected shapes along the bisector of the corner right angle, in the vicinity of the corner, for a square loaded area of width 100 mm and different values of h and k, are depicted in Figs 11a to 11d. (As for the edge-load case, similar plots are available along the edge of the slab [1], but these yield less critical slopes than those obtained from the plots in Fig. 11.) These curves are useful for the designer in order to check the slope caused by a corner load. Based on previous research by Westergaard [11], the following relationship for δ_c was assumed to hold for the present work:

$$\delta_c \propto \left(B_c - \frac{d}{l} \right) \frac{P_c}{kl^2} \qquad (18)$$

where P_c is the total applied load and B_c is a constant. In order to examine the validity of the assumed relationship (18), two square loaded areas of widths 100 mm and 200 mm were considered and the corresponding values of δ_c are shown in Table 1. Based on these results, the validity of relationship (18) was confirmed as illustrated elsewhere [1], and the following expressions were obtained:

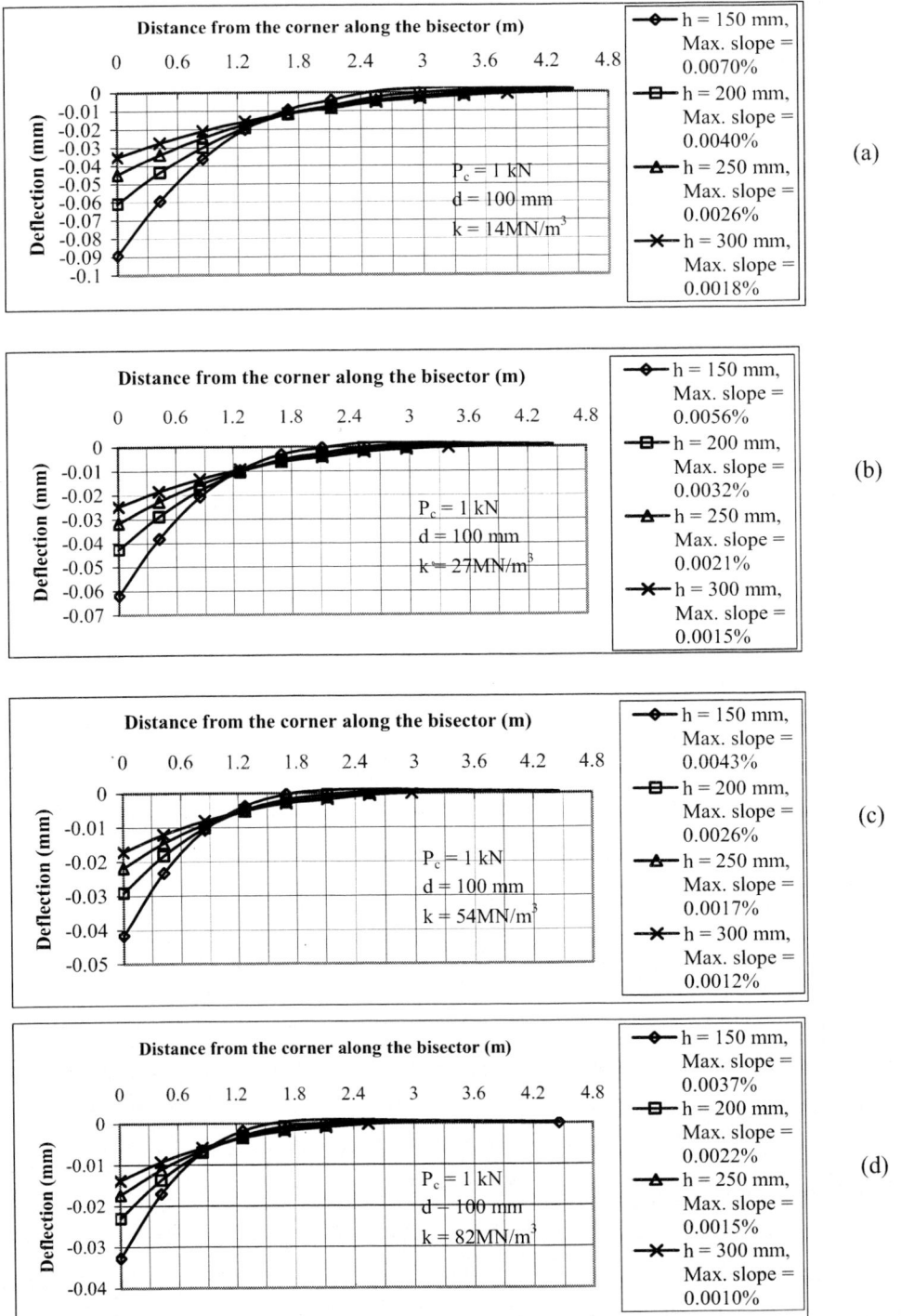

Figure 11. Deflected shape of the slab along the bisector of the corner right angle due to a corner load of width 100 mm for different values of h and k. (Note: origin is at the tip of the corner.)

$$\delta_{c100} = 2.2708\left(0.5775 - \frac{d_{100}}{l}\right)\frac{P_c}{kl^2} \qquad (19)$$

and

$$\delta_{c200} = 1.3518\left(0.9526 - \frac{d_{200}}{l}\right)\frac{P_c}{kl^2} \qquad (20)$$

Similarly to the central-load case, equation (19) for $d = 100$ mm was adopted as the standard-width design formula, and thus the generic expression of δ_c for an arbitrary value of d becomes

$$\delta_c = A_c\left(0.5775 - \frac{d_{100}}{l}\right)\frac{P_c}{kl^2} \qquad (21)$$

where A_c is a constant. Equation (21) was verified for δ_{c200}, when expressed in terms of the standard width d_{100} rather than d_{200}, as shown elsewhere [1], and the ensuing equation is

$$\delta_{c200} = 2.0863\left(0.5775 - \frac{d_{100}}{l}\right)\frac{P_c}{kl^2} \qquad (22)$$

7.1. Effect of the Width of Loaded Area (d)

It was found [1] that the value of δ_c is affected by the distance c between the centre of the patch load and the side of the corner (i.e. the distance from the centroid of the patch load to the corner is $c\sqrt{2}$), and not by the width or the shape of the loaded area (provided that the load is distributed symmetrically about the bisector). Therefore, relationship (22) should be expressed in terms of c rather than d, i.e.

$$\delta_c = A_c\left(0.5775 - \frac{2c_{50}}{l}\right)\frac{P_c}{kl^2} \qquad (23)$$

where $c_{50} = 50$ mm and the corner load is distributed symmetrically about the bisector of the corner right angle. It must be pointed out that, although the numerical outputs of equations (21) and (23) are identical, equation (23) describes the influencing parameters more accurately.

Values of A_c for various values of c (namely c = zero, 25 mm, 75 mm, 125 mm and 150 mm) were obtained, which lead to the following generic equation for δ_c:

$$\delta_c = 2.2708 \left[1.048 - 0.0643 \left(\frac{c}{c_{50}} \right) \right] \left[0.5775 - \frac{2c_{50}}{l} \right] \frac{P_c}{kl^2} \qquad (24)$$

where c is the distance between the centre of the load and the side of the corner, c_{50} = 50 mm and the corner load is distributed symmetrically about the bisector of the corner right angle. Similarly to the central- and edge-load cases, equation (24) was found to be valid for all practical grades of concrete [1].

7.2. Comparison with Existing Experimental Results

The proposed design equation for δ_c was verified in the light of the experimental tests carried out by Child et. al. [18] and Childs and Kapernick [19]. The results of the tests are described in detail elsewhere (Tables 9 and 12 of Childs et. al. [18] and Fig. 16 of Childs and Kapernick [19]). The corresponding theoretical values of δ_c were also calculated based on the Westergaard equation [11], which can be expressed as follows:

$$\delta_c = (1.1 - 0.88 \frac{a_1}{l}) \frac{P_c}{kl^2} \qquad (25)$$

where a_1 is the distance between the corner and the centre of the loaded area measured along the bisector of the corner right angle. Westergaard [11] defined a_1 in terms of the radius of a circular loaded area at the corner a, i.e. $a_1 = a\sqrt{2}$, rather than an independent quantity as he did not comment on whether or not the shape of the loaded area has any effect on the corner deflection. A comparison between the test results, the LFEA-based equation (24) and Westergaard's equation (25) is presented in Table 4. It can be concluded that the LFEA-based and the Westergaard solutions for δ_c are in good agreement with the tests' results, with the LFEA-based solutions being generally somewhat closer to the experimental results than Westergaard's predictions.

7.3. Comparison with the Westergaard Equation for δ_c

A comparison was carried out between the LFEA-based and Westergaard [11] equations for δ_c as described elsewhere [1]. The study was based on square-patch loads of width d (with the sides of the load coinciding with both sides of the corner) ranging from zero to 1000

Table 4. Comparison between δ_c **based on the present LFEA, Westergaard's equation (1925) and some available experimental data**

Experimental data reference	Table 9 in Childs et. al. (1957)		Table 12 in Childs et. al. (1957)	Fig. 16 in Childs and Kapernick (1958)	
E_c (GN/m²)	41	41	41	31	41
h (mm)	152.4	152.4	152.4	203.2	152.4
k (MN/m³)	163	27.17	163	20.37	27.17
c (mm)	106.25	106.25	106.25	135.1	106.25
P_c (kN)	36.92	23.57	26.69	44.48	26.69
l (m)	0.525	0.821	0.525	1.021	0.821
Westergaard δ_c (mm)	0.697	1.208	0.504	1.958	1.367
LFEA δ_c (mm)	0.658	1.213	0.476	1.993	1.373
Test δ_c (mm)	0.533	1.067	0.406	1.524	1.295
(LFEA/Test)%	123%	114%	117%	131%	106%
(Westergaard/Test)%	131%	113%	124%	128%	106%

mm, P_c = 1 kN, and different values of k and h. The results of the comparison study show that both equations give comparable results, with the deflections obtained by the LFEA-based equation being slightly more economical (typical discrepancy of 10%) than Westergaard's for thinner slabs and better soils. However, since the discrepancy is small for practical purposes, it can be concluded that the Westergaard equation for δ_c is both reliable and economical.

8. MAXIMUM DEFLECTION DUE TO SELF-WEIGHT LOAD (δ_{SW})

Based on the results of the sensitivity analyses that were performed to determine the main parameters affecting deflections [1], it was found that the slab self-weight adds a significant downward deflection. Although this deflection does not cause any differential settlement, and therefore can be safely ignored in most cases, equations were obtained in the present LFEA in order to estimate the magnitude of this deflection (δ_{SW}). The main reason is that ignoring δ_{SW} will lead to overestimating the upward deflections caused by patch loads and/or moisture-gradient curling. Thus, it is important to assess the magnitude of δ_{SW} as it helps keep the slab in contact with the subgrade and counteract any potential uplift. Clearly, if the self-weight is solely considered and the subgrade is assumed to behave as a Winkler foundation (which is the assumption made throughout the present LFEA), then the entire slab will deflect uniformly. In fact, the slab will be equally compressed through the thickness and will experience no flexural deformations (these would occur only if the slab were hanging from its edges and were not continuously (uniformly) supported from underneath). This implies that the thicker the slab is, the larger the self-weight deflection that will occur. This is true as long as the load is uniformly applied over the entire surface of the slab and hence every spring, that is assumed to model the subgrade, will carry exactly the same amount of pressure. Therefore, the δ_{SW} is primarily dependent on two parameters: the slab self-weight (i.e. slab thickness) and the modulus of subgrade reaction k, which is a measure of the spring stiffness. It is clear that δ_{SW} is directly proportional to the self-weight load, i.e.

$$\delta_{SW} \propto \gamma h \tag{26}$$

where γ is the unit weight of concrete. Since γ is constant for all types of concrete, equation (26) can be re-written as

$$\delta_{SW} \propto h \tag{27}$$

Finally, the effect of k was introduced by means of a correction factor for the effect of k on δ_{SW} as will be explained below.

A series of parametric studies for the maximum deflection due to the self-weight only were carried out using different values of h and k. The results are shown in Table 5 and thus the linear relationship of equation (27) was verified for different values of h and k [1]. Consequently, the following expression was found for $k = 14$ MN/m^3:

$$\delta_{14} = 0.0017h - 0.000006 \tag{28a}$$

As the value of 0.000006 mm is negligible compared to the rest of the expression, equation (28a) can be reduced to

$$\delta_{14} = 0.0017h \tag{28b}$$

Similarly, for $k = 27$ MN/m^3:

$$\delta_{27} = 0.0009h \tag{29}$$

For $k = 54$ MN/m^3:

Table 5. Maximum deflections (in mm) due to the slab self-weight only

h (mm)	k (MN/m³)	Stiffness of Slab (MNm) $D = \dfrac{E_c h^3}{12(1-\mu^2)}$	Radius of relative stiffness (m) $l = \sqrt[4]{\dfrac{D}{k}}$	Calculated deflection throughout slab (mm)
150	14	9.78	0.914	0.257
	27		0.776	0.133
	54		0.652	0.0667
	82		0.588	0.0439
175	14	15.53	1.026	0.300
	27		0.871	0.156
	54		0.733	0.0778
	82		0.660	0.0512
200	14	23.19	1.134	0.343
	27		0.963	0.178
	54		0.810	0.0889
	82		0.730	0.0585
225	14	33.02	1.239	0.386
	27		1.052	0.200
	54		0.884	0.100
	82		0.797	0.0659
250	14	45.29	1.341	0.429
	27		1.138	0.222
	54		0.957	0.111
	82		0.862	0.0731
275	14	60.28	1.440	0.471
	27		1.222	0.244
	54		1.028	0.122
	82		0.926	0.0805
300	14	78.26	1.538	0.514
	27		1.305	0.267
	54		1.098	0.133
	82		0.988	0.0878

$$\delta_{54} = 0.0004h \tag{30}$$

For $k = 82$ MN/m³:

$$\delta_{82} = 0.0003h \tag{31}$$

Finally, a generic equation to estimate δ_{SW} for any value of k was obtained [1] based on a standard subgrade of $k = 14$ MN/m³, its expression being

$$\delta_{SW} = 0.0017\left(\frac{k_{14}}{k}\right)h \qquad (32)$$

where $k_{14} = 14$ MN/m³. Unlike the case of patch loads, in which E_c affects the value of the maximum deflection, δ_{SW} is independent of E_c. Thus, equation (32) is valid for all concrete grades.

9. MAXIMUM DEFLECTION DUE TO UNIFORMLY-DISTRIBUTED LOAD (δ_{UDL})

In this case, the slab is subject to a uniformly-distributed load q throughout the surface of the slab. The maximum deflection can be obtained by using the formula derived to estimate the maximum deflection for the self-weight load case. In this respect, q can be expressed as the self-weight of a slab of equivalent thickness h_{UDL}. The latter can be calculated from the expression

$$h_{UDL} = \frac{q}{\gamma} \qquad (33)$$

where γ is the unit weight of concrete. Thus, from equations (32) and (33), δ_{UDL} is

$$\delta_{UDL} = 0.0017\left(\frac{k_{14}}{k}\right)h_{UDL} = 0.0017\left(\frac{k_{14}}{k}\right)\left(\frac{q}{\gamma}\right) \qquad (34)$$

where, as defined earlier, $k_{14} = 14$ MN/m³.

10. TOTAL MAXIMUM DEFLECTION (δ_T)

As already explained, the total maximum deflection (δ_T) – due to the cumulative effect of patch, self-weight and UDL loads – should only be considered in order to correct upward deflections due to patch loads and/or curling, and to calculate the total deflection with respect to some absolute datum point (otherwise the effect of δ_{SW} and δ_{UDL} should be ignored for design purposes). In case δ_T is required, it can be calculated from the following expression:

$$\delta_T = \delta_{SW} + \delta_{PL} + \delta_{UDL}$$

$$= \left[0.0017\left(\frac{k_{14}}{k}\right)h\right] + \left[0.0017\left(\frac{k_{14}}{k}\right)\left(\frac{q}{\gamma_c}\right)\right] + \delta_{PL}$$

$$= \left[0.0017\left(\frac{k_{14}}{k}\right)\left(h + \frac{q}{\gamma_c}\right)\right] + \delta_{PL} \tag{35}$$

where δ_{PL} is the maximum deflection due to a patch load at the centre, edge or corner of the slab, which can be determined from the previously obtained equations.

11. CONCLUSION

LFEA-based design guidelines for estimating maximum deflections have been developed. They cover patch loads at the centre, edge or corner of the slab, as well as slab self-weight and UDL applied throughout the slab. The results have shown that the proposed modifications to the long-established (but largely neglected) Westergaard solutions [11, 21], consistently provide more economic estimates of the slab maximum deflections. Moreover, deflected shapes, giving both deflection and slope estimates – due to central, edge and corner patch loads – were obtained, which are vital in helping the designer estimate the effect of the structural loading on the flatness and levelness of the slab. The calculated deflections are valid within the linear range and thus serve to check the usual serviceability requirements which, though widely considered critical, are usually ignored (or are unsatisfactorily treated) in current design guidelines which place emphasis on plastic analysis that, by definition, leave the question of serviceability unanswered.

The present LFEA-based results provide the only data currently available for estimating maximum slopes due to loading, in addition to offering more accurate (and less conservative) deflection estimates. Such serviceability checks are essential in order to ensure that, for the chosen slab thickness, the level of structural loading is acceptable. It must be stressed that deflection and slope adequacy checks for GFS form an integral part in ensuring a safe operation of fork-lift trucks moving on the floor. This is particularly true for floors utilizing Very Narrow Aisle (VNA) fork-lift trucks, which require tight tilt/slope tolerances. Therefore, it is vital not to confuse the level of importance of serviceability checks for GFS (which, themselves, may also affect safety) with that for suspended floors where the serviceability check is just that.

The example in the Appendix considers the case study presented in the preceding article [2], where more accurate stress analysis resulted in considerable thickness savings when compared to results stemming from the more conservative guidelines currently in use: it is interesting that the thinner slab still exhibits acceptable levels of deformation for this particular example, although this need not always be the case, and hence checks such as those illustrated by this example should always be performed once stress checks have proved to be satisfactory. However, it would appear from this example that neither stress nor serviceability considerations prevent the use of thinner slabs in practice although, as stressed in the preceding companion article [2], other criteria (such as factors stemming from environmental conditions and construction techniques) must also be taken into account, and further experimental and field data must be forthcoming on such thinner slabs before they are adopted in practice.

REFERENCES

[1] Abbas A. Analysis and Design of Industrial Ground-Floor Slabs Using the Finite Element Method, Ph.D. Thesis, University of London, 2002.

[2] Abbas A. A.; Pavlović M. N.; Kotsovos M. D. Permissible-stress design of ground-floor slabs, Structures and Buildings, Proc. ICE, 2004, vol. 157, No. 6, 369-384.

[3] Abbas A. A.; Pavlović M. N.; Kotsovos M. D. Elastic analysis of ground-floor slabs under multiple loads, Structures and Buildings, Proc. ICE, 2007, vol. 160, No. 3, 151-164.

[4] Abbas A. A.; Pavlović M. N.; Kotsovos M. D. Aisle-width considerations in ground-floor slabs, Structural Concrete, fib, 2007, vol. 8, No. 3, 125-131.

[5] Abbas A. A.; Pavlović M. N.; Kotsovos M. D. Post-elastic behaviour of Ground-Floor Slabs. Part 1: Plain-concrete slabs, Magazine of Concrete Research, 2007, vol. 59, No. 6, 387-400.

[6] Abbas A. A.; Pavlović M. N.; Kotsovos M. D. Post-elastic behaviour of Ground-Floor Slabs. Part 2: Fabric-reinforced concrete slabs, Magazine of Concrete Research, 2007, vol. 59, No. 6, 401-411.

[7] Concrete Society. Concrete Industrial Ground Floors – A Guide to Their Design and Construction, Crowthorne, 2003, Technical Report 34 (3rd edition).

[8] Chandler J.W.E. Design of Floors on Ground, Cement and Concrete Association, 1982, Technical Report 550 (Publication 42.550).

[9] ACI Committee 302. Guide for concrete floor and slab construction (ACI 302.1R-04), American Concrete Institute, Farmington Hills, Michigan, 2004.

[10] Cement and Concrete Association of Australia. Concrete Industrial Floor and pavement design, North Sydney, 1985, T34.

[11] Westergaard H.M. Stresses in concrete pavements computed by theoretical analysis, Public Roads, 1926, vol. 7, No. 2, 25-35.

[12] Meyerhof G.G. Load-carrying capacity of concrete pavements, J. Soil Mechanics & Foundations Div., Proc. ASCE, 1962, vol. 88, No. SM3, 89-116.

[13] Goldbeck A.T. Thickness of concrete slabs, Public Roads, 1919, vol. 1, No. 12, PASSIM.

[14] Older C. Highway research in Illinois, Trans. ASCE, 1924, vol. 87, 1180-1222.

[15] Kelley E.F. Application of the results of research to the structural design of concrete pavement, Public Roads, 1939, vol. 20, No. 5, 83-104.

[16] Teller L.; Sutherland E. The structural design of concrete pavements. Part 5: An experimental study of the Westergaard analysis of stress condition in concrete pavement slabs of uniform thickness, Public Roads, 1943, vol. 23, No. 8, 167-211.

[17] Losberg A. Pavements and slabs on grade with structurally active reinforcement, J. ACI, Proc. ACI, 1978, vol. 75, No. 12, 647-657.
[18] Childs L.D.; Colley B. E. and Kapernick J.W. Tests to evaluate concrete pavement subbases, J. Highway Div., Proc. ASCE, 1957, vol. 83, No. HW3, 1297-1 to1297-41.
[19] Childs L.D.; Kapernick J.W. Tests of concrete pavements on gravel subbases, J. Highway Div., Proc. ASCE, 1958, vol. 84, No. HW3, 1800-1 to1800-31.
[20] Beckett D. Thickness design method of concrete industrial ground floors, Concrete, 1995, vol. 29, No. 4, 21-23.
[21] Westergaard H.M. Stresses in concrete runways of airports, Proceedings of the 19th Annual Meeting of the Highway Research Board, Washington D. C., National Research Council, 1939, 197-202.

Reviewed by: Dr. R. P. West, Department of Civil, Structural and Environmental Engineering, Trinity College Dublin, Ireland

APPENDIX
COMPARATIVE DESIGN EXAMPLE

1. Introduction

The example presented here shows a comparison between deflections due to single loads at the centre, edge and corner of a plain-concrete ground-floor slab calculated using the currently-existing guidelines (namely Westergaard [11]), as well as the proposed LFEA-based guidelines. In addition, the important issue of slopes is addressed and quantified. The example illustrates the procedures involved in the new guidelines, demonstrates their simplicity, and is a continuation of the design example provided in the earlier companion article that dealt with stresses [2].

2. Problem

Check the deflection/slope adequacy of a ground-floor slab to withstand a fork-lift truck of a maximum wheel load of 13.65 tonnes. Assume that there is no interaction between the wheel loads, and that the radius of the equivalent circular loaded area is 150 mm. The modulus of subgrade reaction is 27 MN/m^3, Young's modulus of elasticity for concrete is 34 GN/m^3 and the Poisson ratio for concrete is 0.15.

3. Solutions

From Abbas [2], the following stress-based minimum thicknesses were calculated:

Design method	LFEA-BASED [2]	TR550 [8]	TR34,3rd Edition [7]
Minimum thickness required (h)	240 mm	300 mm	335 mm

Therefore, the radius of relative stiffness for the LFEA-based case is

$$l = \sqrt[4]{\frac{E_c h^3}{12(1-\mu^2)k}} = \sqrt[4]{\frac{34000*0.24^3}{12(1-0.15^2)*27}} = 1.104 \text{ m}$$

while, according to TR34 (third edition), $l = 1.417$ m.

The TR550 elastic design guide [8] does not provide any guidelines for deflections or slopes, and therefore it will not be considered in the present solution.

3.1 Shape of Loaded Area

For LFEA-based: square at centre, edge and corner.

3.2 Size of Loaded Area

For LFEA-based: width of square $d = \sqrt{\pi(150^2)} = 266$ mm.

For Westergaard: size of load considered only for corner load (radius of circle is $a = 150$ mm and thus $a_1 = 150\sqrt{2}$ mm); Westergaard's solution will be adopted to check the deflection adequacy of TR34 (third edition) slabs.

3.3 Partial Safety Factors Against Collapse

For both LFEA-based and Westergaard methods, all safety factors are taken as 1.0 for serviceability checks.

3.4 Central-Load Case

Calculate maximum deflections and slopes.

i. LFEA-BASED

$d = 266$ mm (i.e. $d > 250$ mm), thus maximum deflection (at centre) due to single load

$$\delta_i = 0.1236 * \left[1.2439 - 0.1038\left(\frac{d}{d_{100}}\right)\right] * \frac{P_i}{kl^2}$$

$$= 0.1236 * \left[1.2439 - 0.1038\left(\frac{0.266}{0.1}\right)\right] * \frac{136.5}{27*1000*1.104^2} * 1000 = 0.496 \text{ mm}$$

Maximum slope due to single load (from Fig. 3, using nearest approximation $h = 250$ mm)

$$s_i = 136.5*0.0001\% = 0.014 \%$$

ii. TR34, 3rd Edition

Maximum deflection at centre due to single load (Westergaard equation [11]) = 0.315mm
No guidelines on slopes provided.

3.5 Edge-Load Case

Calculate maximum deflections and slopes.

i. LFEA-BASED

Maximum deflection (at very edge) due to single load

$$\delta_e = 0.4016\left[1.0376 - 0.0434\left(\frac{d}{d_{100}}\right)\right]\frac{P_e}{kl^2}$$

$$= 0.4016\left[1.0376 - 0.0434\left(\frac{0.266}{0.1}\right)\right]\frac{136.5}{27*1000*1.104^2}*1000 = 1.537 \text{ mm}$$

Maximum slope due to single load (from Fig. 9, using nearest approximation $h = 250$ mm)

$$s_e = 136.5*0.0008\% = 0.11\%$$

ii. TR34, 3rd Edition

Maximum deflection at edge due to single load (Westergaard equation [11]) = 1.090 mm
No guidelines on slopes provided.

3.6. Corner-Load Case

Calculate maximum deflections and slopes.

i. LFEA-BASED

Maximum deflection (at tip of corner) due to single load

$$\delta_c = 2.2708 \left[1.048 - 0.0643 \left(\frac{c}{c_{50}} \right) \right] \left[0.5775 - \frac{2c_{50}}{l} \right] \frac{P_c}{kl^2}$$

The sides of the load coincide with the sides of the corner. Thus $c = \frac{d}{2} = 133$ mm, and

$$\delta_c = 2.2708 \left[1.048 - 0.0643 \left(\frac{0.133}{0.05} \right) \right] \left[0.5775 - \frac{2*0.05}{1.104} \right] \frac{136.5}{27*1000*1.104^2} *1000 = 4.024 \text{mm}$$

Maximum slope due to single load (from Fig.10, using nearest approximation h =250 mm) s_c = 136.5*0.0021% = 0.29 %

ii. TR34, 3rd Edition

Maximum deflection at edge due to single load (Westergaard equation [11]) = 2.352 mm
No guidelines on slopes provided.

4. SUMMARY OF DESIGN EXAMPLE RESULTS

A summary of the results of the present example is presented in the following tables:

4.1 Maximum Deflections

Design method	Centre	Edge	Corner
(i) LFEA-BASED with h = 240 mm	0.496 mm	1.537 mm	4.024 mm
(ii) TR34, 3rd Edition (Westergaard [11] adopted) with h = 335 mm	0.315 mm	1.090 mm	2.352 mm
(i/ii) %	157%	141%	171%

4.2 Maximum Slopes (Guidelines Available Only for LFEA-Based Method)

Design method	Centre	Edge	Corner
(i) LFEA-BASED	0.014%	0.11%	0.29%

5. CONCLUSIONS BASED ON DESIGN EXAMPLE

From the foregoing design example, it can be seen that the LFEA-based minimum allowable thickness of 240 mm (which was found to be adequate in resisting stresses – as shown in Abbas [2] – despite allowing a saving of ~30% over the h =335 mm calculated using TR34 (third edition)) develops acceptable levels of maximum deflections and slopes. It is clear also that, although the TR34-based thicker slab develops smaller maximum deflections – as one would expect given the conservative increase in thickness – such a reduction in deflections appears to be unnecessary, especially when the all-important maximum slope is considered and is found to be adequate in the case of the thinner slab. (It should be pointed that, for the same thickness, the LFEA-based equations consistently provide deflections smaller than those estimated by means of Westergaard's equations, as has been shown in the main body of the article.)

Chapter 8

TOWARDS EFFICIENT ANALYTICAL MODELS FOR SEISMIC ANALYSIS OF MULTISTORIED BUILDINGS

Y. Belmouden[*] and P. Lestuzzi[†]
École Polytechnique Fédérale de Lausanne, Lausanne, Switzerland

ABSTRACT

In this chapter, a novel equivalent planar-frame model with openings is presented. The proposed model was developed to assess the seismic resistance of multistoried structures without resorting to a finite element approach. The model is mainly developed for generation of capacity curves. The model deals with seismic analysis using the Pushover method for masonry and reinforced concrete buildings. It belongs to simple equilibrium analytical models. Its formulation is based on the well-known beam theory (the beam distribution method). The model allows the determination of the member forces (bending moments and shear forces) through three moment equilibrium. Each wall with openings can be decomposed into parallel structural walls made of an assemblage of piers and a portion of spandrels. As formulated, the structural model undergoes inelastic flexural as well as inelastic shear deformations. The mathematical model is based on the smeared cracks and distributed plasticity approach. Both zero moment location shifting in piers and spandrels can be evaluated. The constitutive laws are modeled as bilinear curves in flexure and in shear. A biaxial interaction rule for both axial force - bending moment and axial force – shear force are considered. The model can support any shape of failure criteria. An event-to-event strategy is used to solve the nonlinear problem. Two applications are used to show the ability of the model to study both reinforced concrete and unreinforced masonry structures. Relevant findings are compared to analytical results from experimental, simplified models and finite element models such as Drain3DX and ETABS finite element package.

---

[*] Email addresses: youssef.belmouden@epfl.ch (Y. Belmouden)
[†] pierino.lestuzzi@epfl.ch (P. Lestuzzi)

1. INTRODUCTION

Earthquakes are considered to be the major cause of structural failure of buildings in Europe. Despite their rarity and moderate intensity, earthquakes in the northwest and central Europe have the potential to cause extensive damage and associated financial losses, due to the vulnerability of the local building stock. The mitigation of earthquake hazard involves the collaboration of many specialists with different tasks. One of these topics is structural engineering providing and advancing the knowledge for earthquake resistant construction. However, a problem arises for existing buildings analysis. In this context, in the few last decades, technical advances have been made in seismic engineering and particularly in the seismic vulnerability assessment of existing buildings. The vulnerability assessment focuses on the study of the extent of damage for different earthquake scenarios.

In almost all countries, the majority of the buildings are classified as existing buildings. This is why extensive assessment of such structures is motivated since they have been generally designed to resist gravity loads. Nevertheless, the seismic vulnerability of existing buildings designed against wind loads, is found to be very low.

In practice, the seismic vulnerability assessment of existing buildings is based on the development of simplified analytical models. The need for such models is always motivated by first, the large amount of structures that should be analyzed in a very short time and second, the search for optimal solutions for structural retrofitting.

For vulnerability assessment purposes, the analysis of a large number of existing buildings requires relatively simple approaches that are capable of representing their essential characteristics. The models should be able to evaluate the ultimate strength, maximum displacements and the failure modes. Different models are developed based on analytical and finite element approaches [1]. The analytical models are found to be very simple to use and require lesser amount of data. However they are very limited, particularly for large buildings analysis in terms of structural behavior (coupling effect, distribution of the nonlinearity, modes of failures prediction). The performed analyses show that they are conservative and are not able to represent all features of such buildings [2]. On the other hand, finite element approach is a powerful tool for seismic analysis but it is time consuming and requires a large amount of data. Moreover, refined models based on either discrete or continuum approaches suffer from strong mesh-dependency and require numerous parameters that may not be directly extractable from structural analysis. Hence, these models are very sensitive to the parameter calibration that directly affects the reliability of the results and the analysis stability (lack of convergence, flip-flop occurrence, sudden load falling, and so on). With such methods it is not possible to treat a stock of buildings. Thus, these methods are cumbersome due to the high analytical skills required for their numerical implementation and they are restricted only to practitioners with a high level of knowledge.

A widely used model for structural analysis is the linear (beam-column element) finite element or the equivalent frame model. Despite some limitations in the equivalent frame model, it is very attractive in comparison to complex finite element models [1, 3, 4, 5]. Moreover, they have shown satisfactory results particularly for RC structures. In this context, the proposed model is based on beam-column element and distributed non-linearity approaches. It is adapted to analytical methods without use of finite element method. In the subsequent paragraphs, a novel planar-frame model for seismic vulnerability assessment of existing multistoried buildings is presented.

2. A PLANAR FRAME MODEL FOR STRUCTURAL WALLS WITH OPENINGS

The mathematical model can represent solid walls, frame structural elements (made in beams and columns), coupled walls and perforated walls (or framed walls) [6, 27, 28]. The model can represent different openings. However, the vertical axis should lie through all vertical piers elements as well as for the horizontal axes that should lie through all spandrels. The structural model consists of an assemblage of vertical plane walls with openings that form a single perforated wall. Each structural wall is made of pier elements with or without rigid offsets and a portion of spandrels such that there are two kinds of individual walls: exterior walls and interior walls. The length of these parts of spandrels is equal to the zero moment length, and can be updated at each step depending on the bending moments at the spandrel ends [27, 28].

In the equivalent frame models that are based on finite element method, nonlinear flexural springs (lumped plasticity) are inserted into the model at the ends of the piers and/or spandrel elements. These elements are defined in terms of moment-rotation laws. Translational shear springs are added at each pier and spandrel at mid-points. These springs are expressed in terms of shear force–displacement laws. However, the occurrence of yielding is unlikely along spandrel spans and pier heights. For this reason, nonlinearity should be distributed along the clear pier height and clear spandrel length. Thus, the proposed model is based on the spread nonlinearity approach. Each pier and spandrel can be discretized into a series of slices [7] and each cross-section is considered as homogeneous. The structural element behavior is monitored at the center of the slices [7] while bending moments are evaluated at slice ends.

The mechanical model undergoes flexural as well as shear deformation. In the current formulation, the model only considers a biaxial interaction between axial forces - bending moments (N-M) and axial forces - shear forces (N-V). The so-called shifting of the primary curve technique is used [8]. The axial force is evaluated in a simple manner based on initial axial forces plus vertical shear forces produced in spandrels at joints. A triaxial interaction rule, (N-M-V), is not currently considered. At present, only interaction curves that represent bending moment or shear force interaction with regards to a compressive axial force are considered.

The major features of this model are summarized as follows:

1. All previous attempts to use simplified models based on static equilibrium method, always consider a constant zero moment location (Nuray Aydinoğlu [9], Kilar and Fajfar [10], Lang [11], FEMA356 [12], FEMA306 [13], Paquette and Bruneau [14], and others). The wall formulation herein permits the capture of the coupling effect in elevation due to the nonlinearity distribution in both piers and spandrels. Thus, the zero moment location in both piers and spandrels can be mitigated during the nonlinear analysis.

2. In the current development, the variation of the axial vertical loads are considered for piers only and they are based on an over-simplified approach. The axial loads on piers are updated based on the initial axial forces at each storey plus the shear forces developed on spandrel ends.
3. The nonlinearity is treated using a smeared plasticity approach [7]. Thus, the piers and spandrels are discretized into finite homogenized slices [15]. Variable sections can be specified over either spandrels or piers. In pier elements the axial forces can increase or decrease. In that case, the pier slices can shift either from elastic-to-plastic or from plastic-to-elastic state depending on the axial force distribution.
4. The model can take into account both flexural and shear behavior in the inelastic range. The interaction effect can be defined by using experimental and phenomenological models. These equations are considered as failure criteria that can be defined by points and linear segments. The non linear constitutive model for both flexural and shear behavior is considered as a bilinear envelop curve with a very small post-yield stiffness to avoid numerical problems. The flexural behavior is modeled as a moment-curvature law that is based on an equilibrium statement in a cross-section.
5. The formulation of the model deals with a Pushover analysis. It is based on the well-known event-to-event strategy. A simplified algorithm for systems with interaction effect is presented through an equilibrium correction at each step of calculation. The analysis is performed by a force-controlled technique. The change of sign in a structural element is permitted only in the elastic range. In the inelastic range, this leads to stoppage of the analysis.
6. The structural wall is a planar structure (two-dimensional). However, the sum of all capacity curves, on the basis of the equal top displacement assumption, permits to analyze an entire building and to develop capacity curves.

There are two approaches for modeling axial load variation: (1) based on the axial deformation (extension compression), (2) based on force interaction. As the model is force controlled the second approach was adopted. However, the first one is more convenient for those elements that are subjected to a large difference in forces and stiffnesses along the elements. This is more likely to happen in unreinforced masonry elements than in reinforced concrete elements. This approach necessitates a displacement control formulation of the model. This is beyond of the scope of the present version of the model.

In light of the abovementioned, the simplified model has two major limitations in comparison to a beam formulation (in a finite element approach):

1. Limitation 1: The model does not use a global stiffness matrix (beam element stiffness assemblage, global stiffness matrix including all degree of freedom of the structure),
2. Limitation 2: There is no axial deformation in both piers and spandrels, only rotations and lateral displacements are considered for the simplified model.

Both limitations 1 and 2 are correlated in a finite element method. The use of a global stiffness matrix permits the determination of a global coupling between all degrees of freedom of the discretized structure. The axial load is obtained from axial stiffness (linear elastic in general, EA/L, where A is the cross-section of the finite element, L the length of the finite (beam) element, E the Young modulus of the constitutive material) and axial deformation for a beam element in finite element method. As the axial deformations are omitted, there is no way to calculate axial loads unless an approach based on force equilibrium is used.

In the present model, the analysis of the structure is treated as simple as possible using an analytical approach. In fact, the model treats each wall individually. For each wall, the number of degrees of freedom is equal only to the number of storeys. No intermediate degrees of freedom are considered (along piers and spandrels) as commonly adopted in finite element method (meshing of the structure). Thus, the individual wall behavior is governed by a reduced system of linear equations equal to the number of storeys. The interaction between walls is considered in two levels. First, the axial force redistribution is considered based on force equilibrium in piers-to-spandrels joints using shear forces at pier ends and spandrel ends. Second, an equal top wall-displacement assumption is considered when calculating the Pushover curves.

As stated above, the model formulation is based on the well-known beam theory (the beam distribution method). The model takes into accounts for the equivalent stiffness of piers and spandrels for an individual structural wall (left side, interior and right side wall) at each level. A joint equilibrium permits to consider the piers-to-spandrels coupling effect. The term 'level' means the centre-line (or the neutral axis) of the spandrels between two adjacent storeys that form the pier-to-spandrel joints.

Finally, for capacity curves generation, a simplified approach of the Pushover method based on the event-to-event strategy is adopted [27, 28]. In this method, the event factor for the entire frame subjected to the predefined lateral load pattern. A wall event factor is calculated for the nominal event plus the tolerance [7]. A wall event factor is extracted from the lowest slice event factors. The occurrence of an event depends on the yield surfaces function and yielding limit states for all slices in flexure and in shear.

3. STATIC NON LINEAR ANALYSIS OF A REINFORCED CONCRETE BUILDING

A three-dimensional multistoried building made in RC structural walls is studied [27, 28]. The structure was modeled on both Drain3DX program [16] using a fiber beam element (type 15) and ETABS software [17] using a point hinge beam element. The fiber beam is a nonlinear finite element model in flexure but linear elastic in shear. The behavior of fibers is defined by a stress-strain relationship for both steel and concrete materials. The use of fibers to model cross-sections accounts rationally for axial force – biaxial bending moments. A detailed description of the element is given in the reference including related capabilities, assumptions and limitations [16]. In this application, an elastic perfectly plastic model was adopted for steel fibers. However, a parabolic-rectangular stress block was adopted for concrete material. The floor was modeled as a grid system using the elastic linear beam element type 17 [16]. A validation of the finite element type 15 against experimental results can be found in the reference [2].

On the other hand, ETABS software provides a flexural point hinge finite element model (PHFE) called P-M2-M3. This model considers an interaction between two-way moment curves and axial forces. Since the structural model behaves in the in-plane direction (Z-direction, Fig. 5), the point hinge model performs with a biaxial interaction rule. The hinges are located at the ends of all beam elements, at top and bottom storeys. In the equivalent frame model (EFM), the nonlinear behavior for each slice is defined by a moment-curvature relationship in compression only.

The analysis performed on Drain3DX was force controlled. On the other hand, the analysis performed on ETABS was displacement controlled in the presence of a given lateral load pattern. The adopted load pattern represents the distribution of inertia forces corresponding to the first mode of vibration.

The building studied herein is a six storey torsionally balanced reinforced concrete structure [27, 28]. The floor is of a grid-type with RC slab. In the case studied, the spandrels consist of beam elements representing a grid floor with underbeams. For estimating the stiffness of the floors with underbeams, the effective width of the floor slab is calculated according to the rules suggested by Bachmann and Dazio [18]. All the geometrical and mechanical characteristics of the building (the second moment of inertia of the floor section, the floor load carried by each wall, the normal forces acting on each floor level, and so on) are given in the reference [11, 27, 28].

Three cases are investigated to study the axial force redistribution and the axial force – bending moment interaction rule. They are: (1) rigid floor-type structure with one hundred percent of the floor stiffness, (2) semi rigid-floor type with fifteen percent of the floor stiffness and (3) flexible floor-type with ten percent of the floor stiffness. For each floor-type model, two cases were studied for the EFM and four cases for the PHFE model on ETABS. The case studies are defined as follows:

1. PHFE M1 and M2: Bilinear and elastic-plastic moment-rotation law respectively, without (N-M) interaction,
2. PHFE M3 and M4: Bilinear and elastic-plastic moment-rotation law respectively, including (N-M) interaction,
3. EFM 1 and EFM2: Without and with (N-M) interaction respectively.

In this simulation, the spandrels represent the floor beams that were elastic linear. Thus, the zero moment lengths were chosen to be at the middle of the span length, without rigid offsets, and kept constant during the analysis. The material properties are defined by the tensile strength of concrete, the compressive strength of concrete, the steel maximum strength, the concrete Young modulus and the steel elastic modulus [11, 19].

For the EFM, the walls at each storey are discretized into twenty slices over a storey height. The EFM necessitates moment-curvature laws that are considered idealized for both elastic perfectly plastic and bilinear curves. For RC wall sections, the mechanical properties defining the non linear behavior to be implemented are represented by the yielding moment, the ultimate moment, the first yield curvature, the nominal curvature for a bilinear idealization, and the ultimate curvature [27, 28]. On ETABS, only yielding moments should be specified for the flexural hinges since the yielding rotations are calculated by the program. The axial deformations of the beams are neglected on both ETABS and Drain3DX models.

With regards to the biaxial failure criteria used, the hinge yielding in both PHFE model and slices of the EFM depends closely on the axial load. For low axial loads, the yielding of the sections is delayed, while for high axial loads the yielding is anticipated. These mechanisms are closely related to the axial load redistribution, the floor stiffness and wall coupling. Hence, the global response of the structure is affected (elastic stiffness, structural displacement, damage occurrence).

Numerous capacity curves for the three floor-type models analyzed by Drain3DX, ETABS and the proposed EFM have been displayed. The (N-M) interaction effect increases with the total base shear. The (N-M) interaction has small effect in the first stage of the analysis. As the floor stiffness increases, the force redistribution capacity of the structure increases, the normal forces increase, and then the effect of (N-M) interaction becomes significant. When axial force is still small, the (N-M) interaction is negligible. In other words, the (N-M) interaction rule has no effect for flexible floor-type structures. This application tends to demonstrate the ability of the EFM, in comparison to ETABS's results, to reproduce the interaction between the floor stiffness, the structural wall coupling, the force redistribution, and the failure criteria on the global response of the building.

The order of occurrence of the plastic hinges can be obtained for all the steps of the displacement control from a pushover analysis. The collapse sequence at ultimate state has been extracted [20, 27, 28]. The comparison should be drawn in term of general behavior of the structure, not in terms of exact location of the hinges. Note that, the EFM is compared only to the elastic perfectly plastic PHFE model.

In the first steps of analysis, the diagram of moments (elastic linear diagram), correspond to the relative stiffness between beam-floors and walls. However, as the nonlinearity grows, the wall element stiffness decreases while the beam-floor stiffness remains elastic linear. This means that the shape of the moment diagram tends gradually to a frame-type moment diagram. When a plastic hinge forms at a given wall base in the EFM, the bending moment remains constant at this slice. Generally, as the load increases, the zero moment location shifts to the mid-storey height as explained above. The plastic hinge presented herein corresponds to the step for which the EFM achieves its ultimate state under force control.

For an elastic perfectly plastic moment rotation model without (N-M) interaction, the EFM behaves as a point hinge model with lumped plasticity. The use of force interaction permits modeling the effect of the axial redistribution on the yielding capacity of the structure. Thus, the results show the difference in the number of hinges and the yielded slices when the (N-M) interaction was activated. The results extracted from capacity curves and structural damage assessment point out that the use of force redistribution in simplified models is necessary in particular for existing building analysis more than for structural design.

4. STATIC NON LINEAR ANALYSIS OF AN UNREINFORCED MASONRY BUILDING

The proposed model can be used also for modeling URM structures [27, 28]. The URM piers and spandrels are subdivided into a series of slices. The slices represent a homogeneous brick and mortar one-phase material. As known, the masonry material is a weak isotropic material with very limited ductility. Thus, the softening behavior is very burdensome for computation and causes convergence problems particularly when the analysis is force-controlled. The post-peak behavior with softening is beyond the scope of this model. The yield criteria considered are expressed for flexure and for shear behavior according to the Magenes model (Ref. [21], [22], and [23]). For unreinforced masonry buildings, two constants are required for flexure failure criteria, while nine constants are required for shear failure criteria (Ref. [21], [22], and [23]). The same failure criteria can be found in many other procedures for masonry assessment (Ref. [12], [13], [14], [24] and others]. These criteria deal with elastic perfectly plastic models in terms of moment- rotation and shear force – displacement laws. In this study, the behavior of the spandrel is assumed to be elastic linear in both flexure and shear.

A full-scale two-storey unreinforced masonry tested at the Pavia University was chosen for model validation (Ref. [21], [22], and [23]). This structure has been extensively studied in literature. A remarkable feature of this structure, is that axial loads in piers vary during the experimental test. The variation of axial loads in the considered structure is exploited to study the sensitivity of the model to axial force variation in piers. The structural model is subjected to increasing lateral forces that are applied at the floor levels, keeping a 1:1 ratio between the force at the first and the second floor. In this application the door wall D was chosen as no flange effect is considered. The elastic properties of the structure used in the model such as the maximum compressive strength of a masonry prism orthogonal to the mortar bed, the shear modulus and so on are given in the reference [23, 27, 28].

This structure (Pavia door wall D) was extensively studied with different models:

1. using both a plane-stress nonlinear finite element in a continuum model (F.E.M) and a simplified finite element point hinge model (SAM: Simplified Analysis of Masonry buildings), [22],
2. using a nonlinear eight node solid element (Solid 65, ANSYS Software) with a triaxial yield criteria initially developed for concrete materials and a simplified phenomenological model [5],
3. using a simplified phenomenological model with a diagonal truss-plasticity approach and upper bound plasticity theory [11],
4. using a beam element (a macro element) with additional degree of freedom (TREMURI program) [26].

In the current application, the (N-M) interaction is used. However, the (N-V) interaction was not activated. In fact, it was found that axial forces in the second storey are confined to the first failure mode domain of validity since the variation in axial forces for this storey is very low. The second and third modes of failures are not activated. The first mode of failure corresponds to the corresponding to the shear failure along bed-joints at the end section cracked in flexure. The second mode of failure corresponds to the diagonal cracking at the centre of the panel due to mortar joint failure. The third failure mode represents the diagonal cracking at the centre of the panel due to brick failure. However, despite the variation in axial forces in the first storey, they are confined to the second failure mode domain of occurrence during the analysis. Hence, an elastic perfectly plastic model without (N-V) interaction is used.

The use of rigid offsets is a crucial issue in equivalent frame modeling. The dimensions of rigid offsets in piers are calculated based on an empirical approach proposed by Dolce [25]. In this study, full rigid offsets are considered. The capacity curves (total base shear versus top lateral displacement) are developed for different cases (Table 1).

Table 1. Case studies for both EFM and PHFE models.

Case	Model type	Rigid Zone in Pier	Rigid Zone in Spandrel	(N-M) failure criteria	V Shear effect	Maximum strength ratio (*)
1	PHFE	-	-	-	-	-20.1%
2	PHFE	-	-	×	-	-15.5%
3	EFM	-	-	×	-	-10.7%
4	PHFE	-	-	-	×	-21.7%
5	PHFE	-	-	×	×	-22.4%
6	EFM	-	-	×	×	-20.5%
7	EFM	× (with $2E_m$)	-	×	×	-9.9%
8	EFM	× (with $2E_m$)	-	×	-	+18%
9	EFM	× (with $4E_m$)	-	×	×	-9.9%
10	EFM	× (with $10E_m$)	-	×	×	-9.0%
11	EFM	× (with $2E_m$)	× (with $2E_m$)	×	×	-8.5%
12	EFM	× (with $10E_m$)	× (with $10E_m$)	×	×	-7.1%
13	PHFE	× (with $10E_m$)	× (with $10E_m$)	×	×	-9.3%

Legend: PHFE: Point Hinge Finite Element model, EFM: Equivalent Frame Model, (-) Option considered, (×) Option not considered, (*) The maximum strength ratio=analytical /experimental maximum strengths %, E_m is the masonry Young Modulus.

In the light of the obtained results, the following recommendations are made:

1. The effect of the axial force - bending moment, (N-M) interaction is modeled by the case '1' and '2'. As axial compressive load increases, flexural strength of the piers also increases with regards to the failure criteria.
2. The nonlinear effect of shear mechanism is modeled by cases '3' and '4' in the absence of rigid offsets, and by cases '7', '8' and '9' in the presence of rigid offsets. As expected, the contribution of shear mechanism tends to decrease the capacity of the structure due to the occurrence of shear damage. This feature is successfully captured by the simplified model.
3. The rigid offsets have a significant effect on the global response not only on stiffness, but also on strength capacity of the structure (Ref. [6], [15], and [22]). This is expected as the horizontal element stiffness closely affects the contribution of the frame mechanism to structural response (cases '10', '11' and '12'). The capacity curves obtained from EFM (case '12') versus PHFE model (case '13') are satisfactory.

4. In cases '12' and '13', the resulting capacity curves are close to a certain extent in spite of the smeared approach in the EFM. Both cases '5-6', and '12-13', show the comparison of the modeling performance, including shear effect and (N-M) interaction rule and using either the EFM and the PHFE model with or without rigid zones.

5. CONCLUSION

This paragraph presents a simplified formulation of a novel equivalent frame model. The model permits to consider many relevant features of structural behavior such as structural wall coupling, zero moment location shifting, axial force-bending moment interaction, axial force-shear force interaction, and failure modes prediction. However, in the case of URM buildings, it is well known that smeared crack approach suffers from a few limitations. The smeared crack model is enable to represent effectively the rocking and bed joint sliding mode of failures.

For the development of capacity curves, the obtained results from the proposed model show good agreement with experiment and numerical results. The model has proven its capability to satisfactorily predict the maximum strength. The calculated maximum strengths, in particular for the masonry structure (in the range of nine percent), could be judged as good results since the model is based on simplified approaches in comparison to finite element models. However, the post peak behavior with softening is not yet obtained since the model is force-controlled. Care should be taken when modeling dual buildings as frame-wall structures in particular with respect to the initial zero moment lengths assumption. In all cases, obtained results should be considered from an engineering point of view as is generally done for all simplified existing models.

Further discussion can be made as follows regarding the examples:

a. For the unreinforced masonry structure (second example), it was expected that the resulting damages from the simplified model will be too different than obtained from experiments. Nevertheless, the results have shown only a little difference in damage distribution and in particular in the first storey (the critical storey). Despite the fact that the structural model is implemented on ETABS using a displacement controlled while the simplified model was force controlled. This concludes that despite the simplicity of the present model in comparison to the finite element method (ETABS Software), the difference is in a satisfactory range in spite of the complexity of damage prediction in URM structures. Moreover, the model is two-dimensional without diaphragm effects that are seen in a three-dimensional model.

b. For the reinforced concrete structure (first example), the main shortcoming that is shown in the obtained results between finite element and simplified approaches is that in the last case a symmetrical damage distribution was obtained. This can de readily explained since the first model is a three-dimensional model while the simplified model is two-dimensional. In a three-dimensional model the effect of the diaphragm is considered (wall-to-floor coupling). In the case of a rigid diaphragm with uniform floor rotation around its in-plane axis, the left-side and right-side walls (identical walls) behave in such way that, one side will be subjected to additional compression forces (compression force increasing, uploaded wall in compression) while the other side will be unloaded (compression force decreasing, unloaded wall). Consequently, the damage pattern is unsymmetrical. Consequently, the variation of the axial force is more pronounced in comparison to a structure with a flexible diaphragm. In this example, the floor is a reinforced concrete element and considered as a rigid diaphragm. Moreover, the floor is modeled as linear elastic. If the nonlinearities increase in the walls, the rigid diaphragm effect also increases.

c. This study is the first attempt to apply such an approach for simplified model (simple equilibrium model) rather than finite element. The model tends to incorporate a coupling effect between walls so that the non linear behavior of each individual wall is closely related to the adjacent individual walls (zero moment length update in spandrels). This feature is no longer considered in simplified models commonly used such as FEMA (Federal Emergency Management Agency) method or similar simplified approaches. The capacity curve of a structure, in the simplified methods, is obtained by a sum of individual wall capacity curves without coupling effect (two-dimensional model). But, three-dimensional effects have to be considered only in a finite element approach. In general, the lack of compatibility between tests (numerical and experimental) and the simplified model is due to the concept of the equivalent frame approach. It is well-known that beam formulation is not suited to model planar elements but still largely sufficient when looking for a certain degree of accuracy and global results. Moreover, for the capacity curve generation the equivalent frame model is still attractive and widely used.

Finally, the proposed model is formulated in order to extract capacity curves with damage identification. It can be implemented readily using any programming platform. The model can be used to assess URM structures, RC structures as well as dual structures that are commonly adopted in many countries.

6. REFERENCES

[1] Tzamtzis A. D., Asteris P.G., FE Analysis of complex discontinuous and jointed structural systems: Part 1: Presentation of the method – A state-of-the-art review, *Electronic Journal of Structural Engineering, 1, 2004.*

[2] Belmouden Y., Lestuzzi P., Seismic vulnerability assessment of existing buildings in Switzerland, Applied Computing and Mechanics Laboratory, Structural Engineering Institute, *Ecole Polytechnique Fédérale de Lausanne, Research Report N°6, April 2006.*

[3] Roca P.; Molins C., Marí A. R., *Strength Capacity of Masonry Wall Structures by the Equivalent Frame Method Journal of Structural Engineering, Vol. 131, No. 10*, 2005, pp. -1601–1610.

[4] Salonikios T., Karakostas C., Lekidis V., Anthoine A. *Comparative inelastic pushover analysis of masonry frames, Engineering Structures Journal, Vol. 25, 2005*, pp. 1515-1523.

[5] Kappos A., Penelis G., Drakopoulos C., Evaluation of simplified models for lateral load analysis of unreinforced masonry buildings, *Journal of Structural Engineering, ASCE, Vol. 128, July N°7, 2002*, pp. 890-897.

[6] *ATC-40, Seismic Evaluation and Retrofit of Concrete Buildings, Vol. 1*, Applied Technology Council, Redwood City, California, 1996.

[7] Belmouden Y., Lestuzzi P., Analytical model for predicting nonlinear reversed cyclic behaviour of reinforced concrete structural walls, *Engineering Structures, Vol. 29/7, 2007*, pp. 1263-1276.

[8] ElMandooh K., Ghobarah A., Flexural and shear hysteretic behavior of reinforced concrete columns with variable axial load, *Engineering Structures, Vol. 25, 2003*, pp. 1353-1367.

[9] Nuray Aydinoğlu M., An incremental response spectrum analysis procedure based on inelastic spectral displacements for multi-mode seismic performance evaluation, *Bulletin of Earthquake Engineering, 2003*, pp. 3-36, Vol. 1.

[10] Kilar V., Fajfar P., Simple pushover analysis of asymmetric buildings, *Earthquake Engineering and Structural Dynamics, Vol. 26, 1997*, pp. 233-249.

[11] Lang. K., Seismic vulnerability of existing buildings, *PhD Thesis, Institute of Structural Engineering*, Swiss Federal Institute of Technology, Zürich, Switzerland, February 2002.

[12] *FEMA 356, Prestandard and commentary for the seismic rehabilitation of buildings*, Washington (DC): Federal Emergency Management Agency, 2000.

[13] *FEMA 306, Evaluation of earthquake damaged concrete and masonry wall buildings, basic procedures manual*, Washington (DC): The Partnership for response and recovery, 1999.

[14] Paquette J., Bruneau M., Pseudo-dynamic testing of unreinforced masonry building with flexible diaphragm and comparison with existing procedures, *Construction and Building Materials, Vol. 20, 2006*, pp. 220-228.

[15] Penelis GR.G., An efficient approach for pushover analysis of unreinforced masonry (URM) structures, *Journal of Earthquake Engineering, Vol. 10, N°3*, pp. 359-379, 2006.

[16] Prakash, V., Powell, G.H. and Campbell, S. (1994), *DRAIN3DX: Base Program Description and User Guide*, UCB/SEMM-1994/07, Berkeley: Department of Civil Engineering, University of California, August 1994.

[17] *ETABS, Integrated Building Design Software, Computers and Structures, Inc. Berkeley, California, USA, Version 9, November 2005*.

[18] Bachmann H., Dazio A, A Deformation-Based Seismic Design Procedure for Structural Wall Buildings, *Proceedings of the International Workshop on Seismic Desing Methodologies for the Next Generation of Codes, Bled/Slovenia*, A.A. Balkema, Rotterdam, 24-27 June 1997.

[19] Dazio A., Entwurf und Bemessung von Tragwandgebäuden unter Erdbebeneinwirkung, (in German), *Institut für Baustatik und Konstruktion*, ETH Zürich, Bericht Nr. 254, Birkhäuser Verlag Basel, 2000.

[20] Wilkinson S. M., Hiley R. A., A non-linear response history model for the seismic analysis of high-rise framed buildings, *Computers and Structures, Vol. 84, 2006*, pp. 318-329.

[21] Magenes G., 'A method for pushover analysis in seismic assessment of masonry buildings', *Paper 186612WCEE, 2000.*

[22] Magenes G., Della Fontana A., Simplified non-linear seismic analysis of masonry buildings, *Proc. of the Fifth International Masonry Conference, London, 13th-15th, 1998.*

[23] Magenes G., Calvi G. M., In-Plane response of brick masonry walls, *Earthquake Engineering and Structural Dynamics, Vol. 26, 1997,* pp. 1091-1112.

[24] Tomazevic M., Seismic resistance verification of buildings: Following the new trends, *Seismic Methodologies for the Next Generation of Codes*, P. Fajfar and H. Krawinkler (eds.), Balkema, Rotterdam, 1997.

[25] Dolce M., Schematizzazione e modellazione per azioni nel piano delle pareti (Models for in-plane loading of masonry walls), *Corso sul consolidamento degli edifici in muratura in zona sismica*, Ordine degli Ingegneri, Potenza, 1989.

[26] Galasco A., Lagomarsino S., Penna A., On the use of Pushover analysis for existing masonry buildings, *First European Conference on Earthquake Engineering and Seismology, Geneva, Switzerland, 3-8 September 2006, Paper N°1080.*

[27] Belmouden Y., Lestuzzi P., 'An equivalent frame model for seismic analysis of masonry and reinforced concrete buildings', *Construction Building and Material Journal, 2007 (In Press)*

[28] Belmouden Y., Lestuzzi P., 'An analytical model for capacity curves generation and damage prediction of masonry structures', *International Review of Mechanical Engineering, 2007 (In Press).*

In: Structural Materials and Engineering
Editor: Ference H. Hagy

ISBN 978-1-60692-927-8
© 2009 Nova Science Publishers, Inc.

Chapter 9

DYNAMIC STABILITY OF BEAMS USING A HIGHER ORDER THEORY

Sebastián P. Machado[*] and Víctor H. Cortínez

Centro de Investigaciones de Mecánica Teórica y Aplicada, Facultad Regional Bahía Blanca, Universidad Tecnológica Nacional, Bahía Blanca, Argentina.

ABSTRACT

The dynamic stability of thin walled beams subjected to different sets of boundary conditions is investigated in this presentation. The analysis is based on a seven-degree-of-freedom shear deformable beam theory. The theory is formulated in the context of large displacements and rotations, through the adoption of a displacement field considering moderate bending rotations and large twist. This geometrically non linear formulation is used for analyzing regions of dynamic instability of simply supported, cantilever and fixed-end beams subjected to axial and transverse periodic loads. The influence of shear deformation and inertial effects (corresponding to loading plane) on the unstable regions is analyzed, for mono- and bi-symmetric cross-section beams. This last influence is generally neglected in most of the dynamic instability studies. Such an assumption is valid to a certain extent when the exciting frequency is small in comparison with the free frequency of the loading plane. This is the case frequently assumed for analyzing the dynamic stability of bars subjected to axial excitation. However for transverse excitation, the frequency at which a parametric resonance occurs can be the same order as the natural frequency of the loading plane vibrations.

Ritz variational method is used to reduce the governing equation, the independent displacements vector is expressed as a linear combination of given x-function vectors and unknown t-function coefficients. Regions of dynamic instability of simple and combination resonances are determined by applying Hsu's procedures to the Mathieu equation. The static load parameter and the dynamic amplitude of the excitation load are scaled with the buckling load. This critical value is obtained taking into account the effect of prebuckling deflections. The excitation frequency is also scaled with the lowest frequency value of parametric resonance.

---

[*] For communication: TE: 54-0291-4555220 FAX: 0291 4555311
e-mail: smachado@frbb.utn.edu.ar, vcortine@ frbb.utn.edu.ar

The influence of non-conventional effects is noted in the numerical results. The interaction between the forced vibration and the parametrically excited vibrations on the regions of dynamic instability is significant in some cases. This effect depends on the closeness of the frequency values, i.e. it depends on the stiffnesses ratio between the parametrically excited mode and the vertical mode corresponding to the loading plane.

Key words: Thin-walled beams; shear flexibility; non-linear theory; simple and combination parametric resonance.

1. INTRODUCTION

Thin-walled beams are widely used as structural elements within the fields of civil, aerospace and mechanical engineering, offering a high performance in terms of minimum weight for a given strength. These kinds of members are the most common load-carrying systems in engineering applications. The theory of dynamic stability represents a specific aspect of the stability of the motion, which is related to the vibrations and stability theory of mechanical systems.

It is well known that the spatial stability behavior of thin-walled beam structures is very complex due to the coupling effect of extensional, bending and torsional deformation. Besides, such flexible structures can undergo large displacements and rotations without exceeding their elastic limits. Therefore, a non-linear theory is required for the accurate behavior prediction of such structures. For example, the limitation of the linear stability analysis of beam problems (Vlasov [1]) is the omission of any consideration of the effect of large deflections of the beam. This omission may be sufficiently accurate when the initial deflection of the beam is negligible. In particular, lateral buckling is a relevant phenomenon that involves mechanical complications, since structures may experience large or moderately large deflections and rotations before buckling occurs.

Structural collapse examples, such as the Tacoma Narrows Bridge in 1940 and the loss of dynamic stability of aircraft wings, showed that the instability issue should be considered in the structural design. Dynamic instability of elastic structural elements, such as rods, beams and columns, induced by parametric excitation has been investigated by many researchers. Extensive bibliographies on this subject were given by Evan-Iwanowski [2] and Nayfeh and Mook [3]. Bolotin [4] provided a general introduction to analyze the dynamic stability problems of various structural elements. The Mathieu-Hill equation is obtained in [2, 4] while solving the parametric vibration of a beam subjected to a compressive dynamic force. Nayfeh and Mook [3], used the perturbation method to solve Mathieu-Hill's equation, in order to analyze the behavior of an elastic system under parametric excitation. They established a criterion to yield the transition curves by determining the characteristic exponents in the solution.

In relation to thin-walled beams, Gol'denblat [5] investigated the stability of a compressed thin-walled rod symmetrical about one axis. The problem was reduced to a system of two differential equations. Tso [6] studied the longitudinal-torsional stability, while Mettler [7] and Ghobarah and Tso [8] studied the bending-torsional stability of thin-walled beams. Bolotin [4, 9] and Popelar [10, 11] discussed the dynamic stability of thin-walled beams; typical I and H sections were considered. Hasan and Barr [12] evaluated regions of instability of thin-walled beams of equal angle-section, considering axial and transverse excitation in a cantilever beam.

The development of beam theories usually involves some type of reduction of the three-dimensional constitutive relationships. Often this is accomplished by neglecting the stresses and strains in the transverse directions. In all the above studies shear deformation effect is ignored. Though shear deformations can be small in most civil engineering applications, they may be mainly important in some situations, where bending moments are small compared to shear forces acting on the member. This effect is normally observed in the analysis of beams with small slenderness ratio. M-Y Kim *et al.* [13, 14] and T-W Kim *et al.* [15] analyzed the spatial stability, free vibration and parametric instability behavior of thin-walled frames, taking into account shear deformation effects due to shear forces and warping-torsion. However, in [13, 14] the shear deformable displacement field is introduced based on the semitangential rotations, including second order terms of finite rotations. Machado and Cortínez [16, 17] considered the effect of shear deformation to investigate the stability of composite beam.

A significant amount of research has been conduced in recent years toward the development of non-linear theories of three-dimensional beams. However, many of this theories differ in the order of non-linearity considered in their formulation. For example, second-order displacement field has been used in a formulation of finite element models for 3-D non-linear analysis of beam structures [13, 14, 18-20]. This approximation has several advantages because simplifies the coupling between the displacement and rotations, and so the tangent stiffness matrix (use for the non-linear incremental-iterative analysis) can be simplified. Therefore, this tangent matrix can be decomposed into linear and second-order (non-linear) stiffness matrices. In spite of these advantages, approximations or simplifications that are made in the earlier stages of the derivation may produce the loss of some significant terms in the non-linear strains and in the tangent stiffness matrix. Thus some inaccurate approximations in the coupling between displacement, rotations and their derivates are obtained. The loss of these terms may lead to 'self-straining' caused by superimposed rigid-body motions [21, 22].

To the best of author's knowledge, the dynamic stability analysis of a shear deformable non-linear thin-walled beam was only presented by Machado et al. [23]. The authors developed a second-order beam formulation to study the dynamic stability of composite beams subjected to axial external force. The Galerkin's method was used in order to discretize the governing equation and the Bolotin's method was applied to determine the regions of dynamic instability. The authors demonstrated that the size of the unstable regions can show a discrepancy depending on the influence of the longitudinal inertia.

The purpose of the present investigation is to show the accuracy and efficiency of the proposed formulation to investigate the effect of non-linear coupling on the dynamic stability of thin-walled beams subjected to axial and transverse periodic excitation. The analysis is based on a geometrically higher-order non-linear beam model formulated in the context of large displacements and rotations, through the adoption of a shear deformable displacement field (accounting for bending and warping shear). It should be emphasized here that the necessity of obtaining an accurate bending-torsional coupling arises from the fact that the stability behavior of such structures is influenced by the initial displacements (corresponding to the static load) [16] and by the inertia forces associated to the excitation movement. On the other hand, it has been shown by the authors that the classical and second-order beam theories underpredict and overpredict, respectively, the behavior of these structures [16, 24].

Most of the dynamic stability studies of thin-walled beams have been carried out on the base of linear systems. The effect of geometric nonlinearity has been considered in some few works; among those it is found Bolotin [4], Hasan et al. [12] and Machado et al. [23]. The effect of the interaction between the forced vibration and the parametrically excited vibrations on the unstable regions is generally omitted. In the case of axial excitation, the exciting frequency is generally smaller than the frequency ω_L of the free longitudinal vibrations. However, for beams with small slenderness ratio, the influence of longitudinal vibrations becomes more noticeable, as was shown by Bolotin [4] for solid cross-section non-deformable beams and Machado [23] for thin-walled composite beams. Besides, in the case of transverse excitation, the frequency at which a parametric resonance occurs can be the same order as the natural frequency of the longitudinal vibrations.

The numerical results are obtained for mono- and bi-symmetric open cross-section beam. The length beam and the load height parameter are varied to assess their effects on the dynamic instability behavior. The influence of shear deformation and inertial effects (corresponding to loading plane) on the unstable regions is analyzed. Ritz variational method is used to reduce the governing equation; the independent displacements vector is expressed as a linear combination of given x-function vectors and unknown t-function coefficients. Regions of dynamic instability of simple and combination resonances are determined by applying Hsu's [25] procedures to the Mathieu equation.

2. KINEMATICS

A straight thin-walled beam with an arbitrary cross-section is considered (Fig. 1). The points of the structural member are referred to a Cartesian coordinate system $(x, \bar{y}, \bar{z})$, where the x-axis is parallel to the longitudinal axis of the beam while $\bar{y}$ and $\bar{z}$ are the principal axes of the cross-section. The axes y and z are parallel to the principal ones but having their origin at the shear center (SC), defined according to Vlasov's theory of isotropic beams. Midway through the thickness of each cross-sectional element is the middle surface. A plane perpendicular to the x-axis intersects the middle surface at a curve called the contour. The coordinates corresponding to points lying on the middle line are denoted as Y and Z (or $\bar{Y}$ and $\bar{Z}$). A contour (n,s,x) coordinate system is defined with s following the contour, and n perpendicular to s. This coordinate is introduced on the middle contour of the cross-section system as illustrated in Fig. 2.

$$\bar{y}(s,n) = \bar{Y}(s) - n\frac{dZ}{ds}, \quad \bar{z}(s,n) = \bar{Z}(s) + n\frac{dY}{ds} \tag{1}$$

$$y(s,n) = Y(s) - n\frac{dZ}{ds}, \quad z(s,n) = Z(s) + n\frac{dY}{ds} \tag{2}$$

On the other hand, y_0 and z_0 are the centroidal coordinates measured with respect to the shear center.

$$\begin{aligned} \overline{y}(s,n) &= y(s,n) - y_0, \\ \overline{z}(s,n) &= z(s,n) - z_0. \end{aligned} \tag{3}$$

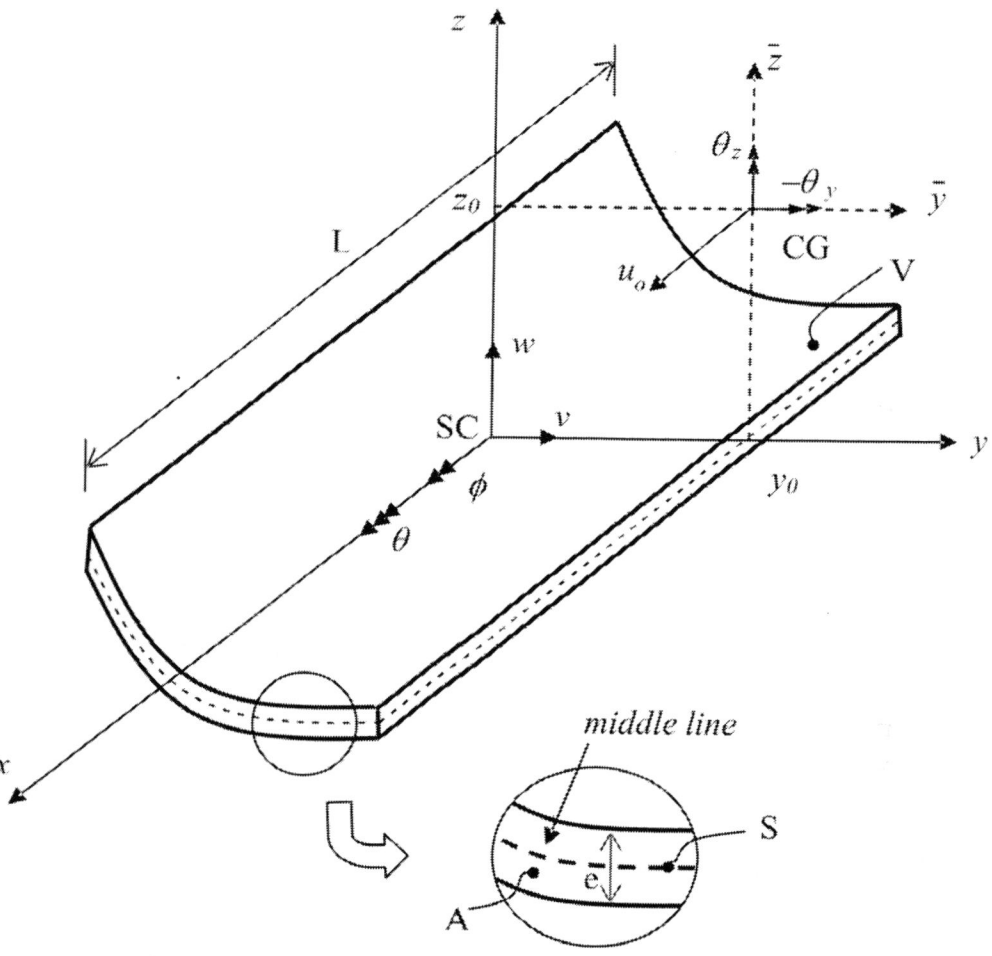

Figure 1. General thin-walled section beam and notation for displacement measures.

The present structural model is based on the following assumptions:

1. The cross-section contour is rigid in its own plane.
2. The warping distribution is assumed to be given by the Saint-Venant function for isotropic beams.
3. Flexural rotations (about the $\overline{y}$ and $\overline{z}$ axes) are assumed to be moderate, while the twist ϕ of the cross-section can be arbitrarily large.
4. Shell force and moment resultant corresponding to the circumferential stress σ_{ss} and the force resultant corresponding to γ_{ns} are neglected.

5. The radius of curvature at any point of the shell is neglected.
6. Twisting linear curvature of the shell is expressed according to the classical plate theory.

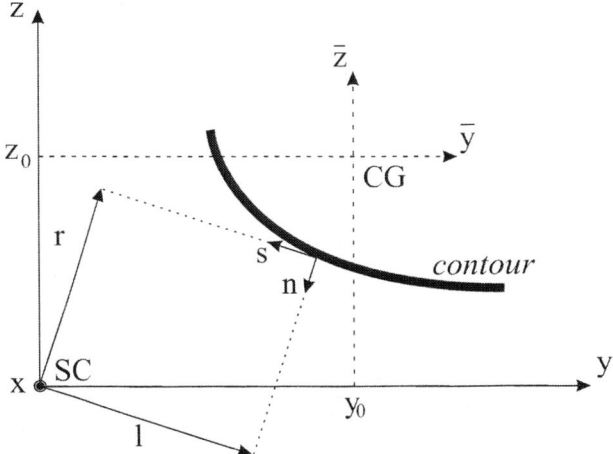

Figure 2. Coordinate system of the cross-section.

2.1. Development of the Displacement Field

According to the hypotheses of the present structural model, the proposed displacement field Eq. (4) is based on the principle of semitangential rotation defined by Argyris [26] to avoid the difficulty due to the noncommutative nature of rotations. In this displacement field, the torsional twist terms ϕ are expressed as trigonometric functions according to hypotheses 3). The displacement field is represented by means of seven degree of freedom corresponding to three displacements (u, v and w), three measures of the rotations (ϕ, θ_y and θ_z) about the shear center axis, $\bar{y}$ and $\bar{z}$ axes, respectively, and a warping variable (θ) of the cross-section. The displacement field is expressed as the following form:

$$u_x = u_o - \bar{y}\left(\theta_z \cos\phi + \theta_y \sin\phi\right) - \bar{z}\left(\theta_y \cos\phi - \theta_z \sin\phi\right) + \omega\left[\theta - \frac{1}{2}\left(\theta_y'\theta_z - \theta_y\theta_z'\right)\right] + \left(\theta_z z_0 - \theta_y y_0\right)\sin\phi,$$

$$u_y = v - z\sin\phi - y(1-\cos\phi) - \frac{1}{2}\left(\theta_z^2 \bar{y} + \theta_z\theta_y \bar{z}\right), \qquad (4)$$

$$u_z = w + y\sin\phi - z(1-\cos\phi) - \frac{1}{2}\left(\theta_y^2 \bar{z} + \theta_z\theta_y \bar{y}\right),$$

where the prime indicates differentiation with respect to x.

This expression is a generalization of others previously proposed in the literature. On the other hand, neglecting the shear flexibility ($\theta_z = v'$, $\theta_y = w'$ and $\theta = \phi'$), approximating $cos\ \phi$ and $sin\ \phi$ by ($1-\phi^2/2$) and ϕ respectively, and conserving non-linear terms up to second-order, the present displacement field coincides with that developed by Fraternali and Feo [27], who formulated a moderate rotation theory of thin-walled composite beams generalizing the infinitesimal theory of sectorial areas by Vlasov [1]. Moreover, the displacement field of the classical Vlasov's [1] theory is obtained when second-order effects are ignored. As a final comparison, taking $\cos \phi = 1$ and $\sin \phi = \phi$ and disregarding the non-linear terms, the displacement field (4) coincides with the one formulated by Cortínez and Piovan [28] for linear dynamics of shear deformable thin-walled beams.

2.2. Warping Displacements

In general, the contour of the cross-section will warp out of its plane as twisting occurs. The warping function ω of the thin-walled cross-section may be defined as:

$$\omega(s,n) = \omega_p(s) + \omega_s(s,n) \tag{5}$$

where, ω_p and ω_s are the contour warping function and the thickness warping function, respectively. They are defined in the form [28]:

$$\omega_p(s) = \frac{1}{m}\left[\int_0^m \left(\int_{s_0}^s [r(s) - \psi(s)]ds\right)ds\right] - \int_{s_0}^s [r(s) - \psi(s)]ds,$$
$$\omega_s(s,n) = -n\, l(s), \tag{6}$$

where m is the length of the cross-sectional thin wall and Ψ is the shear strain at the middle line, obtained by means of the Saint-Venant theory of pure torsion for isotropic beams, and normalized with respect to $d\phi/dx$ [29]. Besides, $r(s)$ represents the perpendicular distance from the shear center (SC) to the tangent at any point of the mid-surface contour, and $l(s)$ represents the perpendicular distance from the shear center (SC) to the normal at any point of the mid-surface contour, as shown in Fig. 2.

$$r(s) = -Z(s)\frac{dY}{ds} + Y(s)\frac{dZ}{ds}, \tag{7}$$

$$l(s) = Y(s)\frac{dY}{ds} + Z(s)\frac{dZ}{ds}. \tag{8}$$

3. THE STRAIN FIELD

The displacements with respect to the curvilinear system (x, s, n) are obtained by means of the following expressions:

$$\bar{U} = u_x(x,s,n), \tag{9}$$

$$\bar{V} = u_y(x,s,n)\frac{dY}{ds} + u_z(x,s,n)\frac{dZ}{ds}, \tag{10}$$

$$\overline{W} = -u_y(x,s,n)\frac{dZ}{ds} + u_z(x,s,n)\frac{dY}{ds}. \tag{11}$$

The three non-zero components ε_{xx}, ε_{xs}, ε_{xn} of the Green's strain tensor are given by:

$$\varepsilon_{xx} = \frac{\partial \overline{U}}{\partial x} + \frac{1}{2}\left[\left(\frac{\partial \overline{U}}{\partial x}\right)^2 + \left(\frac{\partial \overline{V}}{\partial x}\right)^2 + \left(\frac{\partial \overline{W}}{\partial x}\right)^2\right], \tag{12}$$

$$\varepsilon_{xs} = \frac{1}{2}\left[\frac{\partial \overline{U}}{\partial s} + \frac{\partial \overline{V}}{\partial x} + \frac{\partial \overline{U}}{\partial x}\frac{\partial \overline{U}}{\partial s} + \frac{\partial \overline{V}}{\partial x}\frac{\partial \overline{V}}{\partial s} + \frac{\partial \overline{W}}{\partial x}\frac{\partial \overline{W}}{\partial s}\right], \tag{13}$$

$$\varepsilon_{xn} = \frac{1}{2}\left[\frac{\partial \overline{U}}{\partial n} + \frac{\partial \overline{W}}{\partial x} + \frac{\partial \overline{U}}{\partial x}\frac{\partial \overline{U}}{\partial n} + \frac{\partial \overline{V}}{\partial x}\frac{\partial \overline{V}}{\partial n} + \frac{\partial \overline{W}}{\partial x}\frac{\partial \overline{W}}{\partial n}\right], \tag{14}$$

Substituting expressions (4) into (9-11) and then into (12-14), employing the relations (1-3) and (5-8), after simplifying some higher order terms, the components of the strain tensor are expressed in the following form:

$$\begin{aligned}
\varepsilon_{xx} &= \varepsilon_{xx}^{(0)} + n\kappa_{xx}^{(1)}, \\
\gamma_{xs} &= 2\varepsilon_{xs} = \gamma_{xs}^{(0)} + n\kappa_{xs}^{(1)}, \\
\gamma_{xn} &= 2\varepsilon_{xn} = \gamma_{xn}^{(0)},
\end{aligned} \tag{15}$$

where

$$\begin{aligned}
\varepsilon_{xx}^{(0)} &= u_o' + \frac{1}{2}\left(u_o'^2 + v'^2 + w'^2\right) + \omega_p\left[\theta' - \frac{1}{2}\left(\theta_z\theta_y'' - \theta_y\theta_z''\right)\right] + \overline{Z}\left[\left(-\theta_y' - u_o'\theta_y'\right)\cos\phi + \left(\theta_z' + u_o'\theta_z'\right)\mathrm{sen}\phi\right] \\
&+ \overline{Y}\left[\left(-\theta_z' - u_o'\theta_z'\right)\cos\phi - \left(\theta_y' + u_o'\theta_y'\right)\mathrm{sen}\phi\right] + \frac{1}{2}\phi'^2\left(Y^2 + Z^2\right) + \frac{1}{2}\theta_y'^2\overline{Z}^2 + \frac{1}{2}\theta_z'^2\overline{Y}^2 + \theta_y'\theta_z'\overline{ZY} \\
&+ \left(z_0\theta_z' - y_0\theta_y'\right)\mathrm{sen}\phi,
\end{aligned} \tag{16}$$

$$\begin{aligned}
\kappa_{xx}^{(1)} &= -\frac{dZ}{ds}\left[-\left(\theta_z' + u_o'\theta_z'\right)\cos\phi - \left(\theta_y' + u_o'\theta_y'\right)\mathrm{sen}\phi\right] + \frac{dY}{ds}\left[\left(-\theta_y' - u_o'\theta_y'\right)\cos\phi + \left(\theta_z' + u_o'\theta_z'\right)\mathrm{sen}\phi\right] \\
&- l\left[\theta' - \frac{1}{2}\left(\theta_z\theta_y'' - \theta_y\theta_z''\right)\right] - r\,\phi'^2 - \overline{Y}\frac{dZ}{ds}\theta_z'^2 + \overline{Z}\frac{dY}{ds}\theta_y'^2 + \left(\overline{Y}\frac{dY}{ds} - \overline{Z}\frac{dZ}{ds}\right)\theta_y'\theta_z',
\end{aligned} \tag{17}$$

$$\begin{aligned}
\gamma_{xs}^{(0)} &= \frac{dY}{ds}\left[\left(v' - \theta_z - u_o'\theta_z\right)\cos\phi - z_0\frac{1}{2}\left(\theta_z\theta_y' - \theta_y\theta_z'\right) + \left(w' - \theta_y - u_o'\theta_y\right)\sin\phi\right] + (r-\psi)(\phi' - \theta) \\
&+ \frac{dZ}{ds}\left[\left(w' - \theta_y - u_o'\theta_y\right)\cos\phi + y_0\frac{1}{2}\left(\theta_z\theta_y' - \theta_y\theta_z'\right) - \left(v' - \theta_z - u_o'\theta_z\right)\sin\phi\right] + \psi\left[\phi' - \frac{1}{2}\left(\theta_z\theta_y' - \theta_y\theta_z'\right)\right],
\end{aligned} \tag{18}$$

$$\kappa_{xs}^{(1)} = -2\left[\phi' - \frac{1}{2}\left(\theta_z\theta_y' - \theta_y\theta_z'\right)\right], \tag{19}$$

$$\gamma_{xn}^{(0)} = \frac{dY}{ds}\left[\left(w' - \theta_y - u_0'\theta_y\right)\cos\phi + y_0 \frac{1}{2}\left(\theta_z\theta_y' - \theta_y\theta_z'\right) - \left(v' - \theta_z - u_0'\theta_z\right)\sin\phi\right]$$
$$-\frac{dZ}{ds}\left[\left(v' - \theta_z - u_0'\theta_z\right)\cos\phi - z_0 \frac{1}{2}\left(\theta_z\theta_y' - \theta_y\theta_z'\right) + \left(w' - \theta_y - u_0'\theta_y\right)\sin\phi\right] + l(\phi' - \theta). \quad (20)$$

4. VARIATIONAL FORMULATION

Taking into account the adopted assumptions, the principle of virtual work for an isotropic shell may be expressed in the form (Washizu [30]):

$$\iint \left(N_{xx}\delta\varepsilon_{xx}^{(0)} + M_{xx}\delta\kappa_{xx}^{(1)} + N_{xs}\delta\gamma_{xs}^{(0)} + M_{xs}\delta\kappa_{xs}^{(1)} + N_{xn}\delta\gamma_{xn}^{(0)}\right) ds\, dx$$
$$-\iiint \rho\left(\ddot{u}_x\,\delta u_x + \ddot{u}_y\,\delta u_y + \ddot{u}_z\,\delta u_z\right) ds\, dn\, dx$$
$$-\iint(\bar{q}_x\delta\bar{u}_x + \bar{q}_y\delta\bar{u}_y + \bar{q}_z\delta\bar{u}_z)\, ds\, dx - \iint(\bar{p}_x\delta u_x + \bar{p}_y\delta u_y + \bar{p}_z\delta u_z)\big|_{x=0} ds\, dn \quad (21)$$
$$-\iint(\bar{p}_x\delta u_x + \bar{p}_y\delta u_y + \bar{p}_z\delta u_z)\big|_{x=L} ds\, dn - \iiint(\bar{f}_x\delta u_x + \bar{f}_y\delta u_y + \bar{f}_z\delta u_z)\, ds\, dn\, dx = 0,$$

where N_{xx}, N_{xs}, M_{xx}, M_{xs} and N_{xn} are the shell stress resultants defined according to the following expressions:

$$N_{xx} = \int_{-e/2}^{e/2} \sigma_{xx}\, dn, \qquad M_{xx} = \int_{-e/2}^{e/2} (\sigma_{xx}n)\, dn,$$
$$N_{xs} = \int_{-e/2}^{e/2} \sigma_{xs}\, dn, \qquad M_{xs} = \int_{-e/2}^{e/2} (\sigma_{xs}n)\, dn, \qquad N_{xn} = \int_{-e/2}^{e/2} \sigma_{xn}\, dn. \quad (22)$$

The beam is subjected to wall surface tractions $\bar{q}_x$, $\bar{q}_y$ and $\bar{q}_z$ specified per unit area of the undeformed middle surface and acting along the x, y and z directions, respectively. Similarly, $\bar{p}_x$, $\bar{p}_y$ and $\bar{p}_z$ are the end tractions per unit area of the undeformed cross-section specified at $x = 0$ and $x = L$, where L is the undeformed length of the beam. Besides $\bar{f}_x$, $\bar{f}_y$ and $\bar{f}_z$ are the body forces per unit of volume. Finally, denoting $\bar{u}_x$, $\bar{u}_y$ and $\bar{u}_z$ as displacements at the middle line.

Substituting Eqs. (16-20) into Eq. (21) and integrating with respect to s, one obtains the one-dimensional expression for the virtual work equation given by:

$$L_M + L_K + L_P = 0, \quad (23)$$

where L_M, L_k and L_p represent the virtual work contributions due to the inertial, internal and external forces, respectively. Their expressions are given below.

$$L_M = \int_0^L \rho \left[A \frac{\partial^2 u_0}{\partial t^2} \delta u_0 + I_z \frac{\partial^2 \theta_z}{\partial t^2} \delta \theta_z + I_y \frac{\partial^2 \theta_y}{\partial t^2} \delta \theta_y + C_w \frac{\partial^2 \theta}{\partial t^2} \delta \theta + A \frac{\partial^2}{\partial t^2}(v - z_0 \phi) \delta v \right. $$
$$\left. + A \frac{\partial^2}{\partial t^2}(w + y_0 \phi) \delta w + \frac{\partial^2}{\partial t^2}(-Az_0 v + Ay_0 w + I_s \phi) \delta \phi \right] dx, \qquad (24)$$

where A is the cross-sectional area, I_z and I_y are the principal moments of inertia of the cross-section, C_w is the warping constant, I_s is the polar moment with respect to the shear center and ρ is the density.

$$L_K = \int_0^L \left\{ \delta u_0' \left[N + u_0' N - M_z \left(\theta_z' \cos\phi + \theta_y' \sen\phi \right) - M_y \left(\theta_y' \cos\phi + \theta_z' \sen\phi \right) - Q_y \left(\theta_z \cos\phi + \theta_y \sen\phi \right) \right. \right.$$
$$\left. - Q_z \left(\theta_y \cos\phi + \theta_z \sen\phi \right) \right] + \delta v' \left(Q_y \cos\phi - Q_z \sen\phi + v' N \right) + \delta w' \left(Q_z \cos\phi + Q_y \sen\phi + w' N \right)$$
$$+ \delta \theta_z \left[-Q_y (1+u_0') \cos\phi + Q_z (1+u_0') \sen\phi + \frac{1}{2}(Q_z y_0 - Q_y z_0) \theta_y' - \frac{1}{2} T_{sv} \theta_y' - \frac{1}{2} B \, \theta_y'' \right]$$
$$+ \delta \theta_z' \left[-M_z (1+u_0') \cos\phi + M_y (1+u_0') \sen\phi + N z_0 \sen\phi + \frac{1}{2}(Q_y z_0 - Q_z y_0) \theta_y' + \frac{1}{2} T_{sv} \theta_y + \theta_z' P_{zz} + \theta_y' P_{yz} \right]$$
$$+ \delta \theta_y \left[-Q_z (1+u_0') \cos\phi - Q_y (1+u_0') \sen\phi + \frac{1}{2}(Q_y z_0 - Q_z y_0) \theta_z' + \frac{1}{2} T_{sv} \theta_z' + \frac{1}{2} B \, \theta_z'' \right]$$
$$+ \delta \theta_y' \left[-M_y (1+u_0') \cos\phi - M_z (1+u_0') \sen\phi - N y_0 \sen\phi + \frac{1}{2}(Q_z y_0 - Q_y z_0) \theta_z - \frac{1}{2} T_{sv} \theta_z + \theta_z' P_{yz} + \theta_y' P_{yy} \right]$$
$$+ \delta \phi \left[M_y \left((\theta_y' + \theta_z' u_0') \sen\phi + (\theta_z' + \theta_z' u_0') \cos\phi \right) + M_z \left((\theta_z' + \theta_z' u_0') \sen\phi - (\theta_y' + \theta_y' u_0') \cos\phi \right) \right.$$
$$+ Q_y \left((\theta_z - w' + \theta_z u_0') \sen\phi - (\theta_y - w' + \theta_y u_0') \cos\phi \right) + N \left(z_0 \theta_z' - y_0 \theta_y' \right) \cos\phi$$
$$\left. + Q_z \left((\theta_y - w' + \theta_y u_0') \sen\phi + (\theta_z - v' + \theta_z u_0') \cos\phi \right) \right] + \delta \theta_z'' \frac{1}{2} B \, \theta_y - \delta \theta_y'' \frac{1}{2} B \, \theta_z$$
$$\left. + \delta \phi' \left[T_w + T_{sv} + B_1 \, \phi' \right] + \delta \theta' B - \delta \theta \, T_w \right\} dx, \qquad (25)$$

$$L_P = \int_0^L \left\{ -q_x \delta u_0 - q_y \delta v - q_z \delta w - b \, \delta \theta - \delta \theta_z' \frac{1}{2} b \, \theta_y + \delta \theta_y' \frac{1}{2} b \, \theta_z \right.$$
$$+ \delta \theta_z \left[m_z \cos\phi - (m_y + z_0 q_x) \sin\phi + \frac{1}{2} b \, \theta_y' + \frac{1}{2} \lambda_{mx} \theta_y + \lambda_y \theta_z \right]$$
$$+ \delta \theta_y \left[m_y \cos\phi + (m_z + y_0 q_x) \sin\phi - \frac{1}{2} b \, \theta_z' + \frac{1}{2} \lambda_{mx} \theta_z + \lambda_z \theta_y \right]$$
$$+ \delta \phi \left[-m_x \cos\phi - (m_z \theta_z + m_y \theta_y) \sin\phi - (m_y + z_0 q_x) \theta_z \cos\phi \right.$$
$$\left. + (m_z + y_0 q_x) \theta_y \cos\phi + \sin\phi (\lambda_y + \lambda_z + z_0 q_z + y_0 q_y) \right] \right\} dx$$
$$+ \left| -\bar{N} \delta u_0 - \bar{Q}_y \delta v - \bar{Q}_z \delta w - \bar{B} \delta \theta - \delta \theta_z' \frac{1}{2} \theta_y \bar{B} + \delta \theta_y' \frac{1}{2} \theta_z \bar{B} \right.$$
$$+ \delta \theta_z \left[\bar{M}_z \cos\phi - (\bar{M}_y + \bar{N} z_0) \sin\phi + \frac{1}{2} \theta_y' \bar{B} + \theta_y \bar{\lambda}_y + \frac{1}{2} \theta_y \bar{\lambda}_{mx} \right]$$
$$+ \delta \theta_y \left[\bar{M}_y \cos\phi + (\bar{M}_z + \bar{N} y_0) \sin\phi - \frac{1}{2} \theta_z' \bar{B} + \theta_z \bar{\lambda}_z + \frac{1}{2} \theta_z \bar{\lambda}_{mx} \right]$$
$$\left. + \delta \phi \left[-\bar{M}_x \cos\phi - \sin\phi (\bar{M}_z \theta_z + \bar{M}_y \theta_y) - \cos\phi (\bar{M}_y \theta_z - \bar{M}_z \theta_y + \bar{N} z_0 \theta_z - \bar{N} y_0 \theta_y) + \bar{B}_1 \sin\phi \right] \right|_{x=0}^{x=L}, \qquad (26)$$

where q_x, q_y, q_z, m_x, m_y, m_z, b, λ_y, λ_z, and λ_{mx} are resultants of the applied wall surface tractions and $\bar{N}$, $\bar{Q}_y$, $\bar{Q}_z$, $\bar{M}_z$, $\bar{M}_y$, $\bar{B}$, $\bar{M}_x$, $\bar{\lambda}_y$, $\bar{\lambda}_z$, $\bar{\lambda}_{mx}$ and $\bar{B}_l$ represent the resultants of the applied end tractions.

In the present study, the work done by axial and transverse periodic loads in the vertical plane is written as

$$L_P = \int_0^L \left[-q_z(t)\delta w + \delta\phi \, \phi \, e_z q_z(t)\right] dx + \left|-\bar{N}\delta u_0 + \delta\theta_y \, \bar{M}_y(t)\right|_{x=0}^{x=L}, \quad (27)$$

where the distributed load $q_z(t) = q_{z0} + q_{zt} \cos \varpi t$, ϖ is the excitation radian frequency, $q_{z0} = \alpha \, q_{cr}$, $q_{zt} = \beta \, q_{cr}$, α is the static load factor, β is the dynamic load factor, q_{cr} is the buckling load and e_z denotes the eccentricity in z-direction of the applied loads measured from the shear center. In what follows this last one will be called load height parameter. In the same way, $\bar{M}_y(t) = M_0 + M_t \cos\varpi t$ is the bending moment applied in the beam ends. However, when a concentrated load $P(t)$ is applied to the beam at the position ($x = a$), instead of a distributed load, the load $q_z(t)$ in the potential L_P (27) is written as:

$$q_z(t) = P(t)\Delta(x-a), \quad (28)$$

where Δ is the Dirac function.

5. BEAM FORCES AND CONSTITUTIVE EQUATIONS

In the above expressions, the following 1-D beam forces, in terms of the shell stress resultants, have been defined as:

$$N = \int N_{xx} \, ds, \quad M_Y = \int \left(N_{xx}\bar{Z} + M_{xx}\frac{dY}{ds}\right) ds, \quad M_Z = \int \left(N_{xx}\bar{Y} - M_{xx}\frac{dZ}{ds}\right) ds,$$

$$Q_Z = \int \left(N_{xs}\frac{dZ}{ds} + N_{xn}\frac{dY}{ds}\right) ds, \quad Q_Y = \int \left(N_{xs}\frac{dY}{ds} - N_{xn}\frac{dZ}{ds}\right) ds, \quad T_w = \int \left(N_{xs}(r-\psi) + N_{xn}l\right) ds,$$

$$B = \int \left(N_{xx}\omega_p - M_{xx}l\right) ds, \quad T_{sv} = \int \left(N_{xs}\psi - 2M_{xs}\right) ds, \quad (29)$$

where, N corresponds to the axial force, Q_y and Q_z to shear forces, M_y and M_z to bending moments about $\bar{y}$ and $\bar{z}$ axis, respectively, B to the bimoment, T_w to the flexural-torsional moment and T_{sv} to the Saint-Venant torsional moment. In addition, four high-order stress resultants have been defined.

$$B_I = \int \left[N_{xx}\left(Y^2 + Z^2\right) - 2M_{xx}r\right] ds, \quad P_{yy} = \int \left[N_{xx}\bar{Z}^2 + 2M_{xx}\bar{Z}\frac{dY}{ds}\right] ds,$$

$$P_{zz} = \int \left[N_{xx}\overline{Y}^2 - 2M_{xx}\overline{Y}\frac{dZ}{ds} \right] ds, \quad P_{yz} = \int \left[N_{xx}\overline{Y}\,\overline{Z} + M_{xx}\left(\overline{Y}\frac{dY}{ds} - \overline{Z}\frac{dZ}{ds}\right) \right] ds. \qquad (30)$$

In expressions (29) and (30) the integration is carried out over the entire length of the mid-line contour. Assuming the linear elastic material, the force-displacement relations are obtained applying Hook's law. This constitutive law can be expressed in terms of a beam stiffness matrix $[K]$ as defined in Appendix A. Re-arranging the virtual work (25) in terms of the beam forces and moments, the generalized strains can be identified as:

corresponding to N: $\quad ED_1 = u'_o + \dfrac{1}{2}\left(u'^2_o + v'^2 + w'^2\right) + \left(z_0\theta'_z - y_0\theta'_y\right)\operatorname{sen}\phi,$

corresponding to M_y: $\quad ED_2 = \left(-\theta'_y - u'_o\theta'_z\right)\cos\phi + \left(\theta'_z + u'_o\theta'_y\right)\operatorname{sen}\phi,$

corresponding to M_z: $\quad ED_3 = \left(-\theta'_z - u'_o\theta'_y\right)\cos\phi - \left(\theta'_y + u'_o\theta'_z\right)\operatorname{sen}\phi,$

corresponding to B: $\quad ED_4 = \theta' - \dfrac{1}{2}\left(\theta_z\theta''_y - \theta_y\theta''_z\right),$

corresponding to Q_y: $\quad ED_5 = \left(v' - \theta_z - u'_o\theta_z\right)\cos\phi - z_0\dfrac{1}{2}\left(\theta_z\theta'_y - \theta_y\theta'_z\right) + \left(w' - \theta_y - u'_o\theta_y\right)\operatorname{sen}\phi,$

corresponding to Q_z: $\quad ED_6 = \left(w' - \theta_y - u'_o\theta_y\right)\cos\phi + y_0\dfrac{1}{2}\left(\theta_z\theta'_y - \theta_y\theta'_z\right) - \left(v' - \theta_z - u'_o\theta_z\right)\operatorname{sen}\phi,$

corresponding to T_w: $\quad ED_7 = \phi' - \theta,$

corresponding to T_{sv}: $\quad ED_8 = \phi' - \dfrac{1}{2}\left(\theta_z\theta'_y - \theta_y\theta'_z\right),$

corresponding to B_1: $\quad ED_9 = \dfrac{\phi'^2}{2},$

corresponding to P_{zz}: $\quad ED_{11} = \dfrac{\theta'^2_z}{2},$

corresponding to P_{yz}: $\quad ED_{12} = \theta'_y\theta'_z.$ \hfill (31)

These generalized strains correspond to the axial strain (ED_1), the two bending curvatures (ED_2 and ED_3), the torsional curvature (ED_4), the two transverse shear strain (ED_5 and ED_6), the transverse shear strain due to warping (ED_7), the rate of twist and a non-linear term (ED_8 and ED_9), the higher-order axial terms (ED_{10}, ED_{11} and ED_{12}).

6. REDUCED MODEL

In this section the interaction of the various types of motion of thin-walled beams due to non-linear effects is analyzed. Most of the dynamic stability studies of thin-walled beams have been carried out on the base of linear systems. The effect of geometric nonlinearity has been considered in some few works; among those it is found in [4, 12 and 23]. It should be emphasized here that the necessity of obtaining an accurate bending-torsional coupling arises from the fact that the stability behavior of this kind of structures is influenced by the initial displacements (corresponding to the static load) [16] and by the interaction between the forced vibration and the parametrically excited vibrations [23].

The equations of motion of a beam subjected to an axial or transverse excitation are reduced according to the Ritz method. The first step is to obtain the linearized response in the loading plane. This part consists of a term corresponding to the static loading $v_0(x)$ and the other corresponding to the dynamic loading $v_1(x)\cos(\varpi t)$, where it is assumed that $v_1(x) \ll v_0(x)$.

$$\mathbf{v}(x,t) = \mathbf{v}_0(x) + \mathbf{v}_1(x)\cos(\varpi t), \tag{32}$$

where; in the case of axial excitation, $v_0(x)$ correspond to the longitudinal movement; and in the case of transverse excitation in the vertical plane, $v_0(x)$ involves to the displacements w and θ_y.

In order to analyze the dynamic stability of these motions, the displacements are expressed as a linear combination of given x-function vectors $\mathbf{f}_k(x) = \{f_{k1}(x), f_{k2}(x), f_{k3}(x)\}$ and unknown t-function coefficients $q_k(t)$:

$$\mathbf{u}(x,t) = \sum_{k=1}^{n} q_k(t)\mathbf{f}_k(x). \tag{33}$$

The functions $\mathbf{f}_k(x)$ are chosen as eigenfunctions of the linearized equations and boundary conditions. Now, introducing expressions corresponding to the loading plane Eq. (32) and Eq. (33) into Eq. (25), taking variations with respect to the functions q_k and scaling the static load and dynamic amplitude with the critical buckling, one can obtain Eq. (34).

$$\ddot{q}_k + \Omega_k^2 q_k + \beta P_{cr} \eta \cos(\varpi t) \sum_{n=1}^{n} b_{kn} q_n = 0 \quad (k=1,2,\ldots,n.), \tag{34}$$

where $(\Omega_k)^2$ are the system eigenvalues taking into account initial displacements due to the static load, βP_{cr} represent the excitation amplitude load, b_{kn} are the coefficients depending of the eigenfunctions, η is the coefficient that consider the interaction between the forced vibration and the parametrically excited vibrations on the unstable regions. It is important to remark that in Eq. (34) non-linear terms associated with $v_1(x)$ are neglected.

One should notice that the present formulation considers the inertial effects of the loading plane on the parametrically excited vibration modes. This influence is generally neglected in most of the dynamic instability studies. Such an assumption is valid to a certain extent when the exciting frequency is small in comparison with the free vibration frequency ω_p corresponding to the loading plane. This is the case frequently assumed for analyzing the dynamic stability of bars subjected to axial excitation [4, 23]. However for transverse excitation, the frequency at which a parametric resonance occurs can be of the same order as the natural frequency of the loading plane vibrations.

6.1. Transverse Excitation

The methodology explained above is illustrated for the case of simply supported beams subjected to uniform bending $\overline{M}_y(t) = Mo + Mt \cos \varpi t$, where ϖ is the excitation frequency. In this case, the initial displacements are given by the following expressions:

$$w = \frac{Mo}{2EI_y}(Lx - x^2) + \frac{GS_z L\pi}{GS_z \pi^2 - L^2 \rho A \varpi^2} \kappa_\theta Mt \cos(\varpi t)\sin(\pi x/L), \qquad (35)$$

$$\theta_y = \frac{Mo}{2EI_y}(L - 2x) + \kappa_\theta Mt \cos(\varpi t)\cos(\pi x/L), \qquad (36)$$

where

$$\kappa_\theta = \frac{4L}{EI_y \pi^2 + L^2 \left(GS_z - \rho I_y \varpi^2 + \dfrac{GS_z^2 \pi^2}{\varpi^2 L^2 \rho A - \pi^2 GS_z} \right)}, \qquad (37)$$

EI_y is the flexural stiffness and GS_z is the shear stiffness.

Therefore, the initial displacement expressions (w and θ_y) are composed of static and dynamic linear solution. Then, these last Eqs. (35-37) along with Eq. (33) are substituted into Eq. (25). So the resultant expression can be transformed to the following n coupled Mathieu equations Eq. (34). In this example, η has the following expression:

$$\eta = \frac{EI_y \pi^2}{EI_y \pi^2 + L^2 \left(GS_z - \rho I_y \varpi^2 + \dfrac{GS_z^2 \pi^2}{\varpi^2 L^2 \rho A - \pi^2 GS_z} \right)}. \qquad (38)$$

When the influence of the interaction between the forced vibration and the parametrically excited vibrations is omitted $\eta = 1$, which is obtained setting $\varpi = 0$ in the Eq. (38).

6.2. Axial Excitation

In this case a simply supported beams subjected to an axial load $P(t) = P_s + P_d \cos \varpi t$, is considered. The initial displacements in the longitudinal direction are given by the following expressions:

$$u(x,t) = \frac{P_s x}{EA} + \frac{P_d \sin \nu x}{\nu EA \cos \nu L} \cos \varpi t, \quad \text{where} \quad \nu = \varpi \sqrt{\frac{\rho A}{EA}}. \tag{39}$$

As in the previous example, the first right term of the expression (39) represent the influence of the static deformation and the second the inertial influence.

On the other hand, the coefficient η of the Eq. (34), which considers the interaction between the forced vibration and the parametrically excited vibrations, has the following expression:

$$\eta = \frac{\tan \nu L}{\nu L} \frac{1 - \dfrac{\nu^2 L^2}{2\pi^2}}{1 - \dfrac{\nu^2 L^2}{4\pi^2}} \tag{40}$$

7. REGIONS OF INSTABILITY

The different types of unstable boundaries for thin-walled beams subjected to axial or transverse periodic load are studied in this section. The regions of instability for simple and combination resonant frequencies of sum type are determinate by applying Hsu's [25] procedures to the Mathieu Eq. (34). The method combines the method of variation of parameters and the series expansion of the perturbation method into a single treatment. The behavior of the nontrivial solutions in both the stable and unstable cases is deduced with the present analysis. The boundaries of the unstable regions are given as follows:

a. Simple resonance, $\varpi = 2\Omega_k$:

$$2\Omega_k + \frac{\beta \eta b_{kk}}{2\Omega_k} > \varpi > 2\Omega_k - \frac{\beta \eta b_{kk}}{2\Omega_k} \qquad k = 1, 2, ..., N. \tag{41}$$

b. Combination resonance of sum type, $\varpi = \Omega_k + \Omega_j$:

$$(\Omega_k + \Omega_j) - \frac{\beta}{2}\eta \left(\frac{b_{kj} b_{jk}}{\Omega_k \Omega_j}\right)^{1/2} < \varpi < (\Omega_k + \Omega_j) + \frac{\beta}{2}\eta \left(\frac{b_{kj} b_{jk}}{\Omega_k \Omega_j}\right)^{1/2} \qquad k \neq j, \quad k = j = 1, 2, ..., N. \tag{42}$$

8. APPLICATIONS AND NUMERICAL RESULTS

The purpose of this section is to apply the present theoretical model in order to study the dynamic stability of simply-supported, cantilever and fixed-end thin-walled beams, considering open mono- and bi-symmetric cross-sections. The influence of shear deformation and geometrical non-linear coupling is analyzed for both simple and combination resonance. When the beam is excited in the axial (longitudinal) direction, the interactions of this motion on the other flexural or flexural-torsional motions are to be studied. The vibration modes excited parametrically will depend on the symmetry of the cross-section analyzed. Therefore, for this kind of excitation, unstable regions are obtained for mono- and bi-symmetric cross-section beams. In the case of transverse excitation in the vertical plane, this motion (w and θ_y) interacts with flexural-torsional motions (v, θ_z, ϕ and θ) as was explained in section 6. In all the results presented below, the value of the static load parameter is adopted $\alpha = 0.5$, and the excitation frequency ϖ is scaled with the lowest frequency value of parametric resonance (that is the double of the frequency value of the vibration mode first $2\Omega_1$).

8.1. Bisymmetric Open Section

The example considered is a simply supported bisymmetric-I section subjected to an axial excitation. The geometrical and material properties are $L = 6$ m, $h = 0.6$ m, $b = 0.6$ m, $e = 0.03$ m, $E = 200$ GPa and $\mu = 0.3$. When the beam is excited in the axial (longitudinal) direction, the interaction of this movement will depend on the symmetry of the cross-section analyzed. In this example ($y_0 = z_0 = 0$), the system equations are uncoupled. Therefore, there are three main modes of vibration corresponding either to bending or to torsion. In this case, the lowest frequency corresponds to the lateral flexural mode (y-direction), while the highest frequency of vibration corresponds to the vertical flexural mode (z-direction).
Instability regions are shown in the Figs. 3-5, for different beam length. The critical load of the beam corresponding to the flexural mode can be easily obtained by means of expression (43) (as explained by the authors in [31]).

$$P_{cr} = \frac{\pi^2}{L^2} \frac{EI_z GS_y}{GS_y + EI_z \frac{\pi^2}{L^2}} \qquad (43)$$

where EI_z is the flexural stiffness, GS_z and GS_y are shear stiffnesses of a isotropic beam. The definitions of these stiffnesses are given in the Appendix A.

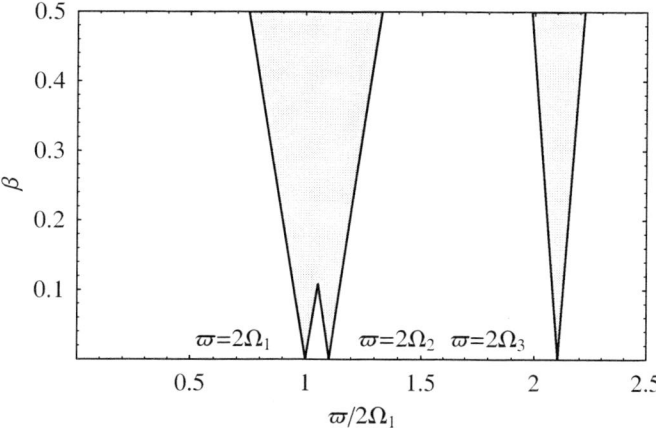

Figure 3. Regions of dynamic instability of a simply-supported beam subjected to an axial periodic force, L = 4 m.

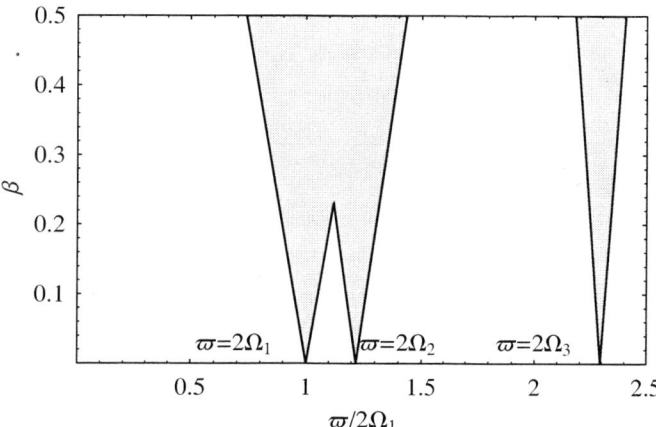

Figure 4. Regions of dynamic instability of a simply-supported beam subjected to an axial periodic force, L = 6 m.

It is observed that the widest unstable region corresponds to the first mode (or to the first frequency of parametric resonance), while the smallest region belongs to the third mode (or third frequency of parametric resonance). The sizes of the unstable regions keep practically constant for the different beam length analyzed. However, in the case of short length beams the main and second unstable regions merge in one larger region. It is due to the closeness of the first and second parametric resonance frequencies.

It is important to mention that the influence of the longitudinal inertia is negligible in all the cases analyzed, due to the exciting frequency is far from the longitudinal natural frequencies of the beam (ω_L).

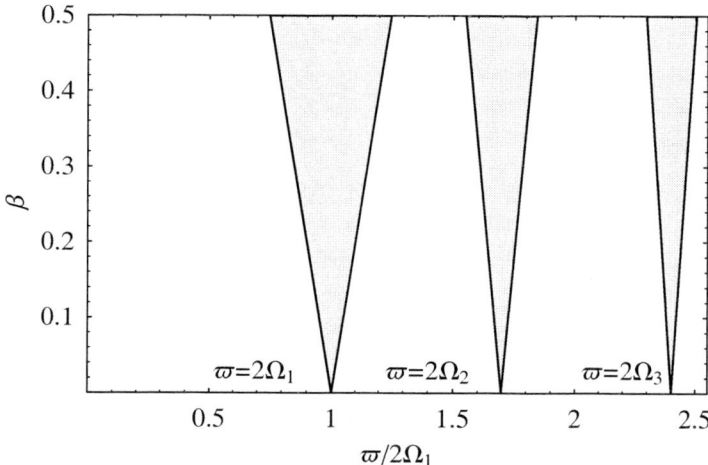

Figure 5. Regions of dynamic instability of a simply-supported beam subjected to an axial periodic force, L = 12 m.

Table 1. Natural frequencies for a bisymmetric beam (*Hz*).

Beam Length	modes	$\alpha = 0$		$\alpha = 0.5$	
		model I	model II	model I	model II
4 m	1	68.19	69.78	48.22	50.47
	2	71.81	73.31	53.21	55.25
	3	112.21	128.57	101.47	119.49
	ω_L	315.87		315.87	
6 m	1	30.79	31.12	21.78	22.23
	2	34.27	34.56	26.47	26.84
	3	54.04	57.81	49.50	53.62
	ω_L	210.56		210.56	
12 m	1	7.77	7.77	5.49	5.49
	2	10.83	10.83	9.33	9.33
	3	14.29	14.29	13.19	13.19
	ω_L	105.29		105.29	

On the hand, the influence of shear deformation on the dynamic behavior of composite beams is analyzed. In Table 1, natural frequencies are given considering two models: the present theory (model I) and neglecting shear flexibility (model II). The shear deformation effect reduces considerably the vibration frequency values when the beam length decreases and it is negligible for slender beams. The effect of shear flexibility on the unstable regions is shown in Fig. 6, considering a beam length L = 4 m. It is observed that the width of the regions does not change for both models. However, when shear deformation is neglected the unstable region moves toward the right, originated by an increase in the parametric frequency values.

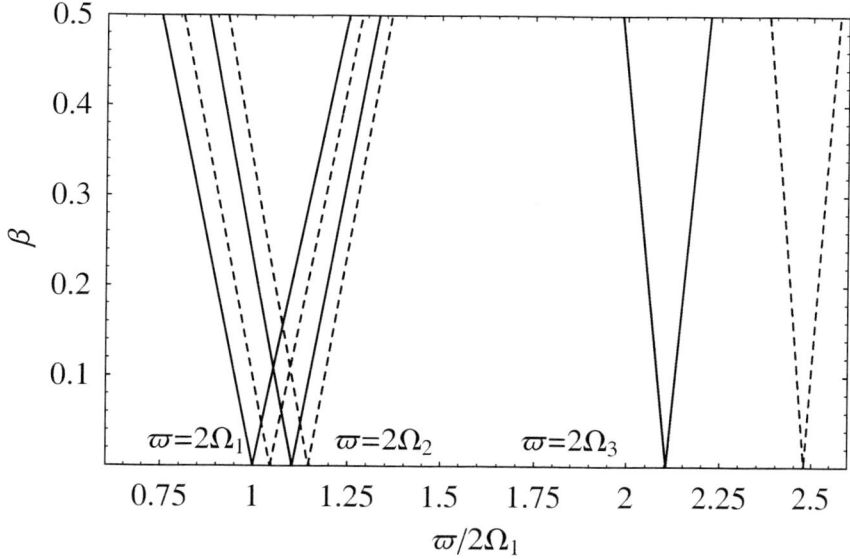

Figure 6. Comparison of the unstable boundaries, (—) present theory, (---) neglecting shear deformation.

The natural frequencies obtained with the present formulation are compared with those obtained from a shell finite elements model using Abaqus [32]. The beam is idealized by 1080 four-node shell elements (S8R). The analysis *STATIC is used in Abaqus as first step of calculus, together with the option NLGEOM to consider the non-linear geometric effect in the pre-loaded state. Then, in the second step, the procedure *Frequency is used for the dynamic analysis of the structure. In Figs. 7, 8 and 9, the three vibration modes of the simply supported beam are shown, obtained with the program of finite elements (Abaqus) and considering a beam length L = 6 m. It can be observed that the present solutions are, in general, in good agreement with those obtained with Abaqus.

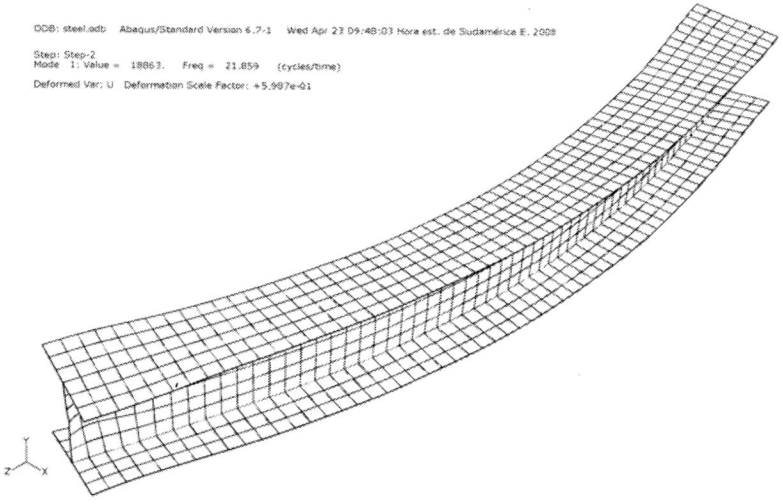

Figure 7. Flexural mode of simply-supported beam subject to axial force ($\alpha = 0.5$), shell model Ω_1 = 21.86 Hz.

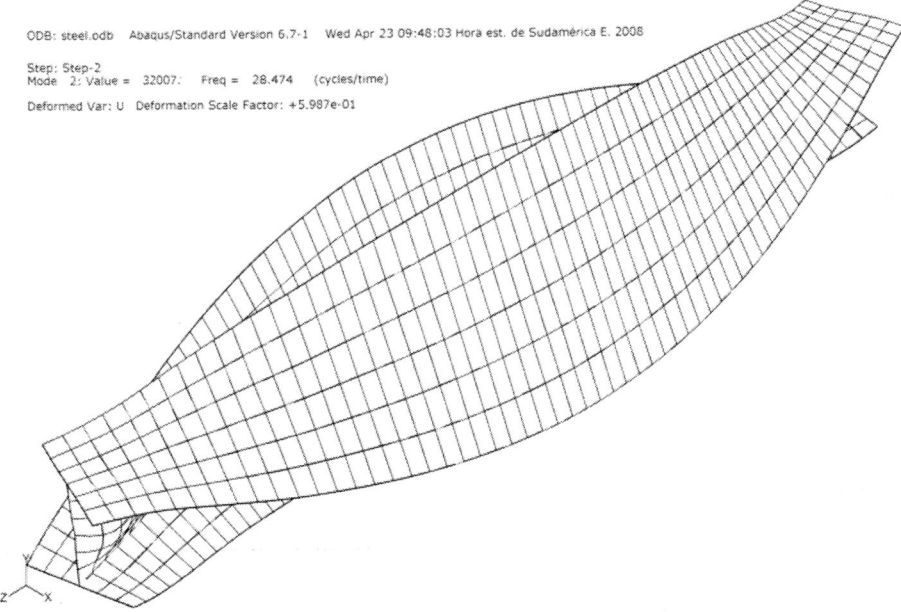

Figure 8. Torsional mode of simply-supported beam subject to axial force ($\alpha = 0.5$). shell model $\Omega_2 = 28.47$ Hz.

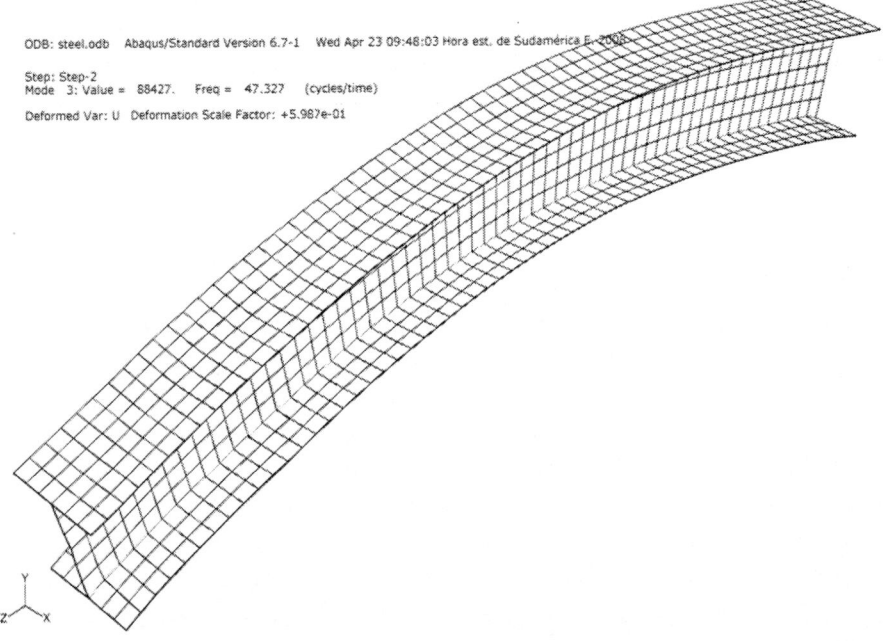

Figure 9. Second flexural mode of simply-supported beam subject to axial force ($\alpha = 0.5$).shell model $\Omega_2 = 47.32$ Hz.

8.2. Monosymmetric Open Cross-Section

The considered example is a monosymmetric channel section, the geometric properties are $L = 6$ m, $h = 0.6$ m, $b = 0.6$ m, $e = 0.03$ m. In this case ($z_0 = 0$), the system of equations corresponding to y-direction (v transversal displacement) are therefore uncoupled. Thus, when the beam is subjected to a longitudinal force, it is possible to excite the beam parametrically in either to bending modes in y-direction (v) or in flexural-torsional vibration mode (w and ϕ). For the cross-section analyzed, the flexural-torsional mode presents the smallest parametric frequency value. Therefore, the excitation frequency is scaled with the value of this last frequency.

Regions of dynamic instability for the channel-beam are shown in the Fig. 10. The first and third instability regions (Ω_1 and Ω_3) correspond to the flexural-torsional mode, while the second region corresponds to the uncoupled flexural mode (Ω_2).

Fig. 10 shows comparative results between the unstable regions obtained by disregarding and considering the influence of the longitudinal vibration. It can be observed that the larger size correspond to the main unstable region ($\Omega_1 = 14.38$ Hz). The second ($\Omega_2 = 40.55$ Hz) and third ($\Omega_3 = 90.71$ Hz) unstable region sizes are smaller when the effect of longitudinal inertia is neglected. However, the third region, corresponding to the second flexural-torsional mode, is wider when this effect is considered. This phenomenon is due to that the parametrically excited frequency Ω_3 is near to the longitudinal natural frequencies of the beam ($\omega_L = 210.58$ Hz). The influence of the longitudinal inertia enlarges the third region, which certainly is composed by two regions, one of them lies near $\varpi = 2\Omega_3$ ($\Omega_3/\Omega_1 = 6.31$); and the second lies near $\varpi = \omega_L$ ($\omega_L/2\Omega_1 = 7.32$).

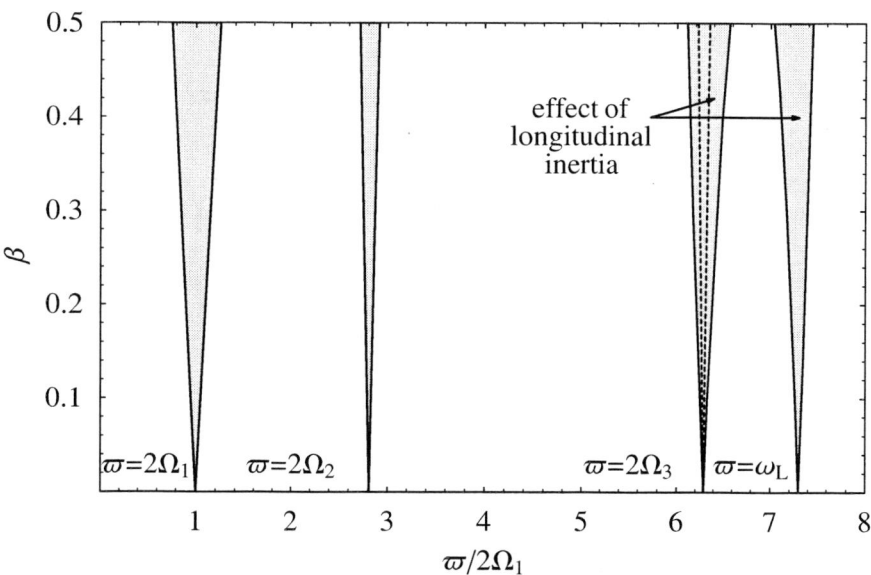

Figure 10. Unstable dynamic regions of a mono-symmetric beam, (—) considering and (- - -) neglecting longitudinal inertia.

8.3. SIMPLY-SUPPORTED BEAM SUBJECTED TO DISTRIBUTED LOAD

In this example a simply supported beam under distributed load $qz(t) = qz_0 + qz_t \cos\varpi t$ is considered for three load positions. The load can be applied to the top flange (case a), at the shear center (case b), and to the bottom flange (case c). The beam considered is a bisymmetric-I section whose geometric properties are $h = 0.6$ m, $b = 0.6$ m, $e = 0.03$ m, $L = 6$m. The analyzed material is the same as the previous examples. The buckling load used to scale the static load $qz_0 = \alpha\, q_{cr}$, corresponding to the flexural-torsional mode, can be easily obtained by means of Eq. (44) (as explained by the authors in [31]):

$$q_{cr} = C_1\gamma EI_z 8\frac{\pi^2}{L^4}\left[-C_2 e_z\gamma + \sqrt{\frac{GS_w GJ + EC_w(GS_w + GJ)\frac{\pi^2}{L^2}}{EI_z\frac{\pi^2}{L^2}\left(GS_w + EC_w\frac{\pi^2}{L^2}\right)} + (C_2 e_z\gamma)^2}\right], \quad (44)$$

$$\gamma = \frac{1}{\sqrt{\left(1-\frac{EI_z}{EI_y}\right)\left(1-\beta\frac{GJ}{EI_y} - \beta\frac{EC_w GS_w\pi^2}{EI_y(GS_w L^2 + EC_w\pi^2)}\right) - \delta\frac{EI_z}{GS_y}\frac{\pi^2}{L^2}\left[1-\frac{GS_y}{GS_z}\left(0.71-\frac{GS_y}{GS_z}0.29\right)\right]}}, \quad (45)$$

where $C_1 = 1.141, C_2 = 0.459, \beta = 0.033$ and $\delta = 0.214$.

Expression (44) also gives the critical load according to the linearized theory, which does not account for the prebuckling deflection, if one takes $\gamma = 1$. Therefore, the presence of the γ coefficient reveals the dependence of the prebuckling effect with respect to the relation between the bending and shear stiffnesses.

On the other hand, the natural frequencies values Ω_k (for $\beta = 0$) are obtained as explained by Machado and Cortínez in [24], taking into account the influence initial displacements.
Instability regions are shown in Figs. 11-13, considering different load heights of a simply-supported beam. The influence of the interaction between the forced vibration and the parametrically excited vibrations on the unstable regions is analyzed in the figures. The unstable boundaries obtained by disregarding this interaction are drawn in dashed lines. The influence of this interaction enlarges the unstable region, which certainly is composed by two or three regions. Therefore, its discard results, inadvertently, in a less critical behavior than in the case of its incorporation.

It is observed that the widest unstable region corresponds to the first mode (or to the first frequency of parametric resonance), while the smallest region depends on the load location. For example, when the load is applied on the top flange and shear center, the smallest correspond to the combination resonance region ($\Omega_1+\Omega_2$). However, when the load is applied on the bottom flange, the second parametric region is the smallest. Besides, for this load condition the second parametric frequency is more distant from the main region ($\Omega_2/\Omega_1 \cong 2$). The natural frequencies in Hz are shown in Table 2, for the unloaded beam $\alpha = 0$ and for $\alpha = 0.5$, considering the dynamic load parameter $\beta = 0$.

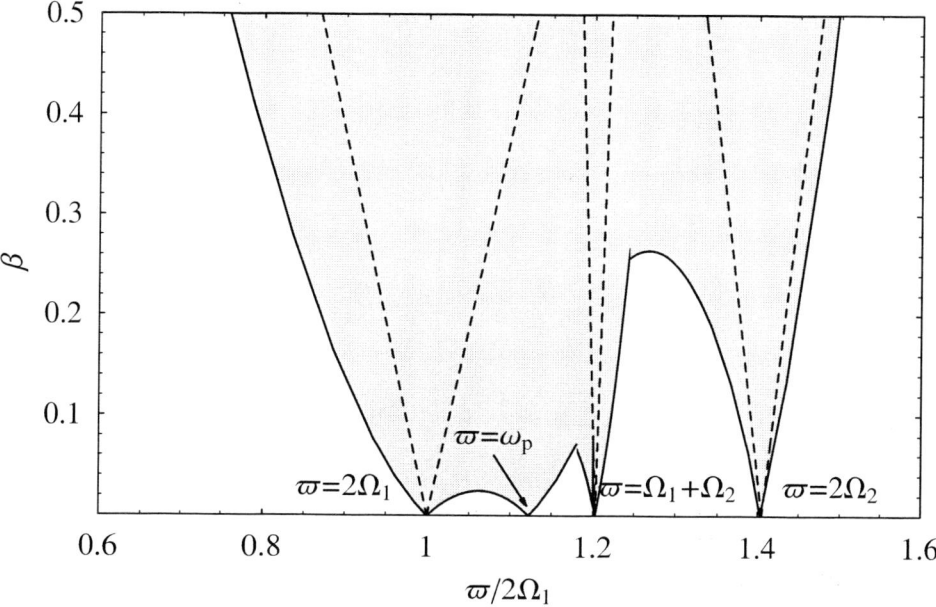

Figure 11. Regions of dynamic instability of a simply-supported beam subjected to distributed load applied on the top flange, (———) present theory, (-----) neglecting interaction with ω_p.

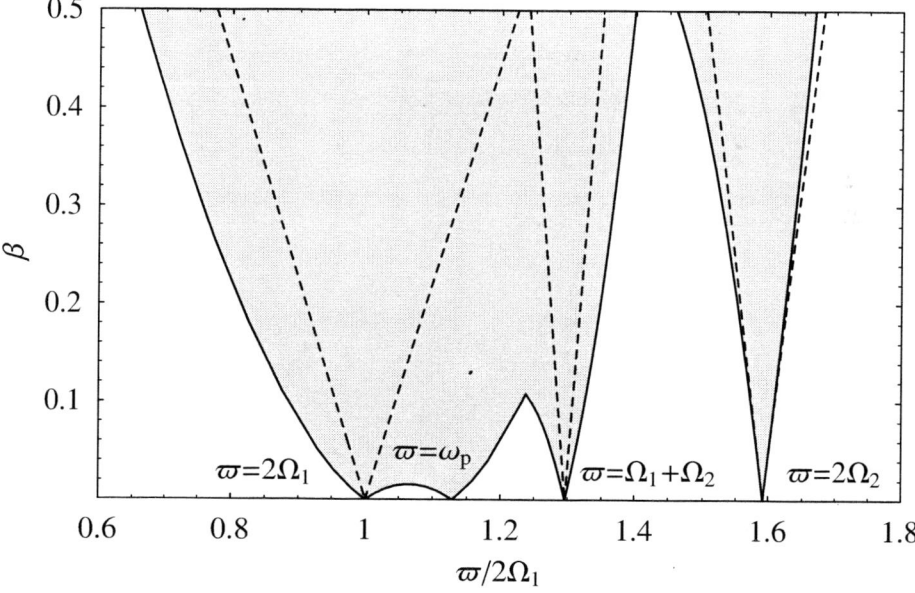

Figure 12. Regions of dynamic instability of a simply-supported beam subjected to distributed load applied on the shear center, (———) present theory, (-----) neglecting interaction with ω_p.

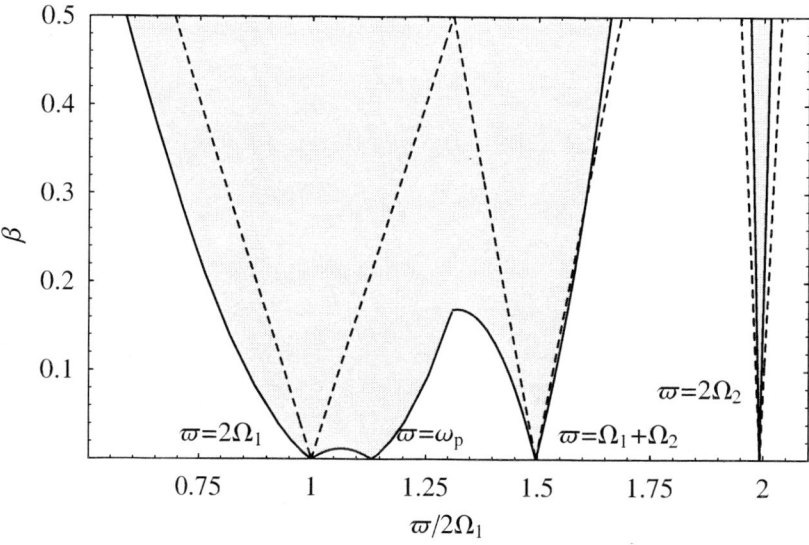

Figure 13. Regions of dynamic instability of a simply-supported beam subjected to distributed load applied on the bottom flange, (———) present theory, (-----) neglecting interaction with ω_p.

In this case, the natural frequency corresponding to the loading plane ω_p is higher than the first parametric resonance frequency $2\Omega_1$. The influence of the vertical inertia enlarges the first region to the right. On the other hand, when the load is applied on the bottom flange (Fig. 13), the main unstable region is largest in comparison with the other load condition (Figs. 11 and 12).

Table 2. Natural frequencies for a simply-supported beam subjected to a distributed load (Hz).

Static load parameter	Load height	1 mode	2 mode	ω_p
$\alpha = 0$		30.79	34.27	53.95
$\alpha = 0.5$	Top	24.07	33.83	53.95
	Shear center	23.96	38.14	53.95
	Bottom	23.86	47.55	53.95

8.4. Cantilever Beam Subjected to End Force

The example considered is a cantilever beam subjected to end force applied on the top flange of its free end. The cross-section properties of the I-beam and the material properties are the same as the previous example. Regions of dynamic instability for the composite beam are shown in Fig. 14. The influence of the interaction between the forced vibration and the parametrically excited vibrations on the unstable regions is analyzed in the figure. The unstable boundaries obtained by disregarding this interaction are drawn in dashed lines. In this case, the natural frequency corresponding to the loading plane (ω_p) is smaller than the first parametric resonance frequency ($2\Omega_I$). The influence of the vertical inertia enlarges the instability region to low frequency values. Therefore, it derives in a large region which certainly is composed by the three parametric regions and the corresponding to in-plane loading vibration mode ($\omega_p = 19.92\ Hz$), for $\beta = 0$. The unstable regions are smaller when the interaction of the forced vibration is omitted, predicting a less critical behavior.

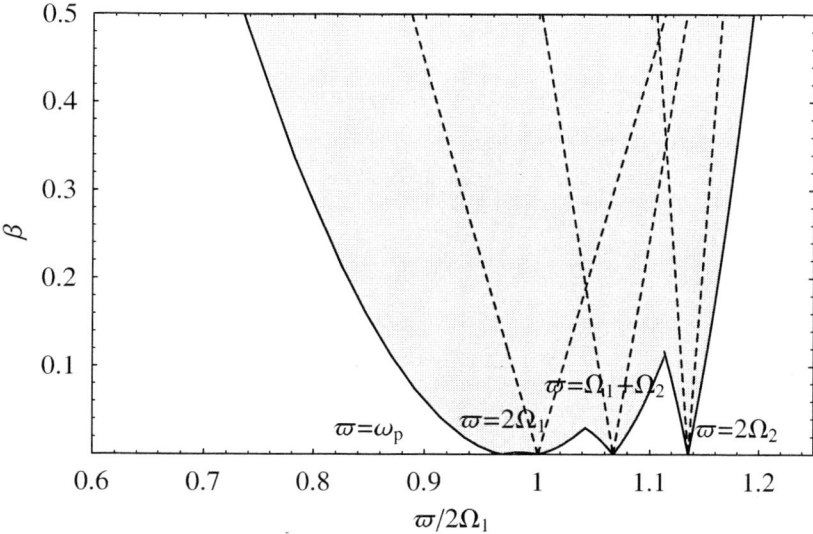

Figure 14. Regions of dynamic instability of a cantilever beam subjected to a concentrated load applied on the top flange, (———) present theory, (-----) neglecting interaction with ω_p.

The natural frequencies in Hz are shown in Table 3, for the unloaded beam $\alpha = 0$ and for $\alpha = 0.5$, considering the dynamic load parameter $\beta = 0$. In this case, the frequencies obtained with the present beam model are compared with those obtained by means of a shell finite element model (Abaqus). It is observed that the present solutions are, in general, in good agreement with those obtained with Abaqus. In Figs. 15 and 16, the deformed configuration of the beam corresponding to the first and second flexural-torsional mode are shown, for $\alpha = 0.5$.

Table 3. Natural frequencies for a cantilever beam subjected to a concentrated load (Hz).

Load height	modes	Present beam model		Abaqus shell model	
		$\alpha = 0$	$\alpha = 0.5$	$\alpha = 0$	$\alpha = 0.5$
top flange	1	11.05	10.28	11.06	10.50
	2	14.90	11.67	14.74	11.26
	ω_p	19.92	19.92	19.99	19.99

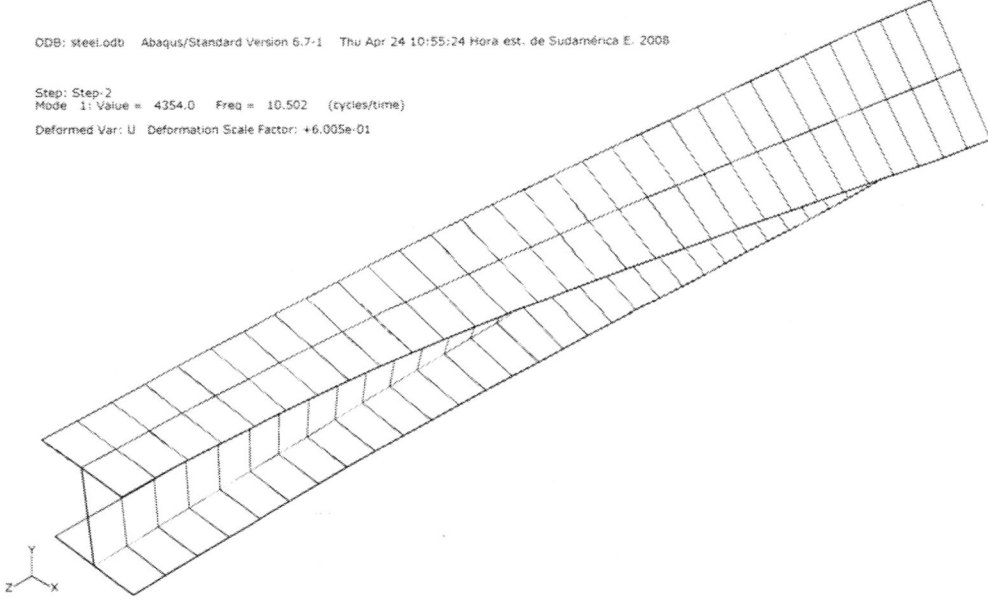

Figure 15. First flexural-torsional mode of a cantilever beam subject to a concentrated force ($\alpha = 0.5$), shell model $\Omega_l = 10.5$ Hz.

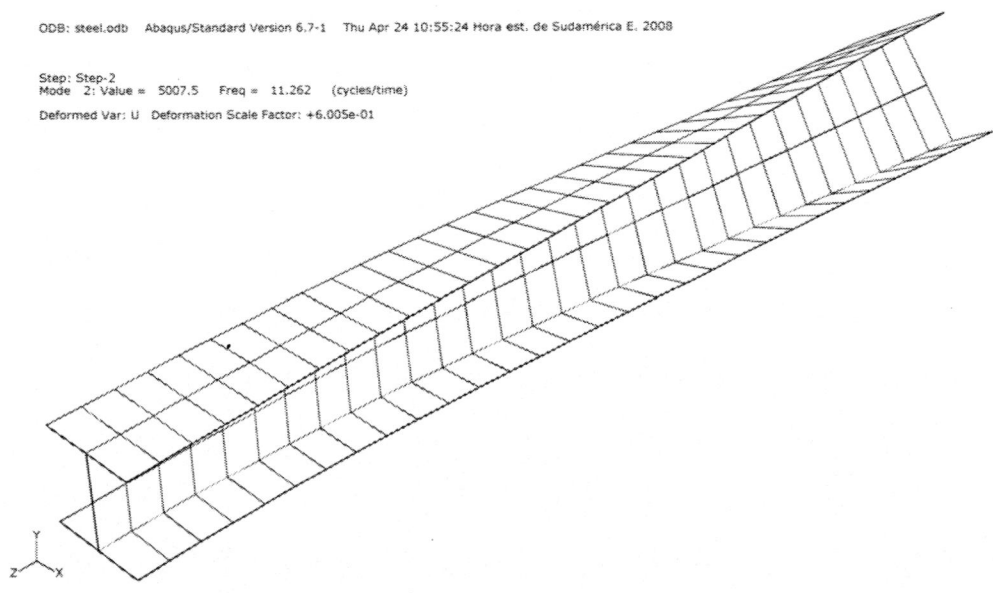

Figure 16. Second flexural-torsional mode of a cantilever beam subject to a concentrated force ($\alpha = 0.5$), shell model $\Omega_l = 11.26$ Hz.

8.5. Fixed-End Beam Subjected to a Concentrated Force

In this example a fixed-end beam is excited transversely by a force at the middle of the span applied in the shear center. The cross-section and the analyzed material are the same as the previous case. In this case, the effect of the static load parameter on the unstable dynamic regions is analyzed.

In Fig. 17-19, the regions of dynamic instability are shown for a static load parameter $\alpha = 0.1$, $\alpha = 0.3$ and $\alpha = 0.5$, respectively. The natural frequencies in Hz are shown in Table 4, considering different static load parameter values and the dynamic load parameter $\beta = 0$.

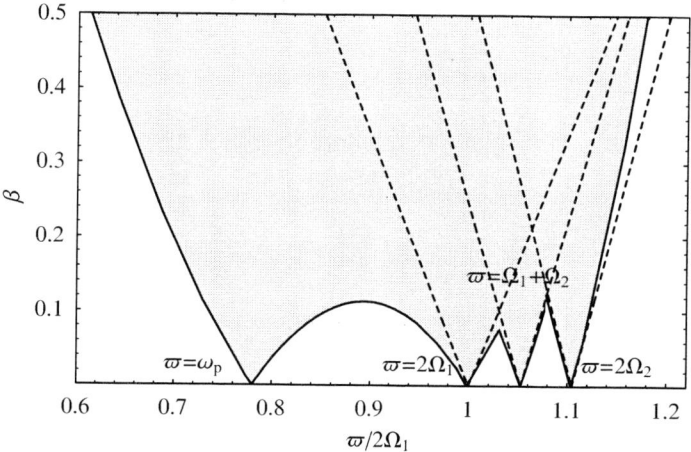

Figure 17. Unstable regions of a fixed-end beam subjected to a concentrated load, $\alpha = 0.1$, (———) present theory, (-----) neglecting interaction with ω_p.

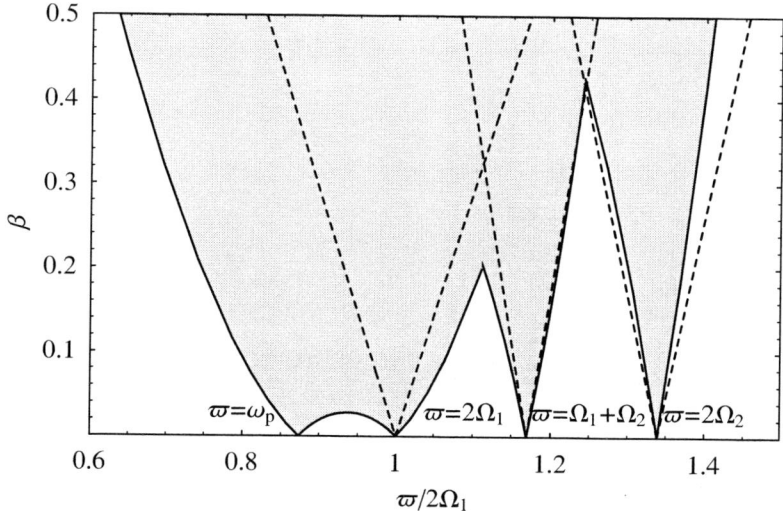

Figure 18. Unstable regions of a fixed-end beam subjected to a concentrated load, $\alpha = 0.3$, (———) present theory, (-----) neglecting interaction with ω_p.

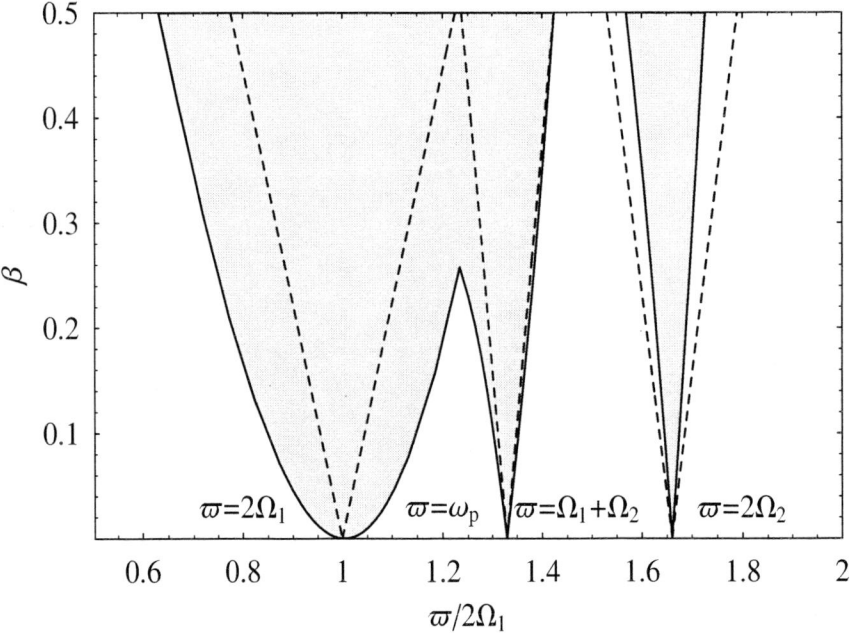

Figure 19. Unstable regions of a fixed-end beam subjected to a concentrated load, $\alpha = 0.5$, (———) present theory, (-----) neglecting interaction with ω_p.

Table 4. Natural frequencies for a fixed-end beam subjected to a concentrated load (Hz).

Static load parameter	1 mode	2 mode	ω_p
$\alpha = 0$	67.50	69.48	101.28
$\alpha = 0.1$	65.32	71.44	101.28
$\alpha = 0.3$	58.61	76.35	101.28
$\alpha = 0.5$	50.23	83.39	101.28

The vibration frequency values corresponding to the first flexural-torsional mode decrease as the static load factor increases. While, the values corresponding to the second mode increase when α increases. This behavior is observed in the unstable regions, where the influence of the vertical inertia enlarges the unstable region to low frequency values. In the case of a small static load parameter ($\alpha = 0.1$), the region of dynamic instability is composed by four regions. This dynamic behavior is due to the closeness of the first and second parametric frequency. On the other hand, when the static load parameter increases ($\alpha = 0.5$), the relation between the parametrically excited frequencies also increases and the third instability region moves away from the main unstable region. It is also observed in this case (Fig. 17) that the parametric frequency ($2\Omega_1 = 100.46$) is hardly the same that the frequency value corresponding to vertical vibration mode ($\omega_p = 101.28\ Hz$), for $\beta = 0$. Anyway, there is an interaction between the forced vibration and the parametrically excited vibrations and this effect enlarges the instability region. The larger unstable region is limited by the frequencies values corresponding to the main parametric and combination resonance ($\Omega_1+\Omega_2$).

9. CONCLUSION

In this paper, the influences of shear deformation and geometrically non-linear coupling on the dynamic stability of thin-walled beam are analyzed. A non-linear beam theory is formulated in the context of large displacements and rotations, through the adoption of a shear deformable displacement field (accounting for bending and warping shear). Regions of dynamic instability are studied for simply-supported, cantilever and fixed-end composite beams subjected to axial and transverse periodic loads. Unstable regions for simple and combination resonant frequencies are expressed in non-dimensional terms and determined by applying Hsu's procedure to the Mathieu equation. The influence of non-conventional effects is noted in the numerical results and can be summarized as follows:

a. The width of the instability regions increases with an increase in the static and dynamic loads.
b. The unstable regions sizes are influenced by the transverse shear effect, when this effect is ignored the boundaries of instability shift to the right. This behavior is originated by an increase in the parametric resonance frequency values and higher differences are observed when the beam length decreases.
c. The interaction between the forced vibration and the parametrically excited vibrations on the regions of dynamic instability is significant in some cases. This effect depends on the closeness of the frequency values, i.e. it depends on the stiffnesses ratio between the parametrically excited mode and the vertical mode corresponding to the loading plane.
d. This influence enlarges the unstable region because the unstable regions are composed by two regions; one of them lies near the parametric excited frequency and the other lies near the natural frequency corresponding to the loading plane, for small values of the dynamic load parameter.
e. In general, the discard of this interaction results, inadvertently, in a less critical behavior for the main parametric and combination unstable regions than in the case of its incorporation.
f. The size of unstable regions depends on the load height parameter, for example the unstable regions are wider when the load is applied on the beam bottom flange.

ACKNOWLEDGMENTS

The present study was sponsored by Secretaría de Ciencia y Tecnología, Universidad Tecnológica Nacional, and by CONICET.

APPENDIX A. CONSTITUTIVE LAW

The constitutive law for a bisymmetric beam is defined in the following form:

$$\{f_g\} = [K]\{\chi\}, \tag{A.1}$$

$$\{f_g\} = \begin{bmatrix} N & M_y & M_z & B & Q_y & Q_z & T_w & T_{sv} & B_1 & P_{yy} & P_{zz} & P_{yz} \end{bmatrix}^T, \tag{A.2}$$

$$\{\chi\} = \begin{bmatrix} ED_1 & ED_2 & ED_3 & ED_4 & ED_5 & ED_6 & ED_7 & ED_8 & ED_9 & ED_{10} & ED_{11} & ED_{12} \end{bmatrix}^T. \tag{A.3}$$

Where $\{f_g\}$ is the vector of generalized forces, $\{\chi\}$ is the vector of the generalized strains and [K] is a symmetric matrix (12 x 12) where the stiffnesses corresponding to the higher-order terms B_1, P_{yy}, P_{zz} and P_{yz} are given by the following contour integrals:

$$K = \begin{bmatrix} EA & 0 & 0 & 0 & 0 & 0 & 0 & 0 & Ek_{1,9} & Ek_{1,10} & Ek_{1,11} & Ek_{1,12} \\ & EI_y & 0 & 0 & 0 & 0 & 0 & 0 & Ek_{2,9} & Ek_{2,10} & Ek_{2,11} & Ek_{2,12} \\ & & EI_z & 0 & 0 & 0 & 0 & 0 & Ek_{3,9} & Ek_{3,10} & Ek_{3,11} & Ek_{3,12} \\ & & & EC_w & 0 & 0 & 0 & 0 & Ek_{4,9} & Ek_{4,10} & Ek_{4,11} & Ek_{4,12} \\ & & & & GS_y & GS_{yz} & GS_{yw} & 0 & 0 & 0 & 0 & 0 \\ & & & & & GS_z & GS_{zw} & 0 & 0 & 0 & 0 & 0 \\ & & & & & & GS_w & 0 & 0 & 0 & 0 & 0 \\ & & & & & & & GJ & 0 & 0 & 0 & 0 \\ & \text{Sym} & & & & & & & EI_R & Ek_{9,10} & Ek_{9,11} & 0 \\ & & & & & & & & & Ek_{10,10} & Ek_{10,11} & 0 \\ & & & & & & & & & & Ek_{11,11} & 0 \\ & & & & & & & & & & & Ek_{12,12} \end{bmatrix}, \tag{A.4}$$

where

$$A = \int dA,$$

$$I_y = \int Z^2 \, dA,$$

$$I_z = \int Y^2 \, dA,$$

$$C_w = \int \omega_p^2 \, dA,$$

$$S_y = \int Y'^2 \, dA,$$

$$S_z = \int Z'^2 \, dA,$$

$$S_w = \int (r-\psi)^2 \, dA,$$

$$S_{yz} = \int Y'Z' \, dA,$$

$$S_{yw} = \int Y'(r-\psi) \, dA,$$

$$S_{zw} = \int Z'(r-\psi) \, dA,$$

$$J = 4 \int dA,$$

$$I_R = \int (Y^2 + Z^2)^2 \, dA,$$

$$k_{1,9} = \int (Y^2 + Z^2) \, dA,$$

$$k_{1,10} = \int \overline{Z}^2 \, dA,$$

$$k_{1,11} = \int \overline{Y}^2 \, dA,$$

$$k_{1,12} = \int \overline{Y}\,\overline{Z} \, dA,$$

$$k_{2,9} = \int \overline{Z}(Y^2 + Z^2) \, dA,$$

$$k_{2,10} = \int \overline{Z}^3 \, dA,$$

$$k_{2,11} = \int \overline{Y}^2 \overline{Z} \, dA,$$

$$k_{2,12} = \int \overline{Y}\,\overline{Z}^2 \, dA,$$

$$k_{3,9} = \int \overline{Y}(Y^2 + Z^2) \, dA,$$

$$k_{3,10} = \int \overline{Y}\,\overline{Z} \, dA,$$

$$k_{3,11} = \int \overline{Y}^3 \, dA,$$

$$k_{3,12} = \int \overline{Y}^2 \overline{Z} \, dA,$$

$$k_{4,9} = \int \omega_p (Y^2 + Z^2) \, dA,$$

$$k_{4,10} = \int \omega_p \overline{Z}^2 \, dA,$$

$$k_{4,11} = \int \omega_p \overline{Y} \, dA,$$

$$k_{4,12} = \int \overline{Y} \omega_p \overline{Z} \, dA,$$

$$I_R = \int (Y^2 + Z^2)^2 \, dA,$$

$$k_{9,10} = \int (Y^2 + Z^2) \overline{Z}^2 \, dA,$$

$$k_{9,11} = \int (Y^2 + Z^2) \overline{Y}^2 \, dA,$$

$$k_{9,12} = \int (Y^2 + Z^2) \overline{Y}\,\overline{Z} \, dA,$$

$$k_{10,10} = \int \overline{Z}^4 \, dA,$$

$$k_{10,11} = \int \overline{Z}^2 \overline{Y}^2 \, dA,$$

$$k_{10,12} = \int \overline{Z}^3 \overline{Y} \, dA,$$

$$k_{11,11} = \int \overline{Y}^4 \, dA,$$

$$k_{11,12} = \int \overline{Y}^3 \overline{Z} \, dA,$$

$$k_{12,12} = \int \overline{Z}^2 \overline{Y}^2 \, dA.$$

(A.5)

REFERENCES

[1] Vlasov, V.Z. (1961). *Thin Walled Elastic Beams*. Jerusalem: Israel Program for Scientific Translation.

[2] Evan-Iwanowski, R.M. (1965). On the parametric response of structures. *Applied Mechanics Review, 18*, 699-702.

[3] Nayfeh, A.H. & Mook, D.T. (1979). *Nonlinear Oscillations*, New York, Wiley.

[4] Bolotin, V.V. (1964). *The Dynamic Stability of Elastic Systems*, San Francisco, Holden-Day.

[5] Gol'denblat, I.I. (1947). *Contemporary Problems of Vibrations and Stability of Engineering Structures*, Moscow, Stroiizdat.

[6] Tso, W.K. (1968). Parametric torsional stability of a bar under axial excitation. *Journal of Applied Mechanics, 35*, 13-19.

[7] Mettler, E. (1962). *Dynamic buckling*, Handbook of Engineering Mechanics, New York, McGraw-Hill.

[8] Ghobarah, A.A. & Tso, W.K. (1972). Parametric stability of thin-walled beams of open section. *Journal of Applied Mechanics, Transactions ASME, 39 Ser E (1)*, 201-206.

[9] Bolotin, V.V. (1953). On the parametric Excitation of Transverse Vibrations. *Collection of Papers: Transverse Vibrations and Critical Velocities, 2*, 5-44.

[10] Popelar, C.H. (1969). Dynamic stability of the flexural vibrations of a thin-walled beam. *International Journal of Solids and Structures, 5*, 549-557.

[11] Popelar, C.H. (1972). Dynamic stability of thin-walled column. *Journal of Engineering, Mechanics Division, Proceedings of the American Society of Civil Engineers, 98*, 657-677.

[12] Hasan, S.A. & Barr, A.D.S. (1974). Non-linear and parametric vibration of thin-walled beams of equal angle-section. *Journal of Sound and Vibration, 31*, 25-47.

[13] Kim, M.Y.; Chang, S.P. & Park H.G. (2001). Spatial Postbuckling analysis of nonsymmetric thin-walled frames. I: Theoretical considerations based on semitangential property. *Journal of Engineering Mechanics, 127*, 769-778.

[14] Kim, S.B. & Kim, M.Y. (2000). Improved formulation for spatial stability and free vibration of thin-walled tapered beams and space frames. *Engineering Structures, 22*, 446-458.

[15] Kim, T.W. & Kim, J.H. (2001). Parametric instability of a cross-ply laminated beam with viscoelastic properties under a periodic force. *Composite Structures, 51*, 205-209

[16] Machado, S.P. & Cortínez, V.H. (2005). Lateral buckling of thin walled composite bisymmetric beams with prebuckling and shear deformation. *Engineering Structures, 27*, 1185-1196.

[17] Machado, S.P. & Cortínez, V.H. (2005). Non-Linear model for stability of thin walled composite beams with shear deformation. *Thin walled Structures, 43*, 1615-1645.

[18] Hasegawa A.; Liyanage, K.K. & Nishino, F. (1986). Spatial instability and nonlinear finite displacement analysis of thin walled members and frames. *Journal of the Faculty of Engineering, University of Tokyo, Series B, 38*, 19-78.

[19] Kitipornchai, S. & Chan, S.L. (1990). Stability and nonlinear finite element analysis. Bull, J.W. Editor. *In Finite Element Applications to Thin walled Structures*. London, Elsevier Applied Science.

[20] Roberts, T.M. (1983). Instability, geometric non-linearity and collapse of thin walled beams. Narayanan, R. Editor. *In Beams and Beam-Columns-Stability and Strength*. London, Applied Science Publishers.

[21] Pi, Y.L. & Bradford, M.A. (2001). Effects of approximations in analysis of beams of open thin walled cross-section - part II: 3-D non-linear behaviour. *International Journal of Numerical Methods in Engineering, 51*, 773-790.

[22] Simo, J.C. & Vu-Quoc, L. (1987). The role of non-linear theory in transient dynamic analysis of flexible structures. *Journal of Sound and Vibration, 119*, 487–508.

[23] Machado, S.P.; Filipich, C.P. & Cortínez, V.H. (2007). Parametric vibration of thin-walled composite beams with shear deformation. *Journal of Sound and Vibration, 305*, 563-581.

[24] Machado, S.P. & Cortínez, V.H. (2007). Free vibration of thin-walled composite beams with static initial stresses and deformations. *Engineering Structures, 29*, 372-382.

[25] Hsu, C. (1963). On the parametric excitation of a dynamic system having multiple degrees of freedom. *Journal of Applied Mechanics, 30*, 367-372.

[26] Argyris, J.H. (1981). An excursion into large rotations. *Computer Methods in Applied Mechanics and Engineering, 32*, 85-155.

[27] Fraternali, F. & Feo, L. (2002). On a moderate rotation theory of thin-walled composite beams. *Composites Part B-Engineering, 31*, 141-158.

[28] Cortínez, V.H. & Piovan, M.T. (2006). Stability of Composite Thin-Walled Beams with Shear Deformability. *Computers & Structures, 84*, 978-990.

[29] Krenk, S. & Gunneskov, O. (1981). Statics of Thin-Walled Pre-twisted Beams. *International Journal of Numerical Methods in Engineering, 17*, 1407-1426.

[30] Washizu, K. (1968).*Variational Methods in Elasticity and Plasticity*. Oxford, Pergamon Press.

[31] Machado, S.P. (2008). Non-linear buckling and postbuckling behavior of thin-walled beams considering shear deformation. *International Journal of Non-Linear Mechanics, 43*, 345-365.

[32] Abaqus (2007). Standard user's manuel, v. 6.7. Printed in the United States of America, Dassault Systèmes.

In: Structural Materials and Engineering
Editor: Ference H. Hagy

ISBN 978-1-60692-927-8
© 2009 Nova Science Publishers, Inc.

Chapter 10

THE PREDICTION OF SURVIVAL PROBABILITIES OF BUILDING STRUCTURES UNDER TRANSIENT EXTREME EXECUTION LOADS

Antanas Kudzys[*]

KTU Institute of Architecture and Construction, Kaunas, Lithuania

ABSTRACT

The effect of random execution (construction) loads, climate actions and their combinations on the probabilistic quality and reliability of structures is discussed. The necessity of using more attention from the probabilistic point of view to the structural reliability prediction of load-carrying members subjected to extreme action (load) effects caused by unfavourable transient site situations is discussed. The methodology of survival probability assessments and predictions of structures during the periods of building carcass erection and mixed construction work is analysed. The design features of survival probability predictions of building structures recommended by Standards SEI/ASCE 37 in the USA and ENV 1991 in Europe for their limit state design are described. The new probabilistic analysis approaches based on the concepts of conventional resistance, coincidence of extreme action effects, random sequence of member safety margins, and transformed conditional probabilities are presented. The prediction of instantaneous and time-dependent survival probabilities of members during their life cycles of carcass erection and mixed construction work is considered. The analysis of partial and total survival probabilities and reliability indices of members as quantitative parameters of their structural quality under execution loads is considered and illustrated by a numerical example. The survival probabilities and reliability indices of precast multiholowcore concrete slabs during construction of masonry walls and mosaic floors for multistorey industrial buildings are under consideration.

Keywords: Execution (construction) loads; Combination of loads; Limit state design, Structural reliability; Safety margin; Series systems; Survival probability; Reliability index.

---

*Correspondence to: E-mail: antanas.kudzys@gmail.com

ABBREVIATIONS

A	Cross-sectional area
d	Duration of the construction load (action)
F	Cumulative distribution function
f	Probability density function
f_y	Yield strength of reinforcement
G	Permanent load (mass)
M	Bending moment
n	Recurrence number of extreme loads (actions)
P	Survival probability
Q	Variable construction load
R	Resistance of members
S	Action (load) effect
T	Thermal and/or shrinkage action
t	Time, period
W	Wind load (action)
X	Main variable
Z	Safety margin of members
z	Lever arm of internal forces
β	Reliability index
γ	Partial factor for loads and actions
δ	Coefficient of variation
λ	Renewal rate of the load (action)
ρ	Coefficient of correlation, correlation factor
σ	Standard deviation
θ	Additional variable as a parameter which contains model uncertainties
Φ	Tabulated distribution function, strength reduction factor

Subscripts

1	Carcass erection period
2	Mixed construction work period
c	Conventional
d	Design value
i	Case of load (action) concurrence
k	Cut of a random process (sequence), characteristic value
m	Mean of a random variable
n	Nominal value
s	Steel bars within the tension zone

1. INTRODUCTION

Standards ASCE-7 (2005) in the USA and EN (1990) in Europe require that building structures shall be designed with appropriate degrees of reliability. The limit state method popularized by these standards are based, respectively, on the load and resistance factor design (LRFD) and the partial safety factor design (PSFD). The structural reliability prediction of load-carrying structures should always be based on the analysis of unfavourable situations which are likely to occur during the construction of buildings and civil engineering works. Due to up-to-date rapid and industrial site work, it is difficult to evade these situations caused by random extreme execution (construction) loads and other actions In this way, extreme site (execution) action effects (bending and torsion moments, axial and shear forces) during construction may be very great and even greater than those caused by use (service) loads. Thus, it is very probable that the high quality of structures may be guaranteed if they have already withstood unfavourable extreme site situations (Saraf et al. 1996). Hence, design loads during construction can govern the design of finished structures. Therefore, designers and other building engineers must pay more attention to the performance criteria, execution loads and actions, strength and stability of structures during their erection.

An intensive loading of structures during construction operations may have not only negative but also some positive effects on their reliability assessment and prediction. Extreme execution loads help structural and resident engineers to convince themselves of the absence of gross human errors in design and construction stages, causing inadmissible deflections and displacements of structures and their members.

Besides, extreme site action effects may be treated as an effective measure in the revised reliability prediction of constructed building members. Therefore, it is impossible to disagree with Melchers (1999) that construction loads acting on a structure during execution periods should be have more attention from the probabilistic point of view. Unfoundedly, these real and urgent recommendations to provide probabilistic approaches in the structural safety analysis for construction conditions are ignored not only in current deterministic and semi-probilistic design codes and standards but are mostly silent in recommendations presented by ISO 2394 (1998) and JCSS 2000 (2000).

Approximately 75-85 % of structural failures may be attributed to human errors (Saffarini 2000) and 30-40 % of these errors are committed at the design stage (Ellingwood 1987). Up to 50 % of failures occur during the execution of structures without site (engineering) inspections (Hall 1988, Stewart 1992). Since execution loads may be treated as proof loads, engineering inspection becomes an effective measure for human error control. According to Yates and Lockley (2002), a fulltime careful inspection during construction work procedures might be the most important factor preventing structural failures and avoiding construction accidents. Moreover, it helps improve a safety climate in site environment (Mohamed 2002). However, a necessity to make the engineering control of construction work stronger may be assessed only by probability-based approaches.

The restriction of loads and avoiding of dangerous situations in sequence of construction period may have decisive effect on the safety of structure during its construction processes. The designer of future permanent structures and the structural engineer of this building must by calculations assess the expediency of restrictions and the reliability of structural members until the construction period is fully completed (Mohamed 2002). Unfortunately, due to a lack of designer experience, it is difficult to know beforehand the intensity of random construction loads that are not fully appreciated by design codes and specified regulations. The deterministic design methods are provided by the necessary code information on analysis models, material and load properties. On the contrary, it is very difficult to use in practice probability-based approaches owing to lack of special (statistical) data. Therefore, design codes need to impose stricter guidelines on the performance of structural analysis (Saffarini 2000). In spite of the efforts made by building engineers and researchers to improve semi-probabilistic analysis methods, the latter do not allow us to assess the structural reliability by quantitative parameters.

In many cases, it would be more expedient to analyse the structural safety of buildings by probability - based methods. The first practical developments towards probabilistic design methods (ISO 2394 1998, JCSS 2000, Vrouwenvelder 2002) are to be welcomed. However, it is difficult to apply these methods in engineering practice due to the great discrepancy among intricate theoretical concepts and practical possibilities of designers, as well as because of some methodological and mathematical features. Probability - based approaches may be acceptable to building engineers only under the indispensable and easy perceptible condition that they may be translated into structural design practice only using unsophisticated and easy perceptible mathematical models. These models may be used as a basis for design of special buildings and structures not completely covered in CEI/ASCE 37 (2002) and EVN 1991-2-6 (1997).

Engineers should possess simplified exact methods for the safety prediction of structures exposed to execution loads and climate actions, because their serviceability cannot be carried out objectively, ignoring structural performance during the construction period (Epaarachchi et al. 2002). However, for the evaluation of extreme execution loads on the reliability index of structures one needs to supplement design codes with statistical data and probabilistic models (Stewart & Rosowsky 1996).

The intention of this paper is to introduce to building engineers new probabilistic approaches and unsophisticated mathematical models in the practical safety analysis of structures subjected to loads and to check the concept of load factors and load combinations for the variability of their transient situation loadings during construction period. This paper is intended for practical use by building engineers knowledgeable in the performance of load-carrying structures. Its provisions are not belonging to loads and other actions caused by gross negligence and error.

2. CONSTRUCTION LOADS IN THE STRUCTURAL STRENGTH DESIGN

2.1. SEI/ASCE 37 Recommendations

Structural failures and collapses in buildings during their construction can be caused not only due to irresponsibility and gross human errors of designers and erectors but also by some conditionalities of vaguely presented and explained directions in design codes and standards. The material presented in SEI/ASCE 37 (2002) recommendations is prepared in accordance with recognized and used in design practice engineering concepts and principles. Therefore, it is expedient to make the acquaintance of minimum design load requirements during construction for building structures designed and constructed to support safety the factored service loads defined in Standard ASCE/SEI 7-05 (2005).

The critical load effect during construction may result from application of uniformly distributed and concentrated one extraordinary erection load or due to recurrent concurrence of two and more loads. The concept of using maximum loads and corresponding load factors in the strength verification of structures during construction is consistent with Standard ASCE 7 recommendations. Besides, in all cases, load-carrying structures, and all parts thereof, should be designed and constructed to support safely the construction loads in load combinations without exceeding the appropriate strength limit states for the structures of buildings and for the safety of working people.

According to SEI/ASCE 37 requirements, the generalized form of the load combinations for strength or limit state design of building structures can be written as:

$$U_{SEI} = \sum_k c_{D,k} D_{n,k} + \sum_i c_{\max} Q_{n,i} + \sum_j c_{APT,j} Q_{n,j} \qquad (1)$$

Here $c_D = \gamma_G$ and $c_{\max} = \gamma_Q$ are the factors for dead loads and maximum value of variable loads; c_{APT} is the load factor for the uncorrelated variable loads at their arbitrary point-in-time (APT) values as the factor for combinations of characteristic (nominal) variable actions; D_n is a nominal variable load; k belongs to all dead and construction material loads; i belongs to all loads occurring at maximum value; j - belongs to all relevant simultaneously occurring variable loads at APT values.

To construction procedure loads Q and their combinations belong a weight of temporary structures, fixed (C_{FML}) and variable (C_{VML}) material loads, personal and equipment load (C_P), horizontal construction loads (C_H), erection and fitting forces (C_F), equipment reactions (C_R), lateral pressure of concrete (C_C) and earth (C_{EH}). In all cases, careful consideration should be given to the placement of stockpiled materials on early-age concrete structures and members of hyperstatic systems with unformed joints. The worstcase load effects caused by erection and fitting forces or equipment reactions must be considered.

The combination rule of construction procedure loads is demonstrated in the numerical example of this paper (see Section 6.2).

The values of construction load factors are from 1.2 (for dead loads) to 2.0 (for erection forces and heavy equipment reactions). The factor for combination value of a variable action or the factor for arbitrary point-in-time value of independent variable actions is equal to 0.5. This value is fairly low while it may be highly unlikely that multiple variable loads will attain their maximum values at the same time of construction period.

The values of factors for construction loads and their combinations depend only on load sources or operation types and probability distribution laws but not on the type of action effects of considered structures. As it is know, construction procedure loads and other actions on a structure and their combinations may be changed during construction. Therefore, I am sure that the time-dependent safety of structures may be objectively assessed and predicted using time-dependent load factors in traditional limit state design approaches or probability-based methods (see Section 6.3).

2.2. ENV 1991-2-6 Recommendations

The features of limit state design of structural members and their systems during execution period of buildings are presented in Part 2-6 of Eurocode 1 (ENV 1997). These features are very close to the directions recommended by Standard SEI/ASCE 37.

Construction loads that have to be considered in transient design situations of buildings include permanent loads (G); working personal with small site equipments (Q_{cs}); movable storage of building materials, precast members and equipment (Q_{cc}); cranes, lifts, vehicles, power installation or heavy control devices (Q_{cd}). All these loads are classified as permanent (dead) or variable, fixed or free, static, quasi-static or dynamic, direct or indirect, uni- or multi-component loads.

The transient design situations should be identified taking into account that all the following construction conditions may very from stage to stage of execution period. Therefore, the probability distribution or/and magnitude of permanent and variable loads may very within the great limits. However, the characteristic (nominal) values of variable construction and environmental actions during execution of buildings may be smaller than those for the limit state verification of structures in persistent service situations.

The characteristic value of self-weight of form-work should be taken equal to 0.5 kN/m² and more. The wind velocity to be not less than 20 m/s at any height over the earth's surface. The simultaneity of particular construction loads during execution stages may be considered as excluded but dangerous transient situations. It is specified in the project specification and appropriate site control measures and are taken into account in the limit state verifications of structures subjected to random joint action effects.

The load combination format for ultimate limit state analysis of structures during construction phases may be presented in the form:

$$U_{EN} = \sum_k \gamma_{G,j} G_{k,j} + \gamma_P P_k + \gamma_{Q1} Q_{k1} + \sum \gamma_{Q,i} \gamma_{O,i} Q_{k,i} \quad (2)$$

Here $\gamma_G = c_D$, $\gamma_Q = c_{max}$ and γ_p are the load factors for characteristic values of permanent (G) and variable construction procedure and environmental (Q) loads, and relevant prestressing force (P); $\psi_o = c_{APT}$ is the factor for combination value of an accompanying variable load or other action.

Unless otherwise specified, for the verification of ultimate limit states of structural resistance by the semi-probabilistic method in Europe or for the strength design method used in the USA, the combinations of execution actions to be considered should be the fundamental combinations with should be the load factors, $\gamma_G = \gamma_Q = \gamma_P$, equal to 1.35 and the factor for load combinations $\psi_o = 0.5$-0.6. Because most actions during execution period cannot be assessed with sufficient accuracy, the simplified verification for building structures may be carried out with the factor for combination value of variable actions $\psi_o = 1$.

The combination of effects caused by execution loads is presented as the unfavourable expression for the strength design of floor slabs by EVN 1991-2-6 (1997) recommendations is presented in Section 6.2.

3. SAFETY MARGINS OF MEMBERS

3.1. The Life Cycle of Carcass Erection

The structural members (beams, slabs, columns, walls) of buildings consist of particular members (sections, connections) representing the only possible failure mode. Structural members consist of two or three particular members and may be treated as auto systems representing multicriteria failure mode. The failure or survival probabilities of structural members and systems may be objectively assessed and predicted only knowing the structural safety parameters of particular members. Usually, the execution loads and other actions are stochastically independent random variables and processes. Due to different in nature and construction technologies, it is expedient to separate two cycles of the times (t_1 and t_2) characterizing the durations of load-carrying carcass erection and mixed site work, respectively. In this way, two different masses of the structures themselves as permanent loads G_1 and their additional components $G_2 = G - G_1$ should be considered. Under the periods of times t_1 and t_2 may be treated the durations of dangerous loading situations that may be shorter as these periods.

The duration of carcass erection life cycles normally is small compared to the design service life of buildings and other construction works. Therefore, environmental load sources (wind, W, temperature changes, T_1, and snow, S) may have very low risk of extreme exposure for loads in this construction life cycle. Besides, the severe environmental actions are seasonal in any short time interval of carcass erection period. Thus, the wind and snow loads during this period may be neglected.

According to full probabilistic approaches and assumptions, the safety margin as the performance function of particular members during the period of time t_1 may be presented in the form:

$$Z_1 = g(\boldsymbol{\theta}, \mathbf{X}_1) = \theta_R R_1 - \theta_G S_{G_1} - \theta_T S_{T_1} - \theta_Q S_{Q_{11}} - \theta_Q S_{Q_{12}} \tag{3}$$

where $\boldsymbol{\theta}$ and $\mathbf{X}_1$ are the vectors of additional and basic variables. The statistical properties of the vector $\boldsymbol{\theta}$ representing uncertainties of models which give the values of resistances and action effects of members, may be presented by its component means $\theta_m = 1.0 - 1.12$ and coefficients of variations $\delta\theta = 0.05 - 0.14$ (JCSS 2000, Stewart 2001, Mirza & MacGregor 1979, Hong & Lind 1996, Ellingwood & Tekie 1999). The effect of $\delta\theta$ on statistical analysis results decreases when the variances of main variables increase. The statistical moments of member resistances and action effects during periods of construction procedures may be calculated using the values $\theta_{Rm} = 1.01$, $\theta_{Tm} = \theta_{Gm} = \theta_{Qm} = 1.0$, $\sigma\theta_R \cong 0.08$,

$\sigma\theta_T = \sigma\theta_G = 0.1$ and $\sigma\theta_Q = 0$ or 0.1, when the probability distribution of extreme action effects obey exponential or other distribution laws, respectively.

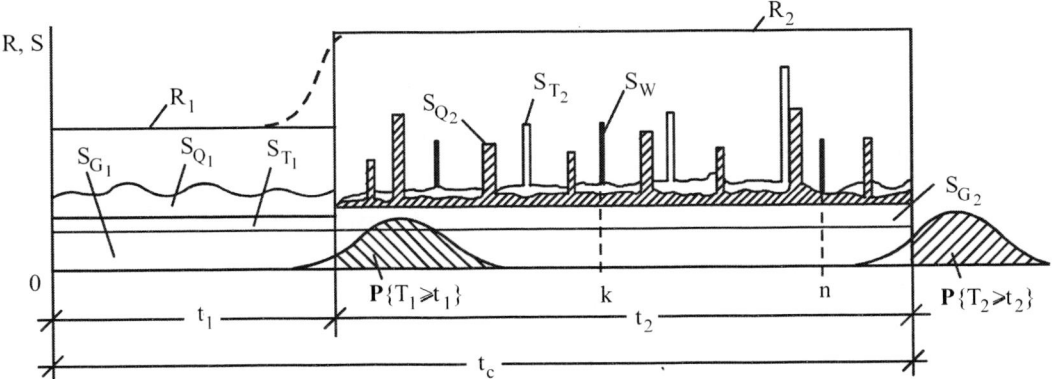

Figure 1. Model for the structural safety prediction of members under carcass erection (t_1) and mixed construction work (t_2) periods

The independent action effects S_{G_1}, S_{T_1} and $S_{Q_{11}}$, $S_{Q_{12}}$ are caused by the mass, G_1, of erected structures, the thermal and/or shrinkage actions, T_1, the fixed and variable placement equipments, erection, fitting and shoring loads, Q_{11} and Q_{12}, respectively. These action effects may be modeled as stationary processes (Fig. 1). Makeshift shores may be required to support the load not only in-situ but also precast concrete and composite structures. The action effect S_{T_1} may be minimized by appropriate detailing provisions. Besides, the effect of low ranges of daily temperatures on construction load combinations is small and may be neglected.

The freshly cast concrete floors of multistorey buildings are supported by propping elements resting on the erected floors below. A relevant load analysis should be considered for the selected critical stage of carcass erection period t_1 (ENV 1991-2-6 1997). The value of self-weight of formworks may taken equal to 0.5 kN/m², unless actual data are available.

3.2. The Life Cycle of Mixed Construction Work

The safety margin process as time-dependent performance function of members during the period of time t_2 may be written as:

$$Z_2(t) = \theta_R R_2 - \theta_G(S_{G_1} + S_{G_2}) - \theta_Q S_{Q_{21}}(t) - \theta_Q S_{Q_{22}}(t) - \theta_T S_{T_2}(t) - \theta_W S_{W_2}(t) \quad (4)$$

where the action effects S_{G_2}, $S_{Q_{21}}$, $S_{Q_{22}}$, S_{T_2} and S_{W_2} are caused by the additional mass G_2 of structures and other permanent loads, the equipment Q_{21}, the loads Q_{22} of concentrated materials and precast elements, the range of ambient temperature T_2 and the monthly maximum wind velocity pressure W_2. The probability distribution of equipment load Q_{21} is close to a normal distribution. The loads Q_{22} and W_2 may be treated as rectangular impulse renewal processes and modeled by the exponential distribution law (JCSS 2000, Quek & Cheong 1992).

For hoisting, launching and other construction stages lasting only a few or several hours, the maximum wind velocity must be defined at the project specification. All structural members including internal walls and partitions must be stabilized during construction to resist the wind loads at any execution stages.

Structural members and other components of buildings can be erected and assembled at environmental temperatures that are very different from their temperatures during service periods. Thermal distortions can be significant during construction when structures are exposed to seasonal ambient temperature variations. Ambient temperature ranges may be very dangerous for structures executed during a summer time and exposed to low temperature actions during severe winter periods or vice versa. According to Raiser (1998), the probability distribution of annual extreme action effects S_{T_2} may be modeled by a Gaussian distribution. Their coefficient of variation may be expressed as:

$$\delta S_{T_2} = \delta T_2 \approx \frac{(\sigma^2 T_{\max} + \sigma^2 T_{\min})^{1/2}}{(T_{\max,m} - T_{\min,m})}, \tag{5}$$

where, $T_{\max,m}$, $T_{\min,m}$ and $\sigma^2 T_{\max}$, $\sigma^2 T_{\min}$ are the means and variances of annual extreme temperatures.

The action effect $S_{Q_2} = S_{Q_{21}} + S_{Q_{22}}$ may be changed into the annual extreme action effect S_S caused by snow pressures, when the structural safety of roof members is considered. However, the probability of concurrence of extreme construction and wind loads during 1-3 years period is small. When the combination of these loads is considered, the lateral wind pressure should be diminished to the values corresponding to the wind speed not more than 18-20 m/s. The mean and standard deviation of the wind velocity pressure the probability distribution law of which may be treated as exponential one should not be less than 200 N/m² (SEI/ASCE 37 2002).

In many cases, the probability distribution laws of extreme values of both Q_{21} and Q_{22} loads may be presented as exponential ones. For the practical sake of the probability-based analysis of structural members, it is expedient to use the conventional bivariate distribution function of two action effects with the mean and variance equal to $S_{Q_2 m} = S_{Q_{21} m} + S_{Q_{22} m}$ and $\sigma^2 S_{Q_2} = (\sigma S_{Q_{21}} + \sigma S_{Q_{22}})^2$.

4. PROBABILISTIC APPROACHES IN STRUCTURAL RELIABILITY ANALYSIS

4.1. Instantaneous Survival Probability

The most unfavourable (extreme) effects of particular members resulting from construction loads or their combinations should be considered in the design phase as sustained $S_s(t)$ or extraordinary $S_e(t)$ values the renewal rate of which is slow or frequent, respectively.

Hall (1988) has suggested to use in structural analysis of particular members the cumulative distribution function of the conventional multivariate action effect as:

$$S_c(t) = \theta_G(S_{G1} + S_{G2}) + \sum_i \theta_Q S_{Qi}(t) \tag{6}$$

The investigations carried out by Stewart (1996) showed that it is not simple to adopt the presented approach in engineering practice due to different probability distribution laws of extraordinary episodic and permanent (dead) or sustained variable loads.

Usually, the probability distributions of permanent and sustained execution and environmental loads and other actions are close to normal and lognormal ones. Therefore, it would be more better for design practice to use the probability distribution of conventional member resistance as:

$$R_c = \theta_R R - \theta_G(S_{G1} + S_{G2}) - \sum_i \theta_Q S_{Qi} \tag{7}$$

In this case, according to the analysis model in Fig. 2, the safety margin process of members may be presented as the finite rank random sequence and written in the form:

$$Z_k = R_c - S_{ek}; \quad k = 1, 2, 3, \ldots, n-1, n \tag{8}$$

where n is the recurrence number of dangerous situations caused by maximum values of single variable loads or combinations of two and more random action effects.

The mean and variance of a random stationary resistance R_c process are:

$$R_{cm} = \theta_{Rm} R_m - \theta_G(S_{G1m} + S_{G2m}) - \theta_{Qm} \sum_i S_{Qim} \tag{9}$$

$$\sigma^2 R_c = \sigma^2(\theta_R R) + \sigma^2(\theta_G S_{G1}) + \sigma^2(\theta_G S_{G2}) + \sum_i \sigma^2(\theta_Q S_{Qi}) \tag{10}$$

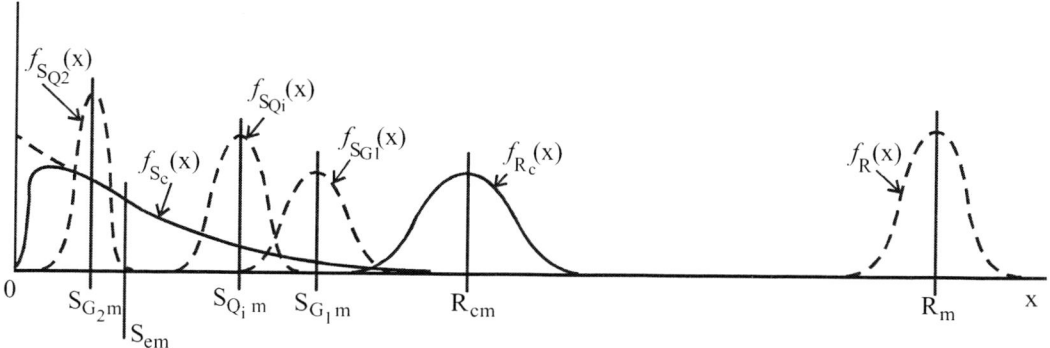

Figure 2. Model for the analysis of instantaneous survival probabilities of particular members using their conventional resistance R_c by Eq. (7) and extraordinary action effects S_e

The instantaneous survival probability of a member at k-th extreme situation (assuming that it was safe at the situations 1, 2, ..., $k-1$) may be expressed as:

$$\mathbf{P}_k = \mathbf{P}\{g[\theta, \mathbf{X}(t_k)] \exists t_k \in [t_1, t_n]\} = \mathbf{P}\{R_c > S_{ek}\} = \int_0^\infty f_{R_c}(x) F_{S_k}(x) \, dx \qquad (11)$$

where $f_{R_c}(x)$ and $F_{S_k}(x)$ are the density and cumulative distribution functions of a conventional resistance and an extreme action effect, respectively. When the probability distribution of extreme action effects belongs to a Gumbel distribution law, the last function may be presented in the form:

$$F_{S_k}(x) = \exp\left[-\exp\left(\frac{S_{ekm} - x}{0.7794 \, \sigma S_{ek}} - 0.5772\right)\right] \qquad (12)$$

where S_{ekm} and σS_{ek} are the mean and standard deviation of an extreme action effect.

When the probability distributions of extreme actions caused by sustained or extraordinary variable loads are close to normal or exponential ones, the instantaneous survival probability $\mathbf{P}_k$ from Eq. (11) of particular members may calculated by the equations:

$$\mathbf{P}_k = \mathbf{P}\{R_c > S_{ek}\} = \Phi\left\{\frac{R_{cm} - S_{ekm}}{(\sigma^2 R_c + \sigma^2 S_{ek})^{1/2}}\right\} \qquad (13)$$

$$\mathbf{P}_k = \mathbf{P}\{R_c > S_{ek}\} = 1 - \Phi\left(-\frac{R_{cm}}{\sigma R_c}\right) - \exp\left(-\frac{R_{cm}}{S_{ekm}} + 0.5\frac{\sigma^2 R_c}{\sigma^2 S_{ek}}\right) \times \left[1 - \Phi\left(-\frac{R_{cm}}{\sigma R_c} + \frac{\sigma R_c}{\sigma S_{ek}}\right)\right] \qquad (14)$$

where $\Phi(\bullet)$ is the cumulative distribution function of the standardized normal distribution; R_{cm}, $\sigma^2 R_c$ and S_{ekm}, $\sigma^2 S_{ek}$ are the means and variances of resistances and extreme action effects of members.

4.2. Time-Dependent Survival Probability

According to probability-based approaches or to the design level III for load-carrying structures, the time-dependent survival probability of structural members may be modeled as series systems with statistically dependent elements as the cuts of random safety margin sequences (Fig. 3).

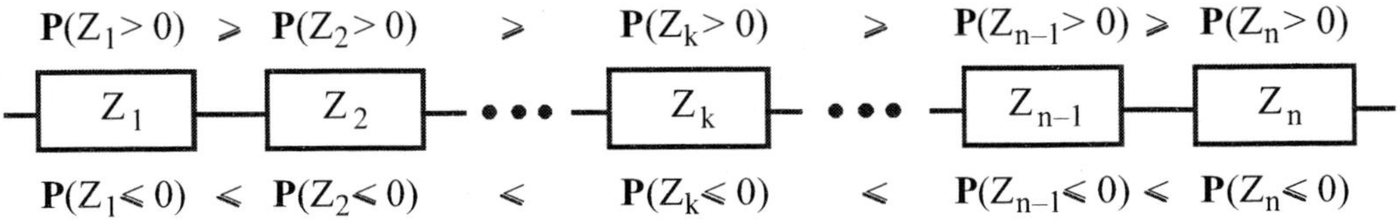

$P(Z_1 > 0) \geqslant P(Z_2 > 0) \geqslant P(Z_k > 0) \geqslant P(Z_{n-1} > 0) \geqslant P(Z_n > 0)$

— Z_1 — Z_2 — $\cdots$ — Z_k — $\cdots$ — Z_{n-1} — Z_n —

$P(Z_1 \leqslant 0) \leqslant P(Z_2 \leqslant 0) \leqslant P(Z_k \leqslant 0) \leqslant P(Z_{n-1} \leqslant 0) \leqslant P(Z_n \leqslant 0)$

Figure 3. Series system with statistically dependent elements

When a structural series system consists of two dependent elements, the probability that either or both of two failures occur may be expressed as:

$$P(Z_2 \leq 0 \cup Z_1 \leq 0) = P(Z_2 \leq 0) + P(Z_1 \leq 0) - P(Z_2 \leq 0)P(Z_1 \leq 0 | Z_2 \leq 0) =$$
$$= P(Z_2 \leq 0) + P(Z_1 \leq 0) - P(Z_2 \leq 0 \cap Z_1 \leq 0) \qquad (15)$$

where $P(Z_2 \leq 0) \geq P(Z_1 \leq 0)$ and $P(Z_1 \leq 0)$ are the probabilities of its failure in the 2 and 1 modes; $P(Z_2 \leq 0 \cap Z_1 \leq 0)$ is the probability that the failure of a system occurs simultaneously in both modes. An evaluation of the second-order intersection probability $P(Z_2 \leq 0 \cap Z_1 \leq 0)$ may be carried out by the rather uncomfortable numerical integration or Monte Carlo simulation methods. However, it is more expedient to use in design practice the unsophisticated method of transformed conditional probabilities (TCPM).

The survival probabilities of rank series systems with two and n elements are:

$$P(Z_2 > 0 \cap Z_1 > 0) = 1 - (Z_2 \leq 0 \cup Z_1 \leq 0) = P(Z_1 > 0)P(Z_2 > 0 | Z_1 > 0) =$$
$$= P(Z_1 > 0)P(Z_2 > 0)\left[1 + \rho_{21}^{x_2}\left(\frac{1}{P(Z_1 > 0)} - 1\right)\right] \qquad (16)$$

$$P\left(\bigcap_{k=1}^{n} Z_k > 0\right) = \prod_{k=1}^{n} P(Z_k > 0)\left[1 + \rho_{21}^{x_2}\left(\frac{1}{P(Z_1 > 0)} - 1\right)\right] \cdots \left[1 + \rho_{k|1..k-1}^{x_k}\left(\frac{1}{P(Z_1 > 0 \cap ... \cap Z_{k-1} > 0)} - 1\right)\right] \times ...$$

$$\ldots \times \left[1 + \rho_{n|1\ldots n-1}^{x_n}\left(\frac{1}{\mathbf{P}(Z_1 > 0 \cap \ldots \cap Z_{n-1} > 0)} - 1\right)\right] \tag{17}$$

Here

$$\rho_{kl} = \rho(Z_k, Z_l) = \sigma^2 R_c / (\sigma Z_k \times \sigma Z_l) \tag{18}$$

$$\rho_k = \rho_{k|1\ldots k-1} = (\rho_{k1} + \rho_{k2} + \ldots + \rho_{k.k-1})/(k-1) \tag{19}$$

are, respectively, the coefficient and factor of correlation of dependent elements that form k-th row of system quadratic correlation matrix;

$$x_k = \left\{\left[4.5 + 30\rho_k^3(1.08 - \rho_k)\right]/(1 - 0.98\rho_k)\right\}^{1/2} \approx \left[(4.5 + 4\rho_k)/(1 - 0.98\rho_k)\right]^{1/2} \approx$$

$$\approx \left[8.5/(1 - 0.98\rho_k)\right]^{1/2} \tag{20}$$

is the bounded index of this conditional factor characterizing its effect on the probability of simultaneous failure of k elements of series systems.

For the sake of simplified but fairly exact probabilistic analysis, the conditional probabilities of higher orders $\mathbf{P}(Z_k > 0 | Z_1 > 0 \cap \ldots \cap Z_{k-1} > 0)$ may be defined as $\mathbf{P}(Z_k > 0 | Z_{k-1})$. Therefore, Eq. (17) may be rewritten as follows:

$$\mathbf{P}\left(\bigcap_{k=1}^{n} Z_k > 0\right) = \prod_{k=1}^{n} \mathbf{P}(Z_k > 0)\left[1 + \rho_{21}^{x_2}\left(\frac{1}{\mathbf{P}(Z_1 > 0)} - 1\right)\right] \times \ldots \times \left[1 + \rho_{k|1\ldots k-1}^{x_k}\left(\frac{1}{\mathbf{P}(Z_{k-1} > 0)} - 1\right)\right] \times \ldots$$

$$\ldots \times \left[1 + \rho_{n|1\ldots n-1}^{x_n}\left(\frac{1}{\mathbf{P}(Z_{n-1} > 0)} - 1\right)\right] \tag{21}$$

For series systems with equireliable and equicorrelated elements, the survival probabilities and correlation factors of system components are: $\mathbf{P}(Z_1 > 0) = \mathbf{P}(Z_{k-1} > 0) = \mathbf{P}(Z_{n-1} > 0) = \mathbf{P}(Z_k > 0)$, $\rho_{21}^{x_2} = \rho_{k,1\ldots k-1}^{x_k} = \rho_{n,1\ldots n-1}^{x_n} = \rho_{kl}^{x_2}$.
From Eq. (21), the survival probability of these systems is:

$$\mathbf{P}\left(\bigcap_{k=1}^{n} Z_k > 0\right) = \mathbf{P}^n(Z_k > 0)\left[1 + \rho_{kl}^{x_2}\left(\frac{1}{\mathbf{P}(Z_k > 0)} - 1\right)\right]^{n-1} \tag{22}$$

where n is the number of elements or safety margin sequence cuts.

4.3. The Analysis of Correlation Factors of Series Systems

The values of correlation factors $\rho_k^{x_k}$ of equicorrelated series systems may be defined from analysis data with real and simplified conditional probabilities presented in Eqs. (17) and (22), respectively. Therefore, the correlation factor of system elements may be expressed, respectively, in the forms:

$$\rho_k^{x_k} = \left[\mathbf{P}\left(\bigcap_{k=1}^{n} Z_k > 0\right) \middle/ \mathbf{P}(Z_k > 0) - \mathbf{P}\left(\bigcap_{k=1}^{n-1} Z_k > 0\right)\right] \middle/ \left[1 - \mathbf{P}\left(\bigcap_{k=1}^{n-1} Z_k > 0\right)\right] \quad (23)$$

$$\rho_k^{x_k} = \left\{ \left[\mathbf{P}\left(\bigcap_{k=1}^{n} Z_k > 0\right) \middle/ \mathbf{P}^n(Z_k > 0)\right]^{1/(n-1)} - 1 \right\} \middle/ \left[\frac{1}{\mathbf{P}(Z_k > 0)} - 1\right] \quad (24)$$

Let us consider the correlation factor values for series systems consisted of four equicorrelated and unequireliable (system 1) or equireliable (system 2) elements. These values are calculated, respectively, from Eqs. (23) and (24), where the survival probabilities of systems with four and three elements are computed by the author using Monte Carlo simulation method (system 1) and by Ahammed and Melchers (2005) using a numerical integration method (system 2).

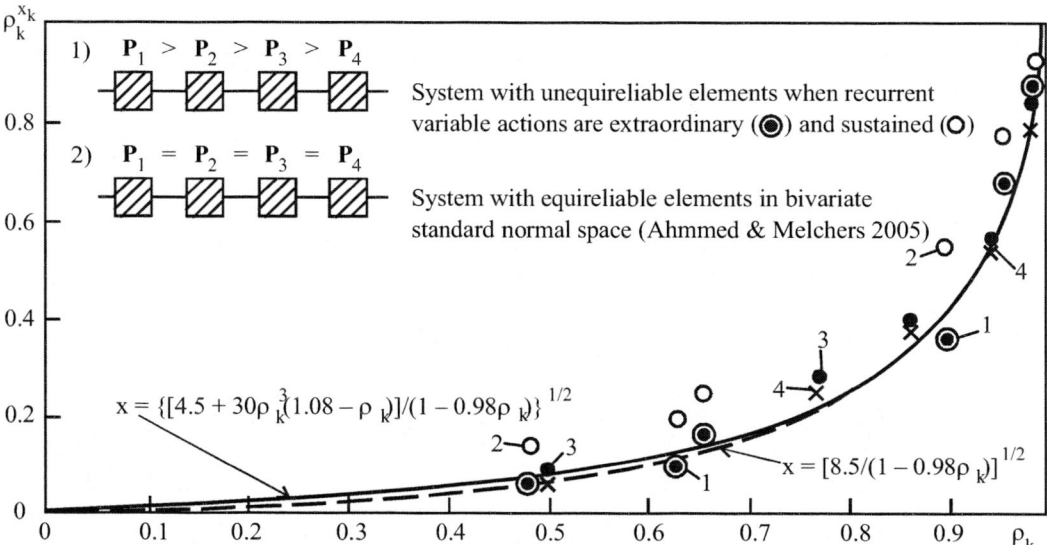

Figure 4. The index correlation factor $\rho_k^{x_k}$ calculated by Eqs. (23) (1, 2, 3) and (24) (4) of two series systems with equicorrelated elements versus their basic values by Eq. (19)

The means and variances of bending resistances and moments of the safety margin sequence of system 1 are: R_{km} = 194, 184, 174, 164 kNm, $\sigma^2 R_k$ = 1600 (kNm)^2; M_m = 60 kNm, $\sigma^2 M = 0, 9, 36, 144, 576, 1296$ (kNm)^2. The probability distribution law of sequence components is a normal for resistance R_k and a normal or a Gumbel for bending moments M caused by recurrent extraordinary and sustained variable actions of considered members. The values of indexed correlation factors $\rho_k^{x_k}$ are presented in Fig. 4 by points 1 and 2.

According to investigations carried out by Ahammed and Melchers (2005), the difference between values of correlation factors of systems with equicorrelated and equireliable elements calculated by Eqs. (23) and (24) is fairly small. This fact is demonstrated by points 3 and 4 in Fig. 4, when the survival probability $\mathbf{P}(Z_k > 0)$ of all elements is equal to 0.99865 and the coefficients of correlation ρ_{kl} of safety margin sequence cuts as system elements are equal to 0.5, 0.766, 0.866, 0.94 and 0.985.

From Fig. 4 one can see that the concept of transformed conditional probabilities is acceptable for the prediction of survival probabilities or series systems. The bounded index x_k of their correlation factors by Eq. (19) may be modeled from Eq. (20).

5. PARTIAL SURVIVAL PROBABILITIES

5.1. The Life Cycle of Carcass Erection

The most unfavourable effects resulting from construction loads and their combinations should be considered in the design phase. The coefficients of variation of members' resistances R_1, action effects S_{G_1} and S_{T_1} are: $\delta R_1 = 0.08$-0.20, $\delta S_{G_1} \approx 0.1$ and $\delta S_{T_1} = 0.1$-0.2. The probability distributions of these random variables are close to normal and lognormal ones (ISO 2394 1998, Ellingwood et al. 1980, EN 1990 2002). Therefore, in the context of the reliability analysis of members during carcass erection periods, Eq. (3) may be presented in the form:

$$Z_1 = R_{c1} - S_{e1} \tag{25}$$

where

$$R_{c1} = \theta_R R_1 - \theta_G S_{G_1} - \theta_T S_{T_1} - \theta_Q S_{Q_{11}} \tag{26}$$

is the conventional resistance of members modeled by a normal distribution;

$$S_{e1} = \theta_Q S_{Q_{12}} = \theta_Q \sum S_{Q_{1i}} \tag{27}$$

is the extreme episodic action effect caused due to one extraordinary execution load or the simultaneity of two and more independent particular construction loads.

The mean value of particular action effects is:

$$S_{Q_im} = \frac{S_{Q_1,k}}{(1+k_{0.95}\delta S_{Q_{1i}})} \quad (28)$$

where S_{Q_ik} and δS_{Q_i} are their characteristic value and coefficient of variation; $k_{0.95}$ is the characteristic fractile factor. Therefore, the mean and variance of the joint action effect may be calculated by the formulae:

$$S_{e1m} = \theta_{Qm}\sum S_{Q_im} \quad (29)$$

$$\sigma^2 S_{e1} = \sum \sigma^2(\theta_Q S_{Q_{1i}}) \quad (30)$$

Handling or placement equipment and many other construction installations belong to the deterministic move-in load category (Karshenas & Ayoub 1994). Therefore, the conventional resistance R_{c1} by Eq. (26) and the action effect S_{e1} by Eq. (27) during concreting operations is assumed to be normally distributed. Thus, the partial survival probability of beams and slabs with fitting-up and stay-in-place moulds may be calculated by Eq. (13), because their instantaneous and time dependent survival probabilities are equal, i.e. $\mathbf{P}_1 = \mathbf{P}\{T \geq t_1\} = \mathbf{P}\{R_{c1} > S_{e1k}\}$.

The independent particular erection and fitting loads belong to the deterministic action category. Due to their combination, random concentration and position on building floors or roofs, these loads may be treated as variable. The distribution of the joint short-term action effect S_{e1} caused by these loads may be considered as an exponential distribution (Mirza & MacGregor 1979, JCSS 2000). Thus, the partial survival probability of members during the period t_1, $\mathbf{P}\{T \geq t_1\}$, may also be calculated analytically by Eq. (14):

5.2. The Life Cycle of Mixed Construction Work

The time-dependent safety margin (4) of members during 1-3-year construction period t_2 (Fig. 1) or during the execution procedure with recurrent extraordinary material and equipment loads may be presented in the form:

$$Z_2(t) = R_{c2} - S_{e2}(t) \quad (31)$$

where

$$R_{c2} = \theta_R R_2 - \theta_G(S_{G_1} + S_{G_2}) - [\theta_Q S_{Q21}] \quad (32)$$

is its conventional resistance modeled by a normal distribution;

$$S_{e2}(t) = \left[\theta_Q S_{Q_{21}}(t)\right] + \theta_Q S_{Q_{22}}(t) + \theta_T S_{T_2}(t) + \theta_W S_{W_2}(t) \tag{33}$$

is the joint extreme episodic action effect caused by statistically independent construction (Q_{21}, Q_{22}), ambient temperature (T_2) and wind (W_2) loads. When the variability of sustained loads Q_{21} is small or considerable, these loads may be treated, respectively, as random stationary or nonstationary processes of the components of Eqs. (32) and (33).

The durations (days a year) and renewal rates (times a year) of these loads are: $d_{Q_{21}} = 1\text{-}2$, $d_{Q_{22}} = 20\text{-}50$, $d_{T_2} = 15\text{-}30$, $d_{W_2} = 1$ and $\lambda_{Q_{21}} = 20\text{-}50$, $\lambda_{Q_{22}} = 2\text{-}5$, $\lambda_{T_2} = 1$, $\lambda_{W_2} = 12$. The recurrence number of two and three concurrent extreme action effects during $t_2 = 1\text{-}3$ – year reference period of mixed construction work may be calculated by the formulae:

$$n_{12} = t_2(d_1 + d_2)\lambda_1\lambda_2 \tag{34}$$

$$n_{123} = t_2(d_1 d_2 + d_1 d_3 + d_2 d_3)\lambda_1\lambda_2\lambda_3 \tag{35}$$

Therefore, the recurrence numbers $n_{Q_{21}Q_{22}} = 2.3\text{-}107$, $n_{Q_{22}T_2} = 0.2\text{-}3.3$, $n_{Q_{22}W_2} = 1.4\text{-}25.2$, $n_{T_2W_2} = 0.5\text{-}3.0$, $n_{Q_{22}T_2W_2} = 0.06\text{-}2.1$.

When recurrent action effects may be treated as intermittent rectangular pulse renewal processes, it is expedient to consider the safety margin as the finite random sequence:

$$Z_{2ik} = R_{c2} - S_{2ik}, \quad k = 1, 2, \ldots, n_i, \quad i = 1, 2, \ldots, m \tag{36}$$

where R_{c2} by (32) is a stationary process and

$$S_{2ik} = \left[\theta_Q S_{Q_{21}k}\right] + \theta_Q S_{Q_{22}k} + \theta_T S_{T_2 k} + \theta_W S_{W_2 k} \tag{37}$$

is the value of the joint action effect at rare cut k of the sequence i.

Due to load combinations, five additional finite random sequences of member safety margins during the execution period t_2 should be considered as:

$$Z_{Q_{21}Q_{22}k} = R_{c2} - \theta_Q S_{Q_{21}k} - \theta_Q S_{Q_{22}k} \tag{38}$$

$$Z_{Q_{22}T_2,k} = R_{c2} - S_{Q_{22}T_2,k} = R_{c2} - \theta_Q S_{Q_{22}k} - \theta_T S_{T_2,k} \tag{39}$$

$$Z_{Q_{22}W_2,k} = R_{c2} - S_{Q_{22}W_2,k} = R_{c2} - \theta_Q S_{Q_{22}k} - \theta_W S_{W_2,k} \tag{40}$$

$$Z_{T_2W_2,k} = R_{c2} - S_{T_2W_2,k} = R_{c2} - \theta_T S_{T_2 k} - \theta_W S_{W_2 k} \tag{41}$$

$$Z_{Q_{22}T_2W_2,k} = R_{c2} - S_{Q_{22}T_2W_2,k} = R_{c2} - \theta_Q S_{Q_{22}k} - \theta_T S_{T_2k} - \theta_W S_{W_2k} \qquad (42)$$

According to Rosowsky and Ellingwood (1992), the annual extreme sum of sustained and extraordinary live loads, $Q_{21}(t) + Q_{22}(t)$, may be modeled as a renewal process described by a Type 1 (Gumbel) distribution of extreme values. Usually, the construction load Q_{21} assumed to be normally distributed and having only an effect on a recurrence number of joint extreme load $Q_{21} + Q_{22}$ calculated by Eq. (34). In this case, Eq. (38) is used in the form:

$$Z_{Q_{22}k} = R_{c2} - \theta_Q S_{Q_{22}k} \qquad (38a)$$

The selection of probabilistic load combinations, including durations and renewal rates of variable construction loads, is based on the features of building types and their construction procedures.

When resistances and action effects may be treated as independent, the instantaneous survival probability of particular members at any cut k of random sequences of safety margins is:

$$\mathbf{P}_{2ik} = \mathbf{P}_{2i}(t_{2k}) = \mathbf{P}\{Z_{2i}(t_{2k}) > 0 \ \exists \ t_{2k} \in [t_{21}, t_{2n}]\} = \mathbf{P}\{R_{c2} > S_{2ik}\} = \int_0^\infty f_{R_{c2}}(x) F_{S_{2ik}}(x) \, dx \qquad (43)$$

where $f_{R_{c2}}(x)$ and $F_{S_{2ik}}(x)$ are the density and cumulative distribution functions of conventional resistance R_{c2} by Eq. (32) and joint action effect S_{2ik} by Eq. (37). This extreme action effect caused by concurrent construction, temperature and wind actions may be assumed to be exponentially distributed. Therefore, its mean and standard deviation may be expressed as equal values, i.e. $S_{2ikm} = \sigma S_{2ik}$. The instantaneous survival probability of members is calculated by Eq. (14).

Usually, the auto correlation coefficient of the execution loads $\rho_{i,kl} = 0$. Therefore, the coefficient of correlation of random sequence cuts is:

$$\rho_{i,kl} = \rho(Z_{2ik}, Z_{2il}) = \text{Cov}(Z_{2ik}, Z_{2il})/(\sigma Z_{2ik} \times \sigma Z_{2il}) = 1/\left(1 + \sigma^2 S_{2ik}/\sigma^2 R_{c2}\right) \qquad (44)$$

where $\text{Cov}(Z_{2ik}, Z_{2il})$ and σZ_{2ik}, σZ_{2il} are auto covariance and standard deviations of the random functions Z_{2ik} and Z_{2il} (Kudzys 1992). Thus, according to the method of transformed conditional probabilities (TCPM) presented in Section 4.2, the partial time-dependent survival probability of members under construction period t_2 at any of extreme load combinations is:

$$\mathbf{P}_{2i} = \mathbf{P}_{2i}\{T_2 \geq t_2\} = \mathbf{P}\left\{\bigcap_{k=1}^{n} Z_{2ik} > 0 \;\exists Z_{2ik} \in [Z_{2i1}, Z_{2in}]\right\} = \mathbf{P}_{2ik}^n \left[1 + \rho_{i,kl}^{x_{ik}}\left(\frac{1}{\mathbf{P}_{2ik}} - 1\right)\right]^{n-1} \quad (45)$$

Here n is the recurrence number of the joint extreme action effect by Eq. (34) or (35); $\mathbf{P}_{2ik}$ by Eq. (43) and x_{ik} by Eq. (20) are the instantaneous survival probability of particular members and the bounded index of correlation factor at cut k of random safety margin sequence.

A strategy for the prediction of the structural safety of members and their systems requires a solution based on the application of integrated survival probabilities. Therefore, according to TCPM approaches, the rank series system of stochastically dependent elements with survival probabilities $\mathbf{P}_{21} > \mathbf{P}_{22} > ... > \mathbf{P}_{2m}$ characterizing its safety at relevant load combinations should be considered (Kudzys 2006).

The correlation factor of rank cuts of a safety margin sequence is:

$$\rho_{i|1...i-1} = (\rho_{i1} + \rho_{i2} + ... + \rho_{i,i-1})/(i-1) \quad (46)$$

Thus, according to TCPM, the survival probability of members at all load combinations during the period t_2 may be calculated by the unsophisticated and fairly exact formula:

$$\mathbf{P}_2 = \mathbf{P}\{T \geq t_2\} = \mathbf{P}\left\{\bigcap_{i=1}^{m} Z_{2i} > 0\right\} = \prod_{i=1}^{m} \mathbf{P}_{2i}\left[1 + \rho_{m|1...m-1}^{x_m}\left(\frac{1}{\mathbf{P}_{m-1}} - 1\right)\right] \times ...$$

$$... \times \left[1 + \rho_{i|1...i-1}^{x_i}\left(\frac{1}{\mathbf{P}_{i-1}} - 1\right)\right] \times ... \times \left[1 + \rho_{21}^{x_2}\left(\frac{1}{\mathbf{P}_1} - 1\right)\right] \quad (47)$$

When the elements are very highly correlated ($\rho \approx 1$) or statistically independent ($\rho \approx 0$), the survival probabilities of members from Eq. (47) are respectively equal to $\mathbf{P}_2 = \mathbf{P}_m$ and $\mathbf{P}_2 = \prod_{i=1}^{m} \mathbf{P}_i$. These results fully comply main requirements of the theory of reliability.

5.3. Reliability Indices of Members

The safety of the structure and people depend on the smallest survival probabilities of particular members exposed to recurrent extreme action effects (Kudzys 2006). The survival probability of these members during the full construction life cycle t_e (Fig. 1) depends on two safety margins as: $Z_1 = R_{c1} - S_{e1}$ and $Z_2 = R_{c2} - S_{e2}$. Here the values of conventional resistances R_{c1} and R_{c2} are given by Eqs. (26) and (32). The extreme action effects S_{e1} and S_{e2} are presented by Eqs. (27) and (37). This total (integrated) survival probability may be introduced as follows:

$$\mathbf{P}_e = \mathbf{P}\{T_e > t_e\} = \mathbf{P}\{Z_1 > 0 \cap Z_2(t) > 0 | Z_1 > 0\} = \mathbf{P}_1 \times \mathbf{P}_2 \left[1 + \rho_{12}^{x_2}(1/\mathbf{P}_{\max} - 1)\right] \quad (48)$$

Here $\mathbf{P}_{\max}$ is the greater value of survival probabilities $\mathbf{P}_1$ by (13) or (14) and $\mathbf{P}_2$ by (45) or (47);

$$\rho_{12} = \sigma^2 R_{c1} / (\sigma Z_1 \times \sigma Z_2) \approx \left[\sigma^2(\theta_R R_1) + \sigma^2(\theta_G S_G)\right] / (\sigma Z_1 \times \sigma Z_2) \quad (49)$$

is the coefficient of correlation of two random safety margins of members the standard deviations of whose are:

$$\sigma Z_1 = \left(\sigma^2 R_{c1} + \sigma^2 S_{e1}\right)^{1/2} \quad (50)$$

$$\sigma Z_2 = \left(\sigma^2 R_{c2} + \sigma^2 S_{e2}\right)^{1/2} \quad (51)$$

For different structures, usually different forms of engineering control exist and they are associated with costs. Therefore, the selection of engineering inspection means is based on the individual approaches. However, in any case this selection is closely related to the reliability indices of members , β, characterizing their structural safety degrees during construction operations.

The reliability indices of members as the inverse Gaussian distributions of their partial $\mathbf{P}_1$, $\mathbf{P}_2$ and total $\mathbf{P}_e$ survival probabilities are defined as:

$$\beta_1 = \Phi^{-1}(\mathbf{P}_1), \quad \beta_2 = \Phi^{-1}(\mathbf{P}_2) \text{ and } \beta_e = \Phi^{-1}(\mathbf{P}_e) \quad (52)$$

Together with the strength, rigidity and stiffness, the reliability indices of structural members may be treated as their quantitative quality parameters. Their target values for members exposed to execution loads are under discussion. The determinations of these values depend on many variables including the expected cost of maintenance and failure. The characteristic survival probabilities 0.999, 0.995 and 0.95 correspond to the reliability indices equal to 3.10, 2.58 and 1.65. When these indices are less than 3.10 and 2.58 or 1.65, it may be expedient, respectively, to use a usual (normal) and tight continuous engineering control of construction procedures or to change these procedures intirely.

6. NUMERICAL EXAMPLE

6.1. Initial Data

Consider the limit state design verifications and the partial and total survival probabilities of precast concrete floor slabs of multistorey industrial buildings with masonry walls during construction. In this period of time, the multihollowcore slabs may be overloaded by brick containers during a wall erection period t_1 (random load Q_{12}) or by an equipment of mosaic floor laying (sustained load Q_{21}) and building materials and installation (extraordinary load Q_{22}) as variable loads during a floor erection period t_2 (Fig. 5).

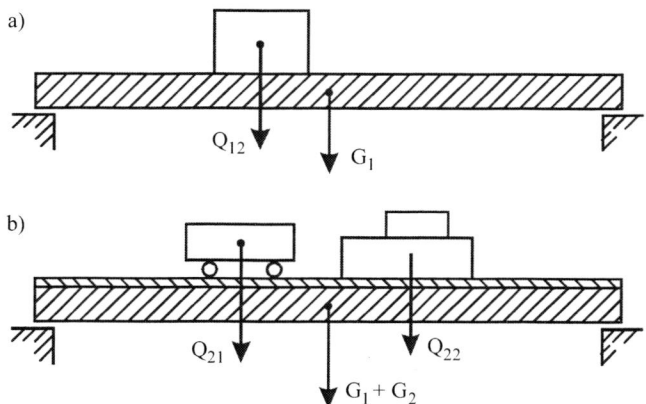

Figure 5. The building slabs subjected to their weight G_1, permanent load G_2, equipment weight Q_{21} and extraordinary loads as brick containers Q_{12} and building materials Q_{22} during stone-work (a) and floor placement (b)

The model uncertainty parameters, means and coefficients of variation of slab bending resistance and bending moments are:

$\theta_{Rm} = 1.01$, $\sigma\theta_R = 0.08$; $\theta_{Gm} = \theta_{Qm} = 1.0$; $\sigma\theta_G = \sigma\theta_Q = 0.10$; $R_{1m} = R_{2m} = R_m = 300$ kNm, $\theta_{Rm}R_m = 1.01 \times 300 = 303$ kNm;

$M_{G_1m} = \theta_{Gm}M_{G_1m} = 90$ kNm, $M_{G_2m} = \theta_{Gm}M_{G_2m} = 14$ kNm, $M_{Q_{12}m} = \theta_{Qm}M_{Q_{12}m} = 34$ kNm, $M_{Q_{21}m} = \theta_{Qm}M_{Q_{21}m} = 29$ kNm, $M_{Q_{22}m} = \theta_{Qm}M_{Q_{22}m} = 24$ kNm;

$\delta R = \delta M_{G_1} = \delta M_{G_2} = 0.1$, $\delta M_{Q_{21}} = 0.2$, $\delta M_{Q_{12}} = \delta M_{Q_{22}} = 1.0$.

Thus, the variances of bending resistance and moments of considered slabs are:

$\sigma^2(\theta_R R) = 1.01^2(0.10 \times 300)^2 + 300^2 \times 0.08^2 = 1494.1$ (kNm)^2,

$\sigma^2(\theta_G M_{G_1}) = 1.0^2(0.10 \times 90)^2 + 90^2 \times 0.10^2 = 162.0$ (kNm)^2,

$\sigma^2(\theta_G M_{G_2}) = 1.0^2(0.10 \times 14)^2 + 14^2 \times 0.10^2 = 3.9$ (kNm)^2,

$\sigma^2 M_1 = \sigma^2(\theta_Q M_{Q_{12}}) = 1.0^2 \times 34^2 + 34^2 \times 0.10^2 = 1167.6$ (kNm)^2,

$$\sigma^2(\theta_Q M_{Q_{21}}) = 1.0^2(0.20 \times 29)^2 + 29^2 \times 0.10^2 = 42.0 \text{ (kNm)}^2,$$

$$\sigma^2 M_2 = \sigma^2(\theta_Q M_{Q_{22}}) = 1.0^2 \times 24^2 + 24^2 \times 0.10^2 = 581.8 \text{ (kNm)}^2.$$

The durations (days a year) and renewal rates (times a year) of variable sustained and extraordinary construction loads during the floor erection period $t_2 = 0.25$ years are: $d_{Q_{21}} = 2$, $d_{Q_{22}} = 60$, $\lambda_{Q_{21}} = 52$, $\lambda_{Q_{22}} = 3$. Therefore, according to Eq. 34, the recurrence number of extreme situations on building site is equal to $n_{Q_{21}Q_{22}} = 6.625$.

6.2. Limit State Design Format

According to the recommendations of strength design method presented in **SEI/ASCE 37**, the precast floor slabs of buildings with masonry walls shall be designed so that their resisting bending moments exceed the action effects caused by factored construction loads in their following combinations:

$$U_{SEI,1} = 1.4G_{1k} + 1.2Q_{11k} + 1.4Q_{12k} \tag{53}$$

$$U_{SEI,2} = 1.2G_{1k} + 1.2G_{2k} + 1.6Q_{21k} + 1.4Q_{22k} \tag{54}$$

Here $G_{1k} = D_{1n} + C_{Dn}$ is a combination of characteristic (nominal) values of main structural (D_{1n}) and construction (C_{Dn}) dead loads; $G_{2k} = D_{2n}$ is an additional characteristic dead load; $Q_{11k} = C_{FML,n}$ is a characteristic value of fixed material loads and erection forces (it does not exist); $Q_{12k} = C_{VML,n}$ and $Q_{22k} = C_{VML,n}$ are characteristic values of variable material loads and erection forces under periods t_1 and t_2; $Q_{21k} = C_{Pn}$ is a characteristic equipment and personal load. Thus, the load combinations (53) and (54) correspond to the loading situations of considered slabs under carcass erection and mixed construction work, respectively.

The characteristic value Y_k of resisting and destroying bending moments are calculated by the formula:

$$Y_k = Y_m(1 \mp k_{0.95}\delta Y) \tag{55}$$

where Y_m is their mean value; $k_{0.95} = 1.65$ or $k_{0.95} = 1.3$ is the 0.95 – fractile (quantile) when the standard distribution of variables is modeled on normal or exponential laws, respectively; δY is a coefficient of variations of variables.

The probability distribution loads G_1, Q_{11}, G_2, Q_{21} and Q_{12}, Q_{22} obey the laws close to normal and exponential ones, respectively. Therefore, the characteristic values of bending resistance and moments are:

$$R_k = 300(1 - 1.65 \times 0.10) = 270 \text{ kNm},$$

$$M_{G_1 k} = 90(1 + 1.65 \times 0.10) = 104.8 \text{ kNm},$$

$$M_{G_2 k} = 14(1 + 1.65 \times 0.10) = 16.3 \text{ kNm},$$

$$M_{e1,k} = M_{Q_{12}k} = 34(1 + 1.3 \times 1.0) = 78.2 \text{ kNm},$$

$$M_{Q_{21}k} = 29(1 + 1.65 \times 0.20) = 38.6 \text{ kNm},$$

$$M_{e2,k} = M_{Q_{22}k} = 24(1 + 1.3 \times 1.0) = 55.2 \text{ kNm}.$$

The structural strength reduction factor of slabs $\Phi = 0.90$. Therefore, their design bending resistance is equal to $R_d = \Phi R_k = 0.90 \times 270 = 243$ kNm. According to the load combinations (53) and (54), the verification of limit states of slabs in design construction situations is carried out by these design conditions:

$$M_{1d} = 1.4 \times 104.8 + 1.4 \times 78.2 = 256.3 \text{ kNm} > 243 \text{ kNm};$$

$$M_{2d} = 1.2 \times 104.8 + 1.2 \times 16.3 + 1.6 \times 38.6 + 1.4 \times 55.2 = 284.4 \text{ kNm} > 243 \text{ kNm}.$$

The verification data show that the sufficient structural safety and integrity of precast slabs during wall and floor construction procedures may be not guaranted. Therefore, the suitable stopgap measures, including temporary bracing of considered slabs should be used. Their safety and serviceability may be achieved also by other measures including normal and tight controls of construction load processes and slab deflections during execution stages.

According to **ENV 1991-2-6** recommendations, the load combinations (53) and (54) may be presented in the form:

$$U_{EN1} = 1.35 G_{1k} + 1.35 Q_{11k} + 1.35 Q_{12k} \tag{56}$$

$$U_{EN2} = 1.35(G_{1k} + G_{2k}) + 1.35 Q_{dom,k} + 0.5 \times 1.35 Q_{ik} \tag{57}$$

Here $Q_{dom,k}$ and Q_{ik} are the characteristic values of dominant and accompanying variable loads.

The partial safety factor for yield strength of reinforcing bars of slabs $\gamma_s = 1.15$. Therefore, the design bending resistance of slabs is equal to

$$R_d = A_s z_s f_{yk}/\gamma_s \approx R_k/\gamma_s = 270/1.15 = 234.8 \text{ kNm}.$$

The verification of limit states of slabs leads to design expressions as:

$$M_{1d} = 1.35 \times 104.8 + 1.35 \times 78.2 = 247.1 \text{ kNm} > 234.8 \text{ kNm};$$

$$M_{2d} = 1.35 \times 104.8 + 1.35 \times 16.3 + 1.35 \times 55.2 + 0.5 \times 1.35 \times 38.6 = 290.2 \text{ kNm} > 234.8 \text{ kNm}.$$

Thus, these analysis results also show that a structural safety of slabs under construction may be infringed not only insignificantly ($M_{1d} > R_d$) but also significantly ($M_{2d} >> R_d$). Therefore, it would be very important to compare the limit state analysis data to the results calculated by probability-based approaches. The sufficient quality of floor slabs may be achieved by normal or tight continuous control of construction procedures and assessed by their quantitative reliability indices.

6.3. Probabilistic Design Format

The means and variances of conventional resistances (resisting moments) of considered slabs expressed by Eqs. (26) and (32) are:

$$R_{c1m} = \theta_{Rm} R_m - \theta_{Gm} M_{G_1 m} = 300 - 90 = 213 \text{ kNm},$$

$$\sigma^2(R_{c1}) = \sigma^2(\theta_R R) + \sigma^2(\theta_G M_{G_1}) = 1494 + 162 = 1656 \text{ (kNm)}^2;$$

$$R_{c2m} = \theta_{Rm} R_m - \theta_{Gm} M_{G_1 m} - \theta_{Gm} M_{G_2 m} - \theta_{Qm} M_{Q_{21} m} = 303 - 90 - 14 - 29 = 170 \text{ kNm},$$

$$\sigma^2(R_{c2}) = \sigma^2(\theta_R R) + \sigma^2(\theta_G M_{G_1}) + \sigma^2(\theta_G M_{G_2}) + \sigma^2(\theta_Q M_{Q_{21}}) = 1494 + 162 + 3.9 + 42 = 1702 \text{ (kNm)}^2.$$

According to Eqs. (14) and (52), the partial survival probability and reliability index of slabs during masonry wall execution processes are:

$$\mathbf{P}_1 = \mathbf{P}_k = 1 - \Phi\left(-\frac{213}{\sqrt{1656}}\right) - \exp\left(-\frac{213}{34} + 0.5\frac{1656}{1167.6}\right) \times \left[1 - \Phi\left(-\frac{213}{\sqrt{1656}} + \frac{\sqrt{1656}}{\sqrt{1167.6}}\right)\right] =$$

$$= 1 - 0.9^711 - 0.003866 \times 0.999973 = 1 - 0.00387 = 0.99613;$$

$$\beta_1 = \Phi^{-1}(0.99613) = 2.66 < 3.10.$$

According to Eq. (44), the coefficient of correlation of cuts of the safety margin sequence $Z_{2k} = R_{c2} - M_{2k}$ during floor erection period is:

$$\rho_{kl} = 1/\left[1 + \sigma^2(\theta_Q M_{Q_{22}})/\sigma^2(\theta_R R_{c2})\right] = 1/(1 + 581.8/1656) = 0.74.$$

Therefore, its bounded index by Eq. (20) is:

$$x_k = [8.5/(1 - 0.98 \times 0.74)] = 5.5616.$$

Thus, the index correlation factor is defined as:

$$\rho_{kl}^{x_k} = 0.74^{5.5616} = 0.1874.$$

According to Eq. (14), the instantaneous survival probability of slabs during mosaic floor laying in building rooms is:

$$\mathbf{P}_{2k} = 1 - \Phi\left(-\frac{170}{\sqrt{1702}}\right) - \exp\left(-\frac{170}{24} + 0.5\frac{1702}{581.8}\right) \times \left[1 - \Phi\left(-\frac{170}{\sqrt{1702}} + \frac{\sqrt{1702}}{\sqrt{581.8}}\right)\right] =$$
$$= 1 - 0.000019 - 0.003623 \times 0.992 = 0.99639.$$

Since the recurrence number of extreme situations $n = n_{Q_{12}Q_{22}} = 6.625$, the survival probability of slabs by Eq. (45) during this period is:

$$P_2 = 0.99639^{6.625}\left[1 + 0.1874(1/0.99639 - 1)\right]^{5.625} = 0.98006.$$

It corresponds to the reliability index $\beta_2 = \Phi^{-1}(0.98006) = 2.05 < 2.58$.

According to Eqs. (50), (51) and (49), the standard deviations and the coefficient of correlation of the safety margins $Z_1 = R_{c1} - S_{e1}$ and $Z_2 = R_{c2} - S_{e2}$ are equal to:

$$\sigma Z_1 = (1656 + 1167.6)^{1/2} = 53.17 \text{ kNm,}$$

$$\sigma Z_2 = (1702 + 581.8)^{1/2} = 47.79 \text{ kNm,}$$

$$\rho_{12} = 1656/(53.14 + 47.79) = 0.652.$$

The bounded index of the correlation factor $\rho_{12}^{x_2}$ by Eq. (20) is:

$$x_2 = \{8.5/(1-0.98 \times 0.652)\}^{1/2} = 4.852.$$

Making use of Eq. (48), the total survival probability of floor slabs under construction is:

$$\mathbf{P}_e = 0.99613 \times 0.98006\left[1 + 0.652^{4.852}(1/0.99613 - 1)\right] = 0.97674.$$

It corresponds to the reliability index $\beta_{12} = \Phi^{-1}(0.97674) = 1.99 < 2.58$. However, 1.99 > 1.65 when the probability of safe construction technology is equal only to 95 %.

Table 1. Reliability indices and indispensable engineering inspection of the slabs

Construction period	Reliability indices	Engineering inspection
t_1 – erection of walls	3.10 > **2.66** > 2.58	normal continuous control
t_2 – erection of floors	2.58 > **2.05** > 1.65	tight —"— —"—
$t_e = t_1 + t_2$	2.58 > **1.99** > 1.65	tight —"— —"—

The results presented in Table 1 show that the selected construction technology may be used but its procedures should be controlled by tight continuous engineering inspection means. The quantitative reliability index helped us assess the safety of construction work fairly objectively.

5. CONCLUSIONS

The execution loads and environmental actions, including their combinations during construction of buildings, are almost in a similar way treated by Standards SEI/ASCE 37 and ENV 1991-2-6 used in the USA and Europe, respectively. Therefore, the relevant semi-probabilistic method of load and resistance factor design (LRFD) and partial safety factor design (PSFD) leads to the same results on the limit state verification of load-carrying structures subjected to random extreme action effects caused by single and recurrent construction loads. However, instead of assessing the residual safety degree and helping to choose stopgap measures for the engineering control of construction loads and behaviors of overloaded executed members in transient situations, the methodology of these methods is dedicated to the analysis of persistent design situations. Therefore, when the action effects of structural members in transient situations under construction are greater than those caused by service loads in persistent situations, the material presented in Sections 2 and 6.2 unwillingly orientate designers and building engineers to consider construction loads in structural design processes as basic variable actions.

It is not complicated to predict the reliability indices of structural members during construction of buildings and the same to ground their engineering (site) control means by probability-based approaches presented in Sections 4 and 6.3. The probability-based assessment of execution (construction) loads and their effects on the structural quality and safety of buildings carried out in a simple and easily perceptible manner should be acknowledged as one of the main tasks facing modern building engineers.

It is expedient to separate in design practice two different in nature execution cycles of buildings defined by structural and technological features of their carcass erection and mixed site work, respectively. Gaussian and lognormal distribution laws may be convenient for the probability distributions of member resistances, permanent, sustained and concreting live variable loads. An exponential distribution is characteristic for loads existing in precast carcass erection period. Annual ambient temperature changes obey a Gaussian distribution law. The episodic extreme action effects caused by concurrent execution, temperature and wind actions may be treated as intermittent rectangular pulse renewal processes and may be assumed to be exponentially distributed. For the sake of simplification of safety prediction of members, it is recommended to base their probabilistic analysis on the concepts of conventional resistances, stochastic sequences and transformed conditional probabilities. Parallel with code deterministic and semi-probabilistic methods, the presented unsophisticated probability-based approaches may stimulate engineers to use probabilistic models in construction practice.

The partial survival probabilities of structural members during carcass erection and mixed construction work of buildings may be calculated by Eqs. (13) and (14) or (45), (47) when their extreme action effects are treated as random single or concurrent variable functions. The total survival probability of these members may be introduced by Eq. (48). These probabilities help us define the reliability indices of load-carrying members, and select objectively relevant execution procedures, construction rates and proper engineering (site) inspection means.

The target values of reliability indices by Eq. (52) of load-carrying members are under discussion among structural engineers. However, it is expedient to use a usual and tight continuous engineering inspection of construction work or to change technology procedures intirely when these indices are less than 3.10 and 2.58 or 1.65, respectively. These values correspond to the survival probabilities equal to 0.999, 0.995, 0.95.

REFERENCES

Ahammed, M. & Melchers, R. E. (2005). New bounds for highly correlated structural series systems: In: *ICOSSAR*, G. Augusti, GJ Schuëller, M Campoli (eds), Millpress, Roterdam: 3656-62.

ASCE/SEI 7-05 (2005). Minimum design loads for buildings and other structures. ASCE/SEI, Reston, USA: 253.

Ellingwood, B. R. (1987). Design and construction error on structural reliability. *Journal of Structural Engineering, 113(2)*: 409-27.

Ellingwood, B. , Galambos, T. V., Mac Gregor, J. G. & Cornell, C. A. (1980). Development of a probability based load criterion for American national standard A58. *NBS special publication 557*.

Ellingwood, B. R. & Tekie, P. B. (1999). Wind load statistics for probability-based structural design. *Journal of Structural Engineering, 125(4)*: 453-63.

EN 1990 (2002). European Committee for Standardization (CEN), Eurocode-Basis of structural design, Brussels: 187.

ENV 1991-2-6 (1997). Eurocode 1: Basis of design and actions on structures – Part 2-6: Actions on structures - Actions during execution. CEN, Brussels: 32 .

Epaarachchi, D. C., Stewart, M. G. & Rosowsky, D. V. (2002). Structural reliability of multistory buildings during construction. *Journal of Structural Engineering, 128(2)*: 205-13.

Hall, W. B. (1988). Reliability of service-proven structures. *Journal of Structural Engineering, 114(3)*: 608-24.

Hong, H. P. & Lind, N. C. (1996). Approximate reliability analysis using normal polynomial and simulation results. *Structural Safety, 18(4):* 329-39.

ISO 2394 (1998). General principles on reliability for structures. Switzerland: 73.

JCSS (2000). Probabilistic model code: Part 1- Basis of design, 12th draft: 62.

Karshenas, S. & Ayoub, H. (1994). Analysis of concrete construction live loads on newly pored slabs. *Journal of Structural Engineering, 120(5):* 1525-42.

Kudzys, A. (1992). Probability estimation of reliability and durability of reinforced concrete structures. *Technika, Vilnius: 143*.

Kudzys, A. (2006). Safety of power transmission line structures under wind and ice storms. *Engineering structures*, 28: 682-89.

Melchers, R. E. (1999). Structural reliability analysis and prediction. *Chichester*, John Wiley & Sons: 437.

Mirza, S. A. & MacGregor, J. G. (1979). Variation in dimensions of reinforced concrete members. *Journal of Structural Engineering*, 105: 751-66.

Mohamed, S. (2002). Safety climate in construction site environments. *Journal of Construction Engineering and Management*, 128(5): 357-84.

Quek, S-T & Cheong, H-F. (1992). Prediction of extreme 3-sec. gusts accounting for seasonal effects. *Structural Safety, Vol. 11, No. 2*: 121-29.

Raizer, V. P. (1998). Theory of Reliability in Structural Design, ACB Publishing House, *Moscow: 302* (in Russian).

Rosowsky, D. & Ellingwood, B. (1992). Reliability of wood systems subjected to stochastic live loads. *Wood and Fiber Science, 24 (1):* 47-59.

Saffarini, H. S. (2000). Overview of uncertainty in structural analysis assumptions for RC buildings. *Journal of Structural Engineering, 126(9):* 1078-85.

Saraf, V. K., Nowak, A. S. & Till, R. (1996), Proof load testing of bridges. *Probabilistic Mechanics and Structural Reliability Proceedings of Seventh Specialty Conference, ASCE*, New York: 526-29.

SEI/ASCE 37-02 (2002). Design loads on structures during construction. ACSE, Reston, USA: 36.

Stewart, M. G. (1992). A human reliability analysis of reinforced concrete beam construction. *Civil Eng. System, 9(3):* 227-47.

Stewart, M. G. (1996). Proof loads, construction area and the reliability of sevice proven structures. *Probabilistic Mechanics and Structural Reliability, ASCE*, Eds. DM Frangopol and MD Grigoriu: 226-29.

Stewart, M. G. & Rosowsky, D. V. (1996). System risk for multi-storey reinforced concrete building construction. *Probabilistic Mechanics and Structural Reliability Proceedings of the Seventh Spaciality Conference. ASCE*, New York: 230-33.

Stewart, M. G. (2001). Effect of construction and service loads on reliability of existing RC structures. *Journal of Structural Engineering, ASCE, Vol. 127 (10):* 1232-36.

Vrouwenvelder, A. C. V. M. (2002). *Developments towards full probabilistic design codes. Structural Safety, Vol. 24 (2-4):* 417-32.

Yates, J. K. & Lockley, E. E. (2002). Documating and analyzing construction failures. *Journal of Construction Engineering and Management, 128(1)*: 8-17.

Reviewed by Prof. M. Holicky, Head of the Department of Structural Reliability, Czech Technical University in Prague, Czech Republic, e-mail: holicky@klok.cvut.cz

INDEX

A

absorption, 49, 50, 51, 55, 56, 57, 58, 59, 60, 61, 62, 74, 75, 76, 79, 80, 81, 82, 83, 84, 85, 86, 89, 90
academic, 51
acceptor, 97
accidents, 329
accounting, 143, 195, 295, 321, 354
accreditation, 235
accuracy, 62, 140, 210, 221, 248, 289, 295, 332
ACI, 228, 237, 241, 272, 273
acid, 3, 143
acoustic, 55
acrylate, 35, 38
acrylic acid, 3
acrylonitrile, 35
activation, 95
adhesion, vii, viii, ix, 1, 3, 7, 9, 13, 22, 24, 31, 33, 41, 135, 136, 137, 143, 144, 146, 147, 148, 149, 151, 152, 153, 154, 155, 156
adjustment, 113, 123
adsorption, 100, 103, 128
aerospace, 294
Ag, 95
age, 46, 55, 56, 59, 86, 331
agent, 8, 9, 35, 38
agents, 3, 7, 8, 54
aggregates, vii, viii, 43, 44, 47, 48, 49, 50, 51, 52, 55, 56, 57, 58, 59, 60, 61, 62, 63, 64, 65, 66, 67, 68, 69, 70, 71, 72, 73, 74, 75, 76, 77, 78, 79, 80, 81, 82, 83, 84, 85, 86, 87, 88, 89, 90
aging, 35, 37, 39, 118
air, 2, 53, 59, 153
airports, 273
algorithm, 282
alloys, viii, 91, 96, 97, 104, 105, 109, 115, 117, 118, 122, 123, 124, 129, 130

alternative, vii, 43, 49, 166, 200, 206, 235
alters, 21
amendments, 235
amide, 8, 14
amine, 8, 9, 14, 33
amplitude, xi, 293, 305, 306
analytical models, x, 279, 280
analytical techniques, 96, 103
antimony, 95
appendix, x, 209, 244
application, 52, 55, 83, 84, 86, 88, 90, 160, 171, 192, 201, 209, 227, 247, 254, 283, 285, 286, 331, 345
ARC, 88
Argentina, 293
argument, 172, 174
arrest, 165
ash, 59, 61, 88
Asia, 157
aspect ratio, ix, 31, 135, 154, 155, 156
asphalt, 46
assessment, 88, 93, 205, 206, 208, 211, 222, 280, 281, 285, 286, 289, 291, 329, 352
assumptions, 96, 98, 139, 144, 283, 297, 301, 333, 354
ASTM, 33, 41, 149, 160, 168, 174, 175
ATC, 178, 194, 195, 196, 197, 201, 202, 212, 217, 237, 239, 290
Athens, 243
atmosphere, 2
attacks, 94
austenitic stainless steels, viii, 91
Australia, 194, 241, 246, 272
Austria, 45, 48
availability, 121
averaging, 147

avoidance, 186

B

barrier, viii, 43, 95, 102, 124, 130
barriers, 52, 53, 54
beaches, 84
beams, x, 138, 144, 148, 153, 228, 231, 233, 236, 281, 284, 293, 294, 295, 296, 297, 298, 299, 305, 306, 307, 308, 310, 311, 321, 324, 325, 333, 342
behavior, vii, 1, 2, 3, 4, 10, 21, 41, 163, 174, 186, 187, 188, 192, 193, 194, 195, 196, 201, 203, 207, 211, 212, 213, 235, 280, 281, 282, 283, 284, 285, 286, 288, 289, 290, 294, 295, 296, 305, 307, 311, 315, 317, 320, 321, 325
Belgium, 45, 46, 48
bending, viii, ix, x, 135, 136, 138, 139, 140, 141, 149, 154, 155, 156, 157, 160, 163, 167, 168, 217, 279, 281, 283, 284, 285, 287, 288, 293, 294, 295, 303, 305, 306, 308, 313, 315, 321, 329, 340, 347, 348, 349, 350
biaxial, x, 217, 279, 281, 283, 284, 285
bleeding, 59
blends, 10, 11, 12, 18, 30, 31, 33
bonding, 3, 7, 24, 33, 143, 152
bonds, 8, 165
boundary conditions, viii, x, 91, 97, 98, 101, 102, 103, 104, 201, 293, 305
bounds, 172, 353
Brazil, 46, 86, 88
Brazilian, 54
breakdown, 162
browser, 232
Brussels, 86, 89, 354
buildings, ix, x, xi, 44, 46, 49, 52, 177, 178, 179, 183, 184, 185, 189, 190, 192, 195, 198, 199, 201, 202, 203, 204, 205, 206, 208, 213, 214, 235, 238, 239, 245, 279, 280, 281, 286, 288, 289, 290, 291, 327, 329, 330, 331, 332, 333, 334, 335, 346, 348, 352, 353, 354
Bulgaria, 91
burning, 3
butadiene, 9, 35
butadiene-styrene, 9

C

cache, 89
CAD, 227
calculus, 311

calibration, ix, 159, 160, 163, 164, 170, 171, 172, 174, 280
campaigns, viii, 43, 56, 57, 59, 60, 61, 70, 80, 83, 85
Canada, 194, 238, 239, 240
capacitance, 108
capillary, 74, 75
carbide, 159, 172
carbides, 159, 161, 172
carbon, vii, 1, 2, 6, 26, 27, 28, 29, 41, 140, 157
carbonization, 2
carboxylic, 35
carrier, 120
case study, 271
cast, 334
cation, 97, 98, 99, 100, 103, 104, 107
CBS, 157
CEE, 242
cell, 149
cement, 50, 52, 54, 58, 59, 84
CEO, 236
ceramic, 49, 50, 51, 52, 53, 59, 60, 61, 63, 87, 89
ceramics, 49, 51, 64
certificate, 235
chemical interaction, 93, 94
chemical properties, 88
chloride, 55, 59, 60, 61, 62, 79, 80, 83, 85
Chloride, 79, 82
chromium, 95, 117, 129
civil engineering, 295, 329
cladding, 95
classes, 55, 61
classical, 137, 139, 296, 298, 299
classification, 49, 51, 54, 57, 80, 205
cleavage, ix, 159, 160, 161, 163, 164, 167, 171, 172, 173, 174
clients, 191, 225
Co, viii, 91, 92, 95, 104, 109, 110, 111, 113, 116, 121, 124, 125, 126
coastal areas, 79
cobalt, 95, 125
codes, 190, 194, 195, 202, 203, 205, 209, 329, 330, 355
coefficient of variation, 335, 342, 348
collaboration, 190, 191, 225, 226, 228, 236, 280
combined effect, 170, 172
commodity, 2, 22
communication, 35, 179, 191, 225, 236, 293
communities, 178
community, ix, 50, 177, 186
compatibility, 13, 191, 289
competition, 27, 213
compilation, 113, 128, 129

complexity, 191, 210, 212, 224, 225, 236, 288
complications, 140, 294
components, 2, 4, 16, 22, 33, 53, 145, 148, 160, 163, 187, 193, 197, 198, 202, 228, 233, 236, 300, 333, 335, 339, 341, 343
composites, vii, viii, 1, 2, 3, 4, 6, 7, 10, 13, 14, 15, 16, 20, 21, 22, 24, 25, 26, 27, 28, 29, 30, 31, 32, 33, 35, 37, 38, 41, 135, 137, 138, 140, 141, 143, 147, 149, 150, 151, 153, 154, 155, 156, 157
composition, viii, 30, 46, 52, 53, 54, 55, 56, 59, 84, 91, 93, 97, 102, 103, 104, 117, 125, 129
compounds, 156
compressive strength, 51, 52, 55, 56, 57, 58, 59, 60, 61, 62, 63, 64, 65, 66, 67, 68, 69, 70, 71, 72, 73, 74, 75, 76, 77, 78, 79, 80, 81, 82, 83, 84, 85, 86, 89, 284, 286
computation, 165, 286
computing, 105, 137
concentration, 7, 97, 98, 99, 101, 103, 104, 107, 108, 120, 123, 124, 125, 128, 130, 188, 342
concrete, vii, viii, x, xi, 43, 44, 46, 49, 50, 51, 52, 53, 54, 55, 56, 57, 58, 59, 60, 61, 62, 63, 64, 65, 66, 67, 68, 69, 70, 71, 72, 73, 74, 75, 76, 77, 78, 79, 80, 81, 82, 83, 84, 85, 86, 87, 88, 89, 90, 186, 187, 217, 219, 234, 239, 240, 244, 245, 246, 248, 253, 256, 260, 262, 266, 268, 270, 272, 273, 279, 282, 283, 284, 286, 288, 290, 291, 327, 331, 334, 346, 354
conduction, 120, 122, 129
conductivity, 97, 120, 123
confidence, ix, 54, 168, 172, 177, 178, 190, 194, 197, 235, 236
configuration, 161, 162, 163, 164, 165, 168, 171, 186, 193, 215, 260, 318
conformity, 205
consensus, 50
constant load, 212
construction, vii, ix, xi, 4, 43, 44, 46, 47, 49, 53, 86, 87, 88, 90, 102, 161, 178, 181, 183, 188, 189, 190, 192, 193, 197, 206, 207, 208, 211, 217, 219, 220, 225, 228, 232, 233, 234, 235, 236, 238, 243, 247, 272, 280, 327, 328, 329, 330, 331, 332, 333, 334, 335, 336, 341, 342, 343, 344, 345, 346, 348, 349, 350, 352, 353, 354, 355
construction and demolition, vii, 43, 44, 86, 87, 88, 90
Construction and demolition, 89
construction materials, 49, 102
construction sites, 49
consumption, 99, 100
contamination, 49, 51, 53

continuity, 103, 205, 209, 235
contractors, 54, 86, 191, 225
control, vii, 1, 3, 31, 178, 181, 188, 197, 228, 282, 285, 329, 332, 346, 350, 352
convergence, 280, 286
copolymer, 9
copolymers, 35
core-shell, 35, 38
correlation, 41, 55, 57, 61, 62, 64, 65, 67, 68, 70, 71, 73, 74, 75, 76, 77, 79, 80, 81, 82, 83, 84, 96, 149, 328, 339, 340, 341, 344, 345, 346, 351
correlation coefficient, 57, 62, 64, 65, 67, 68, 70, 71, 73, 74, 76, 77, 79, 80, 82, 83, 344
correlations, 56, 68
corrosion, viii, 77, 91, 93, 94, 95, 100, 115, 117, 121, 129, 130
costs, 225, 346
coupling, 3, 7, 8, 9, 35, 38, 129, 161, 280, 281, 282, 283, 285, 288, 289, 294, 295, 305, 308, 321
covalent, 8
covalent bond, 8
CPA, 44, 50, 51, 56, 83, 84
crack, 2, 3, 4, 5, 14, 16, 19, 22, 28, 30, 33, 94, 160, 161, 162, 163, 165, 167, 168, 172, 288
cracking, 93, 94, 115, 130, 159, 201, 217, 219, 222, 246, 248, 286
creep, 55, 59, 60, 62, 73, 74, 76, 83, 85
critical behavior, 315, 317, 321
critical value, xi, 4, 7, 293
Croatia, 45
cross-sectional, 5, 22, 296, 299, 302
crystal growth, 102
crystallites, 103, 126
crystallization, 35
crystals, 102, 117, 130
CSR, 38, 39, 40
cumulative distribution function, 336, 337, 338, 344
curing, 54, 58, 59, 61, 71
cycles, 333, 353
cycling, 44, 53
Cyprus, 45
Czech Republic, 45, 48, 355

D

damping, 195, 196, 199, 205, 207, 215
data collection, 56
database, 59
decompression, 88
deep-sea, 54

defects, 2, 49, 97, 98, 99, 100, 107, 120, 129
deficiency, 203, 210, 216
definition, 205, 213, 228, 260, 271
deformability, 50, 84, 85
deformation, viii, ix, x, 3, 4, 10, 13, 14, 15, 16, 17, 22, 24, 31, 35, 38, 135, 136, 138, 139, 140, 149, 150, 154, 155, 156, 160, 161, 162, 172, 193, 197, 199, 212, 213, 222, 271, 281, 282, 283, 293, 294, 295, 296, 307, 308, 311, 321, 324, 325
degradation, 16, 20, 21, 196, 201, 207
degree of crystallinity, 2
degrees of freedom, 282, 283, 325
demand curve, 199, 200
Denmark, 45, 46, 52, 53
density, vii, 1, 2, 49, 50, 51, 55, 56, 57, 58, 59, 60, 61, 62, 63, 80, 81, 82, 83, 84, 85, 86, 87, 122, 302, 328, 337, 344
density values, 56
deposition, 125
deposits, 130
derivatives, 95
designers, 54, 86, 179, 189, 191, 192, 206, 225, 226, 235, 246, 329, 330, 352
detection, 4, 225
detection techniques, 4
developed countries, 46
developing countries, 87
deviation, 33, 328
diaphragm, 288, 290
differential equations, 294
differentiation, 298
diffusion, viii, 79, 85, 91, 92, 97, 99, 100, 101, 102, 103, 105, 107, 109, 110, 112, 114, 117, 118, 120, 121, 122, 124, 125, 126, 127, 128, 130
diffusion process, 102
diffusion rates, 130
directionality, 155
disaster, 236
discomfort, 188
discontinuity, 199
dislocation, 159
dispersion, 61, 62
displacement, ix, x, 5, 14, 25, 177, 178, 183, 187, 188, 193, 194, 195, 197, 198, 199, 200, 201, 202, 203, 206, 207, 209, 210, 211, 212, 213, 214, 216, 217, 219, 221, 222, 223, 224, 235, 236, 237, 241, 281, 282, 283, 284, 285, 286, 287, 288, 293, 295, 297, 298, 304, 306, 313, 321, 324
dissolved oxygen, 123
distortions, 335

distribution, vii, viii, x, 1, 2, 4, 52, 54, 58, 95, 99, 125, 135, 150, 156, 160, 161, 164, 165, 221, 245, 279, 280, 281, 282, 283, 284, 288, 297, 328, 331, 332, 334, 335, 336, 337, 338, 341, 342, 344, 348, 349, 353
distribution function, 328, 335
donor, 97
doped, 105, 110, 117
doping, 96
draft, ix, 177, 178, 203, 205, 206, 208, 209, 214, 215, 221, 222, 235, 236, 239, 354
drainage, 45, 52
drying, 35, 86
ductility, vii, 1, 3, 10, 35, 38, 193, 194, 196, 197, 200, 201, 202, 205, 207, 209, 210, 211, 212, 215, 217, 220, 224, 228, 235, 286
dumping, 52, 53
durability, vii, 43, 49, 50, 51, 55, 58, 62, 76, 79, 83, 84, 85, 86, 88, 89, 90, 192, 236, 248, 354
duration, 109, 199, 203, 333
dynamic loads, 321

E

early warning, 236
earth, 331, 332
earthquake, ix, 177, 178, 179, 180, 181, 183, 187, 188, 189, 190, 191, 192, 193, 194, 197, 198, 201, 202, 206, 210, 211, 213, 214, 235, 238, 280, 290
Eastern Europe, 53
ECM, 87
elastic deformation, 4
elastic fracture, 4
elasticity, 50, 51, 55, 60, 61, 62, 65, 66, 68, 71, 81, 83, 85, 89, 139, 157, 253, 273
elastomers, vii, 1, 2, 9, 22, 40
electric charge, 79
electric field, 97
electric power, 179
electrochemical impedance, 115, 122
electrolyte, 92, 97, 99, 100, 101, 102, 103, 104, 105, 107, 108, 112, 114, 117, 118, 119, 121, 124, 126, 127, 128
electrolytes, 98, 107
electron, 97, 152, 153, 154, 155
elongation, 6, 22, 35, 37, 38
energy, 3, 4, 10, 15, 16, 21, 22, 25, 33, 38, 41, 120, 165, 187, 188, 190, 193, 195, 196, 197, 199, 203, 209, 210
energy parameters, 199
enterprise, 227

environment, vii, 35, 44, 84, 93, 94, 129, 130, 191, 246, 329
environment control, 130
environmental conditions, 93, 272
environmental factors, 93
environmental impact, 44, 53
Environmental Protection Agency, 87
environmental temperatures, 335
epoxy, 9, 35, 140, 157
EPR, 9, 11, 12, 35
equating, 258
equilibrium, x, 92, 107, 108, 110, 128, 279, 281, 282, 283, 289
equilibrium state, 282
estimating, 45, 216, 238, 247, 252, 253, 254, 271, 284
Estonia, 45
estuaries, 84
ethylene, 9, 35
Eurasia, 179
Europe, xi, 45, 46, 47, 86, 89, 280, 327, 329, 332, 352
European Commission, 89, 130
European Union, 45
Eurostat, 45
evolution, 46, 117, 129, 161, 162
examinations, 159
excitation, xi, 206, 235, 293, 294, 295, 296, 303, 305, 306, 308, 313, 324, 325
execution, xi, 54, 327, 329, 330, 332, 333, 335, 336, 341, 342, 343, 344, 346, 349, 350, 352, 353, 354
exercise, 252
experimental condition, 113
exploitation, 131, 132
exposure, viii, 91, 93, 94, 105, 106, 108, 109, 110, 111, 112, 113, 114, 115, 116, 118, 119, 120, 121, 122, 123, 124, 125, 129, 333
extrapolation, 33, 121
extrusion, 2, 16, 19

F

fabric, 245
fabricate, 3
fabrication, 2, 3
failure, ix, x, 2, 3, 4, 147, 159, 161, 164, 165, 166, 168, 169, 170, 171, 183, 186, 187, 188, 212, 213, 236, 279, 280, 282, 285, 286, 287, 288, 333, 338, 339, 346
faults, 180, 183, 208
fax, 135
February, 290

Federal Emergency Management Agency, 239, 289, 290
FEMA, 194, 197, 201, 202, 210, 212, 213, 238, 239, 289, 290
ferrite, 159, 161
FIB, 200, 239
fiber, vii, 1, 2, 3, 6, 7, 8, 9, 13, 14, 15, 16, 19, 20, 21, 22, 24, 25, 26, 27, 28, 29, 30, 31, 35, 38, 39, 40, 283
fiber content, 6, 15, 16, 19, 21, 22, 26, 27, 28, 29, 35, 38
fibers, vii, 1, 2, 3, 6, 7, 9, 10, 11, 12, 18, 19, 21, 22, 24, 25, 26, 27, 28, 30, 31, 33, 35, 38, 41, 283
fibrillar, 13
fibrils, 2
fillers, ix, 54, 136, 156
film, viii, 91, 92, 93, 94, 95, 96, 97, 98, 99, 100, 101, 102, 103, 104, 107, 109, 113, 115, 117, 118, 123, 129, 130
film thickness, 92, 97
films, viii, 91, 93, 95, 96, 98, 99, 106, 109, 110, 111, 113, 116, 117, 119, 121, 123, 124, 125, 126, 129, 130
financial loss, 280
financial support, 236
fines, 50, 51, 52, 53
finite element method, 280, 281, 282, 283, 288
Finland, 45, 48, 91, 131, 132, 133, 175
fire, 192
fire resistance, 192
firms, 53
flatness, ix, 243, 246, 247, 252, 271
flexibility, 211, 225, 235, 294, 298, 311
flexural strength, 287
flooring, 246
flow, 3, 21, 86, 95, 125, 153, 156, 159, 170, 172
flow rate, 125
Foucault, 132
FRA, 44, 50, 51, 52, 56, 83, 84
fracture, vii, ix, 1, 2, 3, 4, 5, 10, 13, 14, 15, 16, 17, 18, 19, 24, 25, 26, 28, 30, 32, 33, 34, 35, 38, 41, 152, 159, 160, 161, 163, 164, 165, 167, 168, 169, 170, 171, 172, 174
fracture stress, 4
fractures, 4
France, 45, 46, 48, 131, 132
freedom, x, 44, 210, 221, 240, 241, 282, 283, 286, 293, 298, 325
friendship, 237
frost, 59
frost resistance, 59
fuel, 95

funding, 130

G

G8, 229
gas, 92, 179
gauge, 149, 235
Gaussian, 335, 346, 353
general knowledge, 152
generalization, 54, 86, 298
generation, x, 99, 100, 279, 283, 289, 291
Geneva, 291
geology, 191, 197
Germany, 45, 46, 48, 53, 242
glass, vii, viii, 1, 2, 3, 5, 6, 8, 9, 11, 12, 13, 16, 18, 20, 22, 26, 27, 28, 33, 38, 39, 41, 49, 135, 149, 150, 151, 152, 153, 154, 155, 156, 157
glass transition, 38
glass transition temperature, 38
goals, 181, 198, 201, 202, 206, 212, 214, 235
government, iv, 225
grades, 51, 256, 262, 266, 270
grading, 58, 59, 84
grafted copolymers, 35
grafting, 35
grain, 120, 122, 129, 159, 165
grain boundaries, 120, 121, 122, 165
graph, 61, 64, 153
graphite, 2
gravity, 40, 225, 247, 280
Greece, 45, 243
groups, 3, 8, 9, 13, 14, 33, 35, 152, 174, 178, 191, 198, 203, 205, 206, 208, 214, 225, 228, 235
growth, viii, 33, 45, 83, 91, 95, 97, 99, 100, 101, 102, 104, 105, 107, 109, 110, 112, 114, 115, 117, 118, 120, 122, 123, 124, 125, 128, 129, 130, 160
growth mechanism, 102, 117, 125, 128
growth rate, 45, 95
guidance, x, 243, 247
guidelines, ix, 177, 178, 186, 187, 190, 194, 201, 203, 225, 228, 243, 245, 246, 247, 271, 273, 274, 275, 276, 330

H

H_2, 115
half-life, 95
hanging, 268
Hawaii, 241
hazards, 194, 208
heat, 2, 102
heating, 21
height, 199, 201, 205, 209, 210, 211, 215, 221, 281, 284, 285, 296, 303, 317, 318, 321, 332
hematite, 125, 128
heterogeneity, 46, 121
heterogeneous, 141
high risk, 198
high temperature, 35, 97, 98, 99
higher quality, 51
high-level, 202
homogeneity, 170
homogenized, 282
homogenous, 147
Honda, 132
Hong Kong, 1, 54, 60
House, 354
human, 188, 329, 330, 354
hybrid, 28, 29, 157
hybridization, 41
hybrids, 16, 28, 29
hydration, 58
hydro, 55
hydrogen, 7, 14, 96
hydrogen bonds, 8
hydrolysis, 35
hydrothermal, 35, 37
hydroxide, 96
hydroxyl, 13, 14, 35
hydroxyl groups, 13, 14
hygienic, 248
hysteresis, 196, 201, 217, 221

I

ice, 354
ICE, 272
idealization, 284
identification, 289
IDR, 207, 213, 218, 219
Illinois, 60, 86, 174, 242, 272
images, 155, 227
immersion, 38, 40, 60, 61, 62, 76, 77, 82, 85, 191
impact energy, 4, 10, 31
impact strength, 3, 9, 10, 12, 22, 30
implementation, 192, 208, 224, 226, 227, 228, 280
impurities, 53, 94
IMS, 116
in situ, 16
inadmissible, 329
incentives, 53, 54
inclusion, 68, 70, 71, 79, 83
independent variable, 331

indication, 74
indicators, 197
indices, xi, 327, 346, 350, 352, 353
industrial, xi, 2, 44, 50, 245, 246, 248, 252, 273, 327, 329, 346
industrial sectors, 2
industry, 44, 49, 89
ineffectiveness, 38
inelastic, x, 187, 188, 194, 195, 196, 197, 199, 205, 207, 210, 213, 215, 220, 221, 279, 282, 289, 290
inert, 2, 44
inertia, 284, 295, 296, 302, 310, 313, 314, 316, 317, 320
information exchange, 179, 192, 206, 236
infrastructure, 245
initiation, 14, 33, 115, 161
injection, 2, 3, 4, 6, 7, 16, 19, 20, 21, 22, 26, 28, 30, 32, 35, 99, 148, 153, 156
injection moulding, 153, 156
injury, 179
inorganic, 2, 6
inorganic filler, 2, 6
inorganic fillers, 2, 6
inspection, 189, 329, 346, 352, 353
inspections, 236, 329
inspiration, 236
instabilities, 140
instability, x, xi, 188, 293, 294, 295, 296, 306, 307, 309, 310, 313, 315, 316, 317, 319, 321, 324
insurance, 190, 236
integration, 188, 236, 304, 338, 340
integrity, 93, 132, 160, 349
interaction, x, xi, 2, 7, 8, 14, 22, 24, 26, 93, 129, 130, 146, 187, 191, 273, 279, 281, 282, 283, 284, 285, 286, 287, 288, 294, 296, 305, 306, 307, 308, 315, 316, 317, 319, 320, 321
interaction effect, 282, 285
interactions, 308
interface, 2, 3, 22, 92, 97, 99, 100, 102, 103, 104, 105, 107, 108, 110, 112, 113, 114, 118, 119, 121, 123, 124, 125, 127, 128, 130, 143, 184, 225, 228
interfacial adhesion, viii, 3, 7, 13, 22, 24, 31, 33, 41, 135, 143, 148, 149, 151, 156
interfacial bonding, 3, 7, 25, 33, 35, 41
internet, 191, 232
interphase, 33
interstitial, 97, 98, 100
interstitials, 100, 125
interval, 333
intrinsic, 44, 55, 62, 63

intrinsic value, 44
ionic, 95, 97, 125, 129
ions, 97, 128, 130
Ireland, 45, 48, 273
iron, 95, 98, 123, 124
island, 179, 183
ISO, 329, 330, 341, 354
isolation, 190, 203
isotropic, 140, 142, 155, 156, 285, 296, 297, 299, 301, 308
Israel, 60, 324
Italy, 45, 48, 53, 87
iteration, 235

J

Japan, 88, 132, 157, 194, 240
Japanese, 132, 240
J_c, 4, 160, 164, 165, 168, 169, 172, 173
Jerusalem, 324
joints, 228, 232, 233, 236, 281, 283, 286, 331
judge, 235
junior high, 181
junior high school, 181

K

kernel, 226, 228, 236
kinetic constants, 103
kinetic parameters, 105
kinetics, viii, 92, 94, 103, 107, 123, 128
King, 194, 198, 208, 240
Kobe, 178, 194

L

L2, 4
labor, 226
laminated, 324
land, 198
landfills, 44, 45, 52
large-scale, vii, 43
Latvia, 45
law, 235, 282, 284, 304, 322, 335, 337, 341, 353
laws, x, 53, 55, 279, 281, 284, 286, 331, 334, 335, 336, 348, 349, 353
legislation, 49, 53, 87
life cycle, xi, 190, 192, 236, 327, 333, 345
life span, 214
life-cycle, 194
life-cycle cost, 194
ligament, 4, 14, 18, 25, 162

likelihood, 164
limitation, 95, 205, 226, 294
limitations, 55, 96, 191, 199, 280, 282, 283, 288
linear, x, xi, 33, 56, 57, 61, 83, 139, 140, 141, 145, 149, 154, 188, 201, 216, 217, 221, 243, 245, 248, 256, 260, 268, 271, 280, 282, 283, 284, 285, 286, 289, 290, 291, 293, 294, 295, 296, 298, 304, 305, 306, 308, 311, 321, 324, 325
linear regression, 56, 57, 61
linear systems, 296, 305
linkage, 33
liquefaction, 181, 182, 183, 198, 208, 214, 235
liquid phase, 102
Lithuania, 45, 327
localised, 93, 94, 130
location, x, 53, 59, 94, 191, 212, 234, 244, 245, 279, 281, 285, 288, 315
lognormal, 336, 341, 353
London, 131, 132, 175, 238, 243, 272, 291, 324, 325
long period, 195
losses, 70, 71, 76, 84, 85
low density polyethylene, 149
low risk, 198, 333
low temperatures, 2, 35
Luxembourg, 45, 89
lying, 33, 296

M

M1, 284
machinery, x, 2, 243
magnetic, 53
magnetite, 108
maintenance, 178, 188, 189, 190, 192, 236, 346
Malta, 45
management, 53, 87, 89, 190, 192, 236
management practices, 89
manganese, 95
manipulation, 145
manufacturing, 2, 245
mapping, 87
market, 47, 53
masking, 58
masonry, x, xi, 44, 46, 50, 83, 88, 279, 282, 285, 286, 287, 288, 289, 290, 291, 327, 346, 348, 350
mass loss, 70, 71
material surface, 94
matrix, vii, 1, 2, 3, 7, 9, 10, 13, 15, 16, 17, 20, 21, 22, 24, 25, 30, 31, 33, 35, 38, 41, 102, 136, 137, 141, 143, 144, 145, 146, 147, 149, 151, 152, 153, 155, 159, 161, 282, 295, 304, 322, 339
measurement, 4
measures, 297, 298, 332, 349, 352
mechanical behavior, vii, 1, 10, 41
mechanical performances, vii, 1, 20, 22, 35
mechanical properties, vii, 1, 2, 3, 6, 8, 10, 13, 20, 22, 28, 30, 31, 40, 43, 68, 83, 88, 90, 93, 137, 143, 147, 149, 284
median, 160, 168, 172
melt, 2, 7, 10, 16, 20, 21, 22, 26, 30, 32, 35, 41
metal oxide, 93
metal oxides, 93
metals, 49, 97, 100, 165
Mexico, 241
microstructure, vii, 1, 9, 40, 88, 118, 129, 137
migration, viii, 91, 99, 101, 103, 122
mining, 45, 52
misconception, 138
misleading, ix, 177
Missouri, 240
mixing, 7, 16, 22, 26, 30, 32, 146, 154
mobility, 120, 125
model system, 138
modeling, 187, 190, 201, 205, 206, 212, 213, 235, 282, 285, 287, 288
models, x, 8, 96, 143, 146, 147, 160, 191, 195, 225, 226, 227, 228, 231, 232, 233, 234, 236, 279, 280, 281, 282, 284, 285, 286, 287, 288, 289, 290, 295, 311, 330, 333, 353
modules, 65, 66
modulus, viii, ix, 2, 6, 9, 10, 20, 21, 22, 27, 28, 30, 31, 35, 37, 38, 41, 50, 51, 55, 60, 61, 62, 65, 68, 71, 83, 85, 89, 135, 136, 137, 138, 139, 140, 141, 142, 144, 145, 146, 147, 148, 149, 150, 151, 152, 153, 154, 155, 156, 157, 160, 161, 162, 164, 174, 244, 252, 253, 268, 273, 283, 284, 286
moisture, 35, 41, 268
molar volume, 92, 101
mold, 22, 23, 28
molecular weight, 21
Monte Carlo, 338, 340
morphology, vii, 1, 9, 10, 13, 24
mosaic, xi, 327, 347, 351
Moscow, 324, 354
motion, 97, 178, 179, 183, 193, 196, 199, 203, 212, 240, 294, 305, 308
moulding, 153, 156
movement, 296, 305, 308
MRA, 44, 50
multimedia, 191
multivariate, 336

N

National Research Council, 190, 241, 273
natural, vii, xi, 43, 44, 57, 60, 64, 191, 293, 296, 306, 310, 311, 313, 315, 316, 317, 318, 319, 321
Nb, 95
negligence, 330
Netherlands, 45, 46, 48, 52, 53
network, 3, 228
New Jersey, 238
New York, 41, 241, 242, 324, 354
New Zealand, 135, 194, 240, 241
Ni, 92, 96, 100, 101, 102, 103, 104, 105, 106, 109, 110, 111, 113, 115, 116, 117, 118, 119, 121, 123, 124, 125, 126
nickel, viii, 91, 96, 100, 104, 105, 109, 115, 117, 118, 122, 123, 124, 129, 130
NiO, 117
nodes, 228
non-linear analysis, 295
nonlinearities, 289
non-linearity, 280, 295, 325
non-renewable, vii, 43, 44
non-uniform, viii, 130, 135, 150, 156
normal, vii, 43, 99, 100, 117, 150, 154, 247, 284, 285, 299, 335, 336, 337, 338, 341, 342, 346, 348, 349, 350, 352, 354
normal distribution, 335, 338, 341, 342
norms, 49, 54, 55, 61, 84
North America, 169
Norway, 45, 48
nuclear, viii, 91, 93, 95, 102, 130, 169
nuclear power, viii, 91, 93, 95, 102, 130, 169
nuclear power plant, viii, 91, 93, 95, 102, 130, 169
nucleation, ix, 159, 160, 174
nuclides, 95, 130
numerical computations, 186
nylon, 6, 7, 8, 9, 11, 12, 16, 18, 19
nylons, 33

O

observations, viii, 19, 91, 107, 129, 163, 172
oil, 157
omission, 247, 294
online, 87, 89
on-line, 294
optical, 16, 22, 23
Optical Emission Spectroscopy, 104
optical micrographs, 16
optimization, 3
orientation, ix, 3, 20, 21, 28, 29, 136, 138, 153, 155, 156, 169
oxidation, 2, 92, 96, 100, 105, 107, 115, 120
oxide, viii, 91, 92, 93, 94, 95, 96, 97, 99, 100, 101, 102, 103, 104, 107, 108, 110, 112, 115, 117, 118, 119, 120, 122, 123, 124, 125, 128, 129, 130
oxide thickness, 95
oxides, viii, 91, 93, 105, 109, 110, 112, 115, 117, 118, 123
oxygen, 97, 99, 100, 104, 123

P

Pacific, 157, 238, 241
PAN, 2
parabolic, 283
parameter, ix, xi, 4, 33, 63, 65, 83, 85, 96, 109, 110, 112, 117, 122, 159, 160, 161, 165, 166, 167, 168, 171, 174, 244, 280, 293, 296, 303, 308, 315, 317, 318, 319, 320, 321, 328
parameter estimates, 96
Paris, 132, 175
particles, vii, 1, 10, 71, 100, 155, 172
particulate matter, 95
partnership, 239
passivation, 94
passive, 94, 97, 98, 101, 115, 129
PBT, 2, 35, 37, 38, 39, 40
pedestrian, 183
peer, 192, 225, 240
peer review, 192
pendulum, 4
per capita, 45, 46
perception, 59
performance indicator, 193, 197, 198, 199, 202
periodic, x, 293, 295, 303, 307, 309, 310, 321, 324
permeability, 59, 60
perturbation, 294, 307
PGA, 190, 193, 203
pH, 93, 128
Philadelphia, 41, 174, 175
physical properties, vii, 35
physico-chemical properties, 88
pitch, 2
planar, x, 138, 279, 281, 282, 289
planning, 190, 225
plants, 53, 54, 169
plaques, 32

plastic, ix, 4, 5, 6, 13, 15, 16, 17, 26, 159, 160, 161, 165, 172, 174, 212, 220, 221, 246, 271, 282, 283, 284, 285, 286
plastic deformation, 13, 16, 17, 160, 161
plastic strain, ix, 159, 160, 172, 174
plasticity, x, 172, 279, 281, 282, 285, 286
plastics, 2, 49
platelet, 157
play, 41, 147
Poisson, 162, 244, 253, 260, 273
Poisson ratio, 260, 273
Poland, 45, 48
police, 53
polyamide, vii, 1, 10, 13
polyamides, 2, 3, 6, 10, 41
polycrystalline, 165
polyester, vii, 1, 35
polyesters, 2
polyethylene, viii, 135, 138, 149, 150
polymer, vii, 1, 2, 3, 9, 10, 11, 12, 21, 22, 24, 25, 30, 32, 40, 141, 143, 144, 152
polymer blends, 30
polymer composites, vii, 1, 2, 33
polymer materials, 30
polymer matrix, vii, 1, 2, 3, 9, 10, 24, 25, 33, 40, 141, 144
polymers, vii, 1, 2, 3, 4, 30, 141
polynomial, 354
polyolefin, 24
polyolefins, 2, 3
polypropylene, vii, 1, 143, 157
poor, 2, 3, 10, 22, 35, 50, 71, 75, 147, 153, 188, 248
population, 183
pore, 102
pores, 100, 102, 103
porosity, 59, 60, 63, 71, 75, 79, 130
porous, 71, 102
Portugal, 43, 48, 53, 59, 84, 86, 87, 88, 89
power, viii, 91, 93, 95, 102, 124, 130, 137, 169, 174, 179, 332, 354
power plant, 93, 124
power plants, 93, 95, 102, 130, 169
pragmatic, 84
precipitation, 100, 102, 117
prediction, xi, 160, 171, 257, 280, 288, 291, 294, 327, 329, 330, 334, 341, 345, 353, 354
predictive model, 95
pre-existing, 110
pressure, 115, 165, 166, 167, 169, 174, 245, 268, 331, 335
prevention, 181, 192, 214

probability, xi, 161, 164, 165, 166, 168, 171, 172, 183, 190, 193, 198, 206, 236, 327, 328, 329, 330, 331, 332, 334, 335, 336, 337, 338, 339, 341, 342, 344, 345, 349, 350, 351, 352, 353
probability density function, 161, 172
probability distribution, 331, 332, 334, 335, 336, 337, 341, 349, 353
producers, 53
production, vii, 41, 43, 44, 45, 46, 47, 48, 49, 51, 52, 53, 54, 58, 59, 79, 84, 89, 100
profit, 44, 54
program, 167, 169, 172, 179, 228, 283, 284, 286, 311
programming, 226, 289
programming languages, 226
propagation, 14, 29, 30, 159, 161
property, iv, 9, 56, 57, 58, 60, 61, 63, 65, 68, 70, 71, 74, 80, 83, 190, 212, 217, 324
proportionality, 254
propylene, 9
protection, 203, 208, 235
protocol, 5
prototyping, 228
PSA, 207, 209, 223
public, 202, 236
pulp mill, 149
pulse, 343, 353

Q

quality control, 49, 53, 54, 190, 192, 206
quantitative estimation, 74

R

RAC, 44, 51, 52, 54, 55, 56, 57, 58, 59, 62, 63, 65, 67, 68, 70, 71, 73, 74, 76, 77, 79, 80, 83, 84, 85, 86
radiation, 93, 125
radius, 244, 252, 258, 266, 273, 274, 298
random, xi, 9, 138, 327, 328, 329, 330, 332, 333, 336, 338, 341, 342, 343, 344, 345, 346, 347, 352, 353
range, viii, x, 5, 15, 30, 40, 57, 58, 91, 112, 121, 151, 164, 167, 169, 171, 192, 195, 236, 243, 246, 248, 255, 271, 282, 288, 334
rapid prototyping, 226
raw material, 53
raw materials, 53
reaction rate, 93, 100
reaction rate constants, 100
reactive groups, 3

reading, 209
reality, 95, 130, 191
reasoning, 107, 146, 225
recovery, 39, 290
recurrence, 336, 343, 344, 345, 348, 351
Recycled Aggregate, v, 43, 49, 52
Recycled Aggregate Concrete, 52
recycling, 44, 46, 49, 51, 53, 54, 85, 86
recycling plant, 53, 54, 86
recycling plants, 53, 54
redistribution, 283, 284, 285
redundancy, 209, 235
regression, 15, 33, 56, 57, 61, 80
regression line, 15, 33, 57, 61, 80
regular, 201, 214, 215, 216
regulation, vii, 43, 190
regulations, 177, 186, 187, 190, 194, 330
rehabilitation, 46, 53, 179, 189, 190, 206, 213, 290
reinforced systems, 155, 156
reinforcement, viii, ix, 2, 3, 7, 10, 16, 21, 77, 85, 135, 136, 137, 138, 141, 143, 144, 146, 147, 148, 149, 150, 151, 154, 155, 156, 184, 186, 187, 188, 189, 220, 228, 229, 230, 232, 233, 234, 236, 273, 328
relationship, ix, 41, 57, 62, 64, 65, 67, 68, 70, 71, 73, 74, 76, 77, 79, 80, 83, 112, 138, 142, 154, 156, 159, 160, 162, 193, 195, 212, 213, 252, 253, 254, 255, 256, 260, 263, 265, 268, 283, 284
relationships, vii, 1, 85, 90, 248, 253, 295
relevance, 96
reliability, ix, xi, 105, 177, 178, 191, 193, 197, 199, 201, 211, 236, 280, 327, 329, 330, 341, 345, 346, 350, 351, 352, 353, 354, 355
repair, 192, 194, 198, 202, 235, 236
replacement rate, 60, 61
replacement ratios, 51, 54, 56, 60, 84
resin, 38
resistance, x, xi, 2, 4, 10, 14, 50, 55, 59, 60, 61, 62, 70, 71, 77, 79, 81, 82, 83, 84, 85, 97, 138, 187, 192, 279, 291, 327, 329, 332, 336, 337, 341, 342, 344, 347, 349, 350, 352
resources, 44
restructuring, viii, 91, 95, 109, 110, 117, 118, 129, 130
retail, 245
retention, 35, 38
rigidity, 217, 222, 346
risk, 93, 190, 193, 198, 203, 235, 236, 354
risk assessment, 190
risk factors, 193
risk management, 190, 236

risks, 192
Ritz method, 305
road map, 87
roads, 272, 273
robotics, 242
robustness, 236
rods, 294
Romania, 45
room temperature, 97
rotations, x, 282, 284, 293, 294, 295, 297, 298, 321, 325
rubber, 9, 16, 18, 19, 30, 31, 33, 35, 38, 41
rubbers, 35
Russian, 354

S

SAC, 194, 239
safety, ix, xi, 53, 115, 177, 178, 179, 180, 192, 194, 198, 202, 210, 214, 235, 271, 274, 327, 329, 330, 331, 333, 334, 335, 336, 338, 339, 340, 341, 342, 343, 344, 345, 346, 349, 350, 351, 352, 353
sample, 4, 33, 109, 110, 116, 140, 149
sampling, 161
sand, 44, 54, 83
saturation, 35, 61, 87, 125
savings, 245, 246, 271
scaling, 160, 161, 162, 163, 164, 305
scanning electron, 152, 153, 154, 155
scatter, ix, 62, 68, 83, 85, 159, 160, 167, 174
school, 181
scientific community, ix, 50, 177, 186
scripts, 228
search, 85, 105, 225, 236, 280
seasonal effects, 354
Seattle, 238, 239
seismic, ix, x, 177, 178, 179, 181, 183, 186, 187, 190, 191, 192, 193, 194, 195, 196, 197, 198, 199, 201, 202, 203, 204, 205, 206, 207, 208, 210, 211, 212, 213, 214, 217, 218, 219, 222, 224, 225, 235, 237, 238, 239, 240, 241, 279, 280, 281, 290, 291
selecting, 142, 193, 194, 198, 201, 226
self, 267
SEM, 10, 13, 16, 17, 24, 31, 32, 152, 153, 155
SEM micrographs, 10, 13, 17, 24, 152
sensitivity, 160, 193, 248, 256, 267, 286
separation, 152, 153, 185, 205, 213, 220
series, 179, 197, 225, 245, 268, 281, 285, 307, 338, 339, 340, 341, 345, 353
service loads, 331, 352, 355
settlements, 245, 247

shape, x, 5, 54, 59, 141, 211, 251, 261, 264, 265, 266, 279, 285
sharing, 151, 237
shear, viii, ix, x, 3, 7, 16, 31, 41, 135, 136, 137, 138, 139, 140, 146, 147, 149, 150, 154, 156, 184, 212, 213, 215, 216, 217, 221, 222, 223, 279, 281, 282, 283, 285, 286, 287, 288, 290, 293, 294, 295, 296, 297, 298, 299, 302, 303, 305, 306, 308, 311, 314, 315, 316, 319, 321, 324, 325, 329
shear deformation, viii, ix, x, 16, 31, 135, 136, 139, 140, 149, 150, 154, 155, 156, 279, 281, 293, 295, 296, 308, 311, 321, 324, 325
shear strength, 136, 137, 146, 147, 212
shell, 297
shock, 169
shores, 334
short period, 205
short-term, 342
sign, 282
silane, 3, 8, 9, 35
silver, 95
similarity, 245
simulation, 105, 109, 187, 191, 201, 205, 206, 212, 225, 284, 338, 340, 354
single crystals, 130
SiO2; 2
sites, 46, 53, 100, 128, 183, 210
skills, 280
skin, 3, 22, 29, 33
Slovakia, 45, 48
Slovenia, 45, 238, 290
socioeconomic, 191
software, 227, 228, 283, 284
soil, 45, 46, 53, 183, 198, 207, 208, 245
soils, 44, 267
solid state, 95, 129, 130
solid waste, 44
solid-state, 102, 117
solubility, 93, 95, 100, 104
sorting, 53
sounds, 236
Spain, 45, 48, 53
spatial, 294, 295, 324
speciation, 108, 128
species, viii, 49, 91, 97, 102, 120, 130
specific gravity, 40
spectrum, 183, 193, 195, 197, 199, 200, 202, 205, 206, 207, 210, 212, 215, 217, 221, 223, 225, 238, 290
speed, 20, 24, 335
sports, 181
springs, 281

sputtering, 105
stability, viii, x, 2, 76, 91, 93, 129, 212, 213, 280, 293, 294, 295, 296, 305, 306, 308, 321, 324, 329
stabilization, 2
stabilize, 2
stages, 115, 225, 236, 295, 329, 332, 335, 349
stainless steel, viii, 91, 96, 100, 104, 105, 109, 117, 118, 122, 123, 129, 130
stainless steels, viii, 91, 96, 104, 105, 109, 117, 118, 122, 123, 129, 130
standard deviation, 335, 337, 344, 346, 351
standards, 41, 149, 194, 211, 220, 247, 327, 329, 330, 352
statistics, ix, 45, 46, 87, 159, 161, 353
steady state, viii, 91, 97, 98, 103
steel, 49, 100, 105, 109, 112, 114, 123, 124, 125, 127, 130, 165, 166, 167, 169, 170, 172, 173, 174, 196, 235, 248, 283, 284
stiffness, vii, ix, 1, 2, 3, 10, 27, 28, 35, 40, 65, 71, 74, 135, 141, 143, 155, 156, 184, 186, 187, 188, 189, 193, 196, 197, 199, 200, 201, 202, 207, 208, 209, 210, 213, 216, 217, 219, 220, 221, 222, 223, 235, 244, 250, 252, 268, 269, 274, 282, 283, 284, 285, 287, 295, 304, 306, 308, 346
STM, 115
stochastic, 160, 353, 354
stock, 148, 280
stoichiometry, 104
storage, x, 243, 245, 332
storms, 354
strain, ix, 2, 4, 31, 32, 34, 38, 39, 120, 137, 140, 141, 159, 160, 161, 162, 163, 164, 167, 171, 172, 174, 217, 283, 299, 300, 305
strains, 295, 304, 305, 322
strength, vii, ix, 1, 2, 3, 6, 7, 8, 9, 10, 20, 21, 22, 27, 28, 30, 31, 33, 35, 37, 38, 39, 41, 50, 51, 52, 54, 55, 56, 57, 58, 59, 60, 61, 62, 63, 64, 65, 67, 68, 70, 74, 76, 79, 80, 81, 82, 83, 84, 85, 88, 89, 92, 97, 112, 114, 122, 126, 127, 136, 137, 146, 147, 165, 166, 186, 187, 188, 193, 194, 195, 197, 198, 199, 201, 202, 207, 209, 212, 213, 219, 220, 222, 223, 235, 243, 246, 280, 284, 286, 287, 288, 294, 328, 329, 331, 332, 333, 346, 348, 349, 350
stress, viii, ix, 3, 4, 5, 6, 7, 10, 16, 22, 30, 33, 35, 38, 89, 93, 94, 115, 130, 135, 136, 137, 138, 140, 141, 147, 150, 156, 159, 160, 161, 162, 163, 164, 165, 166, 167, 168, 169, 170, 171, 172, 173, 174, 187, 188, 243, 271, 272, 273, 283, 286, 297, 301, 303
stress fields, 161

stress intensity factor, 4, 160
stress-strain curves, 30
strikes, 4, 189
structural changes, 97
styrene, 9, 35
substitutes, 125
substitution, 89
summaries, 194
summer, 335
superposition, 206, 211
supervision, 53, 59, 85, 189
supply, 179
surface chemistry, 8
surface modification, 89
surface treatment, 3, 59
surveillance, 189
survival, xi, 327, 333, 337, 338, 339, 340, 341, 342, 344, 345, 346, 350, 351, 352, 353
susceptibility, 76
sustainable development, 88
Sweden, 48, 91
Switzerland, 45, 48, 239, 279, 289, 290, 291, 354
symbols, 99, 115, 119, 120, 123, 139, 147, 165, 167, 168, 171, 173
symmetry, 186, 308
synergistic, 16, 29
synergistic effect, 16
systems, 95, 105, 155, 156, 179, 183, 187, 197, 198, 203, 205, 208, 209, 212, 214, 222, 235, 241, 282, 289, 294, 327, 331, 332, 333, 338, 339, 340, 341, 345, 353, 354

T

Taiwan, v, ix, 177, 178, 179, 180, 183, 187, 189, 190, 192, 195, 196, 202, 203, 206, 237, 238, 240
talc, viii, 135, 138, 149, 153, 154, 155, 156
taxes, 53
technology, 87, 157, 209, 236, 352, 353
TEM, 9, 11
temperature, ix, 20, 21, 38, 40, 41, 92, 93, 98, 99, 100, 108, 110, 112, 121, 159, 160, 164, 165, 168, 171, 172, 173, 174, 333, 334, 335, 343, 344, 353
temperature dependence, 160, 171
tensile, vii, viii, 1, 2, 3, 6, 8, 10, 13, 14, 16, 20, 21, 22, 27, 30, 31, 35, 37, 38, 40, 55, 59, 60, 61, 62, 67, 68, 69, 70, 81, 83, 85, 94, 135, 136, 137, 138, 140, 150, 151, 154, 156, 157, 161, 167, 284

tensile strength, 2, 3, 8, 10, 20, 21, 22, 27, 30, 35, 37, 38, 40, 55, 59, 60, 61, 62, 67, 68, 69, 70, 83, 85, 284
tensile stress, 94
tension, viii, 5, 28, 29, 135, 138, 139, 140, 151, 156, 160, 163, 168, 328
test data, 164
test procedure, 61, 85
thermal aging, 35
thermal stability, 2
thermodynamic, 104
thermodynamic calculations, 104
thermodynamics, 93
thermoplastic, 2, 157
thermoplastics, vii, 1, 2, 3, 138
threatening, 94
three-dimensional, 187, 191, 283, 288, 289, 295
three-dimensional model, 288
threshold, ix, 159, 160, 165, 166, 174
timber, 46, 49
time, viii, xi, 35, 52, 53, 56, 65, 73, 86, 91, 95, 108, 109, 112, 114, 115, 117, 118, 119, 120, 121, 122, 123, 129, 179, 188, 191, 192, 193, 195, 197, 199, 205, 211, 221, 225, 226, 235, 236, 280, 327, 331, 333, 334, 335, 338, 342, 344, 346
time consuming, 226, 280
titanates, 7
titration, 108, 110, 128
Tokyo, 157, 240, 324
tolerance, 283
tort, 102
toughness, vii, ix, 1, 2, 3, 4, 10, 15, 16, 25, 28, 29, 30, 32, 33, 34, 35, 38, 41, 159, 160, 161, 162, 163, 164, 165, 167, 168, 169, 170, 171, 172, 174
toxic, 53
transfer, 3, 7, 22, 35, 41, 92, 97, 113, 143, 147, 236, 245
transformation, 102, 222
transition, ix, 102, 118, 159, 160, 294
transition temperature, 38
transmission, 181, 205, 354
transmission path, 205
transparency, 194, 202, 206
transparent, 192, 199, 203, 235
transport, viii, 91, 95, 97, 98, 99, 100, 102, 103, 105, 107, 109, 110, 116, 121, 126, 127, 129, 130
transportation, 2
trial, 105
trucks, 245, 247, 252, 271
trust, 237

twinning, 159
two-dimensional, 282, 288, 289
two-way, 284

U

UCB, 237, 238, 241, 290
uncertainty, 347, 354
uniaxial tension, 4
uniform, viii, 129, 130, 135, 139, 150, 156, 186, 201, 209, 215, 217, 221, 235, 245, 273, 288, 306
unit transformation, 222
United Kingdom, 45, 46, 48, 53
United States, 194, 240, 325
universal gas constant, 92
universities, 194, 239
updating, 224

V

vacancies, 97, 98, 99, 100, 107
vacuum, 2, 35
validation, 105, 283, 286
validity, 5, 112, 121, 253, 260, 263, 286
variability, 83, 85, 170, 190, 330, 343
variables, 100, 248, 333, 341, 346, 348
variance, 335, 336, 342
variation, 27, 38, 51, 57, 59, 83, 85, 130, 140, 141, 146, 165, 170, 247, 282, 286, 289, 307, 328, 341, 347
vector, xi, 221, 293, 296, 322, 333
vehicles, 332
velocity, 210, 332, 335
versatility, 41
vibration, xi, 284, 294, 295, 296, 305, 306, 307, 308, 311, 313, 315, 317, 320, 321, 324, 325
village, 181
violent, 179
virtual reality, 191
viscoelastic properties, 324
viscosity, 7, 10
visible, 44, 152, 153
visualization, 191, 225, 228
vulnerability, 79, 83, 236, 280, 281, 289, 290

W

waste management, 89

water, 35, 49, 50, 51, 53, 54, 55, 56, 57, 58, 59, 60, 61, 62, 63, 64, 65, 66, 67, 68, 69, 70, 71, 72, 73, 74, 75, 76, 77, 78, 79, 80, 81, 82, 83, 84, 85, 86, 89, 93, 94, 99, 102, 103, 104, 105, 106, 108, 109, 110, 112, 114, 115, 116, 117, 118, 119, 120, 121, 122, 123, 125, 126, 127, 128, 129, 130, 179
water absorption, 49, 50, 51, 55, 56, 57, 58, 59, 60, 61, 62, 63, 64, 65, 66, 67, 68, 69, 70, 71, 72, 73, 74, 75, 76, 77, 78, 79, 80, 81, 82, 83, 84, 85, 86, 89
web, 226, 228
web browser, 226
Weibull, ix, 159, 160, 161, 162, 163, 164, 165, 166, 167, 168, 169, 171, 172, 173, 174
Weibull distribution, 160, 161, 164, 165
weight ratio, 30
wind, 192, 280, 332, 333, 335, 343, 344, 353, 354
windows, 189
winter, 335
wood, viii, 135, 138, 149, 150, 151, 152, 153, 154, 155, 156, 183, 354
workability, 50, 55, 58, 59
workers, ix, 96, 159, 160, 174, 191, 225, 233
World Wide Web, 191, 236
worry, 54
writing, 226
WWW, 191

X

XML, 236
XPS, 104, 106, 115

Y

yield, 4, 9, 10, 22, 31, 149, 162, 165, 170, 171, 186, 237, 263, 282, 283, 284, 286, 294, 350

Z

zinc, 95
zirconium, 95
Zn, viii, 91, 92, 95, 103, 104, 105, 106, 107, 108, 109, 110, 111, 112, 113, 114, 115, 116, 117, 118, 121, 122, 123, 125, 126, 127, 128, 129